SELECTING AN INFERENTIAL STATISTICAL PROCEDURE

Type of design	*Parametric test*	*Nonparametric test*
One sample (when σ_X is known)	z-test (Chapter 10)	none
One sample (when σ_X is not known)	One-sample t-test (Chapter 11)	none
Two independent samples	Independent samples t-test (Chapter 12)	Mann-Whitney U, rank sums test, or one-way chi square (Chapter 15)
Two related samples	Related samples t-test (Chapter 12)	Wilcoxon test (Chapter 15)
Three or more independent samples (one factor)	Between-subjects ANOVA *Post Hoc* test: Tukey's *HSO* protected t-test (Chapter 13)	Kruskal-Wallis H or one-way chi square *Post Hoc* test: rank sums test (Chapter 15)
Three or more related samples (one factor)	Within-subjects ANOVA *Post Hoc* test: Tukey's *HSD* or protected t-test (Appendix C)	Friedman X^2 *Post Hoc* test: Nemenyi's test (Chapter 15)
Two factors	Two-way ANOVA (Chapter 14)	Two-way chi square (Chapter 15)

BASIC STATISTICS FOR THE BEHAVIORAL SCIENCES

BASIC STATISTICS FOR THE BEHAVIORAL SCIENCES

Second Edition

Gary W. Heiman

Buffalo State College

HOUGHTON MIFFLIN COMPANY Boston Toronto

Geneva, Illinois Palo Alto Princeton, New Jersey

Sponsoring Editor: Rebecca J. Dudley
Senior Associate Editor: Jane Knetzger
Senior Project Editor: Charline Lake
Production/Design Coordinator: Jennifer Waddell
Senior Manufacturing Coordinator: Priscilla J. Bailey

Cover design: Janet Theurer, Theurer Briggs Design
Cover image: "Crowd IV," Diana Ong

Printed in the U.S.A.

Library of Congress Catalog Card Number: 95-76952

Student Text ISBN: 0-395-74528-4
Examination Copy ISBN: 0-395-76581-1

123456789-DW-99 98 97 96 95

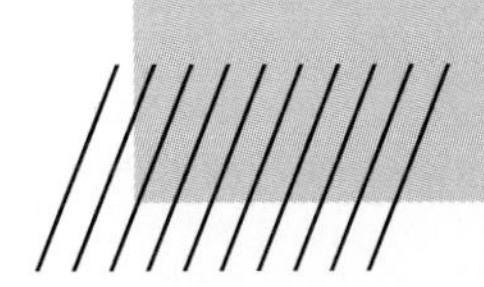

Brief Contents

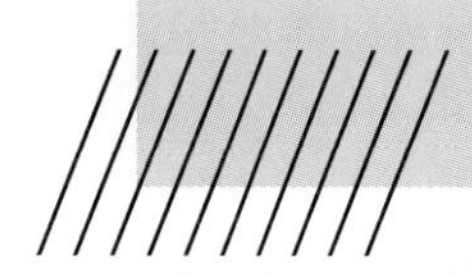

Contents

PART 2

DESCRIPTIVE STATISTICS: DESCRIBING SAMPLES AND POPULATIONS

PART 4

INFERENTIAL STATISTICS

14

Two-Way Analysis of Variance: Testing the Means from Two Independent Variables 414

15
Nonparametric Procedures for Frequency Data and Ranked Data 452

APPENDICES

A

Interpolation 489

B

Additional Formulas for Computing Probability 493

C

One-Way Within-Subjects Analysis of Variance 499

D

Statistical Tables 509

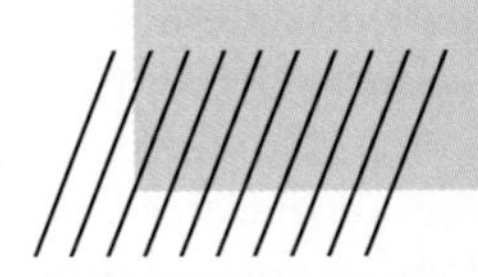

Preface

Many of the undergraduates who enroll in my statistics course have a weak background in mathematics and some degree of "math phobia." By the end of the course, these same students must understand and perform the descriptive and inferential statistical procedures commonly used in behavioral research. The problem is that textbooks often dwell on the remarkable things statisticians can do with statistics and say too little about the things *researchers* commonly do. Such books present a catalog of procedures but do not explain the conceptual purposes of these procedures. Although students can then compute an answer on demand, they do not understand why they should perform the procedure nor what their answer reveals about the data. Therefore, I wanted my students to have a textbook that takes their needs more fully into account: a book that *explains*—clearly, patiently, and with an occasional touch of humor—the way a good teacher does.

My Objectives

In writing this book, I pursued five objectives.

1. Take a conceptual-intuitive approach The text emphasizes the context in which statistics are used to make sense out of data. Each procedure is introduced in the context of a simple study with readily understandable goals. I focus on the purpose of research as examining the relationships between variables, then delineate the procedures for describing and inferring such relationships, and finally return to the conceptual purpose and interpretation of the study. Throughout, I provide students with simplified ways to think about statistical concepts and to see how these concepts translate into practical procedures for answering practical questions.

2. Present statistics within an understandable research context Many of the text's early examples involve simple variables and research questions taken from everyday life, so that students have an intuitive feel for the meaning of the scores and relationships discussed. In later chapters, along with students' developing statistical thinking, examples become more technical and "psychological." Virtually all examples and study questions involve specific variables and research questions, instead of generic data.

3. Deal directly and positively with student weaknesses in mathematics The text presents no formulas or statistical statements without explanation. Formulas are introduced in terms of what they accomplish, and an example of each is worked out completely, step by step. To further reduce the apparent complexity of statistics, I have stressed the similarities among different procedures, showing how, despite slight variations in computations, they have similar components and answer similar questions.

4. Introduce new terms and concepts in an integrated way I have strived to tie each new concept and procedure to previous material, briefly reviewing that material in every possible instance, including in the new chapter openings. Throughout, difficult concepts are presented in small chunks, which are then built into a foundation and later elaborated on.

5. Create a text that students will enjoy as well as learn from To make the text readable and engaging, I repeatedly point out the everyday usefulness of statistics. I have also tried to convey my own excitement about statistics and to dispel the notion that statistics (and statisticians) are boring. One can take a discipline seriously yet still recognize its quirks and foibles and have fun with it.

Organization of the Text

In Part 1, *Getting Started*, Chapter 1 serves as a brief preface for the student and reviews basic math and graphing techniques. Chapter 2 then introduces the terminology, logic, and goals of statistics within the context of behavioral research.

Chapters 3 through 6 make up Part 2, *Descriptive Statistics* (along with a discussion of linear interpolation in Appendix A). I placed the chapter dealing with z-scores (Chapter 6) after the chapters on central tendency and variability (Chapters 4 and 5) so that these building blocks would be fresh in students' minds when discussing z-scores. I included sampling distributions and computing a z-score for a sample mean in this chapter so that students can later see how the logic of inferential statistics is essentially that of computing z-scores.

Part 3, *Describing Relationships*, consists of Chapters 7 and 8, in which correlation and regression are introduced as descriptive procedures, with emphasis on interpreting the correlation coefficient and the variance accounted for. (The point-biserial correlation is included to provide a bridge to measures of effect size in later chapters.) I placed the chapters dealing with correlation and linear regression before the inferential statistical procedures because otherwise it is confusing to introduce these procedures while discussing their inferential tests. Substantial cautions are given, however, about the need for performing inferential procedures on correlation coefficients, and when discussed later, they are presented as a logical variation of significance testing of means.

Part 4, *Inferential Statistics*, begins with Chapter 9, although extensive groundwork is laid in the previous chapters, with strong emphasis on understanding the proportion of the area under the normal curve. Chapter 9 introduces probability and previews hypothesis testing, focusing on using the normal curve to compute probability, with the goal of making decisions about the representativeness of sample means. In Chapter 10, hypothesis testing is formalized using the *z*-test. Chapter 11 presents the single-sample *t*-test, the confidence interval for a single mean, and significance testing of correlation coefficients. Chapter 12 covers two-sample *t*-tests and effect size. Chapter 13 introduces the one-way, between-subjects ANOVA, including *post hoc* tests for equal and unequal *n*'s, eta squared, and omega squared. (The one-way within-subjects ANOVA is described in Appendix C.) Chapter 14 deals with the two-way between-subjects ANOVA, *post hoc* tests for main effects and for unconfounded comparisons in an interaction, as well as graphing and interpreting interactions. Chapter 15 covers the one-way and two-way chi square, as well as the nonparametric versions of all previous parametric tests (with appropriate *post hoc* tests and measures of effect size).

The text is also designed to serve as a reference book for students. To that end, I've included such procedures as the formulas for transforming a raw score into a percentile and vice versa, for the semi-interquartile range, for T-scores, for the F_{max} test, for several types of confidence intervals, and for a rather extensive collection of nonparametric procedures. The instructor can skip the more uncommon procedures, however, without disrupting the discussion of the major procedures.

The text strives to teach students how to interpret their data—not just to report that a result is significant. Thus, I have emphasized such topics as plotting and interpreting graphs and understanding the relationships demonstrated by research. I've also included practical discussions of power and measures of effect size. These discussions occur at the end of a section or chapter so that instructors wishing to skip these topics can easily do so.

Pedagogical Format and Features

A number of features have been built into the book to enhance its usefulness as both a tool for study and a reference.

- New chapter-opening pedagogy encourages review of previous material and provides learning goals.
- "More Statistical Notation" sections introduce new statistical notations at the beginning of the chapter in which they are used, not before, and to reduce student confusion, they are introduced separately from the conceptual issues presented in the chapter.
- Each important procedural point is emphasized by a "STAT ALERT," a summary reminder set off from the text about the calculation or interpretation of a statistic.
- Computational formulas are highlighted throughout the text in color.
- Key terms are highlighted in bold, reviewed in the chapter summary, and listed in the end-of-text glossary. Many mnemonics and analogies are used to promote retention and understanding.
- Graphs and diagrams are thoroughly explained in captions and fully integrated into the discussion.
- "Finally" sections at the end of each chapter provide advice, cautions, and ways to integrate material from different chapters.

- Each "Chapter Summary" provides a substantive review of the material, not merely a list of the topics covered.
- Conceptual and procedural questions, as well as computational problems, are provided at the end of each chapter. Odd-numbered problems (with final and intermediate answers in Appendix E) provide students with a solid review of the material, and even-numbered problems (with answers in the Instructor's Resource Manual) can be used as assigned homework.
- A Summary of Formulas is provided at the end of each chapter, for quick reference. In addition, a list of formulas is shrinkwrapped with each student text for use in closed-book exams.
- A glossary of symbols appears on the inside back cover. New reference tables on the inside front cover provide guidelines for selecting the various descriptive and inferential procedures discussed in the text based on the type of data or research design employed.

New Features in the Second Edition

The first edition was quite well received, and the various reviewers and users suggested little in the way of substantial change. Therefore, my focus was to tighten the conceptual presentations and to incorporate several new explanatory techniques that I developed since the first edition. Throughout, greater emphasis is placed on explaining how to use statistics to mentally "envision" a set of data and to "think" in statistical terms. Further, discussion of the mathematical steps in computing the various procedures has been streamlined, although I have kept the text very friendly to students with math phobia. In addition, I made the following specific changes:

- *New chapter-opening pedagogy* Each chapter now begins by identifying the previously discussed concepts that students should review before continuing, followed by learning goals for the chapter.
- *Expanded practice problems* The number of practice problems at the end of each chapter has increased by fifty percent, with an emphasis on both computational problems and challenging conceptual problems. In addition, practice problems are now provided in the appendices dealing with interpolation, probability, and the one-way within-subjects ANOVA.
- *New software appendix* Appendix F, "Introduction to Statistics Software Packages," provides the student with simple step-by-step instructions for performing all the major statistical procedures discussed in the text using one of three data-analysis software programs: HMSTAT (packaged free with each student text), MYSTAT, and SPSS/PC+ Studentware Plus™. Appendix F provides directions for installing the programs, entering data, running the various procedures, and interpreting the output.

Key changes also have been made to each chapter: Chapter 1 now provides instructions on how students should read a statistics text and study for exams. Chapter 2 has been revamped to emphasize that psychological research is based on the two goals of demonstrating a relationship in sample data and using samples to draw inferences about the population. Chapter 3 includes an improved explanation of using the area under the normal curve to describe scores. Chapter 4 more clearly takes students through the important steps of using a mean score to describe obtained scores, to predict future scores, to convert raw scores to deviation scores, and to estimate the population mean. Chapter 5 now covers using inflection points on a normal distribution to identify the location of the scores that are one standard deviation above or below the mean. Chapter 6 features more discussion of how the normal curve, z-scores, and the central limit theorem allow us to envision the important characteristics of any set of data. Chapters 7 and 8 explain more fully the distinction between a correlational and experimental research design, as well as the similarities and differences in the statistical procedures used in each.

Chapters 9 and 10 clearly set the stage for inferential procedures, with an improved introduction to using a sampling distribution to describe the means we expect if we are representing a particular population and a description of how inferential procedures are ultimately designed to limit Type I and Type II errors. Chapter 11 contains expanded discussion of how researchers maximize statistical power in one-sample experiments and correlational studies. Chapter 12 presents a clearer description of using a squared correlation coefficient to describe effect size and mentions the latest American Psychological Association requirements for reporting effect size with all significant results. Chapter 13 includes an expanded discussion of how the between-groups mean square is an estimate of both error and treatment variance. Chapter 14 features improved graphics showing how the diagram of a two-way design is built from the diagrams of two one-way designs. Chapter 15 has been streamlined, especially in presenting the interpretation of a two-way chi square.

Supplementary Materials

Supporting the text are several ancillaries for students and instructors:

- *Free Data-Analysis Software* A data-analysis program called HMSTAT, custom-tailored to the text by David Abbott of the University of Central Florida, is packaged free with each student text. This menu-driven program can accept and store data, perform all the procedures discussed in the text, and be operated by students with minimal computer background. Directions for using HMSTAT appear in the new Appendix F, prepared by Michael Gayle of the State University of New York at New Paltz, and the Instructor's Resource Manual offers suggestions for integrating computers into the course. The appendix also provides instructions for using MYSTAT and SPSS/PC+ Studentware Plus™ statistical programs.
- *Student Workbook and Study Guide* Additional practice problems are available in the Student Workbook and Study Guide, which I personally revised. Each chapter contains a review of objectives, terms, and formulas, a programmed review, conceptual and computational problems (with answers), and a set of multiple-choice questions similar to those in the Instructor's Resource Manual. A final chapter, called "Getting Ready for the Final

Exam," facilitates student integration of the entire course. Answers to all questions are now provided in each workbook chapter.

- *Instructor's Resource Manual with Test Questions* This supplement, revised by Kay Smith of Brigham Young University, contains approximately 750 test items and problems, as well as suggestions for classroom activities, discussion, and use of statistical software. It also includes answers to the even-numbered end-of-chapter problems from this book. The test items are also available on computer disk for IBM and Macintosh computers.

Acknowledgments

I gratefully acknowledge the help and support of many professionals associated with Houghton Mifflin Company. In particular, I want to thank Michael DeRocco, Rebecca Dudley, Jane Knetzger, and Karen Donovan. A special thanks to Larissa Semanchuk. Finally, I am grateful to the following reviewers who, in evaluating all or parts of this text at one stage or another, provided invaluable feedback.

Bernard C. Beins, Ithaca College
James I. Chumbley, University of Massachusetts, Amherst
Richard P. Deshon, Michigan State University
George M. Diekhoff, Midwestern State University
Susan L. Donaldson, University of Southern Indiana
Warren Fass, University of Pittsburgh at Bradford
Nancy Forger, University of Massachusetts, Amherst
Michael C. Gayle, SUNY New Paltz
Brett W. Pelham, University of California, Los Angeles
Tom Pyle, Eastern Washington University
George A. Raymond, Providence College
Robert A. Reeves, Augusta College
Raymond R. Reno, University of Notre Dame
Kay H. Smith, Brigham Young University
Steven M. Specht, Lebanon Valley College
Lori L. Temple, University of Nevada, Las Vegas
Philip Tolin, Central Washington University
George Whitehead, Salisbury State University
Leonard Williams, Rowan College
Stephanie Gray Wilson, Seton Hall University

Gary W. Heiman

PART 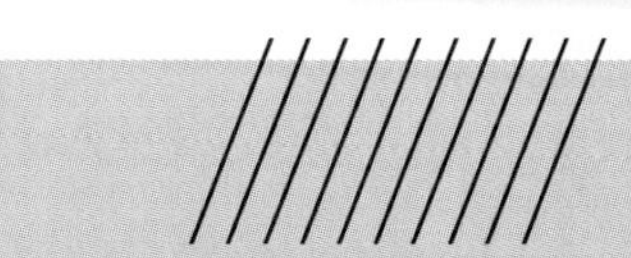

GETTING STARTED

Okay, so you're taking a course in statistics. You probably wonder what's in store for you. Most students know that statistics involve math, but they don't know that studying statistics is much more interesting and educational than merely cranking out a bunch of math problems. A tour through the world of statistics will open up new vistas in how to think, reason, and apply logic, especially when it comes to drawing conclusions from scientific research. You will also learn new ways of simplifying enormous complexities, which will allow you to make confident decisions in seemingly chaotic situations. And statistics will increase your confidence in your own abilities to understand and master an important mental discipline. Moreover, statistics can be fun! Statistics are challenging, there is an elegance to their logic, and you can do nifty things with them. So, keep an open mind, be prepared to do a little work, and you'll be amazed by what happens. You will find that although statistics are a little unusual, they are not incomprehensible, they do not require you to be a math wizard, and they are very relevant to psychology and the behavioral sciences.

The following two chapters provide an overview. The first chapter explains why students in psychology and the behavioral sciences need to learn statistics, and what learning statistics actually involves. The second chapter shows how statistics are generally used in behavioral research.

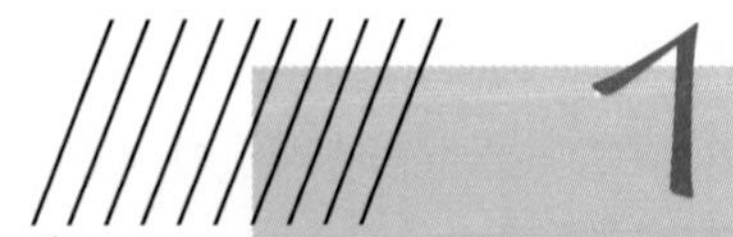

1

Introduction

In this chapter we discuss some common misconceptions about statistics and consider the best way to approach the study of statistics. We also review the math and graphing techniques used in statistics.

As you read this chapter, your goals are simple:

- Learn the general purpose of statistical procedures.
- Develop an effective strategy for learning statistics.
- Prepare yourself for the basic math you'll be using.

SOME ANSWERS TO YOUR QUESTIONS AND CONCERNS ABOUT STATISTICS

Students repeatedly ask the same questions about studying statistics. The answers to these questions may teach you something about psychologists and their use of statistics, as well as relieve any anxiety you may have.

What Are Statistics?

The word *statistics* means different things to different people. To psychologists and behavioral scientists, it is typically a shortened version of the phrase *statistical procedures*, which are computations performed as part of conducting psychological research. The answers obtained by performing certain statistical procedures are also called statistics. Thus, the word **statistics** refers both to statistical procedures and to the answers obtained from those procedures.

What Do Psychologists Do with Statistics?

Psychology and other behavioral sciences are based on empirical research. The word **empirical** means knowledge obtained through observation of events (instead of through intuition or faith). Empirical research involves measurement, and in psychological research we measure behaviors. Such measurement results in numbers, or scores. The scores obtained in research are the **data.** (The word *data,* by the way, is plural, referring to more than one score, so we say "the data are . . .")

Researchers conduct research because they have a research question in mind, and statistics help them to answer the question. For example, to study intelligence, researchers measure the IQ scores of different individuals; to observe how getting rewards influences how often rats press a lever, researchers measure the rats' lever-pressing scores; to learn how children perceive different smells, researchers obtain smell perception data from them; and to study differences in the attitudes of married couples, researchers measure difference scores.

Thus, in any study the researcher ends up with a large batch of data, which must be made manageable and meaningful. Statistical procedures are used to *organize, summarize,* and *communicate* data and then to *conclude* what the data indicate. In essence, statistics help a researcher to make sense out of the data.

But I'm Not Interested in Research; I Just Want to Help People!

Even if you are not interested in conducting research yourself, you still must be an informed user of statistics so that you can understand and learn from other people's research. Let's say that you become a therapist or counselor, and you do not consider yourself a "scientist." You hear of a new therapy that says the way to "cure" people of some psychological problem is to scare the living daylights out of them. This sounds crazy, but what is important is the quality of the research that does or does not support this therapy. As a responsible professional, you would evaluate the research supporting this therapy before you would use it. You could not do so without understanding statistics. However, after you have studied statistics, reading and evaluating such research is relatively easy.

But I Don't Know Anything About Research!

This book is written for the student who has not yet studied the way psychological research is conducted. Whenever we discuss a statistical procedure, we will also discuss simple examples of research that employ the procedure, and this should be enough to get you by. Then, later, when you study research methods, you will know the appropriate statistical procedures to use.

What If I'm Not Very Good at Statistics?

In the grand scheme of things, the application of statistics is one small, although extremely important, step in the research process. Statistics are simply a tool used in the behavioral sciences, just like a wrench is a tool used in the repair of automobile engines. A mechanic need not be an expert wrencher, and a psychologist need not be an expert statistician. Rather, in the same way that a mechanic must understand the correct use of wrenches in order to fix an engine, you must understand the correct use of statistics in order to understand research.

But Statistics Aren't Written in English!

There is no denying that statistics involve many strange symbols and unfamiliar terms. But the symbols and terms are simply the shorthand "code" for communicating statistical results and concepts and for simplifying statistical formulas. A major part of learning statistics is merely learning the code. Think of it this way: in order to understand research you must speak the language, and you are about to learn the language called statistics. Then you will be able to read, understand, and communicate statistical information using the appropriate symbols and terminology. Once you speak the language, much of the mystery surrounding statistics evaporates. In fact, before you're done, you'll understand all of the terms listed in the tables inside the front cover of this book (and all of the symbols listed inside the back cover).

What If I'm Not Very Good at Math?

Although statistics do involve mathematical computations, the math is simple. You only need to know how to add, subtract, multiply, divide, square a number, find a square root, and draw a simple graph. You can perform all statistical calculations with pencil, paper, and a calculator.

What makes statistical procedures *appear* difficult is that most formulas involve a sequence of mathematical operations (first you square the numbers, then you add them together, then you subtract some other number, then you divide by another number, and so on). Working through the formulas is not difficult, but because they are written in a shorthand code it takes a little practice.

So All I Have to Do Is Learn How to Compute the Answers?

No! There is more to statistics than merely crunching numbers through formulas. We will discuss each step in the calculations so that you can see what a formula does and where the answer comes from. But, given the availability of computer programs that perform statistical procedures, the computations are not a big worry. Don't get so carried away with formulas and calculations that you lose sight of the big picture. In the big picture, a statistical answer tells you something about data. Ultimately you want to make sense out of data, and to do that you must compute the appropriate statistic and correctly interpret it. More than anything else, you need to learn *when* and *why* to use each procedure and how to *interpret* the answer from that procedure. Be sure to put as much effort into this as you do into learning how to perform the calculations.

What Should Be My Goal?

Your goal should be to become a competent *user* of statistics, in the same way, for example, that you are a user of telephones. Recognize that you can't be an intelligent user if you do not understand the basic rules of the system. To operate a telephone, you must understand the rules governing long distance and why you must dial different phone numbers to call different people; but you do not need to understand all the inner workings of the telephone network. The same is true in statistics: your task is to learn the rules of the system. Fortunately, statisticians have already developed the internal workings of statistical procedures, so you need only be concerned with applying these procedures in an informed manner.

All Right, So How Do I Learn Statistics?

- Study.
- Think.
- Practice.
- Think.
- Practice some more.

The way to learn statistics is to *do* statistics. Remember, in part you are learning a foreign language called statistics. The way to learn a foreign language is to speak it every day. Therefore, from the outset, try to use precise statistical terminology. In this book, important statistical terms are printed in bold type or are preceded by such phrases as "in statistical language we say . . ." There is also a glossary of terms at the back of the book.

Recognize that because you do not yet speak the language, you will not learn anything if you try to scan a chapter quickly. You must learn to translate the terminology, symbols, and formulas into a verbal description you can understand, and this takes time and effort. Also, accept that this is not a subject you can put off studying until just before an exam and then "cram" the material. If you try this, you won't learn statistics (and your brain will melt). Because you must learn the logic of some rather involved procedures—as well as master the terminology—you should work on statistics a little bit every day. Then you'll get lots of practice, and you'll be able to digest the material in bite-sized pieces. This is the most effective—and least painful—way to learn statistics.

Each new statistical concept builds on previous concepts, so be sure you understand a topic before you move on to the next one. To help you, at the beginning of each chapter I'll tell you the major concepts from preceding chapters that you'll need to already understand. And whenever we discuss a concept covered earlier, there'll be a brief review of the concept or a reference to a previous chapter. Go back and review a topic whenever necessary. (Practice, practice, practice.) Also, statistical concepts are often easier to comprehend if you look at graphs and tables. Take the time to find each table and graph discussed in the text and examine it. It is there to help you. Likewise, every time you encounter a new formula, work through the example presented in the text. Master the formulas and codes at each step, because they often reappear later as part of more complicated formulas.

What Should I Know About This Book?

This book takes an intuitive approach to statistical concepts. Explanations preceded by such phrases as "You can think of it as if . . ." are designed to give your mind something to grasp other than formulas and definitions. Use the examples and analogies to help you remember and understand a concept. However, statistics is a precise discipline, so pay attention to the "official" statistical definition of a concept as well.

The research examples tend to be very simple, and they may give you the impression that every study produces about five scores, and that all scores are nice round numbers like 2 or 5. In fact, real research typically involves many individuals, and the scores may be downright ugly numbers like -104.387. But if you learn to perform a procedure in a simple setting with simple numbers, you'll be able to perform it in a more complex setting.

This book contains several features to direct your attention to important information. At the beginning of each chapter is a list of the major points you should learn from the chapter. After you think you have mastered the chapter, go back and check that you understand all the items in the list. Then most chapters begin with a section titled "More Statistical Notation." This explains the statistical symbols used in the chapter and gives you a chance to familiarize yourself with the new code before you become immersed in the concepts of the chapter. Also, every so often you will see statements labeled "STAT ALERT." These refer to basic concepts you should remember. The last section in each chapter is called "Finally." It contains advice, cautions, and ways to integrate the material from different chapters.

At the end of each chapter is a summary. You should read each statement and determine whether you understand it. There are also practice problems at the end of each chapter, which will help you identify weak spots in your understanding of the material (think of this as a self-test before the real test). After solving each problem, step back and determine whether your answer makes sense. (The correct answer *always* makes sense.) The answers to odd-numbered problems are provided in Appendix E at the back of the book. However, do not cheat yourself by looking at the answer before you have made a serious attempt to solve the problem. Remember, the goal is for *you* to be able to perform these procedures. For quick reference, a list of the formulas discussed is also provided at the end of each chapter.

Your instructor may have chosen the version of this book that comes with a computer program called HMSTAT. To operate the program, you merely tell the computer which procedure you want performed and give it the data; it will then print out the answer. Appendix F, called "Introduction to Statistics Software Packages," explains how to instruct the computer to perform the procedures and gives actual displays from the computer program. The appendix also explains how to use two other widely available applications: the MYSTAT and SPSS/PC+ Studentware Plus™ programs. But remember, the computer is only there to perform computations. Even the fanciest computer cannot tell you *when to use a procedure* or *what the answer means.* That's your job.

REVIEW OF MATHEMATICS USED IN STATISTICS

The remainder of this chapter reviews the basic math operations used in performing statistical procedures. As you will see, there is a system for statistical notation, for rounding and transforming scores, and for creating graphs.

Basic Statistical Notation

Statistical notation is the standardized code for the mathematical operations performed in the formulas, for the order in which operations are performed, and for the answers we obtain.

Identifying mathematical operations We write formulas in statistical notation so that we can apply them to any data. In statistical notation, we usually use the symbol X or Y to stand for each individual score obtained in a study. When a formula says to do something to X, it means to do it to all of the scores we are calling X scores. When a formula says to do something to Y, it means to do it to all of the scores we are calling Y.

The mathematical operations required by the formulas are basic ones. Addition is indicated by the plus sign and subtraction is indicated by the minus sign. We read from left to right, so $X-Y$ is read as "X minus Y." (I *said* this was basic!) This order is important, however, because $10-4$, for example, results in $+6$, but $4-10$ results in -6. With subtraction, always pay attention to what is subtracted from what, and whether the resulting answer is positive or negative: in statistics, the correct answer is often a negative number.

We indicate division by forming a fraction, such as X/Y. The number above the dividing line is called the numerator, and the number below the line is called the denominator (the *d* in denominator stands for "down below"). *Always reduce fractions to decimals,* dividing the denominator *into* the numerator. (After all, 1/2 equals .5, not 2!)

Multiplication is indicated in one of two ways. We may place two components next to each other: XY means "multiply X times Y." Or we may indicate multiplication using parentheses: 4(2) and (4)(2) both mean "multiply 4 times 2."

The symbol X^2 indicates that we should square the score, so if X is 4, X^2 is 16. Conversely, $\sqrt{X}$ means "find the square root of X," so $\sqrt{4}$ is 2. (The symbol $\sqrt{}$ also means "use your calculator.")

Determining the order of mathematical operations Statistical formulas often call for a series of mathematical steps. Sometimes the steps are set apart by parentheses. Parentheses mean "the quantity," so we always find the quantity inside the parentheses first and then perform the operations outside of the parentheses on that quantity. For example, $(2)(4+3)$ indicates that we multiply 2 times "the quantity 4 plus 3." So first we add, which gives us (2)(7), and then we multiply to get 14.

A square root sign also operates on "the quantity," so we may have to compute the quantity inside of the square root sign first. Thus $\sqrt{2+7}$ indicates that we want the square root of the quantity $2+7$; so $\sqrt{2+7}$ becomes $\sqrt{9}$, which is 3.

Most formulas are giant fractions. Pay attention to how far the dividing line is drawn, because the length of a dividing line determines the quantity that is in the numerator and the denominator. For example, we might see a formula that looks like this:

$$\frac{\frac{6}{3}+14}{\sqrt{64}}=\frac{2+14}{\sqrt{64}}=\frac{16}{\sqrt{64}}=\frac{16}{8}=2$$

The longest dividing line indicates that we divide the square root of 64 into the quantity in the numerator. The dividing line in the fraction in the numerator is only under the 6, so we first divide 6 by 3, which is 2. Then we add 14, for a total of 16. In the denominator, the square root of 64 is 8. After dividing, the final answer is 2.

If you become confused in reading a formula, remember that there is an order of precedence of mathematical operations. *Unless otherwise indicated,* perform squaring or the taking of a square root first, then multiplication or division, and then addition or subtraction. Thus, for (2)(4) + 5, we multiply 2 times 4 first and then add 5. For $2^2 + 3^2$, we square first, so we have 4 + 9, which is 13. On the other hand, $(2 + 3)^2$ indicates the quantity 2 + 3 squared, so we work inside the parentheses first; after adding, we have 5^2, which is 25.

Solving equations to find the answer We perform the operations in a formula to find an answer, and we have symbols that stand for that answer. For example, in the formula $B = AX + K$, B stands for the numerical answer we will obtain. Get in the habit of thinking of the *symbols* as quantities. Because B stands for a number, we can discuss whether B is larger than, smaller than, or equal to some other number or symbol.

In each formula, we will find the value of the single term that is isolated on one side of the equals sign, and we will know the values of the terms on the other side of the equals sign. For the formula $B = AX + K$, say that $A = 4$, $X = 11$, and $K = 3$. Now compute B. In working any formula, the first step is to copy the formula and then rewrite it, replacing the symbols with their known values. Thus, we start with

$$B = AX + K$$

Filling in the numbers gives

$$B = 4(11) + 3$$

To keep track of your calculations, it is very important to rewrite the formula again after each step in which you perform *one* mathematical operation. Above, multiplication takes precedence over addition, so we multiply and then write the formula as

$$B = 44 + 3$$

After adding, we have

$$B = 47$$

For simple procedures, you may have an urge to skip rewriting the formula after each step. Don't! That's a good way to introduce errors.

Rounding

Close counts in statistics, so you must carry out your calculations to the appropriate number of decimal places: there is a big difference between $1.64\underline{4}$ and $1.64\underline{6}$. Usually you must round off your calculations at some point. Always carry out calculations so that the *final* answer after rounding has two more decimal places than the original scores. *Do not round off at each intermediate step in the calculations; round off only at the end!* For example, if you have whole-number scores, you will want the final answer to contain two decimal places. Therefore, carry out the intermediate calculations to at least three decimal places, and then round off the final answer to two decimal places.

To round off a calculation to a particular decimal place, mathematicians use the following rules:

> If the number in the next decimal place is 5 or greater than 5, round up. For example, to round to two decimal places, 2.366 is rounded to 2.370, which becomes 2.37.
>
> If the number in the next decimal place is less than 5, round down: 3.524 is rounded to 3.520, which becomes 3.52.

Because there are five digits between 5 and 9 and five digits between 0 and 4, these rules will have us rounding up about 50% of the time and rounding down about 50% of the time.

Recognize that we add zeroes to the right of the decimal point as a way of indicating the level of precision we are using. Rounding 4.996 to two decimal places produces 5, but to show we used the precision of two decimal places, we report it as 5.00.

Transformations

A **transformation** is a mathematical procedure for systematically converting a set of scores into a different set of scores. Adding 5 to each score is a transformation or converting "number correct" into "percent correct" is a transformation. Many statistical procedures are nothing more than involved transformations.

We transform data for one of two reasons. First, if the original scores are awkward, we may perform a transformation to make them easier to work with. For example, if all of the scores contained a decimal, we might transform them by multiplying every score by 10 so that we would have no decimals. Second, we perform a transformation to make different kinds of scores comparable. For example, if you obtained 8 out of 10 on a statistics quiz and 75 out of 100 on an English quiz, it would be difficult to compare the two scores. However, if you transformed the grades to percent correct on each test, you would no longer be comparing apples to oranges.

In statistics, we rely heavily on transformations to proportions and percents.

Proportions A **proportion** is a decimal number between 0 and 1 that indicates a fraction of the total. To transform a number to a proportion, simply divide the number by the total. If 4 out of 10 people pass an exam, then the proportion of people passing the exam is 4/10, which equals .4. If on an exam you score 6 correct out of a possible total of 12 correct, the proportion you have correct is 6/12, which is .5. Conversely, to determine what number constitutes a certain proportion, multiply the proportion times the total. Thus, to find how many questions out of a total of 12 you must answer correctly to get .5 correct, you multiply .5 times 12, and voilà, 6 is .5 of 12.

Percents We can also transform a proportion into a percent. A **percent** is a proportion multiplied by 100. Above, the proportion correct was .5, so you had (.5)(100) or 50% correct. To transform the original test score of 6 out of 12 to a percent, first divide the score by the total to find the proportion and then multiply by 100. Thus, (6/12)(100) equals 50%.

To work in the other direction and transform a percent into a proportion, divide the percent by 100 (50/100 equals .5). To find the original test score that corresponds to a certain percent, transform the percent to a proportion and then multiply the proportion times the total number possible. Thus, to find the number that corresponds to 50% of 12, transform 50% to the proportion .5 and then multiply .5 times 12. Doing all of these steps together, we find that 50% of 12 is equal to (50/100)(12), or 6.

Recognize that percents are whole numbers: think of 50% as 50 of those things called percents. On the other hand, a decimal in a percent is a proportion of *one* percent. Thus, .2% is .2, or two-tenths, of one percent, which is .002 of the total.

Creating Graphs

One type of statistical procedure is none other than plotting graphs. In case it has been a long time since you have drawn one, recall that the horizontal line across the bottom of a graph is called the *X* axis, and the vertical line at the left-hand side is called the *Y* axis. The axes should be drawn so that the height of the *Y* axis is about 60 to 75% of the length of the *X* axis. Such a graph is shown in Figure 1.1. The basic approach in plotting any data is to use pairs of *X* and *Y* scores. Where the two axes intersect is always labeled as a score of 0 on *X* and a score of 0 on *Y*. On the *X* axis, scores become larger positive scores as we move to the *right,* away from zero. On the *Y* axis, scores become larger positive scores as we move *upward,* away from zero.

Say that we measured the height and weight of several people. To plot their scores, we decide to place weight on the *Y* axis and height on the *X* axis. (How we decide

FIGURE 1.1 Arrangement of the *X* and *Y* Axes in a Graph

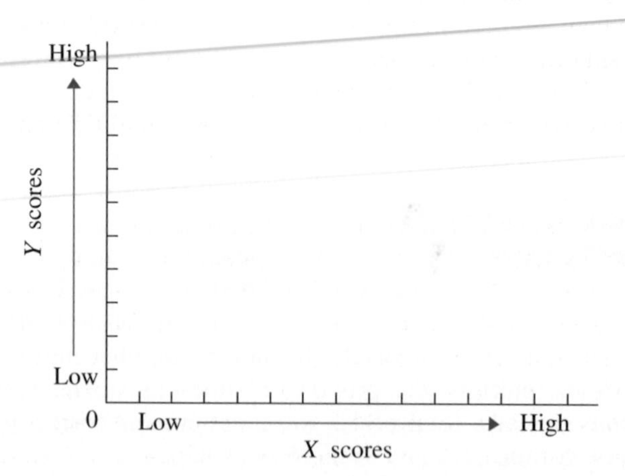

FIGURE 1.2 Plot of Height and Weight Scores

Person	*Height*	*Weight*
Jane	63	130
Bob	64	140
Mary	65	155
Tony	66	160
Sue	67	165
Mike	68	170

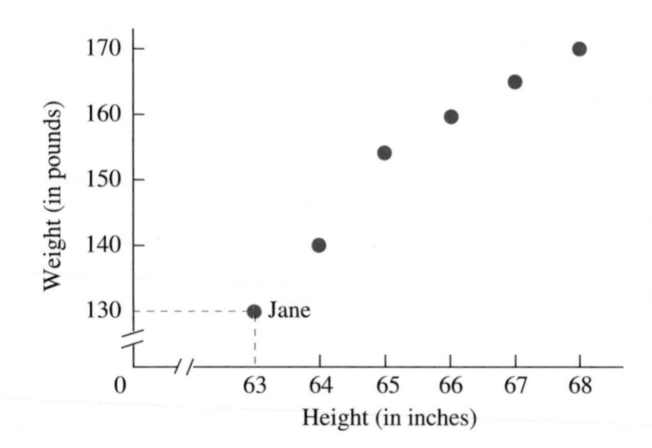

this will be discussed later.) We plot the scores as shown in Figure 1.2. Notice that the lowest height score is 63, and so on the *X* axis the lowest score is 63. Whenever there is a large gap between 0 and the lowest score we are plotting, the axis is compressed using two diagonal lines (//). These diagonal lines indicate that we cut out the part of the *X* axis between 0 and 63 and slid the graph over closer to 0. We did the same thing to the *Y* axis, because the lowest weight is 130.

To fill in the body of the graph, we plot the scores from the table on the left. We start with Jane. She is 63 inches tall and weighs 130 pounds, so we place a dot above the height of 63 and opposite the weight of 130. And so on. Each dot on the graph is called a **data point.** Notice that we can read the graph by using the scores on one axis and the data points. For example, to find the weight of the person who has a height of 67, travel vertically from 67 to the data point and then horizontally back to the *Y* axis: 165 is the corresponding weight.

In later chapters you will learn when to connect the data points with lines and when to create other types of figures. Regardless of the final form of a graph, always label the *X* and *Y* axes to indicate what the scores measure (not just *X* and *Y*), and always give your graph a title indicating what the figure describes.

When creating any graph, it is important to make the spacing between the labels identifying the scores on an axis reflect the spacing between the actual scores. In Figure 1.2, the labels 64, 65, and 66 are equally spaced on the graph, because the difference between 64 and 65 is the same as the difference between 65 and 66. However, in other situations, the labels may not be equally spaced. For example, the labels for scores of 10, 20, and 40 would not be equally spaced, because the distance between 10 and 20 is not the same as the distance between 20 and 40.

Sometimes there are so many different scores that we cannot include a label for each score. The units we use in labeling each axis determine the impression the graph gives. Say that for the above weight scores, instead of labeling the *Y* axis in units of 10 pounds, we labeled it in units of 100 pounds (0, 100, 200), as shown in Figure 1.3. This graph shows the same data as Figure 1.2, but changing the scale on the *Y* axis creates a much flatter pattern of data points. Here we have the misleading impression that regardless of their height, the people

FIGURE 1.3 Plot of Height and Weight Scores Using a Different Scale on the *Y* Axis

Height	*Weight*
63	130
64	140
65	155
66	160
67	165
68	170

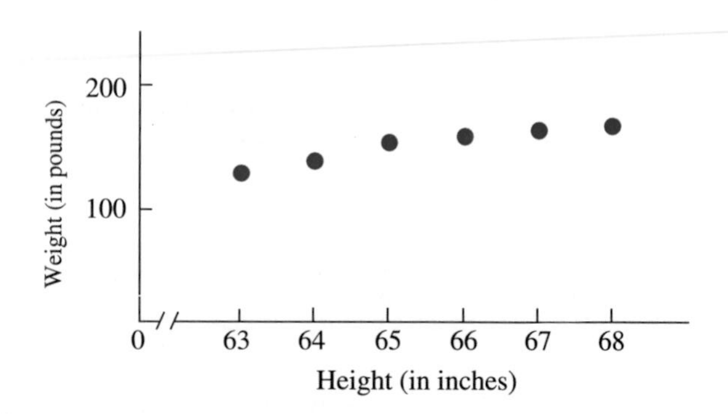

all have about the same weight. However, in looking at the actual scores, we see that this is not the case. Whenever you create a graph, label the axes in a way that honestly presents the data, without exaggerating or minimizing the pattern formed by the data points.

FINALLY

That's all the basic math, algebra, and graphing you need to get started. You are now ready to begin learning to use statistics. In fact, you already use statistics. If you compute your grade average or if you ask your instructor to "curve" your grades, you are using statistics. When you understand from the nightly news that Senator Fluster is projected to win the presidential election or when you learn from a television commercial that Brand X "significantly" reduces tooth decay, you are using statistics. You simply do not yet know the formal names for these statistical procedures or the logic behind them. But you will.

CHAPTER SUMMARY

1. Whether or not they are active researchers, psychologists and behavioral scientists rely on statistical procedures.

2. All *empirical* research is based on observation and involves some form of measurement, resulting in numbers, or scores. These scores are the *data.*

3. The term *statistics* has two meanings. First, it refers to a group of procedures and formulas used to analyze data. Second, it refers to the numerical answers researchers obtain by performing certain statistical procedures.

4. Statistical procedures are used to make sense out of data. We use statistics to organize, summarize, and communicate data and to draw conclusions about what the data indicate.

5. The goal in learning statistics is to know *when* to perform a particular procedure and how to *interpret* the answer.

6. Unless otherwise indicated, the order of mathematical operations is to square or find the square root first, multiply or divide second, and then add or subtract.

7. Perform mathematical operations contained within a set of parentheses first, to find the quantity. Then perform additional operations outside of the parentheses on that quantity. Similarly, perform operations within the square root sign prior to finding the square root of the quantity, and perform operations above and below the dividing line of a fraction prior to dividing.

8. Round off the final answer in a calculation to two more decimal places than are in the original scores. If the digit in the next decimal place is equal to or greater than 5, round up; if the digit is less than 5, round down.

9. A *transformation* is a mathematical procedure for systematically converting one set of scores into a different set of scores. We transform data to make scores easier to work with or to make different kinds of scores comparable.

10. A *proportion* is a decimal between 0 and 1 that indicates a fraction of the total. To transform a score to a proportion, divide the score by the total. To determine the score that constitutes a certain proportion, multiply the proportion times the total.

11. To transform a proportion to a *percent,* multiply the proportion times 100. To transform an original score to a percent, first find the proportion by dividing the score by the total and then multiply by 100. Conversely, to transform a percent to a proportion, divide the percent by 100. To find the original score that corresponds to a certain percent, transform the percent to a proportion and then multiply the proportion times the total.

12. In creating a graph, draw the X and Y axes so that the height of the Y axis is 60 to 75% of the length of the X axis. Scores on the X axis start at zero and proceed to higher positive scores as you read to the *right.* Scores on the Y axis start at zero and proceed to higher positive scores as you read *upward.*

13. A *datapoint* is a dot plotted on a graph to represent a pair of X and Y scores.

14. Place the labels on each axis of a graph so that they accurately reflect the distance between scores and show how the scores change.

PRACTICE PROBLEMS

(Answers for odd-numbered problems are provided in Appendix E.)

1. Why do researchers need to learn statistics?

2. As a student learning statistics, what should be your goals?

3. What does the term *statistical notation* refer to?

4. As part of a study, a researcher measures the IQ scores of a group of college students. What four things will the researcher use statistics for?
5. *a.* To how many places do we round off our final answer after a calculation?
   b. If we are rounding to two decimal places, what are the rules for rounding up or down?
6. For each of the following calculations, to how many places will you round off your final answer?
   a. When measuring the number of questions students answered correctly on a test.
   b. When measuring what proportion of the total possible points students have earned in a course.
   c. When counting the number of people having various blood types.
   d. When measuring the number of dollar bills possessed by each person in a group.
7. If given no other information, what is the order in which we perform mathematical operations?
8. *a.* What is a transformation?
   b. Why do we transform data?
9. What is a proportion and how is it computed?
10. What is a percentage?
11. *a.* What proportion is 5 out of 15?
    b. What proportion of 50 is 10?
    c. One in a thousand equals what proportion?
12. Transform each answer in problem 11 to a percent.
13. How do we transform a percentage to a proportion?
14. The intermediate answers from some calculations based on whole number scores are $X = 4.3467892$ and $Y = 3.3333$. We now want to find $X^2 + Y^2$. What values of X and Y do we use?
15. Round off the following numbers to two decimal places:
    a. 13.7462
    b. 10.043
    c. 10.047
    d. .079
    e. 1.004
16. For $Q = (X+Y)(X^2+Y^2)$, find the value of Q when $X = 3$ and $Y = 5$.
17. Using the formula in question 16, find Q when $X = 8$ and $Y = -2$.
18. For $X = 14$ and $Y = 4.8$, find D:

$$D = \left(\frac{X - Y}{Y}\right)(\sqrt{X})$$

19. Using the formula in question 18, find D for $X = 9$ and $Y = -4$.

20. Of the 40 students in a gym class, 13 played volleyball, 12 ran track (4 of whom did a push-up), and the remainder were absent.

a. What proportion of the class ran track?

b. What percentage played volleyball?

c. What percentage of the runners did a push-up?

d. What proportion of the class was absent?

21. In your stat course, there are three exams: I is worth 40 points, II is worth 35 points, and III is worth 60 points. Your professor defines a passing grade as earning 60% of the points.

a. What score must you obtain on each exam in order to pass it?

b. In total you can earn 135 points in the course. How many points must you earn from the three exams combined in order to pass the course?

c. You actually earn a total of 115 points during the course. What percent of the total did you earn?

22. There are 80 students enrolled in statistics.

a. You and 11 others earned the same number of points. What percent of the class received your grade?

b. Forty percent of the class received a grade of C. How many students received a C?

c. Only 7.5% of the class received a D. How many students is this?

d. A student claims that .5% of the class failed. Why is this impossible?

23. What is a data point?

24. Create a graph showing the data points for the following scores.

X Score	***Y Score***
Student's Age	*Student's Test Score*
20	10
25	30
35	20
45	60
25	55
40	70
45	3

2

Statistics and the Research Process

To understand the upcoming chapter:

- From Chapter 1, understand (1) that we use statistics to make sense out of data, (2) that your goal is to learn when to use a particular statistic and what the answer means, and (3) how to create and interpret graphs.

Then your goals in this chapter are to learn:

- What a relationship between variables is and what is meant by the "strength" of a relationship.
- That when we draw inferences about a population of scores based on a sample of scores, we are actually describing how a behavior occurs in a given situation in nature.
- How random sampling leads to representative or unrepresentative samples.
- When and why descriptive and inferential procedures are used, and what is meant by the terms *statistic* and *parameter.*
- The difference between an experiment and a correlational study, and how to recognize the independent variable, the conditions, and the dependent variable in an experiment.
- The different scales of measurement researchers use.

To understand how psychologists use statistics, you need to first understand why and how they conduct research. As you begin to understand behavioral research, you will see how statistics fit in. In this chapter, we discuss the basics of behavioral research, present an overview of statistical procedures, and get you started in the language of statistics.

THE LOGIC OF SCIENTIFIC RESEARCH

The goal of science is to understand the "laws of nature"—the rules that describe how the universe operates. Behavioral scientists study the laws of nature pertaining to the behavior of living organisms. Implicitly, they assume that there are specific causes or influences that govern every behavior. A law of nature applies to all members of a particular group of living organisms, so the same cause has the same effect, more or less, on the behavior of all members of that group. Thus, the goal of behavioral science is to understand those factors that cause or influence a behavior in a particular group. When psychologists study the mating behavior of sea lions, the development of language in children, social interactions between gorillas, or neural firing in a human's brain, they are ultimately studying the laws of nature.

There are two major components to the logic researchers use to draw conclusions about a law of nature, both involving statistical procedures: we examine our data to learn how the law operates, and we draw conclusions about the group of individuals the law applies to.

The first step in any empirical study is to set up the process of measuring and examining our data.

Variables

The laws of nature are extremely complex, with many factors all combining to cause or influence a behavior. In conducting a particular study, psychologists try to simplify things by examining a law of nature in terms of one specific factor that causes or influences one specific behavior in a specific situation. As we'll see, research and statistics involve a series of translations. First we translate a general hypothesis about a law into specific, measurable observations. Then, after examining our measurements, we translate them back into conclusions about how nature generally operates when it comes to this behavior. For example, cognitive and educational psychologists have proposed various laws involving "learning" and "memory" to describe how people build their knowledge. One component of this process seems to be that a person must repeatedly interact with a set of information in order to learn it. This is a rather global statement, however, which could apply in many situations. Therefore, say we decide to examine this general aspect of learning by translating it into a specific question: Does studying statistics cause you to learn them?

We have thus begun to simplify nature, by translating a law into something, X, that causes or influences some behavior, Y. To complete our translation, we must decide precisely what we mean by "studying" and how we will measure it, and what we mean by "learning" and how we will measure it. In the language of research, the factors we measure that cause or influence behaviors—as well as the behaviors themselves—are called variables. A **variable** is anything that, when measured, can produce two or more different values. A few of the variables found in psychological research include your age, race, gender, and intelligence; your personality type, attitudes, or political or religious affiliation; how anxious, angry, or aggressive you are; how attractive you find someone; how long or hard you will work at a task; how accurately you recall a situation; or how long it takes you to make a particular decision. Other variables measure how loud a noise is; how much you are paid to do something; the type or amount of information occurring in a situation, and whether it is presented auditorily or visually; whether a scary event occurs; how potent a drug is that you consume; and how friendly someone acts toward you.

We can separate variables into two general categories. If a score indicates the amount of a variable that is present, the variable is called a **quantitative** variable. A person's height, for example, is a quantitative variable because a score indicates the quantity of height that is present. Some variables, however, cannot be measured in amounts. Instead, a score classifies an individual on the basis of some characteristic. Such variables are called **qualitative, or classification,** variables. A person's gender, for example, is a qualitative variable, because the "score" of male or female indicates a *quality,* or category.

For our study, we might measure "studying" using such variables as how much effort is put into studying or the number of times a chapter is read, but say we select the variable of the number of hours spent studying for a statistics test. We might measure "learning" by measuring how well statistical results can be interpreted or how quickly a particular procedure can be performed, but say we select the variable of grades on the statistics test.

Relationships Between Variables

Once the variables are selected, then the most basic assumption in scientific research is this: if Y is influenced by or otherwise related to X by a law of nature, then *different* amounts or categories of Y will occur when *different* amounts or categories of X occur. Thus, if nature relates those mental activities we call studying to those mental activities we call learning, then different amounts of learning should occur with different amounts of studying. For us to learn about this law of nature, we will observe and measure some students and *see* if different amounts of our learning variable do occur with different amounts of our studying variable.

Thus, we translate natural events and behaviors into scores on our variables. Then we examine a law of nature by examining the mathematical relationship between our variables. A **relationship** occurs when a change in one variable is accompanied by a consistent change in another variable. Because we measure individuals' scores on variables, a relationship is a *pattern* in which certain scores on one variable are paired with certain scores on another variable. As the scores on one variable change, the scores on the other variable change in a consistent manner. For example, rumor has it that there is a relationship in which the more you study, the higher your test grade will be, but the less you study, the lower your grade will be.

What might this relationship look like? Say that we asked some students how long they studied for a test and their subsequent grades on the test. We might find the scores shown in Table 2.1. These scores form a relationship, because as the scores on the variable of study time change (increase), the scores on the

TABLE 2.1 Scores Showing a Relationship Between the Variables of Study Time and Test Grades

Student	*Study time in hours*	*Test grades*
Jane	1	F
Bob	1	F
Sue	2	D
Tony	3	C
Sidney	3	C
Ann	4	B
Rose	4	B
Lou	5	A

variable of test grades also change in a consistent fashion (also increase).[1] Further, when the scores on the study time variable do *not* change (for example, Jane and Bob both studied for 1 hour), the scores on the grade variable do not change either (for example, they both received F's). In statistics, we often use the term *association* when talking about relationships. In the same way that your shadow's movements are associated with your movements, low study times are associated with low test grades and high study times are associated with high test grades.

As researchers, we want to understand such a relationship, and a major use of statistical procedures is to examine the scores in a relationship and the pattern they form. The simplest relationships fit either the pattern "the more you *X*, the *more* you *Y*" or the pattern "the more you *X*, the *less* you *Y*." Thus, the saying "the bigger they are, the harder they fall" describes a relationship, as does that old saying "the more you practice statistics, the less difficult they are." Relationships may also form more complicated patterns where, for example, more *X* at first leads to more *Y*, but beyond a certain point even more *X* leads to *less* *Y*. For example, the more you exercise, the better you feel, but beyond a certain point more exercise leads to feeling less well, as pain, exhaustion, or death sets in.

Although the above examples reflect relationships involving quantitative variables, we can also study relationships involving qualitative variables. In such a relationship, as the category or quality changes, scores on the other variable change in a consistent fashion. Consider, for example, that men typically are taller than women. If you think of male and female as "scores" on the variable of gender, then this is a relationship, because as gender scores change (going from male to female), height scores change in a consistent fashion (decrease). We can study any combination of qualitative and quantitative variables in a relationship.

1. The data presented in this book are a work of fiction. Any resemblance to real data is purely a coincidence.

Strength of a relationship In Table 2.1, there is a perfectly consistent association between hours of study time and test grades: all those who studied the same amount received the same grade. In the real world, however, all of the people who study the same amount of time will not receive the same test grade. (Life is not fair.) Consistency is not all or nothing, however, so a relationship can be present even if the association between scores is not perfectly consistent. There can be some *degree* of consistency so that as the scores on one variable change, the scores on the other variable *tend* to change in a consistent fashion. For example, Table 2.2 presents a relationship between the variables of number of hours spent studying and number of errors made on a test. To some degree, higher scores on the study time variable tend to be associated with lower scores on the error variable. However, every increase in study time is not matched perfectly with a decrease in errors, and sometimes the same studying score produces different error scores. The consistency in a particular relationship is called its strength: The **strength of a relationship** is the extent to which one value of the Y variable is consistently associated with one and only one value of the X variable. It is the *degree of association* between the variables.

Relationships are not perfectly consistent because of individual differences. The term **individual differences** refers to the fact that no two individuals are identical and that differences in genetic make-up, experience, intelligence, personality, and many other variables all influence behavior in a given situation. It is because of individual differences that a particular law of nature operates in *more or less* the same way for all members of a specific group; other laws govern how an individual's characteristics combine to determine how a particular law operates. Thus, for example, the hours that you study is only one of a number of factors that influence your test performance. Your intelligence, aptitude, and motivation also play important roles. Because students exhibit individual differences in these characteristics, they will each be influenced differently by the same amount of studying, and thus they will behave differently on a test as a result. Because individual differences produce differences in test scores for a particular study time, scores will be only somewhat consistently associated with study times.

TABLE 2.2 Scores Showing a Relationship Between Study Time and Number of Errors on Test

Student	*Study time in hours*	*Number of errors on test*
Amy	1	12
Joe	1	11
Cleo	2	11
Jack	2	10
Terry	3	9
Chris	4	9
Sam	4	8
Gary	5	7

L

Mathematically, the scores from two variables can produce any strength of relationship, from perfectly consistent association to no association. However, in actual psychological research, individual differences produce relationships that at best contain only some degree of association. Theoretically, perfectly consistent relationships are possible, but they do not occur in the real world. Therefore, it is never enough to merely say that we have observed a particular pattern in a relationship: we must also determine the strength of the relationship.

> ***STAT ALERT*** In statistics and research, we are concerned not only with the existence of a relationship but also with the strength of the relationship, or with the degree of association in it.

No relationship When there is no consistent association between two variables, there is no relationship. For example, there is (I think) no relationship between the number of chocolate bars people consume each day and the number of times they blink each minute. If we measured the scores of individuals on these two variables, we might have the data shown in Table 2.3. Here there is no consistent change in the scores on one variable as the scores on the other variable change. Instead, the same blinking scores tend to show up for each chocolate bar score. Notice that, because there is no relationship, *differences* in the amount of chocolate consumed are not associated with consistent differences in blinking. Ultimately, when we look for a relationship, we look for a pattern of differences among the scores: for each different score on one variable, there should tend to be a different group of scores on the other variable. Much of statistics boils down to trying to identify such differences.

To summarize then, when there is a relationship, certain scores on one variable tend to be associated with certain scores on the other variable, and when the scores on one variable change, there tends to be a consistently different group of scores on the other variable.

Graphing relationships It is important that you be able to recognize a relationship and its strength when looking at a graph. It is not a coincidence that in a graph we have the X and Y axes and that we refer to a relationship as involving an X and Y variable with X and Y scores. But how do we decide which variable

TABLE 2.3 Scores Showing No Relationship Between Number of Chocolate Bars Consumed per Day and Number of Eye-Blinks per Minute

Student	*Number of chocolate bars consumed per day*	*Number of eye-blinks per minute*
Mark	1	20
Ted	1	22
Ray	2	20
Denise	2	23
Maria	3	23
Irene	3	20

to call X or Y? In any study, we implicitly ask this question: For a *given* score on one variable, what scores occur on the other variable? The "given" variable is the X variable (plotted on the X axis), and the other variable is the Y variable (plotted on the Y axis). Thus for our study we ask, "For a given amount of study time, what test grades occur?" So study time is the X variable and test grades is the Y variable.

Notice that once you have identified your X and Y variables, there is a special way of communicating the relationship between them. We describe a relationship using this general format: "Scores on the Y variable change *as a function of* changes in the X variable." Thus so far we have discussed relationships involving "higher test grades as a function of greater study times" and "number of eye blinks as a function of amount of chocolate consumed." Likewise, if you hear of a study titled "Differences in Career Choices as a Function of Personality Type," you know that the researcher looked to see how Y scores that measure career choices consistently changed as X scores that measure personality types changed.

STAT ALERT The "given" variable in a study is designated the X variable, and we describe a relationship using the format "changes in Y as a function of changes in X."

As we saw in Chapter 1, we can create a graph by plotting the data points formed when particular X scores are paired with particular Y scores. Just as you read a sentence from left to right, you read a graph from left to right along the X axis, simultaneously observing the pattern of change in the Y scores. In essence you ask, "As the scores on the X axis increase, what happens to the scores on the Y axis?" Figure 2.1 shows the data points from four sets of data.

In Graph A, we have our original test-grade and study-time data from Table 2.1. Here, the pattern is such that as the X scores increase, the Y scores also increase. Further, because everyone who obtained a particular value of X obtained the same value of Y, the graph shows that there is perfectly consistent association.

Graph B shows test errors as a function of number of hours studied from Table 2.2. Here the pattern is that increasing X scores are associated with decreasing values of Y. Further, because there are different values of Y at an X score, this relationship is weaker and less than perfectly consistent.

Say that Graph C shows the relationship between Y scores reflecting different career choices and X scores reflecting different personality types. There is again a pattern reflecting a relationship where, as the X scores increase, the Y scores also increase. However, here a relatively wide range of different career-choice scores are paired with each personality-type score, reflecting an even weaker relationship than in Graph B.

Graph D shows the eye-blink and chocolate-bar data from Table 2.3. We see no consistent pattern of change in Y scores; more or less the same values of Y are associated with each value of X. As here, whenever a graph shows an essentially flat pattern—so that Y scores tend to neither increase nor decrease with increasing X scores—the graph reflects zero association and no relationship.

FIGURE 2.1 Plots of Data Points from Four Sets of Data

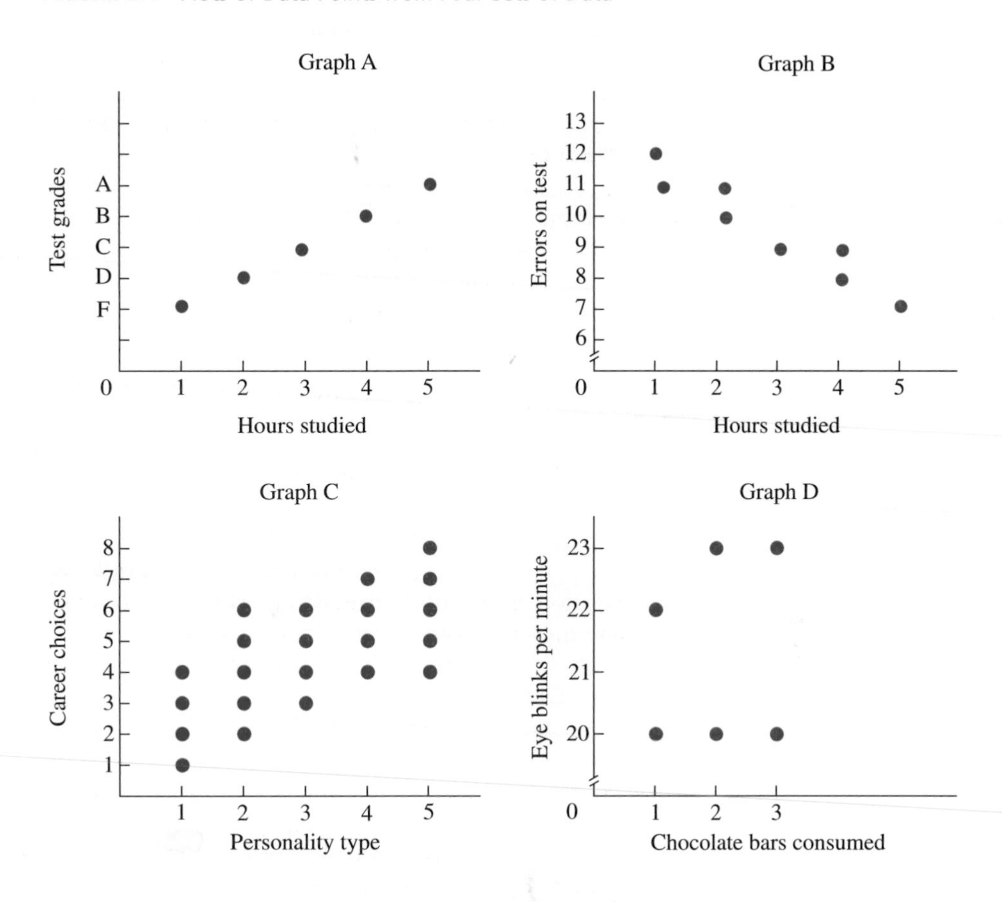

Using Relationships to Discover Laws

Now you can see how relationships form part of the logic of scientific research. The goal of science is to discover the laws of nature, and a relationship between the variables in a study is a telltale sign that a law of nature is at work. As in our original study-time research, if we think we have identified a law of nature, we translate it into variables we can measure. Then we conduct research to see if there is a relationship between these variables. We use statistical procedures to examine the scores, determine the degree of association, and understand the mathematical relationship. Then, because the scores reflect behaviors and events, we translate the mathematical relationship between the scores into the relationship between the behaviors and events they reflect. Ultimately, then, we will have learned something about how the universe operates. For example, we have observed that more hours spent studying is related to better test performance. Thus we have observed *one instance,* at least, where it looks like, in nature, more of the mental activities we call studying do result in more of the mental activity we call learning. However, because we did not observe that eye-blink scores are related to chocolate-bar scores, we have no evidence that the activities involved in eating chocolate bars influence the behavior of blinking. That, more or less, is the logic for interpreting scientific data.

> ***STAT ALERT*** The focus of all scientific research is the study of relationships.

Drawing conclusions about a relationship, however, is only one component of the logic of research. Combined with it is the logic of sampling from a population.

SAMPLES AND POPULATIONS

Recall that any law of nature applies to a specific group of individuals (all mammals, all humans, all male white rats, all four-year-old English-speaking children). The entire group to which the law applies is called a **population.** The population contains all possible members of the group, so we usually consider a population to be infinitely large. Although ultimately scientists discuss the population of *individuals,* in statistics we talk of the population of *scores,* as if we had already measured the behavior of everyone in the population in a particular situation. Thus, you can think of a population as the infinitely large group of all possible scores we would obtain if we could measure the behavior of everyone of interest in a particular situation.

Of course, to measure an infinitely large population would take roughly forever. Instead, therefore, we study a sample of the population. A **sample** is a relatively small subset of a population that is intended to represent, or stand in for, the population. It is the sample or samples that we measure in a study and the scores from the sample(s) constitute the data. The individuals measured in a sample are called the **subjects.** Again, however, as with a population, in statistics we discuss a sample of scores as if we had already measured the subjects in a particular situation.

Notice that the definitions of a sample and a population depend on your perspective. Say that we measure the students in your statistics class on some variable. If these are the only individuals we are interested in discussing, then we have measured the population of scores. On the other hand, if we are interested in the population of all college students studying statistics, then we have collected a sample of scores to represent that population. But if we are interested in both the population of college males studying statistics and the population of college females studying statistics, then the males in the class are one sample and the females in the class are another sample, and each sample is intended to represent its respective population. And there is one other way to view a sample and a population. One or more scores from *one* student can be considered a sample representing the population of all possible scores which that student might produce. Thus, a population is any complete group of scores that would be found in a particular situation, and a sample is a subset of those scores that we actually measure in that situation.

Drawing Inferences About a Population

The logic behind samples and populations is this: because we usually cannot measure the entire population, we use the scores in a sample to *infer* or to *estimate* the scores we would expect to find if we could measure the whole population. Essentially we treat the observations in our sample as rather interchangeable with any other observations we might obtain from the population: any group of subjects—any sample—*should* give scores similar to any other group's scores. Therefore, by observing our subjects, we are observing the equivalent of all other potential subjects. All potential subjects *are* the population. Thus, it *should* be true that as the sample goes, so would the population go, if we could measure it completely. So, for example, if a sample of business executives produces certain motivation scores in a particular situation, then the population of all business executives should score in a similar way in the same situation. Or, if a sample of five-year-old children obtains certain aggressiveness scores, then all five-year-olds should show about the same level of aggressiveness in an identical situation. And remember that scores reflect behavior; by translating the scores in a sample into the behaviors they reflect, we can infer the behavior of the population. Thus, when the nightly news predicts who will win the presidential election based on the results of a survey, researchers are using a sample to represent a population. The scores from the sample survey (usually containing about 1200 voters) are used to infer the voting *behavior* of the population of over 50 million voters.

In the same way, if our sample of study-time and test-score data shows a particular relationship, we want to infer that a similar relationship would be found in the population. Say that in our research we obtained the relationship between study time and test errors shown in Figure 2.2. Inferring the relationship that would be found in the population means that if we could measure every student

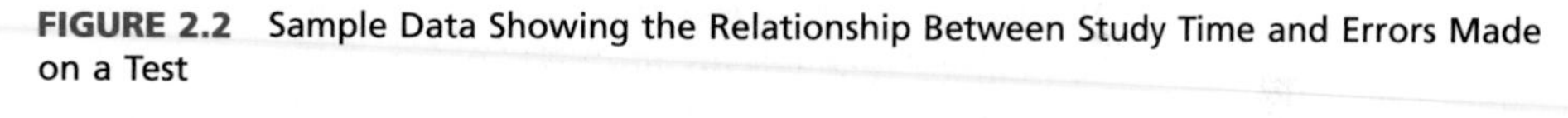

FIGURE 2.2 Sample Data Showing the Relationship Between Study Time and Errors Made on a Test

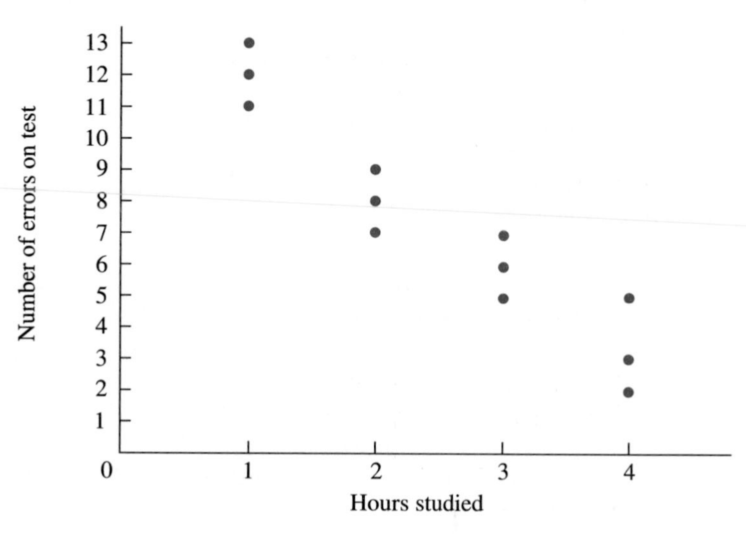

on our variables and then graph that data, we expect that we should see a similar graph. Then we can infer that every time the studying variable changes, we would obtain a different *population* of error scores. Thus, because our sample who studied 1 hour made between 13 and 11 errors on the test, we expect that, if all students studied for 1 hour, they should produce a population consisting of error scores generally between 13 and 11. However, our sample who studied for 2 hours produced scores of between 10 and 7 errors. Therefore we expect that if we went back and had all students study for 2 hours, we should obtain a different population of scores that would consist of scores between 10 and 7 errors. Similarly, we expect that studying for 3 hours should produce a population of scores consisting mainly of scores between 7 and 5 errors, and so on.

Now you can see how research combines the logic of relationships with that of sampling from the population. We originally hypothesized that, in nature, interacting with information is related to learning it. To determine if this is true, we must first find a relationship in our sample data where, as scores reflecting study time change, scores reflecting exam performance also change. But this will provide only *one instance* in which nature seems to operate as we think. Our goal is to draw the inference that a similar relationship would be found for *all* students in the population—that more or less for all students, higher studying scores are associated with better test performance. Then, by translating the scores back into the behaviors and events they reflect, we essentially create a picture of how nature works. Ultimately we infer that for all students there is something about the mental activities we call learning that does relate to the mental activities we call studying, such that greater learning generally occurs with greater studying. By concluding that greater learning occurs with greater studying, we will have come full circle, having obtained evidence that this law of nature does operate as we originally proposed.

Thus, the ultimate goal of research is to describe the relationship found in the population, so that we can infer that as everyone's score on one variable changes, his or her score on the other variable tends to change in a particular fashion. But remember that we do not actually observe *everyone*! The problem is that our sample data provides only one instance—one example—of the larger population. As we see next, the major potential flaw in this logic is that our sample may not be a "good" example of the population.

Representativeness of a Sample

For us to draw accurate conclusions about a population, our sample must be "representative" of that population. In a **representative sample,** the characteristics of the sample accurately reflect the characteristics of the population. Statistically, the characteristics of the population include such things as how often each score occurs, what the highest and lowest scores are, and what the average score is. Thus, if the average score in the population is 50, then the average score in a representative sample will be around 50. If 30% of the scores in the population are 45, then around 30% of the scores in a representative sample will be 45, and so on. To put it simply, a representative sample accurately represents the population.

Whether a sample is representative depends on how we select the sample. From a statistical standpoint, the most important aspect of creating representative samples is random sampling.

Random Sampling **Random sampling** is a method of selecting a sample in which (1) all possible scores in the population have the same chance of being selected for a sample and (2) all possible samples have the same chance of being selected. Because we obtain scores from subjects, random sampling means that we are unbiased, or unselective, in choosing our subjects, so that all members of the population have an equal chance of being selected. To create a random sample from a population, we select subjects based simply on the luck of the draw. We might place the names of all potential subjects on slips of paper in a very large hat, stir them up, and then draw the names of the subjects to be in our sample. Or we might use a computer to generate random numbers and select subjects with corresponding social security numbers. Any way we do it, we are as unbiased as possible in selecting subjects, and we try to select from all segments of the population of interest. We may be selective in choosing the population, but then we use random sampling to select the sample from that population. For example, if I want to study only 12-day-old babies, I will randomly select from the population of 12-day-old babies to create my sample.

A random sample *should* turn out to be representative of the population, because random sampling allows the characteristics of the population to occur in the same way in the sample. For example, say that the population of students at your college are all standing in a field, and 60% of them are female. Random sampling is analogous to randomly and blindly wandering through that field. Out of a sample of 100 people you encounter, about 60 of them should be female, because that's how often females are out there. Likewise, if 30% of the individuals in the population have a score of 45, then 30% of the subjects we select through random sampling should also have a score of 45. In the same way, random sampling should produce a sample having all of the characteristics of the population.

At least we *hope* it works that way!

Unrepresentative Samples I keep saying that a random sample "should" be representative, but this is a *very* big should! Nothing forces a random sample to be representative of a particular population. The trouble is that whether a sample is representative is determined by which subjects we select to be in the sample, and which subjects we select is determined by random chance. Therefore, whether a sample is representative depends on random chance. We can, just by the luck of the draw, obtain a very unusual sample of subjects who are not at all representative of most subjects in the population. For example, 30% of the individuals in the population may have a score of 45, but through random sampling we may not select *any* subjects who have this score, or we may select *only* subjects who have this score. Likewise, although the population average may be 50, the sample average may be something very different because of the individuals in the sample we happen to select.

Thus, random sampling is a double-edged sword. We use random sampling to produce a representative sample, and usually it works pretty well. But random sampling may backfire on us, producing a very unrepresentative sample, in which case the characteristics of the sample do not match the characteristics found in the population. Therefore, we can never take a sample at face value, automatically assuming that it accurately represents the population.

The representativeness of a sample is the major potential flaw in research. If our sample is unrepresentative, then our observations are not interchangeable with any other observations we might obtain, our sample's scores are not similar to any other sample's scores, and so as our sample goes is *not* how the population would go, if we could measure it. To see the problem this creates, return to Figure 2.2 on page 25 and consider this: By chance, we might have unknowingly selected only poor students who studied for 1 hour, so their scores may be unrepresentative of the scores of the typical student who studies for 1 hour. Or, we may have selected only exceptionally good students who studied for 4 hours, so their scores do not represent the typical student who studies for 4 hours. And so on. If the sample of scores at each study time does not accurately represent the population of scores we'd find at that study time, then the *relationship* in our sample data does not accurately represent the relationship we'd find in the population. If we could study the population of all students, we might find a relationship very different from this one, or worse, we might find *no* relationship!

Thus, if our samples are unrepresentative, the entire logic of research falls apart: if as the sample goes is *not* how the population goes, then we do not create an accurate picture of nature, and any evidence we think we have for a particular law of nature is misleading and our conclusions are wrong. Yet society uses the conclusions from scientific research to create such serious things as drugs and surgical techniques, treatments for homicidal psychopaths, bridges, airplanes, and nuclear bombs! We'd *really* like our conclusions about how nature works to be accurate. Therefore, whether a sample relationship accurately represents the relationship found in the population is one of the most important concerns we have when interpreting research. But, the problem is that we can never *know* whether a sample is representative or not. We would have to measure the entire population to see how well the sample and population matched. But if we could measure the entire population, we would not need the sample to begin with! To rescue us from this dilemma, we use statistical procedures to help us decide what inferences we can safely draw from our sample data. So, after we have used statistics to understand the data and relationship observed in our sample, we always stop and perform other procedures that allow us to decide whether the sample relationship would actually be found in the population. Only then do we proceed to infer that the sample data accurately describes how nature operates.

USING STATISTICAL PROCEDURES TO ANALYZE DATA

Recall that statistics help us make sense out of our data, and now you can see what "making sense" means. First, because we conduct research to demonstrate a relationship, we use statistical analysis to determine whether a relationship is present. In our study-time research, we want to first determine whether there is a relationship between subjects' error scores and the amount of time they studied.

Second, making sense of the data means describing the scores and the relationship we find in a study. We want to know how many errors are associated with a particular amount of study time, how much errors decrease with increased study time, how consistently errors decrease, and so on. Third, if we understand a relationship, then we should be able to use a subject's score on one variable to accurately predict his or her score and corresponding behavior on some other variable. If we know how many hours subjects study, we should be able to accurately predict the number of errors they make on the test. Thus, we use statistical procedures to analyze data so that we may (1) determine whether a relationship is present, (2) describe the relationship and the scores in it, and (3) predict the scores on one variable using the scores on another variable.

As we've seen, however, we are always talking about two things: the sample data we have collected and the population of scores being represented. Therefore, we separate statistical procedures into two categories: descriptive statistical procedures, which deal with samples, and inferential statistical procedures, which deal with populations.

Descriptive Statistics

Recall that individual differences ensure that there is no such thing as a perfectly consistent relationship. Therefore, in real data, there will be many different scores in each sample. Researchers are invariably confronted with a mind-boggling array of different numbers that may have a relationship hidden in it. The purpose of descriptive statistics is to bring order to this chaos. **Descriptive statistics** are procedures for organizing and summarizing data so that we can communicate and describe the important characteristics of the data. (Descriptive statistics are used to describe data, so when you see *descriptive,* think *describe.*)

How do descriptive statistics work? In our study-time research, of the three subjects who studied for 1 hour, one made 11 errors, one made 12 errors, and one made 13 errors. To simplify the data, we can say that this sample produced around 11 to 13 errors. In fact, split the difference: on average close to 12 errors are associated with 1 hour of study time. By saying that this sample produced close to 12 errors, we have used descriptive statistics. We summarized the data by reducing the results to one number (12). We communicated an important characteristic of the data by saying that the scores are around 12 (and not around 33). We described another important characteristic of the data by stating that the scores are *close* to 12 (as opposed to being spread out around 12).

Statisticians have developed descriptive procedures that enable us to answer five basic questions about the characteristics of a sample:

1. *What scores did we obtain?* To answer this question, we organize and present the scores in tables and graphs.

2. *Are the scores generally high scores or generally low scores?* We can describe the scores with one number that is the "typical" score.

3. *Are the scores very different from each other, or are they close together?* As we shall see, there are mathematical ways of describing "close."

4. *How does any one particular score compare to all other scores?* Certain statistical transformations allow us to easily compare any score to the rest of the scores.
5. *What is the nature of the relationship we have found?* We can summarize the important characteristics of a relationship and use this information to predict scores on one variable if we know a score on another variable.

Parts 2 and 3 of this book discuss the specific descriptive procedures used to answer these five questions. When we have answered these questions, we have described the important characteristics of a sample.

Realize, however, that even though we compute a precise mathematical answer, it will not precisely describe every score in the sample. (In our study, everyone who studied 1 hour did not score 12.) Because of those individual differences again, a particular law of nature operates in *more or less* the same way for all members of a group, and so the best we can do is to summarize the general tendency or trend in the scores from a group. In any summary, less accuracy is the price we pay for less complexity and less quantity, so descriptive statistics always imply "around" or "more or less."

Inferential Statistics

After we have answered the five questions above for our sample, we want to answer the same questions for the population of scores represented by our sample. Thus, although technically descriptive statistical procedures are used to describe samples, the logic of descriptive statistics is also applied to populations. Of course, because we usually cannot measure the scores in the population, we must *estimate* the description of the population, based on the sample data.

However, we cannot automatically assume the sample reflects the population. Remember that the inferences we draw about a population will be accurate only if the sample is representative of the population. Although we saw that random sampling should produce representative samples, we also saw that it may not. Therefore, before we draw any conclusions about the population of scores, we must first perform inferential statistics. **Inferential statistics** are procedures that allow us to decide whether to conclude that the sample data accurately represent a particular relationship in the population. As the name implies, the focus of inferential procedures is to help us make accurate *inferences* about the scores and relationship represented by our sample.

How do we do this? Whether a sample is representative depends on chance, on the luck of the draw of which subjects are selected for the sample. Although we never know if random chance produces an unrepresentative sample, we have a good idea of how chance operates. As we shall see, inferential statistics involve *models* of how nature produces random chance events. Using such models, we can decide at least whether it is *likely* that the sample data are representative of a particular relationship in the population. If the sample data are deemed representative, then we assume that each sample gives a reasonably accurate estimate of the corresponding population. Then we use our descriptive statistics, computed from a sample, as the basis for our estimates of the scores that would be found in the population. Thus, as a result of our inferential procedures, we may estimate that everyone who studies for 1 hour will in fact make around 12 errors, and so on.

Part 4 of this book discusses inferential statistics in great detail. Until that time, simply think of inferential procedures as ways of deciding whether our sample data are "believable": should we believe that we would find similar data, forming a similar relationship, in the population?

STAT ALERT Descriptive statistics summarize the sample data, and inferential statistics allow us to decide what inferences we should draw about the population.

Statistics and Parameters

In all research, then, we set up a situation in which we measure the scores of a randomly selected sample. We use descriptive procedures to describe the sample, and then we use inferential procedures to make estimates about the corresponding population of scores we expect we would find if we could measure the entire population. From our description of the population, we may be able to infer that a relationship exists in the population and thereby learn something about a law of nature.

So that we know when we are describing a sample and when we are describing a population, statisticians use the following system. A number that describes a characteristic of a *sample* of scores is called a **sample statistic** or simply a **statistic.** Different statistics describe different characteristics, and the symbols for different statistics are different letters from the English alphabet. On the other hand, a number that describes a characteristic of a *population* of scores is called a **population parameter** or simply a **parameter.** The symbols for different parameters are different letters from the Greek alphabet. For example, if we compute your bowling average from a sample of your scores, we are computing a descriptive statistic. The symbol for a sample average is a letter from the English alphabet. If we then estimate the average of the population of all your bowling scores based on the sample average, we are estimating a population parameter. The symbol for a population average is a letter from the Greek alphabet.

THE CHARACTERISTICS OF A STUDY

The preceding discussion applies to virtually all psychological research. We will see a variety of descriptive procedures that serve the general purpose of describing the data, as well as a variety of inferential procedures that are used to help us draw inferences about the relationship in the population. Recall that a big part of your job in this course is to learn when to use each particular statistical procedure. Which specific descriptive or inferential procedure we employ within a specific study is determined by three things. First, you must decide what it is you want to know—what question about the characteristics of the sample or population do you want to answer? Then, your choice of procedures depends on two aspects of the study being conducted: the specific research design being employed and the characteristics of the scores being measured.

Research Designs

Once a researcher has identified the variables and relationship he or she will seek to observe and the sample and population his or her conclusions will be based on, the specifics of the research study must be determined. However, there are many ways to set up, or "design," a particular study. A **research design** is the way in which a study is laid out: how many samples there are, how the subjects are tested, and the other specifics of how a researcher goes about demonstrating a relationship. As we shall see, different research designs require different statistical procedures. Therefore, your job as a beginning statistics student is to learn to recognize the characteristics of different designs. Research designs can be broken into two major types because essentially researchers have two ways of demonstrating a relationship: experiments and correlational studies.

Experiments

Not all research is an experiment. Technically, an **experiment** involves a procedure in which the researcher actively changes or manipulates one variable, measures the scores on another variable, and keeps all other variables constant. The logic of an experiment is extremely simple, and you use it every day. For example, if you believe a certain light switch controls a particular light, you flip that switch, keep all other switches constant, and see whether the light comes on. In an experiment, if we believe there is a relationship in which variable X influences behavior Y, we produce different amounts of X, keep all other variables constant, and see whether the scores on variable Y change in a consistent fashion.

For example, say that we conduct an experiment to examine the relationship between the variables of amount of time spent studying statistics and number of errors made on a statistics test. We decide to compare 1, 2, 3, and 4 hours of study time, so we randomly select four samples of students. We give one sample of subjects 1 hour of study time, administer the test, and then count the number of errors each subject makes. We give another sample 2 hours of study time, administer the test, and then count the errors, and so on for our remaining samples. If we understand the laws of nature governing learning and test taking, then as we increase the length of time subjects study, their number of errors should decrease.

There are names for the components of any experiment, and you will use them daily in research and statistics.

The independent variable An **independent variable** is a variable that is directly changed or manipulated by the experimenter. Implicitly, it is the variable that we think causes a change in the other variable. In our studying experiment, we manipulate how long subjects study, because we think that longer studying causes fewer errors. Thus, amount of study time is the independent variable. Or, in an experiment to determine whether eating more chocolate causes people to blink more, amount of chocolate consumed would be the independent variable, because the experimenter manipulates the amount of chocolate a subject eats. You can remember the independent variable as the variable that the experimenter manipulates *independent* of what the subject wishes. (In our study, some subjects studied for 4 hours whether they wanted to or not.)

Technically, a true independent variable is manipulated by doing something *to* subjects. However, there are many behavior-influencing variables that an experimenter cannot change by doing something to subjects. For example, we might hypothesize that growing older causes a change in some behavior. Yet, we can't randomly select subjects and then *make* some of them 20 years old and some of them 40 years old. In such situations, the experimenter must manipulate the variable in a different way. Here we would randomly select one sample of 20-year-olds and one sample of 40-year-olds. Similarly, if we wanted to examine whether a qualitative variable such as gender was related to some behavior, we would select a sample of females and a sample of males. In our discussions, we will *call* such variables independent variables, because the experimenter controls them by controlling a characteristic of the samples. Technically, though, such variables are called *quasi-independent variables.*

Thus, the experimenter is always in control of the independent variable, either by determining what is done to a sample or by determining a characteristic of the subjects in each sample. In essence, a subject's "score" on the independent variable is determined by the experimenter. In our examples, subjects in the sample that studied 1 hour have a score of 1 on the study-time variable, and subjects in the 20-year-old sample have a score of 20 on the age variable.

An independent variable is the *overall* variable a researcher examines; it is potentially composed of many different amounts or categories. From these, the researcher will actually test subjects under only certain conditions of the independent variable. A **condition** is a specific amount or category of the independent variable that creates the specific situation under which the subjects' scores on some other variable are measured. Thus, although our independent variable above was amount of study time—which could be any amount—our conditions of that variable involved only 1, 2, 3, or 4 hours of study. Likewise, if we compare the errors of males and females, then "male" and "female" are each a condition of the independent variable of gender. A condition is also known as a **level** or a **treatment:** by having subjects study for 1 hour we determine the specific "level" of studying that is present, and this is one of the ways we "treat" our subjects.

The dependent variable If a relationship exists, then as we change the conditions of the independent variable, we will observe a consistent change in subjects' scores on the dependent variable. The **dependent variable** is the variable that is measured under each condition of the independent variable. Scores on the dependent variable are presumably caused or influenced by the independent variable, so scores on the dependent variable *depend* on the conditions of the independent variable. In our studying experiment, the number of errors on the test is the dependent variable, because we believe that the number of errors depends on how long subjects study. Or, if we manipulate the amount of chocolate subjects consume and then measure their eye blinking, eye blinking is our dependent variable. Because we measure subjects' scores on the dependent variable, it is also called the *dependent measure.* Notice that we always ask, "What scores occur on the dependent variable for a *given* amount of the independent variable?" Thus, we always ask, "Are there consistent changes in the dependent variable *as a function of* changes in the independent variable?"

STAT ALERT The independent variable is always manipulated by the experimenter, and the dependent variable is always what the subjects' scores measure.

Drawing conclusions from experiments After we have conducted our experiment involving amount of study time and number of test errors, we need to examine our data. Table 2.4 shows a useful way to diagram the design of an experiment, label the components, and organize the data. Each column is a condition of the independent variable—study time—under which some subjects were tested. Each number in a column is a subject's score on the dependent variable of number of test errors. Is there a relationship here? Yes. Why? Because as subjects' scores on the variable of amount of study time change (increase), their scores on the variable of number of test errors also tend to change (decrease) in a consistent fashion. (To see this relationship more easily, go back and look at the graph of these data in Figure 2.2.)

As we will see, there are special descriptive statistics that are commonly used to summarize scores and describe and understand the relationship found in an experiment. Then, as usual, we want to draw inferences about the population of all subjects with respect to our variables. Therefore we use special inferential statistics with experiments to decide whether the different samples of scores found under our conditions are likely to represent different populations of scores that would be found if we included everyone from the population in our experiment. Then, if our sample statistics in our study-time experiment tell us that students who study for 1 hour score close to 12 errors, we can infer that if all college students study for 1 hour, we would find a population parameter showing close to 12 errors. But if our statistics indicate that studying for 2 hours produces around 8 errors, then we would expect a different population parameter indicating around 8 errors for 2 hours of study. And so on.

On the other hand, if we cannot conclude that there are different populations of error scores associated with different amounts of study time, then we cannot

TABLE 2.4 Diagram of an Experiment Involving the Independent Variable of Number of Hours Spent Studying and the Dependent Variable of Number of Errors Made on a Statistics Test

Each column contains subjects' scores measured under one condition of the independent variable.

	Independent variable: number of hours spent studying			
	Condition 1: 1 hour	*Condition 2: 2 hours*	*Condition 3: 3 hours*	*Condition 4: 4 hours*
Dependent variable: number of errors made on a statistics test	13 12 11	9 8 7	7 6 5	5 3 2

infer a relationship in the population. And we have not learned about how a law of nature operates.

Thus, from a statistical perspective, the goal of any experiment is to show that there is a different population of scores on the dependent variable associated with each condition of the independent variable.

The problem of causality Essentially, the logic a researcher uses in an experiment is this: "If I do this or that to subjects in terms of one variable, it should *make* subjects behave and score in a particular way on the other variable." Therefore, when we conclude that an experiment shows a relationship in the population, we can discuss the relationship as if changing the independent variable "causes" the scores on the dependent variable to change. However, it is important to realize that we cannot definitively prove that X causes Y. Implicitly we recognize that some other variable may actually be the cause. In our studying experiment, for example, perhaps the subjects who studied for 1 hour had headaches and the actual cause of high error scores was not lack of study time, but headaches. Or perhaps the subjects who studied for 4 hours were more motivated, and this motivation produced lower error scores. Or perhaps some subjects cheated, or perhaps the moon was full, or who knows what! Although we try to eliminate and control these other variables through well-designed research, we are never completely certain that we have done so. Therefore, in any relationship we demonstrate, there is always the possibility that some unknown variable is actually causing the scores to change.

This is especially true when we examine a quasi-independent variable. For example, if we find that a sample of males has different scores than a sample of females, we *cannot* claim that the sex of the subjects causes differences in scores. This is because not only would the samples differ in terms of the gender variable, but coincidentally the females would tend to differ from the males along a host of other variables, including height, hair length, amount of makeup, interests, and attitudes. Any one of these variables might actually be causing the differences in scores.

Recognize that statistics are not a solution to this problem. There is no statistical procedure that will prove that one variable causes another variable to change. Think about it: how could some formula written on a piece of paper "know" what causes certain scores to occur? How could a statistical result prove any statement you make about the real world?

STAT ALERT Statistics don't prove anything!

An experiment merely provides evidence, like the evidence in a court of law. How the experiment was conducted and how well the experimenter controlled the variables are part of the evidence supporting a certain conclusion. Statistical results are additional evidence to support the conclusion. Such evidence helps us to *argue* for a certain point of view, but it is not "proof." As scientists, we always know there is the possibility that we are wrong.

Correlational Studies

Sometimes we do not manipulate any variables and instead design the research as a correlational study. In a **correlational study,** we simply measure subjects' scores on two variables and then determine whether there is a relationship. Originally, for example, we used a correlational approach to study the relationship between amount of study time and number of test errors: We simply asked a random sample of students how long they studied for a particular test and what was their grade on the test. Then we determined whether a relationship exists, asking, "Did subjects who studied longer receive higher grades?" Likewise, in a different study we would have a correlational design if we measured subjects' career choices and their personality type, asking, "Is career choice related to personality type?" Or say that "in the wild" we simply observe many different aspects of the behavior of some subjects (say wolves) and then ask, "Which aspects are related?" (Perhaps the aggressiveness of males is related to the number of females they attract.)

Drawing conclusions from correlational studies As usual, our first goal is to understand and summarize the relationship in our sample data; and we have particular descriptive procedures used with correlational designs. Then, as in an experiment, the goal of a correlational study is to infer the relationship we expect would be found in the population; for this there are certain inferential procedures that help us decide whether the data are likely to represent such a relationship. Thus, as in our studying experiment, our correlational study will, we hope, allow us to conclude that such a relationship would be found in the population, so that for all students who study for a certain amount of time there is one population of scores, and for those students who study for a different amount of time there is a different population of scores.

Again, the problem of causality Our statistical procedures for correlational studies, as with experiments, do not prove that changes in X cause changes in Y. In fact, in a correlational study, the researcher does not actively manipulate or control either variable, so a correlational study does *not* fit the logic of "If I do this or that to subjects in terms of one variable, it should make subjects score in a particular way on the other variable." Rather, this design fits the logic of "I wonder if scores on the two variables *naturally* change together in a consistent fashion?" Because we don't do anything that even *might* cause a variable to change, the cause of this change is unknown. Therefore, even more than in an experiment, there are many variables that may actually cause scores to change. Thus, we can *never* say that, based on a correlational study, changes in one variable *cause* the other variable to change. Changes in X might cause changes in Y, but we have demonstrated the relationship in such a way that we have no convincing evidence that this is the case. At most, all we can say with confidence is that there is a relationship, or association, between the scores on the two variables. (We will explore this issue further when we discuss correlation in Chapter 7.)

Types of Variables

Whether we conduct an experiment or a correlational study, we measure subjects on variables, so we end up with a batch of scores. Depending on the variables, the numbers that comprise the scores can have different underlying mathematical characteristics. The particular mathematical characteristics of a variable determine which particular descriptive or inferential procedure you should use. Therefore, part of your job is to learn to recognize the characteristics of the data that indicate which procedures are appropriate. You must always pay attention to two important characteristics of your variables: the type of measurement scale involved and whether the scale is continuous or discrete.

The four types of measurement scales Numbers mean different things in different contexts. The meaning of the number 1 on a license plate is different from the meaning of the number 1 in a race, which is different still from the meaning of the number 1 in a hockey score. The kind of information that scores convey depends on the *scale of measurement* that is used in measuring the variable. There are four types of measurement scales: nominal, ordinal, interval, and ratio.

With a **nominal scale,** each score does not actually indicate an amount; rather, it is used simply for identification, as a name. (When you see *nominal,* think *name.*) License plate numbers and the numbers on the uniforms of football players reflect a nominal scale of measurement. In research, a nominal scale is used to identify the categories of a qualitative, or classification, variable. For example, we cannot perform any statistical operations on the words *male* and *female.* Therefore, we might assign each male subject a 1 and each female subject a 2. However, because this scale is nominal, we could just as easily assign males a 2 and females a 1, or we could use any other two numbers. Because we assign numbers in a nominal scale arbitrarily, they do not have the mathematical properties normally associated with numbers. For example, as used on this scale, the number 1 does not indicate more than 0 yet less than 2 as it usually does.

When a variable is measured using an **ordinal scale,** the scores indicate rank order: the score of 1 means the most or least of the variable, 2 means the second most or least, and so on. (For *ordinal,* think *ordered.*) In psychological research, ordinal scales are used, for example, to rank subjects in terms of their aggressiveness or to have subjects rank the importance of certain attributes in their friends. Each score indicates an amount of sorts, but it is a relative amount. For example, relative to everyone else being ranked, you may be the number 1 student, but we do not know how good a student you actually are. Further, with an ordinal scale there is not an equal unit of measurement separating each score. In a race, for example, first may be only slightly ahead of second, but second may be miles ahead of third. Also, there is no number 0 in ranks (no one can be "zero-ith").

When a variable is measured using an **interval scale,** each score indicates an actual amount, and there *is* an equal unit of measurement separating each score: the difference between 2 and 3 is the same as the difference between 3 and 4. (For *interval,* think *equal* interval.) Interval scales include the number 0, but it is not a "true" zero. It does not mean zero amount; it is just another point on the scale. Because of this feature, an interval scale allows negative numbers. Temperature (measured in centigrade or Fahrenheit) is an interval scale. A measurement of zero degrees does not mean that zero amount of heat is present; it only means that there is less than 1 degree and more than -1 degree. Interval scales are often used with quantitative variables measured by psychological tests, such as intelligence or personality tests. Although a score of zero may be possible, it does not mean zero intelligence or zero personality. Note that with an interval scale, it is incorrect to make "ratio statements" about the amount of a variable at one score relative to the amount at another score. For example, at first glance it seems that 4 degrees centigrade has twice as much heat as 2 degrees. However, if we measure the *same* physical temperatures using the Fahrenheit scale, 2 and 4 degrees centigrade are about 35 and 39 degrees Fahrenheit, respectively, so now the one amount of heat is not twice that of the other. (Essentially, if we don't know the true amount of a variable that is present at a score of zero, then we don't know the true amount that is present at any other score.)

Only with a **ratio scale** do the scores reflect the true amount of the variable that is present, because the scores measure an actual amount, there is an equal unit of measurement, *and* 0 truly means that zero amount of the variable is present. Therefore, ratio scales cannot include negative numbers, and only with ratio scales can we make ratio statements, such as "4 is twice as much as 2." (So for *ratio,* think *ratio!*) In psychological research, ratio scales are used to measure quantitative variables, such as the number of errors made on a test, the number of friends someone has, or the number of calories consumed in a day.

To help you remember the four scales of measurement, Table 2.5 summarizes their characteristics.

Discrete and continuous scales In addition to being nominal, ordinal, interval, or ratio, a measurement scale is either continuous or discrete. A **continuous scale**

TABLE 2.5 Summary of Types of Measurement Scales

Each column describes the characteristics of the scale.

	Type of measurement scale			
	Nominal	*Ordinal*	*Interval*	*Ratio*
What does the scale indicate?	Quality	Relative quantity	Quantity	Quantity
Is there an equal unit of measurement?	No	No	Yes	Yes
Is there a true zero?	No	No	No	Yes
How might the scale be used in research?	To identify males and females as 1 and 2	To judge who is 1st, 2nd, etc., in aggressiveness	To convey the results of intelligence and personality tests	To state the number of correct answers on a test

allows for fractional amounts; it "continues" between the whole-number amounts. With a continuous scale, decimals make sense. Age is a continuous variable because it is perfectly intelligent to say that someone is 19.6879 years old. (In fact, age is a continuous ratio variable.) To be continuous, a variable must be at least theoretically continuous. For example, intelligence tests are designed to produce whole-number scores. You cannot obtain an IQ score of 95.6. But theoretically an IQ of 95.6 makes sense, so intelligence is a theoretically continuous interval variable.

On the other hand, some variables involve a **discrete scale,** and then the variables can only be measured in whole-number amounts. Here, decimals do not make sense. Usually, nominal and ordinal variables are discrete. In addition, some interval and ratio variables are discrete. For example, the number of cars someone owns and the number of children someone has are discrete ratio variables. It sounds strange when the government reports that the average family has 2.4 children and owns 1.78 cars, because these are discrete variables being treated as if they were continuous. (Imagine a .4 child driving a .78 car!)

There is a special type of a discrete variable. When there can be only two amounts or categories of the variable, it is a **dichotomous** variable. Pass/fail, male/female, and living/dead are examples of dichotomous variables.

> ***STAT ALERT*** Whether a variable is continuous or discrete and whether it is measured using a nominal, ordinal, interval, or ratio scale are factors used to determine which statistical procedure to apply.

FINALLY

The terms and logic introduced in this chapter are used throughout the scientific world. Psychologists and behavioral scientists thoroughly understand such terms as relationship, independent and dependent variable, condition, and descriptive statistic. These terms are a part of their everyday vocabulary, and they think using these terms. For you to understand research and apply statistical procedures (let alone understand this book), you too must learn to think in these terms. The first step is to always be careful to use the appropriate terminology.

As you proceed through this course, however, don't let the terminology and details obscure your ultimate purpose. Statistics are only one of the tools involved in research. Keep things in perspective by remembering the overall logic or flow of research, which can be summarized as the following five steps:

1. Based on a hypothesized law of nature, we design either an experiment or a correlational study to measure variables and possibly observe a relationship in the scores we collect from our sample.
2. We use descriptive statistical procedures to understand our scores and the relationship they form. The relationship in our sample data is at least one instance that provides evidence that nature works in a certain way.
3. We use inferential procedures to decide whether our sample represents the scores and relationship that we would find if we could study everyone in the population.

4. Because scores reflect behaviors and events, by describing the scores and relationship that would be found in the population, we are actually describing how the behavior of all members of a particular group operates in a particular situation.

5. Because a law of nature governs the behavior of all the members of a particular group in a particular situation, when we describe the behavior of the population, we *are* describing how a law of nature operates.

CHAPTER SUMMARY

1. The goal of psychological research is to discover the laws of nature governing behavior. To do this, we study the relationships between variables.

2. A *variable* is anything that, when measured, can produce two or more different values. Variables may be *quantitative,* measuring a quantity or amount, or *qualitative,* measuring a quality or category.

3. A *relationship* occurs when a change in scores on one variable is associated with a consistent change in scores on another variable.

4. The term *individual differences* refers to the fact that no two individuals are identical. Because of individual differences, relationships can have varying *strengths,* showing only some *degree* of consistent association between the scores on two variables.

5. The "given" variable in any study is always designated the X variable, and we phrase our description of a relationship using the format "changes in Y *as a function of* changes in X."

6. The large group of all individuals to which a relationship applies is known as the *population.* In statistics, the population is the entire group of scores that we wish to describe. The subset of the population that is actually measured is the *sample,* and the individuals in a sample are the *subjects.*

7. In conducting research, we select subjects using *random sampling.* For a sample to be random, all possible scores in the population must have the same chance of being selected and all possible samples must have the same chance of being selected.

8. Random sampling is used to produce *representative samples.* Representative means that the characteristics of the sample accurately reflect the characteristics of the population.

9. *Descriptive statistics* are procedures used to organize, summarize, and describe sample data. *Inferential statistics* are procedures that allow us to decide whether sample data represent a particular relationship in the population.

10. A *statistic* is a number that describes a characteristic of a sample of scores. The symbols for statistics are letters from the English alphabet. A sample statistic is used to infer or estimate the corresponding *population parameter.*

A parameter is a number that describes a characteristic of a population of scores. The symbols for parameters are letters from the Greek alphabet.

11. Which particular descriptive or inferential procedure we use depends in part on our *research design,* the particular way in which a study is laid out.

12. In an *experiment,* a relationship is demonstrated by the experimenter's manipulating or changing the *independent variable* and then measuring scores on the *dependent variable.* Each specific amount or category of the independent variable used in an experiment is known as a *condition, treatment,* or *level.*

13. In a *correlational study,* neither variable is actively manipulated. Scores on both variables are simply measured as they occur, and then the relationship is described.

14. In any type of research, if a relationship is observed, it may or may not mean that changes in one variable *cause* the other variable to change.

15. Which particular descriptive or inferential procedure we use also depends on the *scale of measurement* used to measure the variables. We may use (1) a *nominal scale,* in which numbers name or identify a quality or characteristic; (2) an *ordinal scale,* in which numbers indicate a rank order; (3) an *interval scale,* in which numbers measure a specific amount, but with no true zero; or (4) a *ratio scale,* in which numbers measure a specific amount and 0 indicates truly zero amount.

16. Which procedure we use also depends on whether a variable is *continuous,* in which case decimals make sense, or *discrete,* in which case decimals do not make sense. A *dichotomous variable* is a special type of discrete variable that has only two amounts or categories.

PRACTICE PROBLEMS

(Answers for odd-numbered problems are provided in Appendix E.)

1. How can you recognize when a relationship exists between two variables?

2. What are the two components of the logic of using research results to learn about the laws of nature?

3. Why can't a scientist expect to observe a perfectly consistent relationship between variables?

4. What is the difference between an experiment and a correlational study?

5. In an experiment, what is the difference between the independent variable and the conditions of the independent variable?

6. In an experiment, what is the dependent variable?

7. What is random sampling?

8. *a.* Why do random samples occur that are representative of the population?
b. Why do unrepresentative samples occur?

9. What are descriptive statistics used for?
10. What are inferential statistics used for?
11. *a.* What is the difference between a statistic and a parameter?
    b. What types of symbols are used for statistics and parameters?
12. Using the words "statistic" and "parameter," how do we describe a relationship in a population?
13. A student, Poindexter, conducted a survey. In his sample, 83% of females employed outside the home would rather be in the home raising children. After performing all the necessary statistical analyses, he concluded that "the statistical analyses prove that most working women would rather be at home." What is the problem with this conclusion?
14. In study A, a researcher gives a sample of subjects various amounts of alcohol and then observes any decrease in their ability to walk. In study B, a researcher notes the various amounts of alcohol that subjects drink at a party, and then observes any decrease in their ability to walk.
    a. Which study is an experiment and which is a correlational study. Why?
    b. Which study will be best for showing that drinking alcohol causes an impairment in walking? Why?
15. Another student, Foofy, conducts a survey of the beverage preferences of college students on a random sample of students. Based on her findings, she concludes that most college students prefer sauerkraut juice to other beverages. What statistical argument can you give for not accepting her conclusions?
16. In each of the following research projects, identify the independent variable, the conditions of the independent variable, and the dependent variable:
    a. A researcher studies whether scores on a final exam are influenced by whether background music is soft, loud, or absent.
    b. A researcher compares freshmen, sophomores, juniors, and seniors with respect to how much fun they have while attending college.
    c. A researcher investigates whether being first-born, second-born, or third-born is related to intelligence.
    d. A researcher examines whether length of daily exposure to a sun lamp (15 minutes versus 60 minutes) accounts for differences in self-reported depression.
    e. A researcher investigates whether being in a room with blue walls, green walls, red walls, or beige walls influences aggressive behavior in a group of adolescents.
17. What are the two aspects of a study to consider when deciding on the particular descriptive or inferential statistics that you should employ?
18. *a.* Define the four scales of measurement.
    b. Rank order the four scales of measurement, from the scale that provides the most specific and precise information about the amount of a variable present to the scale that provides the least specific information.

19. For the following data sets, which sample or samples have a relationship present?

Sample A		Sample B		Sample C		Sample D	
X	*Y*	*X*	*Y*	*X*	*Y*	*X*	*Y*
1	1	20	40	13	20	92	71
1	1	20	42	13	19	93	77
1	1	22	40	13	18	93	77
2	2	22	41	13	17	95	79
2	2	23	40	13	15	96	74
3	3	24	40	13	14	97	71
3	3	24	42	13	13	98	69

20. In which sample in problem 19 is there the strongest degree of association? How do you know?

21. Below are graphs of data from three studies. Which depict a relationship? How do you know?

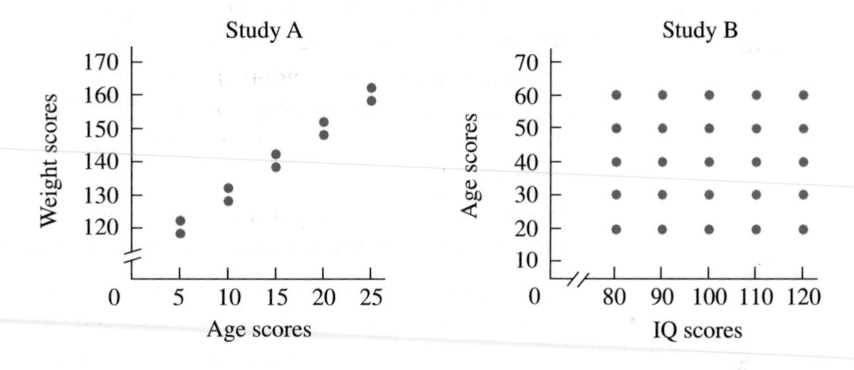

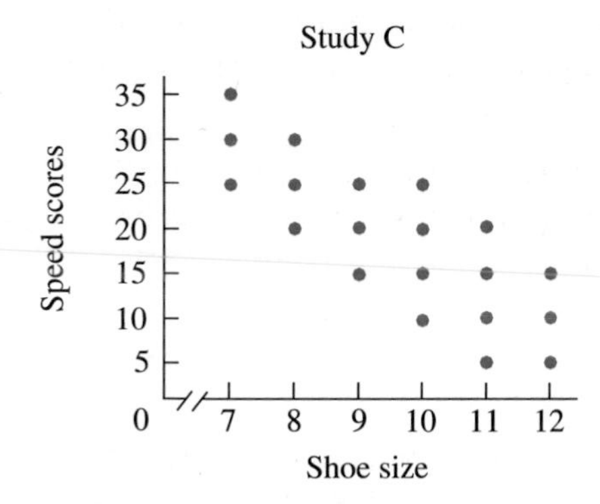

22. Which study in problem 21 demonstrates the greatest degree of association between the variables? How do you know?

23. In problem 21, why is each relationship a telltale sign that a law of nature is at work?

24. *a.* Poindexter says that Study A in problem 21 examines age scores as a function of weight scores. Is he correct?

b. Poindexter also claims that in Study C, the researcher is asking, "For a given shoe size, what speed scores occur?" Is he correct?

c. Let's say the studies in problem 21 were conducted as experiments. In each, which variable is the independent (or quasi-independent) variable and which is the dependent variable?

25. In the chart below, identify the characteristics of each variable.

Variable	***Qualitative or quantitative***	***Continuous, discrete, or dichotomous***	***Type of measure-ment scale***
gender	Qual		
academic major	Qual		
time	Qty		
restaurant ratings	Qty		
speed	Qty		
money	Qty		
position in line	Qty		
change in weight	Qty		

PART 2

DESCRIPTIVE STATISTICS: DESCRIBING SAMPLES AND POPULATIONS

So, we're off! In research we attempt to discover the laws of nature by looking for a relationship between variables. We use statistical analysis as a tool to look with. In Chapter 2 we saw that descriptive procedures are used to answer five questions about a sample of data:

1. What scores did we obtain?

2. Are the scores generally high scores or generally low scores?

3. Are there large differences between the scores, or are there small differences between the scores?

4. How does any one particular score compare to all other scores?

5. What is the nature of the relationship we have found?

Using the answers from these questions, we then estimate the answers to the same questions about the population that is represented by the sample. In each of the four chapters of Part 2, we will discuss the procedures used to answer each of the first four questions above. In Part 3 we will answer the final question.

3

Frequency Distributions and Percentiles

To understand the upcoming chapter:

- From Chapter 1, be sure you can calculate proportions and percents.
- From Chapter 2, understand the four types of measurement scales and the phrase "as a function of."

Then your goals in this chapter are to learn:

- How simple frequency, relative frequency, cumulative frequency, and percentile are computed, and what each tells you.
- How the different types of frequency tables, bar graphs, histograms, and polygons are created.
- What normal, skewed, bimodal, and rectangular distributions are, and how to interpret them.
- How the proportion of the total area under the normal curve corresponds to the relative frequency of scores.

We call the scores we measure in a study the *raw scores:* the scores are "uncooked" and not yet "digestible." Descriptive statistics help us cook down the raw scores into an organized and interpretable form.

As we've seen, one way to organize raw scores is to display them in graphs and tables. A basic rule for you as a researcher is to always create a table or graph. As the saying goes, "A picture is worth a thousand words," and nowhere is this more appropriate than in organizing a large group of scores. Such organization is imperative if you are to make sense out of your data. Also, descriptive statistical procedures allow you to communicate your results to others, and a table or graph is often the most efficient way to do this. Finally, we look for a relationship in our data, and an organized table or graph allows us to see it.

Before we get to examining the relationship between the scores of two variables, however, we must first summarize the scores on each *individual* variable. As usual, which statistical procedure we employ in a given situation depends first upon what we wish to know about our data. Presenting the scores in a graph or table helps us answer our first question about data: what scores did we obtain? In fact, buried in any batch of scores are two important questions: What scores occurred, and how often did each score occur? As we will see in this chapter, we can answer these questions simultaneously, by organizing the data in one of four ways: using each score's "simple frequency," its "relative frequency," its "cumulative frequency," or its "percentile."

But first . . .

MORE STATISTICAL NOTATION

By constructing a table or graph, we create a distribution. A **distribution** is the general name for any organized set of data. We organize scores so that we can see the pattern formed by the scores. In statistical language, the pattern formed by the scores is the way the scores are *distributed* in the sample or population.

In most statistical procedures we must count how *many* scores we have. There is an important symbol to represent this number: N stands for the number of scores in a set of data. (When you see N, think *Number.*) An N of 10 means that we have 10 scores, or $N = 43$ means that we have 43 scores. In statistical terminology, N is the *sample size:* N indicates how big a sample is in terms of the number of scores it contains. Note however, that N stands for the total number of scores, *not* the number of different scores. For example, if the 43 scores in a sample are all the same number, N still equals 43. When we have one score for each subject, N corresponds to the number of subjects in the sample. Get in the habit of treating the symbol N as a quantity itself so that you will understand such statements as "the N subjects in the sample" or "increasing N" or "this sample's N is larger than that sample's N."

We are also concerned with how often each score occurs within a set of data. How often a score occurs is the score's **frequency,** symbolized by f. Also learn to treat f as a quantity: one score's f may be larger than another score's f, we can add the f's of different scores, and so on. As we'll see, there are several ways to describe a score's frequency, so we often combine frequency and f with other terms and symbols.

CREATING SIMPLE FREQUENCY DISTRIBUTIONS

There are several ways to answer the question "Which scores occurred and how often did each occur?" The most common way is to create a simple frequency distribution. A **simple frequency distribution** shows the number of times each score occurs in a set of data. The symbol for a score's simple frequency is simply f. To find f for a score, count how many times that score occurs in the data. If three subjects scored 66, there are three scores of 66 in the data, so the frequency of 66 (its f) is 3. Creating a simple frequency distribution involves counting the frequency of every score in the data.

Presenting Simple Frequency in a Table

To see how we present a simple frequency distribution in a table, let's begin with the following sample of raw scores. (Perhaps these scores measure some deep psychological trait, or perhaps they are something silly like the number of chocolate Bing-Bongs each subject eats in a day; it makes no difference.)

14	14	13	15	11	15	13	10	12
13	14	13	14	15	17	14	14	15

There are only 18 scores here, but in this disorganized arrangement, it is difficult to make sense out of them. Watch what happens, though, when we arrange them into a simple frequency table, as shown in Table 3.1. We create the table by counting the number of times each score occurred. Then we create a score column and an f column. Notice that the score column begins with the highest score in the data at the *top* of the column. Below that in the column are all *possible* whole-number scores in decreasing order, down to the lowest score that occurred. Thus, although no subject obtained a score of 16, we still include it.

Now we can easily see the frequency of each score and discern how the scores are distributed. We can also determine the combined frequency of several scores by adding together their individual f's. For example, in Table 3.1, the score of 13 has an f of 4 and the score of 14 has an f of 6. The combined frequency of 13 and 14 is $4 + 6$, or 10. Notice too that there are 18 scores, so N equals 18. If we add together all the values in the f column, the sum will equal 18.

STAT ALERT The sum of all individual frequencies always equals ***N***.

TABLE 3.1 Simple Frequency Distribution Table

The left-hand column identifies each score, and the right-hand column contains the frequency with which the score occurred.

Score	*f*
17	1
16	0
15	4
14	6
13	4
12	1
11	1
10	1
Total:	$18 = N$

The 1 subject who obtained 17, plus the 0 subjects scoring 16, plus the 4 subjects who scored 15, and so on will equal the *N* of 18. As a check on any frequency distribution you create, add up the frequencies. If the sum of the individual frequencies does not equal *N*, you made a mistake.

That is how we create a simple frequency distribution. Such a distriubtion is also called a *regular frequency distribution* or a plain old *frequency distribution.*

Graphing a Simple Frequency Distribution

We create graphs because they provide an easy way to communicate the overall distribution of a set of scores. A simple frequency distribution essentially shows the relationship between each score and the frequency with which it occurs. We ask, "For a given score, what is its corresponding frequency?" and then we observe changes in frequency *as a function of* changes in the scores. Therefore, in creating a graph, we place the scores on the *X* axis and the frequency of the scores on the *Y* axis. Then we look to see how the frequency changes as the scores increase.

> ***STAT ALERT*** A graph of a frequency distribution always shows the scores on the *X* axis.

Recall from Chapter 2 that a variable may involve one of four types of measurement scales—nominal, ordinal, interval, or ratio. As shown in the following sections, the type of scale involved determines whether we graph a frequency distribution as either a bar graph, a histogram, or a polygon.

Bar graphs Recall that in nominal data, each score is merely a name for some category, and in ordinal data, each score indicates rank order. We graph a frequency distribution of nominal or ordinal scores by creating a bar graph. A **bar graph** is a graph in which we draw a vertical bar centered over each score on the *X* axis. *In a bar graph, adjacent bars do not touch.*

Figure 3.1 shows two bar graphs of simple frequency distributions. Here the frequency table is included so that you can see how each score was plotted, but usually we do not include the table. The upper graph shows the nominal variable of political affiliation of subjects, where a score of 1 indicates Republican, a 2 indicates Democrat, a 3 indicates Socialist, and a 4 indicates Communist. The lower graph shows ordinal data reflecting the frequencies with which a certain baseball team has ranked in the top four positions nationally in the last 20 years. In both graphs, the height of each bar corresponds to the score's frequency.

The reason we crate bar graphs here is that, in both nominal and ordinal scales, no equal unit of measurement separates the scores. The space between the bars indicates this fact. Recall that interval and ratio scales do have an equal unit of measurement between scores, so these scales are not plotted using bar graphs. Instead, we have two ways of graphing such scores, depending upon how many *different* values of scores—the range of scores—the data includes.

Histograms When plotting a frequency distribution containing a small range of interval or ratio scores, we create a histogram. A **histogram** is similar to a

FIGURE 3.1 Simple Frequency Bar Graph for Nominal and Ordinal Data

The height of each bar indicates the frequency of the corresponding score on the X axis.

Nominal variable of political affiliation	
Score	*f*
4	1
3	3
2	8
1	6

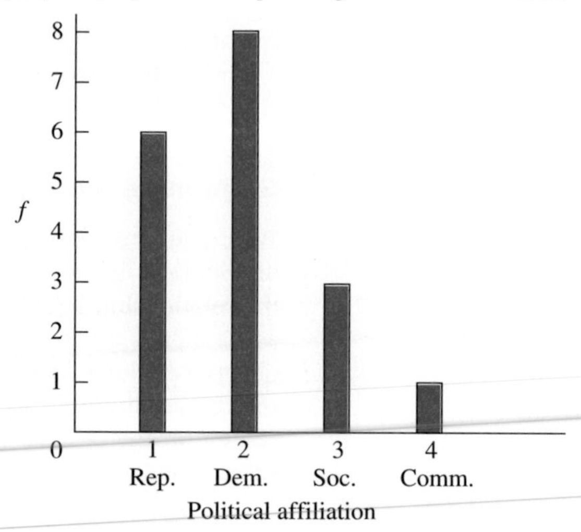

Ordinal variable of baseball team rankings	
Score	*f*
4	3
3	8
2	4
1	5

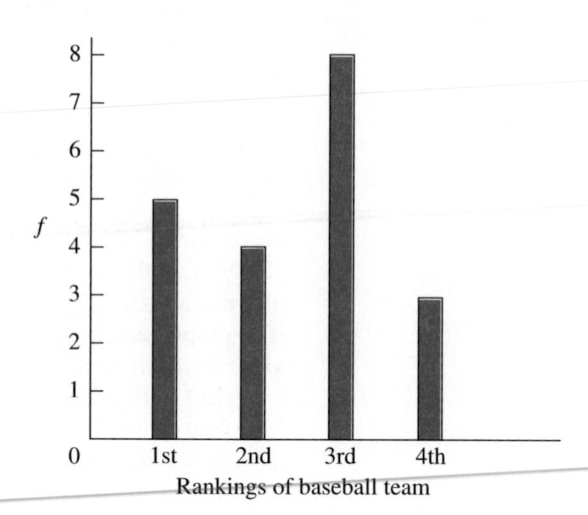

bar graph except that *in a histogram adjacent bars touch.* The absence of a space between the bars in a histogram signals an equal unit of measurement between scores. Histograms are especially appropriate when the interval or ratio variable is discrete (that is, there can be no fractions). For example, say that we measured a sample of subjects on the discrete ratio variable of number of parking tickets received, and we obtained the data in Figure 3.2. The histogram for these data appears to the right of the scores. Again, the height of each bar indicates the corresponding score's frequency.

Usually we don't create a histogram when we have a large range of different scores (say if subjects had from 1 to 50 parking tickets). The 50 bars would need to be very skinny, so the graph would be difficult to read. Likewise, sometimes

FIGURE 3.2 Histogram Showing the Simple Frequency of Parking Tickets in a Sample

Score	*f*
7	1
6	4
5	5
4	4
3	6
2	7
1	9

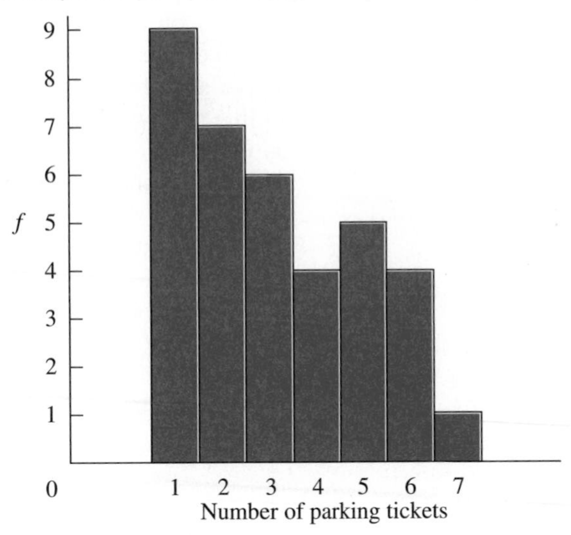

we plot more than one sample of scores on the same graph, and even with a small range of scores, overlapping histograms are hard to read. Instead, in such situations we create a frequency polygon.

Frequency polygons To construct a **frequency polygon** we place a data point over each score on the *X* axis at the height on the *Y* axis corresponding to the appropriate frequency. Then we connect the data points using straight lines. To illustrate this, Figure 3.3 shows our parking ticket data plotted as a frequency polygon.

FIGURE 3.3 Simple Frequency Polygon Showing the Frequencies of Parking Tickets in a Sample

Score	*f*
7	1
6	4
5	5
4	4
3	6
2	7
1	9

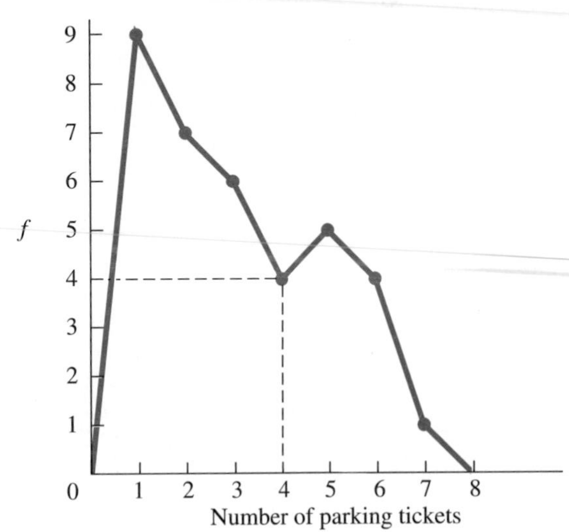

Notice that, unlike a bar graph or histogram, a simple frequency polygon includes on the X axis the next score above the highest score in the data and the next score below the lowest score (in Figure 3.3, scores of 0 and 8 are included). These added scores have a frequency of 0, so the polygon touches the X axis. In this way we create a complete geometric figure—a polygon—with the X axis as its base.

Often in statistics you must read the frequency of a score directly from the polygon. To do this, locate the score on the X axis and then move upward until you reach the line forming the polygon. Then, moving horizontally, locate the frequency of the score. For example, as shown by the dashed line in Figure 3.3, the score of 4 has an f equal to 4.

STAT ALERT The height of the polygon above any score corresponds to the frequency with which that score occurred.

TYPES OF SIMPLE FREQUENCY DISTRIBUTIONS

In statistics we have special names for polygons having certain characteristic shapes. Each shape comes from an idealized frequency distribution of an infinite population of scores. By far the most important frequency distribution is the normal distribution. (This is the big one, folks.)

The Normal Distribution

Figure 3.4 shows the polygon of the ideal theoretical normal distribution. For reference, assume that these are test scores from the population of college students. Although specific mathematical properties define this polygon, in general it is a bell-shaped curve. However, don't call it a bell curve! Because this shape occurs

FIGURE 3.4 The Ideal Normal Curve

Scores farther above and below the middle scores occur with progressively lower frequencies.

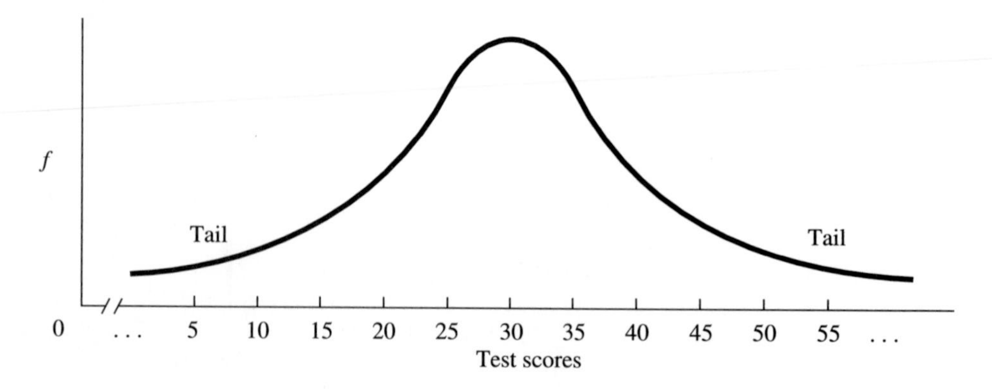

so often, the polygon itself is called the *normal distribution* or the *normal curve,* or we say that the scores are *normally distributed.*

To help you intepret the normal curve (or any polygon for that matter), imagine that you are flying in a helicopter over a parking lot. The X and Y axes are laid out on the ground, and the entire population of subjects is present (it's a very big parking lot). Those subjects who received a particular score stand in line in front of the marker for their score on the X axis. The lines of subjects are packed so tightly together that, from the air, all you see is a dark mass formed by the tops of many heads. If you painted a line that went behind the last subject in line at each score, you would have the outline of the normal curve. This view is shown in Figure 3.5.

Thus, you can think of the normal curve as a solid geometric figure made up of all the subjects and their different scores. When we move vertically up from any score to the height of the curve and then read off the corresponding frequency on the Y axis, it is the same as counting the number of people in line at the score. Either way we know how many people obtained that score, which is the score's f. Likewise, we might, for example, read off the simple frequencies on the Y axis for the scores between 30 and 35 and, by adding them together, obtain the frequency of scores between 30 and 35. We'd get the same answer if we counted the number of people in line above each of these scores and added them together. And if we read off the simple frequencies on the Y axis for all scores and add them together, we will have the total number of scores (our N). This is the same as counting the number of subjects standing in all lines in the parking lot.

As you can see from Figures 3.4 and 3.5, the normal distribution has the following characteristics. The score with the highest frequency is the middle score between the highest and lowest scores (the longest line of subjects in the parking lot is at the score of 30). The normal curve is *symmetrical,* meaning that the left half (containing the scores below the middle score) is a mirror image of the right half (containing the scores above the middle score). As we proceed away from the middle score toward the higher or lower scores, the frequencies

FIGURE 3.5 Parking Lot View of the Ideal Normal Curve

The height of the curve above any score reflects the number of subjects obtaining that score.

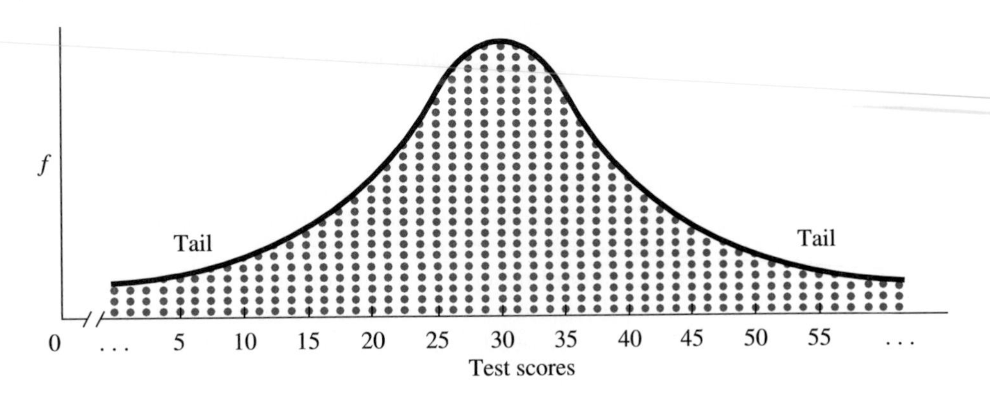

at first decrease only slightly. As we proceed farther from the middle score, each score's frequency decreases more drastically, with the highest and lowest scores having relatively very low frequency.

In statistics, the scores that are relatively far above and below the middle score of any distribution are called the **extreme scores.** In a normal distribution, the extreme scores have a relatively low frequency. In the language of statistics, the far left and right portions of a normal curve containing the relatively low frequency, extreme scores are called the **tails** of the distribution. In Figures 3.4 and 3.5, the tails are roughly below the score of 15 and above the score of 45.

Because the ideal normal curve represents a theoretical infinite population of scores, it has several characteristics that are not found with polygons created from actual sample data. First, with an infinite number of scores, we cannot label the Y axis with specific values of f. (Simply remember that the higher the curve, the higher the frequency.) Second, the theoretical normal curve is a smooth curved line. There are so many different whole-number and decimal scores that we do not need to connect the data points with straight lines. The individual data points form a solid curved line. Finally, there is no limit to the extreme scores in the ideal normal curve: regardless of how extreme a score might be, theoretically such a score will sometimes occur. Thus, as we proceed to ever more extreme scores in the tails of the distribution, the frequency of each score decreases, but there is never a frequency of zero so the theoretical normal curve approaches but never actually touches the X axis.

Before you proceed, be sure that you are comfortable reading the normal curve. Can you see in Figure 3.4 that the most frequent scores are between 25 and 35? Do you see that a score of 15 has a relatively low frequency and a score of 45 has the same low frequency? Do you see that there are relatively few scores in the tail above 50 or in the tail below 10? Above all, you must be able to see this in your sleep:

On a normal distribution, the farther a score is from the central score of the distribution, the less frequently the score occurs.

Overlapping normal distributions Sometimes we have two or more overlapping normal distributions plotted on the same set of X and Y axes. For example, say that we wish to compare the populations of males and females on the variable of height. Figure 3.6 shows two idealized normal distributions of the data. Generally males tend to be taller than females, but the overlapping parts of the polygons show that some males and females are the same height and some females are taller than some males. (If this were a parking lot full of males and females, then in the overlapping portions of the curve, members of one sex would be standing on the shoulders of those of the other sex.) In examining overlapping distributions, simply ignore one distribution when looking at the other: we count only males when looking at the male distribution and only females when looking at the female distribution. The height of each polygon at a score indicates the frequency with which the score occurred in that distribution. Thus, for example, the score of 69 inches occurs more frequently in the male distribution than in the female distribution.

FIGURE 3.6 Overlapping Distributions of Male and Female Height

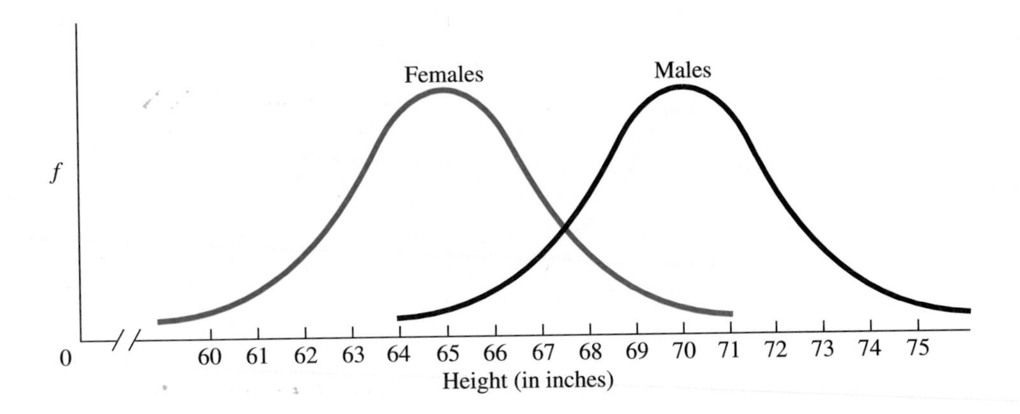

The normal curve model The pattern reflected in a normal distribution is very common in psychological research, occurring when measuring such variables as creativity, intelligence, or personality. In each case, most people tend to score at or slightly above or below a middle score, with progressively fewer people scoring at more extreme low or high scores. Of course, scores from the real world never conform exactly to the precise mathematical normal curve. Nonetheless, they are often close enough to forming the ideal normal curve for us to use it as a model. We noted in Chapter 2 that inferential statistics are based on models of how nature operates. Our most common model is the **normal curve model.** As we'll see, we often say that we assume a population of scores fits the normal curve model and is normally distributed. By this we mean that the actual distribution of scores in the population may not perfectly fit the ideal normal curve, but that it is close enough to treat it as if it does. Then, because we have precise mathematical ways of describing the normal curve, we also have a system for describing the distribution of actual scores.

Notice what the normal curve model enables us to do. Recall that the goal of research is to describe the laws of nature so that we understand how everyone will behave in a particular situation. In statistical terms, this means describing the entire population of scores we would find in a particular situation. Saying that the population forms a normal distribution is a major part of describing the population, so we are well on the way to our goal.

Approximations to the normal distribution We have a statistical term for describing a distribution that is an approximately normal distribution. Consider the three curves in Figure 3.7. The word *kurtosis* refers to how skinny or fat a distribution is. Curve A is generally what we think of when we discuss the ideal normal distribution; such a curve is called mesokurtic (*meso* means middle). Curve B is skinny relative to the ideal normal curve, and it is called leptokurtic (*lepto* means thin). Leptokurtic distributions occur when only a few scores around the middle score have a relatively high frequency. On the other hand, Curve C is fat relative to the ideal normal curve, because there are many different scores

FIGURE 3.7 Variations of Bell-Shaped Curves

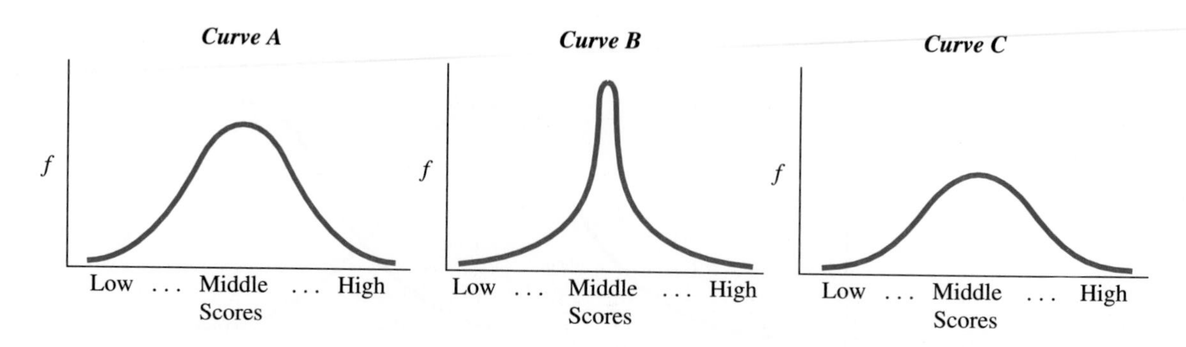

around the middle score that each have a relatively high frequency. Such a curve is called platykurtic (*platy* means broad or flat).

These terms help us to describe various normal distributions. For statistical purposes, however, as long as we have a reasonably close approximation to the normal curve, differences in shape are not all that critical.

Other Common Frequency Polygons

The distribution from every variable does not conform to a normal distribution. When a distribution does not fit the normal curve, it is called a *nonnormal* distribution. The three most common types of nonnormal distributions are *skewed, bimodal,* and *rectangular* distributions.

Skewed distributions A **skewed distribution** is similar in shape to a normal distribution except that it is not symmetrical: the left half of the polygon is not a mirror image of the right half. A skewed distribution has only one pronounced tail. As shown in Figure 3.8, a distribution may be either negatively skewed or positively skewed, and the skew is where the tail is.

FIGURE 3.8 Idealized Skewed Distributions

The direction in which the distinctive tail is located indicates whether the skew is positive or negative.

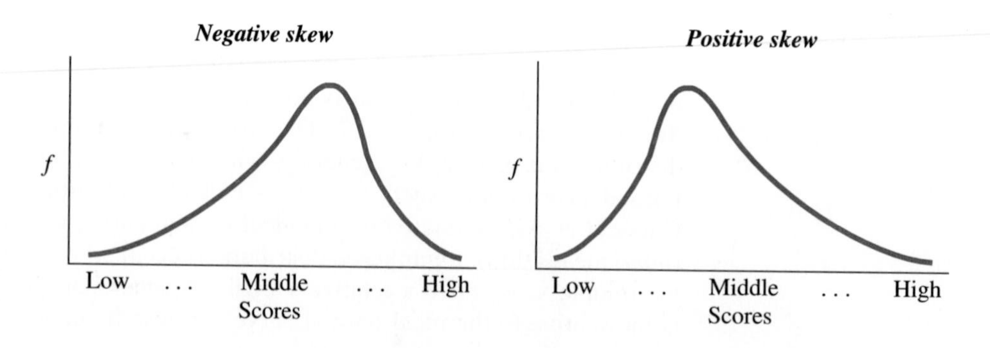

A **negatively skewed distribution** contains extreme low scores that have low frequency, but does not contain extreme high scores with corresponding low frequency. The left-hand polygon in Figure 3.8 shows an idealized negatively skewed distribution. This pattern might be found, for example, if we measured the running speed of professional football players. Most would tend to run at higher speeds, but with a relatively few linemen lumbering in at the slower speeds. To recognize that such a curve is negatively skewed, remember that the pronounced tail is over the lower scores, sloping toward zero, toward where *negative* scores would be.

On the other hand, a **positively skewed distribution** contains extreme high scores that have low frequency, but does not contain extreme low scores with corresponding low frequency. The right-hand polygon in Figure 3.8 shows a positively skewed distribution. This pattern might be found, for example, if we measured the time subjects take to react in a given situation. Most frequently, "reaction time" scores will tend to be rather low, but every once in while a subject will "fall asleep at the switch," requiring a large amount of time and thus producing a high score. To recognize that such a curve is positively skewed, remember that the tail slopes away from zero, toward where the higher, *positive* scores are located.

> ***STAT ALERT*** Whether a skewed distribution is negative or positive corresponds to whether the distinct tail slopes toward or away from where the negative scores would be located.

Bimodal and rectangular distributions An idealized bimodal distribution is shown in the left-hand side of Figure 3.9. A **bimodal distribution** is a symmetrical distribution containing two distinct humps where there are relatively high frequency scores. At the center of each hump is a score that occurs more frequently than the surrounding scores, and technically the center scores in each hump have the same frequency. Such a distribution would occur with test scores, for example, if most students scored around 60 or 80, with fewer students failing or scoring in the 70's or 90's.

The right-hand side of Figure 3.9 presents a rectangular distribution. A **rectangular distribution** is a symmetrical distribution shaped like a rectangle. There

FIGURE 3.9 Idealized Bimodal and Rectangular Distributions

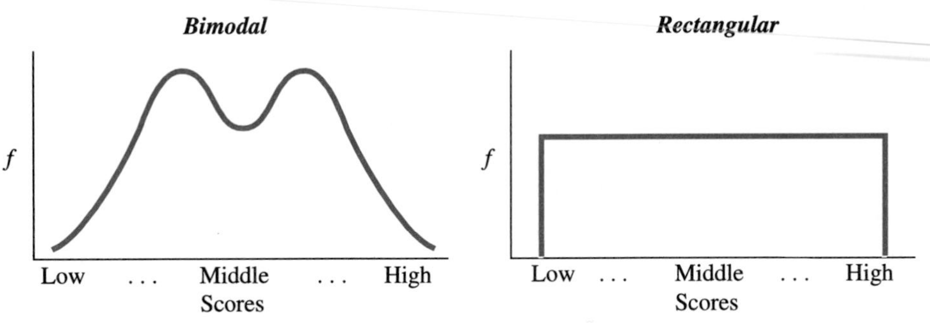

are no discernible tails, because the extreme scores do not have relatively low frequencies. Such a distribution occurs whenever the frequency of all scores is the same.

Distributions of real data versus ideal distributions If you're wondering why you need to know the names of the previous ideal distributions, it's because we use descriptive statistics to describe the important characteristics of a sample of data. One important characteristic is the shape of the frequency distribution which the data forms, so we apply the names of the previous distributions to sample data as well. However, real data are never pretty, and the distribution of a sample of scores will tend to be a bumpy, rough approximation to the smooth idealized curves we've discussed. Figure 3.10 shows several frequency distributions of sample data, as well as the corresponding labels we might use. (Notice that we can apply these names even to choppy histograms or bar graphs.) We generally assume that the sample represents a population that more closely fits the corresponding ideal polygon: We expect that if we measured the population, the additional scores and their corresponding frequencies would "fill in" our sample curve, smoothing it out to form the ideal curve.

We shall return to simple frequency distributions throughout the remainder of this book. However, counting each score's simple frequency is not the only thing we do in statistics.

FIGURE 3.10 Simple Frequency Distributions of Sample Data with Appropriate Labels

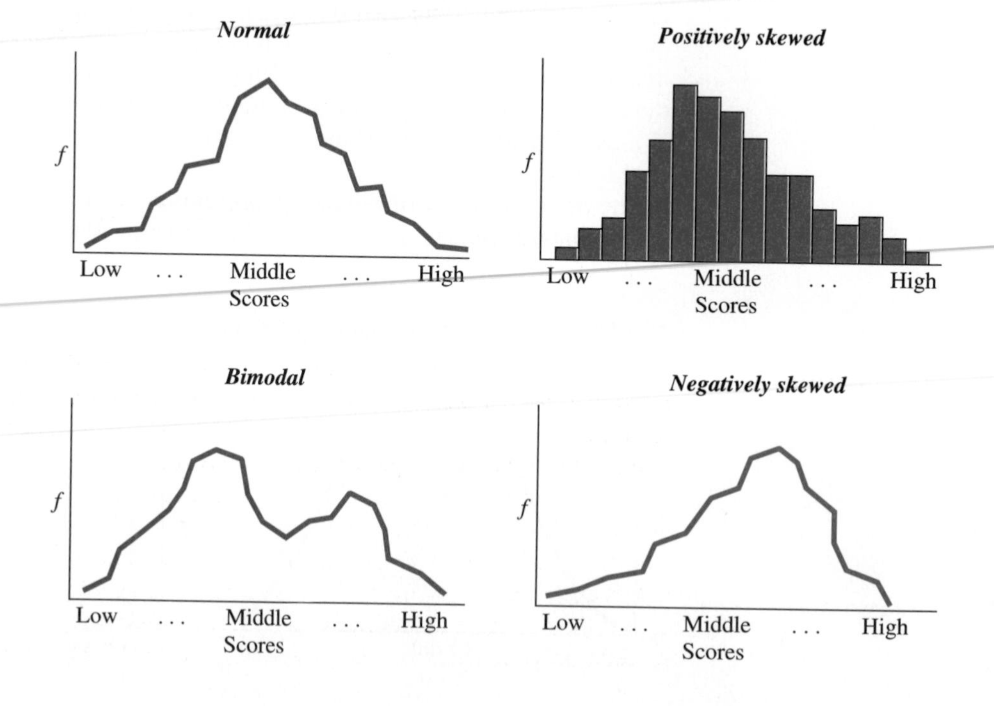

CREATING RELATIVE FREQUENCY DISTRIBUTIONS

Another way to organize scores is to transform each score's simple frequency into a relative frequency. **Relative frequency** is the proportion of the total N made up by a score's simple frequency. While simple frequency is the number of times a score occurs in the data, relative frequency is the proportion of time the score occurs in the data. The symbol for relative frequency is *rel. f.*

Why do we compute relative frequency? We are again asking how often certain scores occured, but relative frequency is often easier to interpret—*to make sense out of*—than simple frequency. For example, the finding that a score has a simple frequency of 60 is difficult to interpret, because we have no frame of reference. However, we can easily interpret the finding that a score has a relative frequency of .20, because this means that the score's f is .20 of the total N; the score occurred .20 of the time in the sample.

Here is your first statistical formula.

THE FORMULA FOR COMPUTING A SCORE'S RELATIVE FREQUENCY IS

$$rel.\ f = \frac{f}{N}$$

This formula says that to compute the relative frequency of a score, divide that score's frequency by the total N.

For example, say that out of an N of 10 scores, the score of 7 has a simple frequency of 4. What is the relative frequency of 7? Using the formula, we have

$$rel.\ f = \frac{f}{N} = \frac{4}{10} = .40$$

The score of 7 has a relative frequency of .40, meaning that 7 occurred .40 of the time in the sample.

Presenting Relative Frequency in a Table

A distribution based on the relative frequency of the scores is called a **relative frequency distribution.** To create a relative frequency table, first create a simple frequency table, as we did previously. Then add a third column labeled "*rel. f*" to show the relative frequency of each score.

As an example, say that we asked a sample of mothers how many children they each have, and then we compiled the results as shown in Table 3.2. To compute *rel. f*, we need N, which is the total number of scores in the sample. Although there are 6 scores in the score column, N is not 6. As shown by the sum of the f's, there are a total of 20 scores in this sample. The relative frequency for each score is the f for that score divided by N. The score of 1, for example,

TABLE 3.2 Relative Frequency Distribution of Number of Children

The left-hand column identifies the scores, the middle column shows each score's frequency, and the right-hand column shows each score's relative frequency.

Score	f	rel. f
6	1	.05
5	0	.00
4	2	.10
3	3	.15
2	10	.50
1	4	.20
Totals:	20	1.00 = 100%

has $f = 4$, so the relative frequency of 1 is 4/20, or .20. Thus, in our sample, .20 of the subjects have 1 child.

We can also determine the combined relative frequency of several scores by adding the individual relative frequencies together. For example, in Table 3.2, a score of 1 has a relative frequency of .20, and a score of 2 has a relative frequency of .50. Therefore, the relative frequency of 1 and 2 is .20 + .50, or .70; these scores occurred .70 of the time, so mothers having 1 or 2 children compose .70 of our sample.

You may find that working with relative frequency is easier if you transform the decimals to percents. (Remember that officially relative frequency is a proportion.) If we convert relative frequency to percent, we have the percent of time that a score or scores occurred. To transform a proportion to a percent, multiply the proportion times 100. Above we saw that .20 of the scores were the score of 1, so (.20)(100) = 20%: 20% of all scores were 1. To transform a percent back to a relative frequency, divide the percent by 100.

To check your work, remember that the sum of all the relative frequencies in a distribution should equal 1.0: all scores together should constitute 1.0, or 100%, of the sample.

STAT ALERT Relative frequency is interpreted as the proportion of time that certain scores occur in a set of data.

Graphing a Relative Frequency Distribution

As with simple frequency, we graph relative frequency with a bar graph if the scores are from a nominal or ordinal scale, and with a histogram or polygon if the scores are from an interval or ratio scale. Figure 3.11 presents examples using the relative frequency distribution we created in Table 3.2. These graphs are drawn in the same way as corresponding graphs of simple frequency except that here the Y axis reflects relative frequency so it is labeled in increments between 0 and 1.0.

Finding Relative Frequency Using the Normal Curve

When our data form a normal distribution, an extremely important statistical procedure is to determine relative frequency directly from the normal curve. One reason for visualizing the normal curve as the outline of a parking lot full of people is so that you think of the normal curve as forming a solid geometric figure having an area underneath the curve. What we consider to be the total space occupied by subjects in the parking lot is, in statistical terminology, the **total area under the curve.** This area represents the combined total frequency of all scores. We can also find the area of any portion of the normal curve. We take a vertical "slice" of the polygon above certain scores and the area of this portion of the curve is the space occupied by the subjects obtaining those particular scores. This area represents the combined frequency for these scores. We then compare this area to the total area, to determine the **proportion of the total area under the curve.** Now here's the important part:

> **The proportion of the total area under the normal curve at certain scores corresponds to the relative frequency of those scores.**

Here's an example. Figure 3.12 shows a normal curve with the "parking lot view." A vertical line is drawn through the middle score of 30, and so .50 of the parking lot is to the left of the line. Because the complete parking lot contains all of our subjects, a slice that is .50 of it contains 50% of our subjects. (We can ignore those relatively few subjects who are straddling the line.) Subjects are

FIGURE 3.11 Examples of Relative Frequency Distributions (Using Data in Table 3.2)

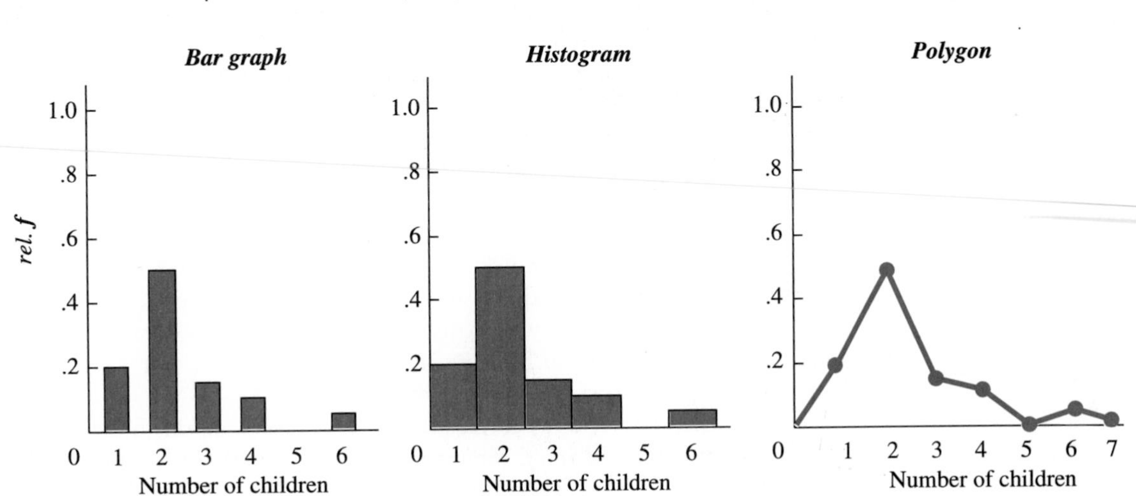

FIGURE 3.12 Normal Curve Showing .50 of the Area Under the Normal Curve

The vertical line is through the middle score, so 50% of the distribution is to the left of the line and 50% is to the right of the line.

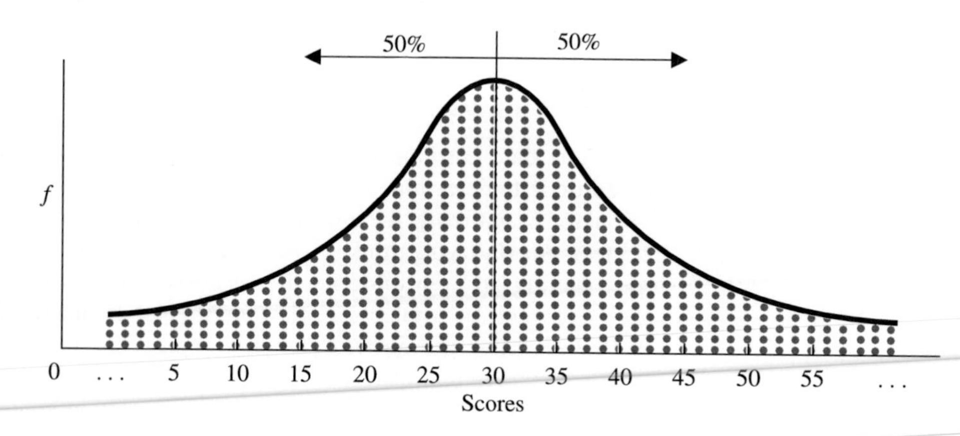

standing in the left-hand part of the lot because they received scores of 29, 28, and so on, so in total, 50% of our subjects obtained scores below 30. Because 50% of our subjects had these scores, in other words, the scores below 30 occurred 50% of the time in these data. So the scores below 30 have, altogether, a relative frequency of .50.

In the same way, we can identify any proportion of the total area under the curve. Think of a proportion of the area under the curve as the proportion of subjects who are standing in that part of the parking lot. The proportion of subjects standing in that part of the parking lot is equal to the proportion of time that subjects obtained those scores out of all the scores in the distribution. The proportion of time that certain scores occur out of all scores *is* relative frequency.

Of course, statisticians don't fly around in helicopters, eyeballing parking lots. But the same principle applies. Say that by using a ruler and protractor, we determine that in Figure 3.13 the total area under the curve—the entire polygon—occupies an area of 6 square inches on the page. To find the area of any geometric figure, we consider its height and width. The area under the curve depends on the curve's height, which is due to the simple frequency of each score, and its width, which is due to spanning adjacent scores. Thus, the total area under the curve corresponds to the total of all frequencies for all scores, which is our *N*. Say that we also determine that the area under the curve between a score of 30 and 35 covers 2 square inches. This area, too, is due to the frequencies of the scores found there. Therefore, the total frequency of scores between 30 and 35 constitutes 2 out of the 6 square inches created by the frequency of all scores, so these scores constitute two-sixths, or 33%, of the total distribution. In other words, the scores between 30 and 35 constitute 33% of our *N*, so they occur 33% of the time and have a relative frequency of .33.

Because area corresponds to frequency, we would obtain the same answer if we used our formula for *rel. f.* First we would add together the simple frequencies of every score between 30 and 35. Then, dividing the sum by *N*, we would again find that the relative frequency is .33. However, the advantage of using area

FIGURE 3.13 Finding the Proportion of the Total Area Under the Curve

The complete curve occupies 6 square inches, with scores between 30 and 35 occupying 2 square inches.

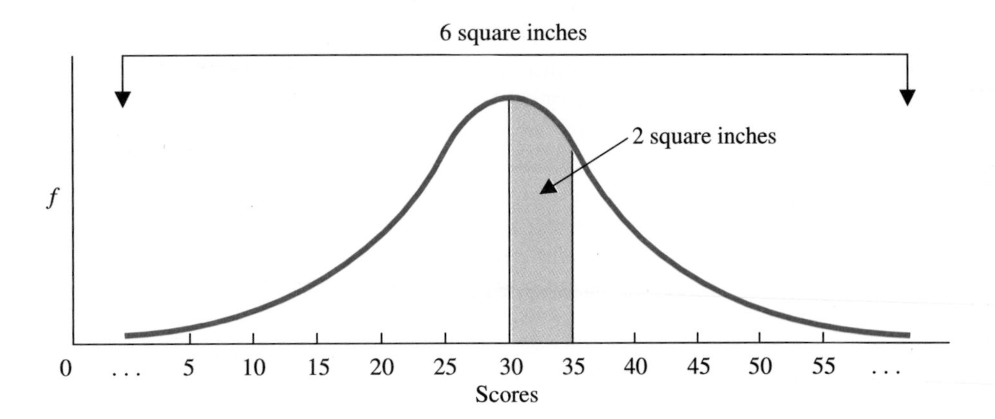

under the curve is that we can get the answer without knowing the N or the simple frequencies of these scores. Whatever the N might be and whatever the actual frequency of each score is, we know that the area these scores comprise is 33% of the total area, and that's all we need to know to determine their relative frequency. This is especially advantageous because, as we'll see in Chapter 6, statisticians have created a system for easily finding the proportion of the total area under any part of the normal curve, which means we can easily determine the relative frequency for any group of scores within a normal distribution. (No, you won't need a ruler and a protractor.) Until that time, simply remember this:

> ***STAT ALERT*** Total area under the normal curve corresponds to the times that all scores occur, so a proportion of the total area corresponds to the proportion of time certain scores occur, which is their relative frequency.

CREATING CUMULATIVE FREQUENCY DISTRIBUTIONS

Sometimes we want to know not only how often a particular score occurred, but simultaneously its standing relative to other scores in the data. Knowing that 10 people received a grade of 80 on an exam or that .15 of the class received an 80 may not be as informative as knowing that 30 people scored above 80 or 60 people scored at or below 80. When we seek such information, the convention in statistics is to count from *lower* scores, computing cumulative frequency. **Cumulative frequency** is the frequency of all scores at or below a particular score. The symbol for cumulative frequency is *cf.* The word *cumulative* implies accumulating. To compute a score's cumulative frequency, we accumulate, or add, the simple frequencies for all scores below that score and then add the score's frequency, to get the frequency of scores at or below the score.

Presenting Cumulative Frequencies in a Table

To create a cumulative frequency table, first create a simple frequency table. Then add a third column labeled "*cf.*" As an example, say that we measured the ages of a sample of adolescents (as part of a study of the relationship between age and the stress of becoming an adult). To summarize the age scores, we create the distribution in Table 3.3. To compute cumulative frequency, begin with the *lowest* score. In Table 3.3, we see that no one scored below 10 and 1 subject scored 10, so we put 1 in the *cf* column opposite 10 (there is one subject who is 10 years of age or younger). Next, there were two scores of 11. We add this *f* to the *cf* for 10, so the *cf* for 11 is 3 (there are 3 subjects at or below the age of 11). Next, no one scored 12 and three subjects scored below 12, so the *cf* for 12 is also 3 (there are still 3 people at age 12 or below). In the same way, the *cf* of each score is the frequency for that score plus the cumulative frequency for the score immediately below it.

As a check on any cumulative frequency distribution you create, verify that the *cf* for the highest score equals *N*: all of the *N* subjects obtained either the highest score or a score below it.

Graphing a Cumulative Frequency Distribution

Usually it makes sense to compute cumulative frequency only for interval or ratio data, and the convention is to create a polygon. Figure 3.14 shows the cumulative frequency polygon of the distribution from Table 3.3.

TABLE 3.3 Cumulative Frequency Distribution of Age Scores

The left-hand column identifies the scores, the center column contains the simple frequency of each score, and the right-hand column contains the cumulative frequency of each score.

Score	*f*	*cf*
17	1	19
16	2	18
15	4	16
14	5	12
13	4	7
12	0	3
11	2	3
10	1	1
Total:	19	

FIGURE 3.14 Cumulative Frequency Polygon Showing the Cumulative Frequencies of the Scores in Table 3.3

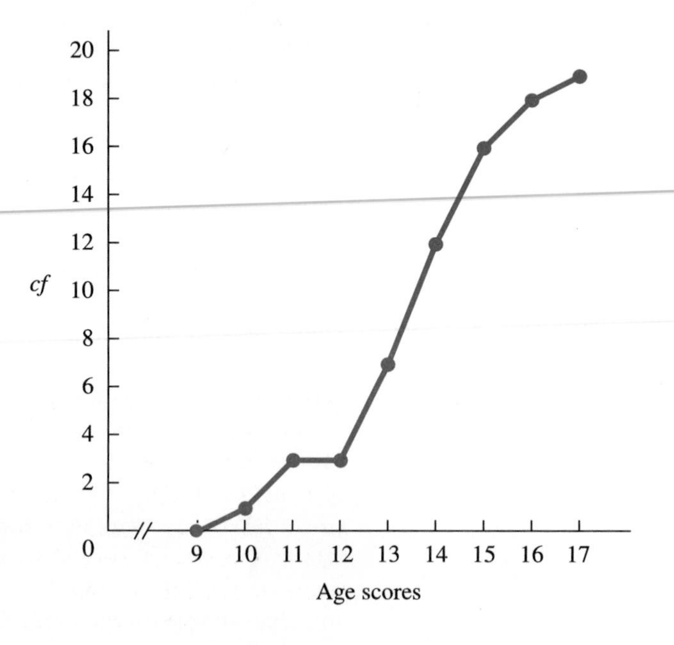

In a cumulative frequency distribution, the *Y* axis is labeled "*cf.*" As in plotting simple frequency, we include on the *X* axis the next score below the lowest score in the data: 10 is the lowest score in the data, but 9 is included on the graph. Unlike the polygon of a simple frequency distribution, however, a cumulative frequency distribution does not include the next score above the highest score in the data.

Cumulative frequencies never decrease. There cannot be fewer subjects having a score of 13 or below than received a score of 12 or below. Therefore, when you plot cumulative frequency, as the scores on the *X* axis increase, the height of the polygon must either remain constant or increase.

COMPUTING PERCENTILE

There is one other way to answer the question of how often various scores occurred. We saw that the simple frequency of a score may be difficult to interpret because we have no frame of reference. Therefore, we transformed it into relative frequency, because a proportion (or percent) of the time a score occurs may be easier to interpret than the number of times it occurs. Likewise, cumulative frequency may be difficult to interpret. To decide whether 60 people scoring at or below 80 is a lot or a little depends on how many people took the test. But if I say that 98% of the scores were at or below 80, you now have a better understanding of just how the test scores are distributed and what a score of 80 reflects. By computing such a percentage, we are computing the score's percentile. A **percentile** is the percent of all scores in the data that are at or below a certain score. While cumulative frequency indicates the *number* of subjects who obtained a particular score or below, a percentile indicates the *percentage* of subjects who obtained a particular score or below. Thus, for example, if a person scored at the 25th percentile, we know that 25% of all subjects scored at or below that person's score.

We have two approaches for determining the percentile of a particular score, or for finding the score that is at a particular percentile: we can use the area under the normal curve, or we can use formulas.

Finding Percentile Using the Area Under the Normal Curve

Earlier we saw that when our data produce a normal distribution, we can use the proportion of the area under the normal curve to compute relative frequency. We can also use the proportion of the area under the curve to compute percentile. On a graph, lower scores are always placed to the left of a particular score. Thus, a percentile for a given score corresponds to the percent of the total area under the normal curve that is to the *left* of the score. For example, on the distribution of test scores in Figure 3.15, .5 or 50% of the curve is to the left of the middle score of 30. Because scores to the left of 30 are below it, 50% of the distribution is below 30 (50% of the subjects are standing to the left of the line in the parking

FIGURE 3.15 Normal Distribution Showing the Area Under the Curve to the Left of Selected Scores

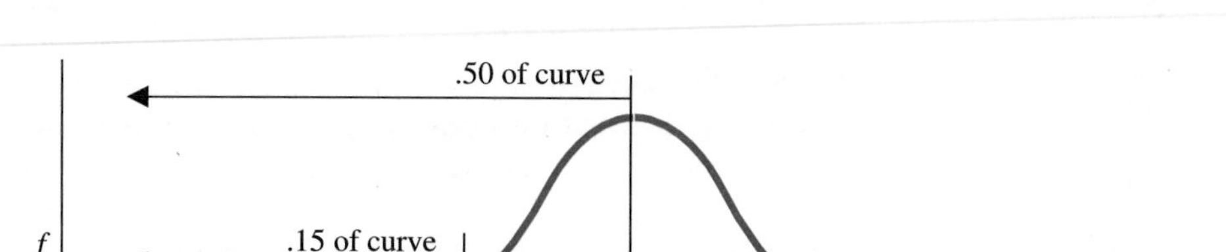

lot, and all their scores are less than 30). Thus, the score of 30 is at the 50th percentile.

Likewise, to find the percentile for a score of 20, we would find the percent of the total curve that is to the left of 20. If we find that 15% of the distribution is to the left of it, then 20 is at the 15th percentile. We may also work the other way, finding the score at a given percentile: if we were seeking the score at the 85th percentile, we would measure over until 85% of the area under the curve is to the left of a certain point. If, as shown in Figure 3.15, the score of 45 is at that point, then 45 is the score at the 85th percentile.

Notice that we make a slight change in our definition of percentile when we are using the normal curve. Technically, a percentile is the percent of scores *at* or below a certain score. However, we use the normal curve to describe large samples or populations, so we can treat those subjects who scored at the score as a negligible portion of the total (remember that we ignore those relatively few subjects who are straddling the line.) Then a percentile is the percent of all scores that are *below* a certain score. Thus, in Figure 3.15, since the score of 30 is at the 50th percentile, we treat this as meaning that 50% of the scores are below 30 and 50% are above it. However, if we are describing a small sample, we should not say that 50% of the scores are above the 50th percentile and 50% are below it: those subjects scoring *at* the 50th percentile might actually constitute, say, 10% of the sample, which is not a negligible amount. And if we say that 50% are below, 50% are above, and 10% are at the score, we have the impossible total of 110% of the sample! Therefore, with small samples, percentile is defined and calculated as the percent of scores at or below a particular score.

Using the normal curve to compute percentile will be reasonably accurate only if we have a rather large sample or a population that closely fits the normal curve. If we have only a few scores, or the data forms a nonnormal distribution, we instead use formulas for directly calculating percentile. Before we can discuss these formulas, however, you must first understand the procedures for creating grouped frequency distributions.

CREATING GROUPED FREQUENCY DISTRIBUTIONS

A rule of thumb for any type of frequency table is that there should be between about 8 and 18 rows in the table. Fewer than 8 scores tend to produce a very small, often unnecessary table, while more than 18 scores tends to produce a very large, inefficient table. In the previous examples we presented the *f, rel. f* or *cf* for each individual score, so we created **ungrouped distributions.** When there are too many individual scores to produce a manageable ungrouped distribution, we create a grouped distribution. In a **grouped distribution** we combine different scores together into small groups and then report the total *f, rel. f,* or *cf* of each group.

To see how this is done, say that we measure how anxious subjects get when making a speech by giving them an anxiety questionnaire immediately afterward. We obtain the following anxiety scores:

3	4	4	18	4	28	26	41	5	40	4	6	5
18	22	3	17	12	26	4	20	8	15	38	36	

The scores are between a low of 3 and a high of 41, spanning 39 possible different scores. You can count them on your fingers, or you can calculate the number of values spanned between any two scores by using this formula:

$$\text{Number of values} = (\text{High score} - \text{Low score}) + 1$$

Thus, 41 minus 3 equals 38, plus 1 equals 39, indicating that, including the 41 and the 3, there is a span of 39 values between 41 and 3.

In creating a grouped distribution, the first step is to decide how to group the scores so that each group spans the same range of scores. To facilitate this, we can operate as if the sample contained a wider range of scores than was actually in the data. For example, we will operate as if the above scores were from 0 to 44, spanning 45 scores. This conveniently allows us to create nine groups, each spanning 5 scores. Then we create the grouped distribution shown in Table 3.4.

The group labeled "0–4" contains the scores 0, 1, 2, 3, and 4, while "5–9" contains scores 5 through 9, and so on. Each group is called a *class interval,* and the number of values spanned by every class interval is called the *interval size.* Here we have used an interval size of 5, meaning that each group spans five scores. (For example, using the above formula, $(4 - 0) + 1 = 5$.) We typically choose an interval size that is easy to work with (such as 2, 5, 10, or 20), rather than something unfriendly (like 17). Also, we choose an interval size that will result in between 8 and 18 class intervals.

Notice several things about the layout of the score column in Table 3.4. First, each class interval is labeled with the low score on the left. Second, the low score in each interval is a whole-number multiple of our interval size of 5. Third, every class interval has the same interval size, including the highest and lowest intervals. (Even though the highest score in the data is only 41, we have the complete interval of 40–44.) Finally, the intervals are arranged so that we proceed to higher scores as we proceed toward the top of the column.

TABLE 3.4 Grouped Distribution Showing *f*, *rel. f*, and *cf* for Each Group of Scores

The left-hand column identifies the lowest and highest score in each class interval.

Scores	*f*	*rel. f*	*cf*
40–44	2	.08	25
35–39	2	.08	23
30–34	0	.00	21
25–29	3	.12	21
20–24	2	.08	18
15–19	4	.16	16
10–14	1	.04	12
5– 9	4	.16	11
0– 4	7	.28	7
Total:	25	1.00	

To complete the table, we find the *f* for each class interval by summing the individual frequencies of all scores in the group. In the original raw data above, there were no scores of 0, 1, or 2, but there were two 3's and five 4's. Thus, the 0–4 interval has a total *f* of 7. For the 5–9 interval, there were two 5's, one 6, no 7's, one 8, and no 9's, so the 5–9 interval has a total *f* of 4. And so on.

We compute the relative frequency for each interval by dividing the *f* for the interval by *N*. Remember, *N* is the total number of raw scores (here 25), not the number of class intervals. Thus, for the 0–4 interval, *rel. f* equals 7/25, which is .28.

We compute the cumulative frequency for each interval by counting the number of scores in the data that are at or below the *highest* score in the interval. Begin with the lowest interval. There are 7 scores of 4 or below, so the *cf* for interval 0–4 is 7. Next, *f* is 4 for the scores between 5 and 9, and adding the 7 scores below the interval produces a *cf* of 11 for the interval 5–9. Likewise, the *cf* for each interval is the *f* for that interval plus the *cf* for the interval immediately below it.

We can summarize the steps in creating a grouped distribution as follows:

1. Select a low score and a high score that span all the scores in the sample and that allow the scores to be divided into intervals having the same interval size.

2. Arrange the intervals from highest to lowest, and label each interval with the lowest and highest scores in the interval, putting the lowest score on the left.

3. Compute the frequency and relative frequency for each interval based on the scores that fall in the interval. List as the cumulative frequency for each interval the cumulative frequency of the highest score in that interval.

Recognize that in deciding how to organize a particular distribution, you must weigh the advantages and disadvantages of grouping the data. The disadvantage is that grouped distributions present the individual scores less precisely than do ungrouped distributions. For example, the above 0–4 interval has an *f* of 7, so all we know from the table is that 7 subjects scored somewhere between 0 and 4. If we had chosen a larger interval size, say 10, then we would have even less precision. The advantage, however, is that grouped distributions shrink a large number of different scores into a manageable form, so that we can comprehend the overall distribution better than if we examined a mass of individual scores.

Real Versus Apparent Limits

What if one of the scores in the above example were 4.6? This score seems too large to be in the 0–4 interval, but too small to be in the 5–9 interval. To allow for such scores, we must consider the "limits" of each interval. The upper and lower numbers we use to identify each interval in the score column of a frequency table are called the *apparent upper limit* and the *apparent lower limit,* respectively. However, apparent limits always imply another type of limit, called the *real limit.* The left-hand portion of Table 3.5 shows the apparent limits for the previous grouped data, and the right-hand portion gives the implied real limits for each interval.

As you can see, (1) each real limit is halfway between the lower apparent limit of one interval and the upper apparent limit of the interval below it, and (2) the lower real limit of one interval is always the same number as the upper real limit of the interval below it. Thus, 4.5 is halfway between 4 and 5, so 4.5 is the lower real limit of the 5–9 interval and the upper real limit of the 0–4 interval. Also, notice that the difference between the lower real limit and the upper real limit always equals the interval size (9.5 − 4.5 = 5).

Real limits eliminate the gaps between intervals, so now we can place all scores. A score such as 4.6 falls in the interval 5–9, because it falls between 4.5

TABLE 3.5 Real and Apparent Limits

The apparent limits in the left-hand column imply the real limits in the right-hand column.

Apparent limits (lower–upper)	*imply*	*Real limits (lower–upper)*
40–44	→	39.5–44.5
35–39	→	34.5–39.5
30–34	→	29.5–34.5
25–29	→	24.5–29.5
20–24	→	19.5–24.5
15–19	→	14.5–19.5
10–14	→	9.5–14.5
5– 9	→	4.5– 9.5
0– 4	→	−0.5– 4.5

and 9.5. If we have scores equal to a real limit (such as two scores of 4.5), we put half in the lower interval (between –0.5 and 4.5) and half in the upper interval (4.5–9.5). If one such score is left over, we flip a coin to pick the interval.

Notice that the principle of real limits also applies to ungrouped data. Implicitly, each individual score is actually a class interval with an interval size of 1. Thus, when we label the score column in an ungrouped distribution with the score of 6, we are writing both the upper and the lower apparent limits. However, the lower real limit for this interval is 5.5, and the upper real limit is 6.5.

Graphing Grouped Distributions

Grouped distributions are graphed in the same way as ungrouped distributions, *except* that we label the X axis differently.

To graph grouped simple frequency or grouped relative frequency, label the X axis using the *midpoint* of each class interval. To find the midpoint, multiply .5 times the interval size, and add the result to the lower real limit. In our example the interval size was 5, which multiplied times .5 is 2.5. For the interval 0–4, the lower real limit was $-.5$. Adding 2.5 to $-.5$ yields 2. Thus, we use the score of 2 on the X axis to represent the class interval 0–4. Similarly, for the interval 5–9, 2.5 plus 4.5 is 7, so we represent this interval using the score of 7.

As usual, for nominal or ordinal scores we create a bar graph and for interval or ratio scores we create a histogram or polygon. Figure 3.16 presents a histogram and polygon for the grouped simple frequency distribution created back in Table 3.4. The height of each bar or data point corresponds to the total simple frequency of all scores in the class interval. We plot a relative frequency distribution in the same way, except that the Y axis is labeled in increments between 0 and 1.0.

FIGURE 3.16 Grouped Frequency Polygon and Histogram

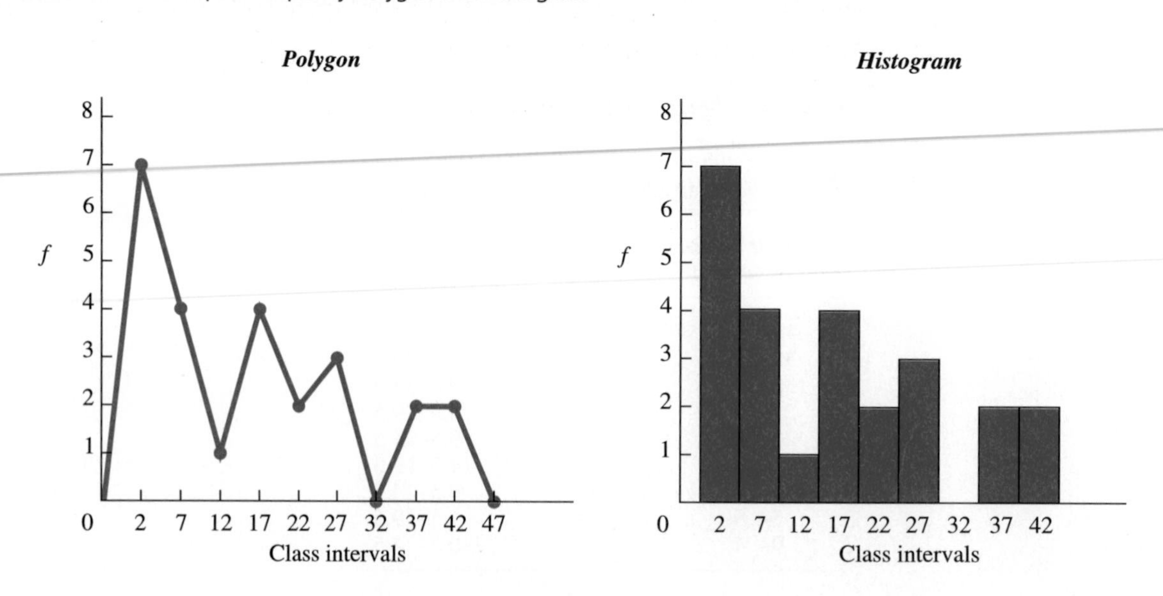

FIGURE 3.17 Grouped Cumulative Frequency Polygon

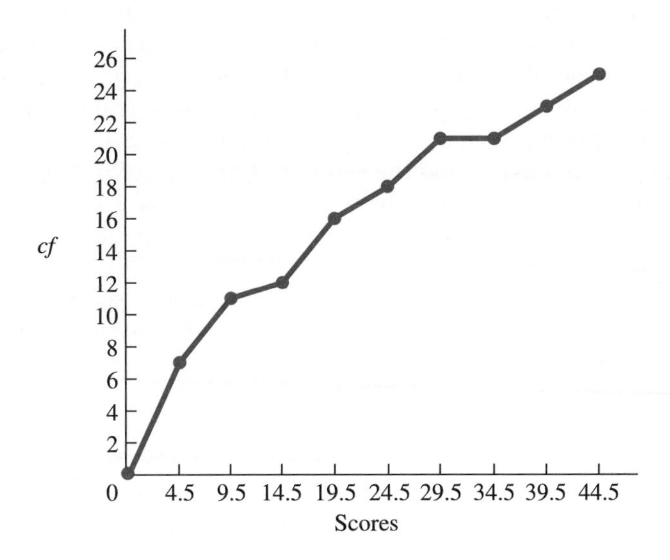

Figure 3.17 presents the grouped cumulative frequency polygon for the preceding data. To graph a grouped cumulative frequency distribution, label the X axis using the *upper real limit* of each interval. Thus, for example, the interval 0–4 is represented at 4.5 on the X axis, and the interval 5–9 is at 9.5. Then create a polygon in which each data point is the *cf* for a group.

Now that you understand real limits, we can discuss our final topic: the formulas for computing the score that corresponds to a given percentile, and for computing the percentile of a given score.

Determining the Score at a Given Percentile

We compute percentiles from a cumulative frequency distribution. As an example, say that we have collected scores from laboratory rats, where each score reflects the number of minutes required for the rat to find the end of a maze. The cumulative frequency distribution for these data is presented in Table 3.6.

TABLE 3.6 Cumulative Frequency Distribution of Maze-Running Times for Laboratory Rats

Score	*f*	*cf*
5	1	10
4	1	9
3	2	8
2	3	6
1	3	3
	$N = 10$	

Say that we wish to determine the score at the 50th percentile. We ask, "50% of the scores are at or below which score?" To find the score at a particular percentile, we find the score that has a *cf* that corresponds to that particular percentage of *N*. Here we are looking for the score with a *cf* that is 50% of *N*. Since *N* is 10, the score at the 50th percentile is the score having a *cf* of 5.

The trouble is that no score in the table has a *cf* of exactly 5. The score of 1 has a *cf* of only 3, and the score of 2 has a *cf* of 6. Obviously, the score having a *cf* of 5 is somewhere between the scores of 1 and 2. As this illustrates, the percentile we seek may not correspond to one of the scores that actually occurred in the sample.

To compute a percentile, we first treat the scores as if they were from a continuous variable that allows decimals. Then we look at the real limits. Looking at the real limits for the scores of 1 and 2, we see

Scores	***f***	***cf***
1.5–2.5	3	6
.5–1.5	3	3

Because the score we seek has a *cf* of 5, we want a score above 1.5 (with a *cf* of only 3), so the score we seek is in the interval 1.5–2.5. Thus, we will proceed into this interval far enough beyond 1.5 to accumulate a *cf* of 5. We assume that the frequency in an interval is evenly spread throughout the interval, so that, for example, if we go to a score that is halfway between the upper and lower limits, we accumulate one-half of the frequency in the interval. Conversely, if we accumulate one-half of the frequency in an interval, we assume we are at the score that is halfway between the upper and lower limits. In our problem, the score we seek has a *cf* of 5, so we want the score above 1.5 that increases the *cf* by 2. If we went to a score of 2.5, we would accumulate an additional *f* of 3, increasing the *cf* by 3, which is too much. We want an *f* of 2 out of the 3, so we want two-thirds of the total frequency in the interval. To accumulate two-thirds of the frequency in the interval, we go to the score that is two-thirds of the way between the lower and upper limits. To find the score that is two-thirds of the way between 1.5 and 2.5, we multiply two-thirds, or .667, times the interval size of 1, which gives us .667. Then, adding .667 to 1.5 takes us to the score of 2.17. Thus, 2.17 is two-thirds of the way through this interval so we assume it has a *cf* of 5. Therefore 2.17 is at the 50th percentile. We conclude that 50% of our rats completed the maze in 2.17 minutes or less.

Luckily, we have a formula that accomplishes everything above all at once.

THE FORMULA FOR FINDING THE SCORE AT A GIVEN PERCENTILE IS

$$\text{Score} = \text{LRL} + \left(\frac{\text{target } cf - cf \text{ below interval}}{f \text{ within interval}}\right)(\text{interval size})$$

In English, the formula says that we need these components.

1. target *cf*: the cumulative frequency of the score we seek. To find it, we transform the percentile we seek into a proportion and then multiply the proportion by *N*. The interval containing this *cf* contains the score we seek. (In our example above, the target *cf* is 5.) You must find the target *cf* before you can find any other component.
2. LRL: the lower real limit of the interval containing the score we seek. (In the example above, it is 1.5.)
3. *cf* below interval: the cumulative frequency for the interval below the interval containing the score we seek. (In the example, it is 3.)
4. *f* within interval: the frequency in the interval containing the score we seek. (In the example, it is 3.)
5. interval size: the interval size used to create the frequency distribution. (Above, it is 1.)

Table 3.7 shows where in the original cumulative frequency distribution you would find the components needed to determine the score at the 50th percentile.

We put these numbers into the formula for computing the score at a given percentile:

$$\text{Score} = \text{LRL} + \left(\frac{\text{target } cf - cf \text{ below interval}}{f \text{ within interval}}\right)(\text{interval size})$$

which becomes

$$\text{Score} = 1.5 + \left(\frac{5 - 3}{3}\right)(1)$$

We must first deal with the fraction. After subtracting 5 − 3, we have 2/3, which is .667. So

$$\text{Score} = 1.5 + (.667)(1)$$

TABLE 3.7 Cumulative Frequency Distribution Showing Components for Computing the Score at the 50th Percentile

	Score	f	cf	
	4.5–5.5	1	10	
	3.5–4.5	1	9	
	2.5–3.5	2	8	← target *cf* of 5 is in this interval
LRL →	1.5–2.5	3 ← *f* within interval	6	
	.5–1.5	3	3	← *cf* below interval

↑ interval size = 1

After multiplying, we have

$$\text{Score} = 1.5 + .667$$

So finally,

$$\text{Score} = 2.17$$

Again, the score at the 50th percentile in these data is 2.17.

Although this example has an interval size of 1, we can use the above formula for any grouped distribution having any interval size.

Finding a Percentile for a Given Score

We can also work from the opposite direction when we have a score in mind and wish to determine its percentile. To find the percentile for a given score, we find the *cf* of the score within the interval, plus the *cf* below the interval, and then determine the percent of scores that are at or below the score.

Mathematically we accomplish this using the following formula.

THE FORMULA FOR FINDING THE PERCENTILE OF A GIVEN SCORE IS

$$\text{Percentile} = \left(\frac{cf \text{ below interval} + \left(\dfrac{\text{score} - \text{LRL}}{\text{interval size}} \right) \left(\begin{matrix} f \text{ within} \\ \text{interval} \end{matrix} \right)}{N} \right)(100)$$

This formula requires the following components:

1. score: the score for which we are computing the percentile.
2. *cf* below interval: the cumulative frequency of the interval below the interval containing the score.
3. LRL: the lower real limit of the interval containing the score.
4. *f* within interval: the frequency in the interval containing the score.
5. interval size: the interval size used to create the grouped distribution.
6. *N*: the total number of scores in the sample.

Say that we want to use this formula to find the percentile of the score of 4 in our rat data. We find the components for the formula as shown in Table 3.8.

We then put these numbers into the formula for finding the percentile of a given score:

$$\text{Percentile} = \left(\frac{cf \text{ below interval} + \left(\dfrac{\text{score} - \text{LRL}}{\text{interval size}} \right) \left(\begin{matrix} f \text{ within} \\ \text{interval} \end{matrix} \right)}{N} \right)(100)$$

TABLE 3.8 Cumulative Frequency Distribution Showing Components for Computing the Percentile of a Given Score of 4

Score	f	cf
4.5–5.5	1	10 ← N
3.5–4.5	1	9
2.5–3.5	2	8 ← cf below interval
1.5–2.5	3	6
.5–1.5	3	3

score of 4 is in this interval → 3.5–4.5

LRL → 3.5

interval size = 1

f within interval

which becomes

$$\text{Percentile} = \left(\frac{8 + \left(\frac{4.0 - 3.5}{1}\right)(1)}{10}\right)(100)$$

Working on the fraction in the numerator first, we find that 4.0 minus 3.5 is .5, which divided by 1 is still .5, so we have

$$\text{Percentile} = \left(\frac{8 + (.5)\,(1)}{10}\right)(100)$$

Multiplying .5 by 1 gives .5, so we have

$$\text{Percentile} = \left(\frac{8 + .5}{10}\right)(100)$$

After adding, we have

$$\text{Percentile} = \left(\frac{8.5}{10}\right)(100)$$

and after dividing, we have

$$\text{Percentile} = (.85)(100)$$

Finally, the answer is

$$\text{Percentile} = 85$$

Thus, the score of 4.0 in the above distribution is at the 85th percentile, so 85% of the rats completed the maze in 4 minutes or less.

FINALLY

In this chapter you've learned descriptive techniques for describing scores using simple frequency, relative frequency, cumulative frequency, or percentile. Recognize that you *have* learned when to use some statistics here and what their answers mean. All are procedures whose answers indicate how often certain scores in the data occur, but each provides a slightly different perspective that allows you to interpret the data in a slightly different way. As usual, which particular procedure you should use is determined by the scale of measurement reflected by the scores. Likewise, whether you create a bar graph, histogram, or polygon is determined by the measurement scale (and the number of different scores in the data). Beyond this, which procedure you should use is determined simply by which provides the most information. However, as a researcher, you may not automatically know which is the best technique for a given situation. So, use the trial-and-error method: try everything, and then choose the technique that most accurately and efficiently summarizes the data for your purposes.

As an aid to learning statistics, start drawing graphs. In particular, draw the normal curve. When you are working problems or taking tests, draw the normal curve and indicate where the low, middle, and high scores are located. Being able to see the frequencies of the different scores will greatly simplify your task.

CHAPTER SUMMARY

1. The number of scores in a sample is symbolized by N.
2. A *simple frequency distribution* organizes data by showing the frequency with which each score occurred. The symbol for *simple frequency* is f. The sum of the individual f's of all the scores in the sample equals N.
3. For ungrouped distributions, when graphing a simple frequency distribution, plot the scores along the X axis and the frequency of the scores along the Y axis. If the variable involves a nominal or ordinal scale, create a *bar graph,* in which the adjacent bars do not touch. If the variable involves an interval or ratio scale and there are relatively few different scores, create a *histogram.* It is similar to a bar graph except that adjacent bars do touch. If there are many different scores from an interval or ratio variable or there is more than one sample of scores, create a *polygon:* place a data point above each score at the corresponding frequency and then connect adjacent data points with a straight line. Also include the score above the highest score and below the lowest score.
4. A *normal distribution* forms a symmetrical, bell-shaped curve known as the *normal curve.* In it extreme high and low scores occur relatively infrequently, scores closer to the middle score occur more frequently, and the middle score occurs most frequently. The low frequency, extreme low and extreme high scores are in the *tails* of the distribution.
5. The *normal curve model* assumes that the frequency distribution for a population of scores generally fits the normal curve.

6. A *negatively skewed distribution* is a nonsymmetrical distribution containing low frequency, extreme low scores, but not containing corresponding low frequency, extreme high scores. A *positively skewed distribution* is a nonsymmetrical distribution containing low frequency, extreme high scores, but not containing corresponding low frequency, extreme low scores.

7. A *bimodal distribution* is a symmetrical distribution containing two areas where there are relatively high frequency scores.

8. A *rectangular distribution* is a symmetrical distribution in which the extreme scores do not have relatively low frequencies.

9. The *relative frequency* of a score, symbolized by *rel. f*, is the proportion of time that the score occurred in a distribution. A relative frequency distribution is graphed in the same way as a simple frequency distribution except that the Y axis is labeled in increments between 0 and 1.0.

10. Relative frequency corresponds to the *proportion of the total area under the normal curve.* If the area under the curve above a score or scores constitutes a certain proportion of the total area, then the frequency of the score or scores will constitute that same proportion of the total N.

11. The *cumulative frequency* of a score, symbolized by *cf*, is the frequency of all scores at or below the score. A cumulative frequency polygon is graphed in the same way as a simple frequency polygon except that the Y axis is labeled cumulative frequency and we do not include the score above the highest score.

12. *Percentile* indicates the percent of all scores at or below a given score. On the normal curve, the percentile of a score is the percent of the area under the curve to the left of the score.

13. In an *ungrouped distribution*, the *f*, *rel. f*, or *cf* of each individual score is reported.

14. In a *grouped distribution*, we combine different scores into small groups and then report the total *f*, *rel. f*, or *cf* for each group. Each group of scores is called a *class interval*, and the range of scores in the interval is called the *interval size*. The lowest score shown in each class interval is the lower apparent limit, and the highest score shown in a class interval is the upper apparent limit.

15. The lower real limit of an interval is the score that is halfway between the lower apparent limit and the upper apparent limit of the interval below. The upper real limit of an interval is the score that is halfway between the upper apparent limit and the lower apparent limit of the interval above.

16. When graphing a grouped simple frequency distribution or a grouped relative frequency distribution, label the X axis using the *midpoint* of each class interval. When graphing a grouped cumulative frequency distribution, label the X axis using the *upper real limit* of each interval.

PRACTICE PROBLEMS

(Answers for odd-numbered problems are provided in Appendix E.)

1. What do each of the following symbols mean?
   - *a.* N
   - *b.* f
   - *c.* $rel.\ f$
   - *d.* cf
2. What type of frequency graph should you create when counting each of the following?
   - *a.* the males and females at a college
   - *b.* the different body weights reported in a statewide survey
   - *c.* the number of sergeants, lieutenants, captains, and majors in an army battalion
   - *d.* the people falling into one of eight salary ranges
3. What is the difference between graphing a relationship as we did in Chapter 2 and graphing a frequency distribution?
4. What does a rectangular distribution show about the relationship between frequency and the different scores?
5. What does it mean when a score is in one of the tails of the normal distribution?
6. A professor observes that a distribution of test scores is positively skewed. What does this tell the professor about the difficulty of the test?
7. In reading psychological research you encounter the following statements. Interpret each one.
   - *a.* "The IQ scores were approximately normally distributed."
   - *b.* "A bimodal distribution of physical agility scores was observed."
   - *c.* "The distribution of the patients' memory scores was severely negatively skewed."
8. From the data 1, 4, 5, 3, 2, 5, 7, 3, 4, 5, Poindexter created the following frequency table. What five things did he do wrong?

Score	*f*	*cf*
1	1	0
2	1	1
3	2	3
4	2	5
5	3	8
7	1	9
	N = 6	

9. *a.* How is percentile defined in a small sample?

   b. How is percentile defined for a large sample or population when calculated using the normal curve?

10. *a.* On a normally distributed set of exam scores, Poindexter scored at the 10th percentile, so he claims that he outperformed 90% of his class. Why is he correct or incorrect?

    b. Because Foofy's score had a relative frequency of .02, she claims she had one of the highest scores on the exam. Why is she correct or incorrect?

11. *a.* What is the difference between a score's simple frequency and its relative frequency?

    b. What is the difference between a score's cumulative frequency and its percentile?

12. What is the difference between how we use the proportion of the area under the normal curve to determine a score's relative frequency and how we use it to determine a score's percentile?

13. The following shows the distribution of final exam scores in a large introductory psychology class. The proportion of the total area under the curve is given for two segments.

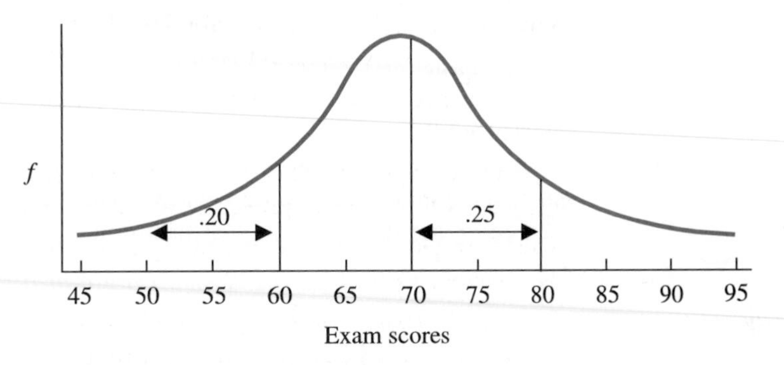

    a. Order the scores 45, 60, 70, 72, and 85 from most frequent to least frequent.

    b. What is the percentile of a score of 60?

    c. What proportion of students scored between 60 and 70?

    d. What is the percentile of a score of 80?

14. What is the advantage and disadvantage of using grouped frequency distributions?

15. Organize the scores below in a table showing simple frequency, relative frequency, and cumulative frequency.

49	52	47	52	52	47	49	47	50
51	50	49	50	50	50	53	51	49

16. What is the percentile for the score of 51 in question 15?

17. What score in question 15 is at the 50th percentile?

18. Organize the scores below in a table showing simple frequency, cumulative frequency, and relative frequency.

16 11 13 12 11 16 12 16 15
16 11 13 16 12 11

SUMMARY OF FORMULAS

1. *The formula for computing a score's relative frequency is*

$$rel.\ f = \frac{f}{N}$$

where f is the score's simple frequency and N is the number of scores in the sample.

2. *The formula for finding the score at a given percentile is*

$$\text{Score} = \text{LRL} + \left(\frac{(\text{target } cf - cf \text{ below interval})}{f \text{ within interval}}\right)(\text{interval size})$$

where "target cf" is the cumulative frequency of the score we seek, "LRL" is the lower real limit of the interval containing the score we seek, "cf below interval" is the cumulative frequency for the interval below the interval containing the score we seek, "f within interval" is the frequency in the interval containing the score we seek, and "interval size" is the interval size used to create the frequency distribution.

3. *The formula for computing the percentile of a given score is*

$$\text{Percentile} = \left(\frac{cf \text{ below interval} + \left(\frac{\text{score} - \text{LRL}}{\text{interval size}}\right)\left(\begin{matrix} f \text{ within} \\ \text{interval} \end{matrix}\right)}{N}\right)(100)$$

where "score" is the score for which we are computing a percentile, "cf below interval" is the cumulative frequency for the interval below the interval containing the score, "LRL" is the lower real limit of the interval containing the score, "f within interval" is the frequency in the interval containing the score, "interval size" is the interval size used in creating the distribution, and N is the total number of scores in the sample.

4

Measures of Central Tendency: The Mean, Median, and Mode

To understand the upcoming chapter:

- From Chapter 2, you should understand the logic of "statistics" and "parameters" and the difference between an independent and a dependent variable.
- From Chapter 3, understand when to create bar graphs versus polygons, how to interpret polygons, and how a percentile is calculated using the area under the curve.

Then your goals in this chapter are to learn:

- How measures of central tendency are used to describe data.
- What the mean, median, or mode indicates and when each is appropriate.
- How a sample mean is used to describe both individual scores and the population of scores.
- What is meant by "deviations around the mean" and what they convey about each score's location and frequency in a normal distribution.
- How to interpret and graph the results of an experiment.

The graphs and tables discussed in Chapter 3 are important tools for presenting data. After all, the purpose of descriptive statistics is to describe the important characteristics of a set of data, and the type of distribution it forms *is* one important characteristic. However, graphs and tables are not the most efficient way to summarize the characteristics of a distribution. Instead, we can compute individual numbers that each summarize and describe an important characteristic of a sample or population of scores. Which characteristic we describe depends upon what it is we want to know about the data.

In this chapter we discuss the important statistics and parameters called measures of central tendency. A measure of central tendency allows us to answer the question "Are the scores generally high scores or generally low scores?"

MORE STATISTICAL NOTATION

Recall that the symbol X stands for a subject's raw score. If a formula requires us to perform a particular mathematical operation on X, then we perform that operation on each score in a sample. Also, recall that parentheses mean "the quantity." We first perform the operations inside the parentheses and then perform any operations outside of the parentheses on the quantity.

A new important symbol is Σ, the Greek capital letter S, called sigma. Sigma is used in conjunction with a symbol for scores, so you will see such notations as ΣX. In words, ΣX is pronounced **"sum of X"** and literally means to find the sum of the X scores. Thus, ΣX for the scores 5, 6, 9, and 7 is 27, and in code we would say $\Sigma X = 27$. Notice that we do not care whether each X is a different score. If the scores are 4, 4, and 4, then $\Sigma X = 12$.

This chapter also introduces a symbol used in performing certain transformations on scores. The symbol K stands for a constant number. This symbol is used when we add the same number to each score or when we multiply, divide by, or subtract a constant.

Now, on to central tendency.

WHAT IS CENTRAL TENDENCY?

To understand central tendency, you need to alter your perspective of what a score indicates. You should think of a score as indicating a *location* on a variable. For example, if I am 70 inches tall, do not think of my score as indicating that I have 70 inches of height. Instead, think of me as being located on the variable of height at the point marked 70 inches. Think of any variable as an infinite continuum, a straight line, and think of a score as indicating a subject's location on that line. Thus, as shown in Figure 4.1, my score locates me at the address

FIGURE 4.1 Locations of Individual Scores on the Variable of Height

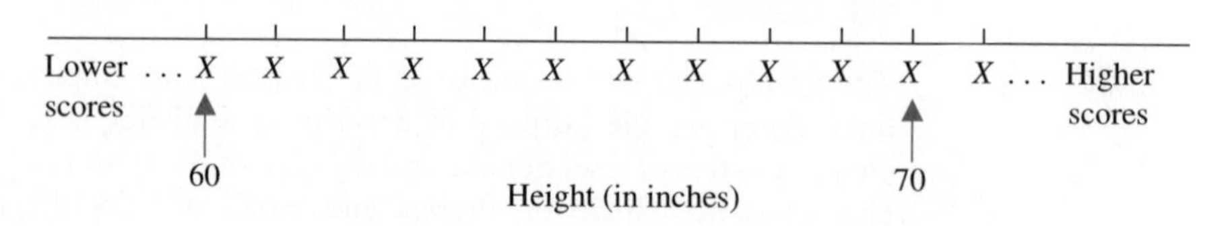

labeled 70 inches. If my brother is 60 inches tall, then he is located at the point marked 60 on the height variable. The idea is not so much that he is 10 inches shorter than I am, but rather that we are separated by a *distance* of 10 units—in this case, 10 "inch" units. In statistics, scores are locations, and the difference between any two scores is the distance between them.

From this perspective, a frequency polygon for a sample or population shows the location of each score. Again visualize the parking lot view of the normal curve: subjects' scores determine *where* they stand. A high score puts them on the right side of the lot, a low score puts them on the left side, and a middle score puts them in a crowd in the middle. Further, if two distributions contain different scores, then the *distributions* have different locations on the variable. Figure 4.2 shows the polygons for two samples of height scores, one consisting of low scores and the other consisting of higher scores.

We began this chapter by asking, "Are the scores generally high scores or generally low scores?" From the above perspective you can see that we are actually asking, "*Where* on the variable is the distribution located?" A **measure of central tendency** is a score that summarizes the location of a distribution on a variable. In essence, the score is used to describe *around* where most of the distribution is located. This means that ideally the distribution is centered around the measure of central tendency, with the majority of the scores close to this score. Thus, the purpose of a measure of central tendency is to indicate where the *center* of the distribution *tends* to be located. It provides us with one score that serves as a reasonably accurate address for the entire distribution.

Thus, in Sample A in Figure 4.2, we see that the most frequent scores are in the neighborhood of 59, 60, and 61 inches, so we can summarize this sample by saying that the majority of the scores are located at around 60 inches. In Sample B, the distribution tends to be centered around the score of 70 inches.

Notice that the above example illustrates how we use descriptive statistics: from them we obtain an idea of what's in a set of data and can envision the important aspects of the distribution *without* looking at graphs or tables of all of the individual scores. If a researcher told us only that one normal distribution is centered at 60 and the other is centered around 70, we can envision the essence

FIGURE 4.2 Two Sample Polygons on the Variable of Height

Each polygon indicates the locations of the scores and their frequencies.

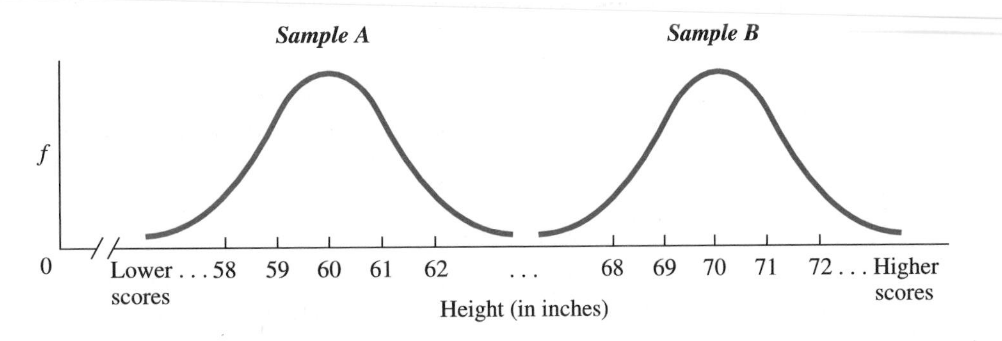

of the information presented in Figure 4.2. Although we lose some detail and thus some accuracy, we do answer our question about whether the scores in each sample are generally high scores or generally low scores. We will see other statistics that add more to our mental picture of a distribution, but measures of central tendency are at the core of summarizing a distribution.

STAT ALERT The first step in summarizing any set of data is to compute the appropriate measure of central tendency.

The trick is to compute the measure of central tendency that allows us to *accurately* envision where most scores in our data are actually located. Recall that one of your overall goals is to learn when to use a particular statistical procedure. As we shall see, we decide which measure of central tendency to calculate by first considering the scale of measurement we used, so that our summary makes sense given the nature of our scores. Second, we consider the shape of the frequency distribution that our scores produce, to ensure that the measure of central tendency does accurately summarize the distribution.

There are three common measures of central tendency. In the following sections we will first discuss the mode, then the median, and finally the most common measure of central tendency, the mean. With this background, we will then discuss how we use measures of central tendency to interpret experiments.

THE MODE

If we are trying to describe where most of the scores tend to be located, then one way to summarize the distribution is to use the one score that occurs most frequently. The most frequently occurring score is called the **modal score** or simply the **mode.** Consider the test scores 2, 3, 3, 4, 4, 4, 4, 5, 5, and 6. The score of 4 is the mode, because its frequency is higher than that of any other score in the sample. You can see how the mode summarizes this distribution from the polygon in Figure 4.3. Most of the scores are at or around 4. Notice that Figure 4.3 is roughly a normal curve, with the highest point on the curve over the mode. When a polygon has one hump, such as on the normal curve, the distribution is called **unimodal,** indicating that one score qualifies as the mode.

There may not always be a single mode in a set of data. For example, consider the test scores 2, 3, 4, 5, 5, 5, 6, 7, 8, 9, 9, 9, 10, 11, and 12. Here two scores, 5 and 9, are tied for the most frequently occurring score. These scores are plotted in Figure 4.4. In Chapter 3 we saw that such a distribution is called **bimodal,** meaning that it has two modes. Describing this distribution as bimodal and identifying the two modes does summarize where most of the scores tend to be located, because most scores are either around 5 or around 9.

Uses of the Mode

The mode is typically used to describe central tendency when the scores reflect a nominal scale of measurement. Recall that a nominal scale is used when we

FIGURE 4.3 A Unimodal Distribution

The vertical line marks the highest point on the distribution, thus indicating the most frequent score, which is the mode.

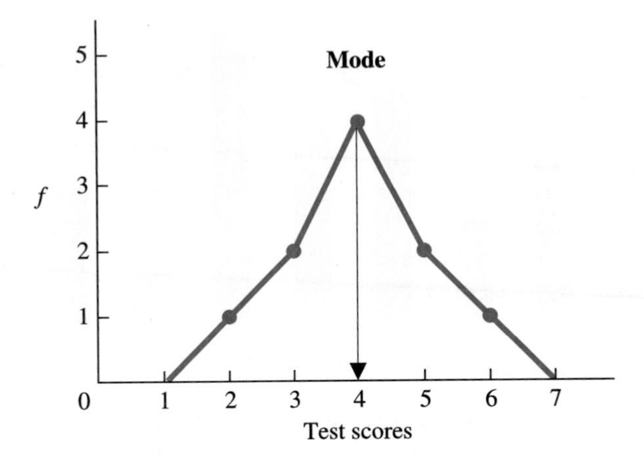

categorize subjects on the basis of a qualitative, or classification, variable. For example, say that we asked subjects their favorite flavor of ice cream and counted the number of responses in each category. Such data might produce the bar graph shown in Figure 4.5. A useful way to summarize such data would be to indicate the most frequently occurring category. Reporting that the modal response was a preference for category 5, "goopy chocolate," would be very informative. When our data are ordinal, interval, or ratio, it is useful to know the mode, but we can usually use other, better measures of central tendency.

FIGURE 4.4 A Bimodal Distribution

Each vertical line marks one of the two equally high points on the distribution, so each indicates the location of one of the two modes.

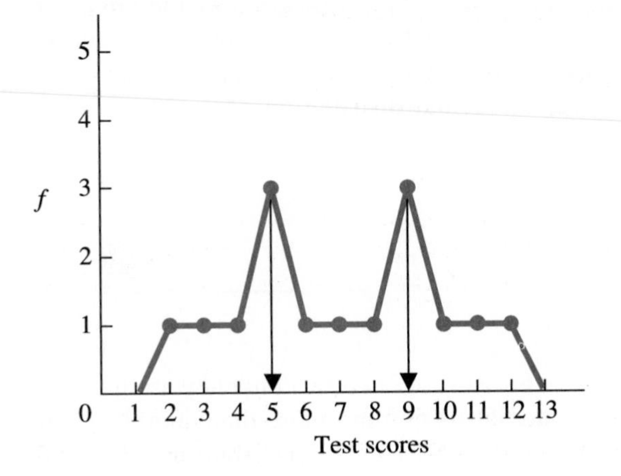

FIGURE 4.5 Bar Graph Showing the Frequencies of Preferred Ice Cream Flavors

The mode is flavor 5, "Goopy Chocolate."

Problems with the Mode

Two potential problems arise when using the mode to describe central tendency. One problem is that, depending on the distribution we obtain, many scores may have the same highest frequency. In distributions with more than two modes, we begin to fail to summarize the data. In the most extreme case, we might obtain a rectangular distribution with scores such as 4, 4, 5, 5, 6, 6, 7, and 7. Here the mode should not be determined. Depending on your perspective, either there is no mode or all scores are the mode. Either way, the mode will not summarize the data.

A second problem is that the mode does not take into account any scores other than the most frequent score(s), so it ignores much of the information in the data. An *accurate* summary should reflect all of the important information, so that we can accurately envision the corresponding distribution. For example, say that I tell you the mode in a sample is 7. This gives you an idea of the entire distribution, but it may give you the *wrong* idea. The actual data might consist of 7, 7, 7, 20, 20, 21, 22, 22, 23, and 24. Although 7 is the single most frequently occurring score, the problem is that the majority of the sample is not located at 7 or even around 7: most of the scores are up there between 20 and 24. Thus, as usual, we must examine the actual distribution of our data, because employing the mode may not accurately summarize around where *most* scores in the entire distribution are located.

THE MEDIAN

Often a better measure of central tendency is the median score, or simply the median. The **median** is nothing more than another name for the score at the 50th percentile. Recall that 50% of the distribution is at or below the score at the 50th percentile. Thus, if the median is 10, then 50% of all scores are either the score of 10 or a score less than 10. (Note that the median will not always be one of the actual scores that occurred.) The median is typically a better measure of central tendency than the mode, because (1) only one score can be the median and (2) the median will usually be around where most of the scores in the distribution tend to be located. The symbol for the median is usually its abbreviation, Mdn.

As we saw in Chapter 3, with a large sample or population the 50th percentile is the score that separates the lower 50% of the distribution from the upper 50% of the distribution (ignoring those subjects at that score). For example, we can draw a line on a normal curve such that .50 of the area is to the left of the line. This is shown in Graph A of Figure 4.6. Because 50% of the scores are to the left of the line, they are below the score at the line. Therefore, the score at the line is the 50th percentile. In other words, the score is the median.

In fact, the median is the score below which .50 of the area of *any* shaped polygon is located. For example, in the positively skewed distribution in Graph B of Figure 4.6, the vertical line is drawn so that .50 of the area under the curve is to the left of the line and .50 is to the right of the line. Envisioning this as a parking lot, we can see that 50% of our subjects are standing to either side of the line. The scores of those subjects standing to the left of the line constitute the lower 50% of all scores. Therefore, the score at the line is the median.

We have two ways of calculating the median. As in Graph A of Figure 4.6, when the scores form a normal or any other *symmetrical* distribution (where the left half mirrors the right half), we can calculate the median as the score that is halfway between the highest and lowest scores in the distribution. Because the distribution is symmetrical, this middle score splits the polygon in half, with .50 of the area under the curve always below the middle score, so that 50% of the scores are below it. (Notice that in a perfect normal distribution, the median is also the most frequently occurring score, so it is the same score as the mode.)

Sometimes, however, we do not calculate the median using the above approach. When a distribution is not symmetrical, we have no easy way to determine the

FIGURE 4.6 Location of the Median in a Normal Distribution (A) and in a Skewed Distribution (B)

The vertical line indicates the location of the median, with one-half of the distribution on each side of it.

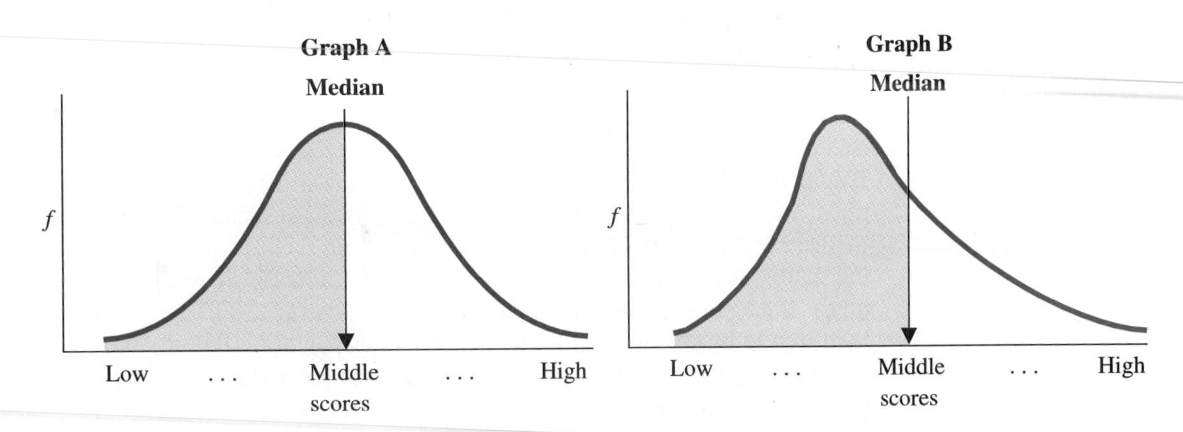

point below which .50 of the area under the curve is located. (In Graph B of Figure 4.6, one-half of the *area* of the skewed distribution is not at the score midway between the highest and lowest scores.) Further, using the area under the curve on any polygon does not precisely determine the median score, because we ignore that portion of the subjects who scored *at* that score. In a small sample, that portion may not be a negligible amount, so we obtain only a rough estimate of the true median score. Therefore, when we have data that does not form a symmetrical distribution or when we seek the precise median (such that 50% of the scores are at or below that score), we use our second procedure: We directly calculate the 50th percentile from our scores using the formula discussed in Chapter 3 for finding the score at a given percentile (on page 72). (Most computer programs employ this formula, providing the easiest solution.)

Uses of the Median

The median is not used to describe nominal data: to say, for example, that 50% of our subjects preferred "Goopy Chocolate" or below is more confusing than it is informative. On the other hand, the median is often the preferred measure of central tendency when the data are ordinal (rank-ordered) scores. For example, say that a group of students ranked how well a college professor teaches. If we report that the professor's median ranking was 2, we communicate that 50% of the students rated the professor as number 1 or number 2.

As we shall see in a later section, the median is also the most appropriate measure of central tendency when interval or ratio scores form a very skewed distribution. In addition, for any distribution, the median may be one of the measures of central tendency we compute, because it's nice to know where the 50th percentile is located.

Problems with the Median

In computing the median we still ignore some information in the data. The median reflects only an accumulation of the frequencies of lower scores until we have 50% of the distribution, without considering the mathematical values of these scores or of the scores in the distribution above the 50th percentile. Therefore, the median is usually not our first choice for describing the central tendency of most distributions.

THE MEAN

By far the all-time measure of central tendency in psychological research is the mean score, or simply the mean. The **mean** is the score located at the exact mathematical center of a distribution. Although technically we call this statistic the arithmetic mean, it is what most people call the average. You compute a mean exactly the same way you compute an average: add up all the scores and then divide by the number of scores you added. Unlike the mode or the median, the mean does not ignore any information in the data. In adding up all of the scores, you include the value of each score, as well as the number of times it occurs.

The symbol for a *sample* mean is $\overline{X}$. It is pronounced "the sample mean" (not "bar X": bar X sounds like the name of a ranch!). As with other symbols, get in the habit of thinking of $\overline{X}$ as a quantity itself, so that you understand statements such as "the size of $\overline{X}$" or "this $\overline{X}$ is larger than that $\overline{X}$."

To compute $\overline{X}$, recall that the symbol meaning "add up all the scores" is ΣX and the symbol for the number of scores is N. Then,

THE FORMULA FOR COMPUTING A SAMPLE MEAN IS

$$\overline{X} = \frac{\Sigma X}{N}$$

As an example, take the scores 3, 4, 6, and 7. The sum of X is all of the scores added together, so $\Sigma X = 20$. There are four scores, so N is 4. Thus $\overline{X} = 20/4 = 5$. Saying that the mean of these scores is 5 tells us that the exact mathematical center of this distribution is located at the score of 5. (As this example illustrates, the mean may be a score that does not actually occur in the data, but it is still the location of the center of the distribution.)

What is the exact mathematical center of a distribution? We can think of the mathematical center of a distribution as its balance point. Visualize a polygon as a teeter-totter on a playground. A score's location on the teeter-totter corresponds to its location on the X axis. The left-hand side of Figure 4.7 shows the scores 3, 4, 6, and 7 sitting on the teeter-totter. The mean score of 5 is the point that balances the distribution. The right-hand side of Figure 4.7 shows how the mean is the balance point even when all of the scores do not have the same frequency (the score of 1 has an f of 2). Here the mean is 4 (because $\Sigma X/N = 20/5 = 4$), and it balances the distribution.

Uses of the Mean

Computing the mean is appropriate when getting the "average" of our scores makes sense. Therefore, we do not use the mean when describing nominal data. For example, say we are studying political affiliation and we arbitrarily assign

FIGURE 4.7 The Mean as the Balance Point of a Distribution

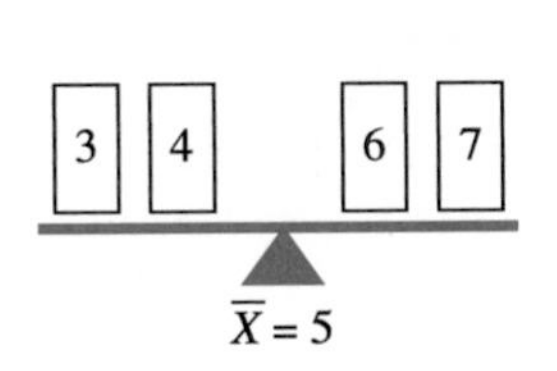

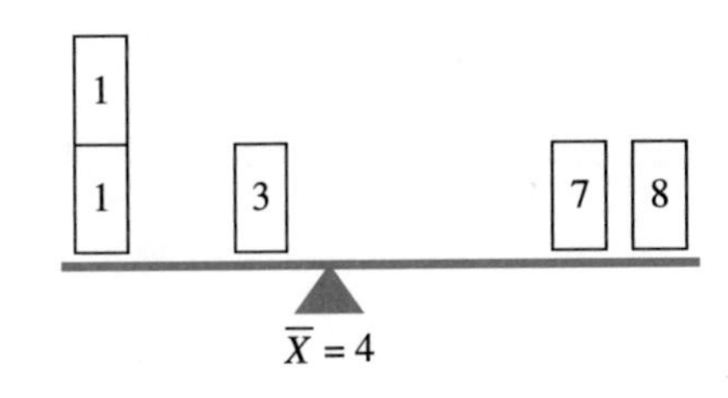

a 1 if a subject is a Democrat, a 2 if a Republican, a 3 if a Socialist, and so on. It is meaningless to say that the mean political affiliation was 2.3; the mode or percentages would be much more informative. Likewise, we usually prefer the median when describing ordinal data (because, for example, it is strange to say that on average runners came in 5.7th in a race). This leaves us with using the mean to describe interval or ratio data (where there is a consistent unit of measurement and zero). The mean is especially useful when the variable is at least theoretically continuous (decimals make sense).

In addition to needing interval or ratio scores, however, the appropriateness of using the mean depends on the shape of the distribution. Recall that our goal is to identify the point around which most of the scores in the distribution tend to be located. The mean is simply the mathematical center of the distribution. Therefore, we must be sure that the mathematical center of the distribution is also the point around which most of the scores are located. This will be the case, and the mean will be most accurate when we have a *symmetrical* and *unimodal* distribution. For example, say that we are studying the intelligence of cats by timing how long it takes them to escape from a puzzling maze. Our data are 1, 2, 3, 3, 4, 4, 4, 5, 5, 6, and 7 minutes, which form the roughly normal distribution shown in Figure 4.8. The mean is appropriate here because the balance point is also that point *around* which *most* of the scores are located. The mean score of 4 is located both at the midpoint of the distribution and around where the most frequently occurring scores are located. Thus, the mean of 4 accurately summarizes that most of the cats took around 4 minutes to escape.

Notice that, as Figure 4.8 illustrates, the score of 4 is also the middle score between the highest and lowest scores. Previously we saw that the median score in any symmetrical distribution is the middle score between the highest and lowest scores. Because the mean is the middle score in a symmetrical distribution, and because the middle score divides the area under the curve in half, the mean score is always also the median score on a symmetrical distribution.

FIGURE 4.8 Location of the Mean on a Distribution Formed by the Escape Times 1, 2, 3, 3, 4, 4, 4, 5, 5, 6, and 7

The vertical line indicates the location of the mean score, which is the balance point of the distribution.

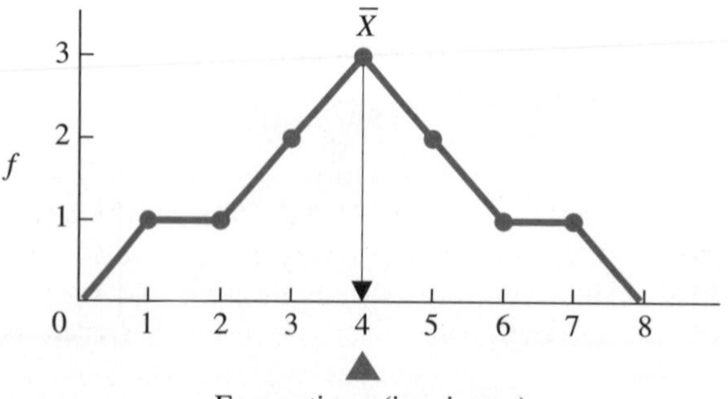

The ultimate symmetrical distribution is the normal distribution. Therefore, we use the mean as our measure of central tendency whenever we are summarizing a normal or approximately normal distribution. Notably, on a perfect normal distribution all three measures of central tendency are located at the same score. As in Figure 4.8, the mean score of 4 splits the area under the curve in half, so 4 is also the median score. At the same time, the middle score has the highest frequency in a normal distribution, so 4 is also the mode.

If a distribution is only roughly normal, then the mean, median, and mode will be close to but not exactly the same score. In such a situation, you might think that any one of the measures of central tendency would be close enough. Not true. Because the mean uses all of the information in the data, the mean is the preferred measure of central tendency. Further, most inferential statistical procedures are based on the mathematical properties of the mean. Therefore, the rule for you as a researcher should be that the mean is the preferred statistic to use with interval or ratio data unless it clearly provides an inaccurate description of the central tendency of the distribution.

STAT ALERT Whenever we have an approximately normal distribution of interval or ratio scores, we describe the central tendency of the distribution by computing the mean.

Problems with the Mean

The mean does not accurately describe a highly skewed distribution. To understand this, consider what happens to your grade average if you obtain one low grade after receiving many high grades. Your average drops like a rock. The low score produces a negatively skewed distribution, and the mean gets pulled away from where most of your grades are, toward that low grade. What hurts, of course, is then telling someone your average, because it's misleading. It sounds like all of your grades are relatively low, while only you know that you have that one zinger. For precisely this reason, we do not use a mean score to summarize highly skewed distributions.

The mean is pulled toward the tail of a skewed distribution because mathematically it must balance the distribution. For example, say that our previous cat subjects produced the escape-time scores of 1, 2, 2, 2, 3, and 14, forming the positively skewed distribution shown in Figure 4.9. Without the 14, the scores would form a symmetrical distribution with a mean of 2, right around where most scores are located. However, because that one exceptionally slow cat has the extreme score of 14, the mean is pulled away from the low scores. We can picture the distribution as a teeter-totter with a bunch of little guys (1's, 2's, and 3's) sitting at one end, trying to balance the big guy (the 14) sitting way at the other end. To keep the teeter-totter level, the balance point must be located at the mean of 4. The problem is that a measure of central tendency is supposed to describe where most of the scores tend to be located. But, as in Figure 4.9, there are no scores located at 4, and most scores are not around 4. The mean is where the mathematical center is, *but in a skewed distribution the mathematical center is not the point around which most of the scores tend to be located.*

FIGURE 4.9 Location of the Mean on a Skewed Distribution Formed by the Time Scores 1, 2, 2, 2, 3, 14

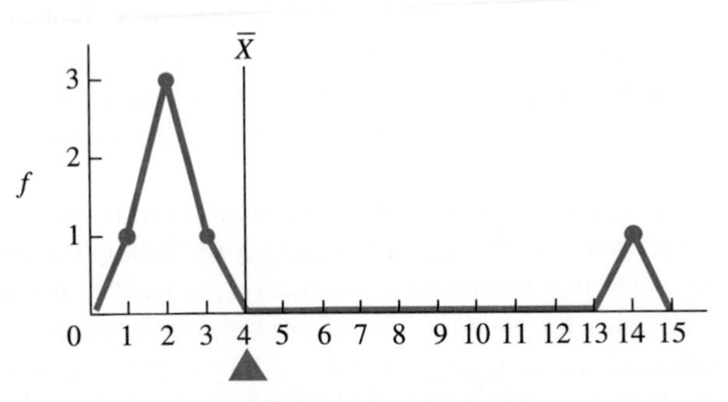

The solution is to use the median when summarizing a very skewed distribution. Figure 4.10 shows the relative positions of the mean, median, and mode in skewed distributions. The mode is the most frequent score, the median cuts the area in half, and the mean is the balance point. When the distribution is positively skewed, the mean is larger than the median. When the distribution is negatively skewed, the mean is less than the median. In both cases, the mean is pulled toward the extreme tail of the distribution and does not accurately describe the central tendency of the entire distribution. Likewise, the mode tends to be toward the side away from the extreme tail, so most of the distribution is not centered around the mode either. However, the median is not thrown off by extreme scores occurring in only one tail of the distribution, because it does not take into account the actual values of the scores. Thus, of the three measures, the median most accurately reflects *around* where most of the scores tend to be located.

It is for the above reasons that the government typically uses the median to summarize such skewed distributions as that of yearly income or the price of houses. For example, the median income in the United States is around $35,000

FIGURE 4.10 Measures of Central Tendency for Skewed Distributions

The vertical lines show the relative positions of the mean, median, and mode.

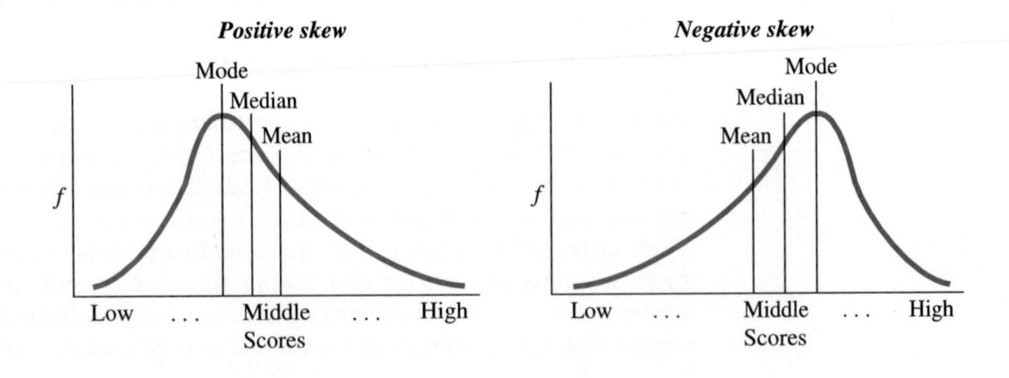

a year. But there are a relatively small number of corporate executives, movie stars, professional athletes, rock stars, and the like who make millions! If we were to average these extreme high incomes with everyone else's, the mean income would work out to be close to $50,000. However, because most incomes are not located at or around $50,000, the median is a much better summary of the distribution, reflecting the many lower incomes instead of being drawn off by the few extremely high incomes.

Believe it or not, we've now covered the basic measures of central tendency. We can summarize the preceding discussion of using central tendency to determine whether a sample contains generally high scores or generally low scores as follows:

1. We employ the mode with nominal (categorical) data or with a distinctly bimodal distribution of any type of scores.
2. We employ the median with ordinal (ranked) scores or with a very skewed distribution of interval/ratio scores.
3. We employ the mean with a symmetrical, unimodal distribution of interval/ratio scores.

It is a coincidence of how the laws of nature operate that most psychological variables produce a roughly normal distribution of interval or ratio scores. Therefore, more often than not, we summarize the data in psychological research using the mean. Because the mean is used so extensively, we will delve further into its characteristics and uses in the following sections.

TRANSFORMATIONS AND THE MEAN

Recall that we perform transformations on raw scores so that the scores are easier to work with or so that we can compare scores from different variables. The simplest transformation is to add, subtract, multiply, or divide each score by a constant. This brings up the burning question of how transformations affect the mean.

If we add a constant value, K, to each raw score in a sample, the new mean of the transformed scores will equal the old mean of the raw scores plus K. For example, the scores 7, 8, and 9 have a mean of 8. If we add the constant 5 to each score, we have 12, 13, and 14. The new mean is 13. The old mean of 8 plus the constant 5 also equals 13. Thus, the rule is that *new* $\overline{X}$ = *old* $\overline{X} + K$. The same logic applies for other mathematical operations. If we subtract K from each score, *new* $\overline{X}$ = *old* $\overline{X} - K$. If we multiply each score by K, *new* $\overline{X}$ = (*old* $\overline{X}$)K. If we divide each score by K, *new* $\overline{X}$ = (*old* $\overline{X}$)/K.

The above rules also apply to the median and to the mode. Whenever we use a constant, we merely change the location of each score on the variable by K points. We move the entire distribution to higher scores or to lower scores, so we also move the address of the distribution by a corresponding amount.

DEVIATIONS AROUND THE MEAN

To understand why the mean is the mathematical center of any distribution, you must understand that, *in total,* the mean is just as far from the scores above it as it is from the scores below it. The distance separating a score from the mean is called the score's **deviation,** indicating the amount the score **deviates** from the mean. A score's deviation is equal to the score minus the mean.[1] In symbols, a deviation is the quantity $(X - \bar{X})$.

> ***STAT ALERT*** Always subtract the mean *from* the raw score when computing the score's deviation.

Thus, if the sample mean is 47, a score of 50 deviates by +3, because 50 − 47 is +3. A score of 40 deviates from the mean of 47 by −7, because 40 − 47 = −7. Notice that a score's deviation consists of a number and a sign. A positive deviation indicates that the score is greater than the mean. A negative deviation indicates that the score is less than the mean. The size of the deviation (regardless of its sign) indicates the distance the score lies from the mean. The larger the deviation, the farther the score is from the mean. Conversely, the smaller the deviation, the closer the score is to the mean. A deviation of 0 indicates that the score is equal to the mean.

When we determine the deviations of all the scores in a sample above and below the mean, we find the *deviations around the mean.* The **sum of the deviations around the mean** is the sum of all differences between the scores and the mean. Here's why the mean is the mathematical center of any distribution:

The sum of the deviations around the mean always equals zero.

For example, consider the symmetrical distribution containing the scores 3, 4, 6, and 7, which has a mean of 5. The upper portion of Table 4.1 shows how we compute the deviations around the mean for these scores. The lower portion of Table 4.1 shows the deviations for the skewed distribution containing the scores 1, 2, 2, 2, 3, and 14, which has a mean of 4. As you can see, in each sample, the sum of the deviations is zero. In fact, for *any* sample of scores you can imagine, having a distribution of any shape, the sum of the deviations around the mean will always equal zero. This is because the sum of the positive deviations equals the sum of the negative deviations, so the sum of all deviations is zero. Thus the mean is the center of a distribution, because, in the same way that the center of a rectangle is an equal distance from both sides, in total, the mean is an equal distance from scores above and below it.

Many of the formulas we will eventually use involve something similar to finding the sum of the deviations around the mean. The statistical code for finding the sum of the deviations around the mean is $\Sigma(X - \bar{X})$. We always start inside parentheses, so we first find the deviation for each score: $(X - \bar{X})$. The Σ indicates that we then find the sum of the deviations. Thus, as in the upper portion of Table 4.1, $\Sigma(X - \bar{X}) = -2 + -1 + 1 + 2$, which equals zero.

1. In some textbooks, the lowercase letter x is the symbol for the amount a score deviates from the mean, and thus $x = (X - \bar{X})$.

TABLE 4.1 Computing Deviations Around the Mean

The mean is subtracted from each score, resulting in the score's deviation.

Score	*minus*	*Mean score*	*equals*	*Deviation*
3	−	5	=	−2
4	−	5	=	−1
6	−	5	=	+1
7	−	5	=	+2
			Sum =	0

Score	*minus*	*Mean score*	*equals*	*Deviation*
1	−	4	=	−3
2	−	4	=	−2
2	−	4	=	−2
2	−	4	=	−2
3	−	4	=	−1
14	−	4	=	+10
			Sum =	0

USING THE MEAN TO INTERPRET RESEARCH

It is important to understand that the sum of the deviations equals zero, because then you understand that only the mean score is *literally* "more or less" the score that all subjects obtained: individual scores may be higher or lower than the mean, but those that are higher balance out those that are lower. Because of this, the mean provides researchers with a very useful tool for interpreting research. As we will see in the following sections, a sample mean is used in four ways: to describe any score in a sample, to predict additional scores, to describe a score's location within a distribution, and to draw inferences about the population.

Using the Mean to Describe Individual Scores in a Sample

Because the mean is in the mathematical center of a distribution, it is, more or less, the *typical* score in the distribution. This implies that if all of the scores in the sample were the same, they would all be the mean score. Therefore, the mean score is our best summary for describing any individual score.

For example, think about how you describe the individual grades of your friends who have a B average in college. They may not always get B's, but you operate as if they do. Thus, if asked what you think they received in a particular course, you'd estimate a B. For every other course, you'd also estimate a B. Likewise, if the test average for a sample of students is 80, we think of everyone in the class as receiving around 80. Without actually examining every score, our best estimate for each and every individual would be the score of 80.

Of course, in describing each individual's score this way, we will sometimes be wrong. The amount of error we will have when describing any single score is the amount that the mean differs from the actual score the subject obtained. In symbols, this error is equal to $(X - \overline{X})$. We have already seen that $(X - \overline{X})$ is the amount a score deviates from the mean. The reason we use the mean score to describe each individual score is that the *total* of our errors when using the mean to describe everyone in the sample is the sum of these deviations, or $\Sigma(X - \overline{X})$, *which always equals zero.* For example, the test scores 70, 75, 85, and 90 have a $\overline{X}$ of 80. One student, Quasimodo, scored the 70. We would estimate he scored 80, so we would be wrong by -10. But another student, Attila, scored the 90; by estimating an 80 for him, we would be off by $+10$. In the same way, the errors in our description for the entire sample will cancel out so that our total error is zero.

If we used any single score other than the mean, our total error would be *greater* than 0. If, for example, we described the above scores using a score of 75 or 85, the sum of the deviations would be $+20$ or -20, respectively. Having a total error of $+20$ or -20 is not as good as having a total error of zero. A total error of zero means that, *over the long run,* we will overestimate by the same amount that we underestimate. A basic rule of statistics is that if we can't have a summary number that accurately describes every score, then the next best step is to have a number that more or less describes every score, with the same degrees of more and of less. (There is an old joke about two statisticians shooting targets. One hits one foot to the left of the target, and the other hits one foot to the right of the target. "Congratulations," one says. "We got it!") Likewise, if we cannot perfectly describe every score, then we want our errors—our over- and underestimates—to balance out to zero. Only the mean score provides us with that capability.

Using the Mean to Predict Individual Scores

Recall that as part of understanding the laws of nature, we want to be able to predict the behavior we would see in a given situation. Predicting a behavior translates into predicting the scores we would find in that situation. Because the mean is the typical score we have previously found in that situation, the mean is our best prediction of any other, unknown scores that would be found in that situation.

For example, think about the grades you *predict* for your friends who have a B average in college. Because they have been B students, and you assume nothing has happened that drastically changed them, your best guess is that they will continue to be B students. You treat any past course as essentially interchangeable with any future course, so you predict that they will receive a B in statistics and in every other course they will complete. In the same way, if the test average for a sample of students is 80, then we assume that the subjects in our sample are essentially interchangeable with similar individuals who were not in our sample. Assuming nothing drastic has happened that changes the behavior, we'd predict that anyone who missed the test would also score around 80. Essentially, because the mean is the typical score of our subjects, we assume that it also typifies the score of anyone else who might have been a subject.

Of course, our predictions will be wrong sometimes, because every subject will not score at the mean score. The error in our predictions is the difference between the mean score we predict for subjects and the actual scores they would have obtained. On the one hand, we don't know what these scores would be. But on the other hand, all subjects are essentially interchangeable, so we expect the differences between the mean and the scores that these subjects *would* have obtained to be about the same as the differences between the mean and the scores that the subjects in our sample *did* obtain. Thus, we can determine how well the mean predicts the scores in our sample of known scores (as if we didn't already know them) and use this to estimate how well the mean would predict any other, unknown or future scores. (In statistics, we always estimate the amount of error in predicting unknown scores based on how well we can predict the scores in some known sample.)

The reason we use the mean score to predict any individual score is because, again, our total error will equal zero. The amount of error when using the mean score to predict any score in our sample is the difference between the mean and the actual score the subject obtains, or again $(X - \bar{X})$. The total error in predicting all the scores in a sample is the sum of these deviations, or $\Sigma(X - \bar{X})$, which equals zero. Thus, for example, if we predict a score of 80 (the mean) for Quasimodo and then learn he scored at 70, our prediction is wrong by -10. But by predicting 80 for Attila, who scored 90, we are off by $+10$. In the same way, the errors in our predictions for the entire sample will cancel out so that our total error is zero. Based on this, we assume that other subjects will behave similarly to Quasimodo, Attila, and the rest, so that by using the mean to predict any unknown scores, our total error will also be zero.

Of course, this is not the whole story. Although the *total* error in our predictions will equal zero, a prediction for any *individual* score may be off by a country mile. In later chapters we will see how to reduce our error in predicting each individual score. For now, simply remember that *unless you have additional information about the scores,* the mean is the best score to use when predicting or describing scores. In either case, our total error, considering both over- and underestimates, will balance out to zero.

Using the Mean to Describe a Score's Location in a Distribution

In computing the deviations around the mean, we have simply performed a transformation on the raw scores (we subtracted K from each score, where $K = \bar{X}$). Recall that we perform transformations so that scores are easier to interpret. Because the mean is at the center of the distribution, a deviation score is a transformation that easily communicates a score's position relative to where most of the other scores are located.

Say that we give subjects a test of creativity and obtain the raw scores 1, 2, 3, 3, 4, 4, 4, 5, 5, 6, and 7. Look at the approximately normal distribution these scores form, shown in Figure 4.11 on the next page. Below each score is its deviation. Since a positive deviation occurs when the raw score is larger than the mean, a positive deviation indicates that the score is located to the right of the mean on the distribution. Conversely, a negative deviation indicates that the

FIGURE 4.11 Frequency Polygon Showing Deviations from the Mean

The first row under the X axis indicates the original creativity scores, and the second row indicates the amounts the raw scores deviate from the mean.

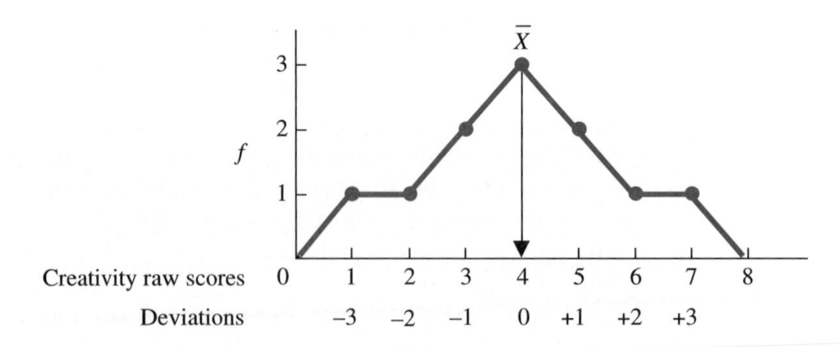

creativity score falls to the left of the mean. Further, the larger the deviation, whether positive or negative, the farther from the mean the raw score lies. Thus, by examining a score's deviation, we can determine the score's location relative to the mean and, indirectly, the score's location relative to the rest of the distribution.

The deviation scores in a normal distribution also tell us about the frequencies of the corresponding raw scores. The larger the deviation, whether positive or negative, the farther into the tail of the distribution the raw score lies. As you know, scores lying farther in the tail of a normal distribution occur less frequently. Thus, the larger a deviation, the less frequently the corresponding raw score occurs. Also, notice that the larger a score's deviation, the less frequently the *deviation* occurs. Above, the raw score of 7 produces the deviation of +3. Because 7 is an extreme score that occurs only once, its deviation is extreme and also occurs only once. In fact, the frequency of any score's deviation will equal the frequency of that score. As shown in Figure 4.11, whether we label the X axis using scores or deviations, we have the same frequency polygon.

The advantage of using deviation scores, however, comes when we wish to interpret any individual score. A raw score by itself may be meaningless because as psychologists, we usually do not have an implicit, intuitive understanding of how a single score should be viewed in the grand scheme of things in nature. If, for example, I tell you that your raw score on a normally distributed creativity test is 6, you won't know if your score is good, bad, or indifferent, because you have no frame of reference. However, if I say that your score produced a deviation score of +2, you can then at least interpret your score *relative* to the rest of the distribution. By envisioning the graph in Figure 4.11, you know that a positive deviation of +2 indicates that you are above average (which with creativity is good), you know that you are to the right of and therefore above the 50th percentile (which is also good), and you know that you are out there in the direction of the less common scores, where the most creative people are (so that's good too). Conversely, if these scores reflected the number of blunders on a test, you'd know you were out there among those who made the highest number of blunders (and that's not so good). In either case, at least you'd have a better idea of how to interpret—*make sense of*—your score, and that is the purpose of statistics.

We will elaborate on deviations around the mean in subsequent chapters. For now, remember that on the normal curve, the larger the deviation (whether positive or negative), the farther the raw score is from the mean and thus the less frequently the score and its deviation occur.

Using the Sample Mean to Describe the Population Mean

Recall that research ultimately is designed to describe the entire population of scores we would find in a given situation if we could measure all the scores. Populations are unwieldy, so we summarize a population using a measure of central tendency. Here is what we want to know: If we examined the population, around which point would most of the scores be located?

Because we usually have interval or ratio scores that we assume form at least an approximately normal distribution, we usually describe the population using the mean. Recall that parameters are numbers that describe the important characteristics of a population and that they are symbolized using letters from the Greek alphabet. The mean score of a population is one such parameter, and it is symbolized by the Greek letter μ, pronounced "mew." Thus, saying that μ equals 143 is the same as saying that the population mean is 143. We use the symbol μ simply to show that we are talking about a population mean, as opposed to $\overline{X}$, which shows that we are talking about a sample mean. However, a mean is a mean, so a population mean has all the characteristics of a sample mean:

1. μ is the arithmetic average of all the scores in the population.

2. μ is the score at the mathematical center of the distribution, so μ is the balance point.

3. The sum of the deviations around μ is zero.

Previously we saw that these characteristics make the mean the best summary statistic when we are describing or predicting any individual score and that they are the basis for determining any score's relative location in the distribution. For the same reason, in the absence of any additional information, we treat the value of μ as the typical score in the population, as the best value to use when predicting any individual score in the population, and as the basis for describing the relative location of any score in the population.

How do we determine the value of μ? If we know all the scores in the population, then we compute μ using the same formula we used to compute $\overline{X}$:

$$\mu = \frac{\Sigma X}{N}$$

Of course, usually we think of a population as being infinitely large and unmeasurable, so we cannot directly compute μ. Instead we estimate the value of μ based on the mean of a random sample. If, for example, a sample's mean in a particular situation is 99, then our best guess is that if we could actually compute it, the population μ in that situation would also be 99. We can make such an inference because the logical explanation for obtaining a sample mean of 99 in the first place is that the population mean is 99. After all, a population with a mean of 99 is the one that is most likely to produce a sample having a mean of 99. In such a population, most subjects score at or close to 99, so in a random sample we are very likely to repeatedly run into subjects who score around 99. Thus, we are very likely to obtain a sample mean of 99. On the other hand, if the population mean is 4,000, then relatively few subjects score around 99, and so we are very unlikely to obtain an entire sample that scores around 99. Thus, wherever most scores in a sample are located should be where most scores in the corresponding population are located, so a random sample mean *should be* a good estimate of the population μ. (This assumes, of course, that the sample is representative of the population.)

> ***STAT ALERT*** Because the mean score is at the center of a normal distribution, it is the best summary score to use when thinking about the scores in either a sample or in its corresponding population.

Now that we have discussed the characteristics of sample and population means, we can discuss how they come together in an experiment.

SUMMARIZING THE RESULTS OF EXPERIMENTS

As we discussed in Chapter 2, in an experiment we obtain a sample of scores on the dependent variable for each condition of the independent variable. If a relationship exists between the independent variable and the dependent variable, then as the conditions of the independent variable change, there will be a consistent change in the scores on the dependent variable: the scores will have certain values under one condition and different values under the other conditions.

Say that we are interested in the relationship between the amount of information that subjects must remember and the number of errors they make when recalling the information. Specifically, we predict that subjects will make more errors in recalling a long list of words than in recalling a short list. We conduct an overly simplistic experiment to demonstrate the relationship between the length of the list to be remembered and the number of errors produced during recall. We randomly select three samples of subjects. Subjects in one sample read a list containing 5 words and then recall it. Subjects in another sample read a 10-item list and then recall it, and those in the third sample read a 15-item list and then recall it. What is our independent variable? It is the length of list to be remembered, with the three conditions being 5-, 10-, and 15-item lists. Our dependent variable is the number of errors subjects make in recalling their list. If a relationship exists, then as we change the length of the list, there will be a consistent change in subjects' error scores.

Say we study this relationship using the unrealistically small N of 3 subjects per sample. The error scores for the three conditions appear in Table 4.2. It appears that there is a relationship here, because there tends to be a different and higher set of error scores associated with each condition. Most experiments, however, involve much larger N's, and with many different scores it is difficult to detect a relationship by looking at the raw scores. But that's why we have

TABLE 4.2 Numbers of Errors Made by Subjects Recalling a 5-, 10-, or 15-Item List

Independent variable: Length of list.

Condition 1: 5-item list	*Condition 2: 10-item list*	*Condition 3: 15-item list*
3	6	9
4	5	11
2	7	7

measures of central tendency: so that we can summarize the scores and at the same time simplify the relationship. Thus, our first step is always to compute a measure of central tendency for the scores in each condition.

Summarizing a Relationship Using Measures of Central Tendency

As we've seen, which measure of central tendency we should compute depends on the characteristics of our scores. In an experiment our scores reflect the *dependent variable.* Therefore we choose the mean, median, or mode depending upon (1) the scale of measurement used to measure the dependent variable and (2) for interval or ratio scores, the shape of the distribution they form. In determining the shape of the distribution, we consider that the sample data may not produce a recognizable shape but that ultimately we want to describe the population anyway. Therefore, we usually rely on how the scores are assumed to be distributed in the population. How do we know how a population is distributed? We do not perform an experiment in a vacuum, so from previous research, books, and other sources we can determine the assumed shape of the population.

> ***STAT ALERT*** The measure of central tendency we employ in an experiment is determined by the type of *dependent* scores we collect.

In our experiment above, the number of recall errors is a ratio variable that is assumed to form an approximately normal distribution, so we will compute the mean score for each condition. In fact, most dependent variables in psychological research are interval or ratio variables that are assumed to be normally distributed, so computing the mean score for each condition is the predominant method of summarizing experiments.

We compute the mean of each condition in Table 4.3 by computing the mean in each column. Examining these means, we quickly recognize evidence of a relationship:

Condition 1: 5-item list	*Condition 2: 10-item list*	*Condition 3: 15-item list*
$\overline{X} = 3$	$\overline{X} = 6$	$\overline{X} = 9$

Interpreting this experiment is simply a translation process: By looking at the above means, we can envision the kinds of distributions that produced them back in Table 4.2. Thus, recalling a 5-item list resulted in one distribution of scores, located around the mean of 3 errors; recalling a 10-item list resulted in a different distribution, located around 6 errors; and recalling a 15-item list resulted in a different distribution, located around 9 errors. Thus, the means indicate that a relationship exists, because as the conditions of the independent variable change (from 5 to 10 to 15 items), the scores on the dependent variable also tend to change in a consistent fashion (from around 3, to around 6, to around 9 errors, respectively). Further, we can use the mean score to describe the individual scores in each condition, as we have done previously. In condition 1, for example, we'd describe every subject as if he or she made about 3 errors, we would predict that any other subject in this condition would make about 3 errors, and we would interpret any subject's actual score by determining its deviation from the mean of 3.

Recall that in an experiment we have a relationship when the scores on the dependent variable change *as a function of* changes in the conditions of the independent variable. Because the mean score reflects the typical score for each condition, we also have a relationship when the mean scores on the dependent variable change as a function of changes in the conditions of the independent variable. Further, our experiment "works" because it demonstrates that, literally, list length is a variable that makes a *difference* in individual recall scores and therefore in the mean scores. As researchers we often communicate that we have found a relationship simply by saying that we have found a difference (between the means). If we find no difference, we have not found a relationship.

STAT ALERT An experiment shows a relationship whenever the means from two or more conditions have different values.

The above logic also applies to studies in which it is appropriate to compute the median or mode. For example, say that we conduct a study in which the dependent variable is political party affiliation, to see if it changes with the subjects' year in college. Political parties are categories involving nominal scores, so the mode is the appropriate measure of central tendency. We might see that freshmen most often claim to be Republican, but the mode for sophomores is Democrat, for juniors Socialist, and for seniors Communist. These data reflect a relationship because they indicate that as college level changes, political affiliation tends to change (with upperclassmen typically becoming more liberal). Likewise, say we learn that the median income for freshmen is lower than the median income for sophomores, which is lower than that for upperclassmen. This tells us that the location of the corresponding distribution on the variable of income is different for each class, so we know that the income "scores" of individual subjects must be changing as their year in college changes.

So that we can easily see a relationship when it is present, we summarize experiments by creating graphs.

Graphing the Results of an Experiment

Recall that in creating a graph, we place the "given" variable on the *X* axis. In an experiment, the given variable is always the independent variable. (Earlier, for example, we asked, "For a given list length, what error scores occurred?") Therefore, the convention for graphing the results of an experiment is to place the independent variable on the *X* axis and the dependent variable on the *Y* axis. Then we read the graph from left to right, asking, "As the conditions of the independent variable change, what happens to the scores on the dependent variable?"

To create a graph of the results of an experiment, we have two decisions to make. Because we want to simplify the data, we do not plot the individual dependent scores found in each condition. Rather, on the *Y* axis we plot the value of the measure of central tendency we've computed for each condition. Thus, if we have not already done it, we must identify the type of dependent variable we have measured and compute the appropriate measure of central tendency. Notice that a potential for confusion is present when using our formulas. When we create a frequency distribution, the scores that subjects produce are placed on the *X* axis, and so our formulas are written using *X*'s. However, now the dependent scores that subjects produce are placed on the *Y* axis.

> ***STAT ALERT*** Although in our formulas we use the symbol *X* to stand for each subject's score on the dependent variable, we graph such scores on the *Y* axis.

Our second decision is to select the type of graph we will create. As we see below, we create either a *line graph* or a *bar graph.* The type of graph we create is determined by the characteristics of our *independent variable.*

Line graphs Whenever an independent variable is an interval or ratio variable, we connect the data points with straight lines. This type of graph is called a **line graph.**

In our recall experiment above, the independent variable of length of list involves a ratio scale, because it has an equal unit of measurement and a true zero. Therefore, we create the line graph shown on the left in Figure 4.12. First we label the *X* and *Y* axes with the specific variables (not "independent variable" and "dependent variable"). Note that the label on the *Y* axis in Graph A is *mean* recall errors, because in this experiment it is appropriate to compute a mean of the dependent scores in each condition. The numbers on the *X* axis correspond to the values of the independent variable in the various conditions. Then we place a data point above the 5-item condition opposite 3 errors, because the mean error score for the 5-item list was 3. Similarly, we place a data point above the 10-item condition at 6 errors, and a data point above the 15-item condition at 9 errors. Then we connect adjacent data points with straight lines. We use straight lines with interval or ratio data because we assume that the relationship continues in a straight line between the points shown on the *X* axis. For example, we assume that if there had been a 6-item list, the mean error score would have fallen on the line connecting the means for the 5- and 10-item lists.

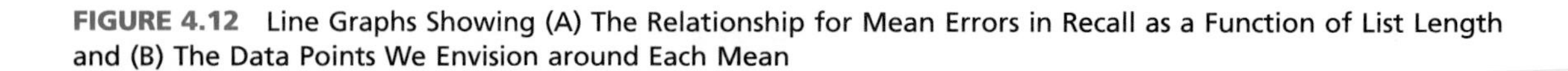

FIGURE 4.12 Line Graphs Showing (A) The Relationship for Mean Errors in Recall as a Function of List Length and (B) The Data Points We Envision around Each Mean

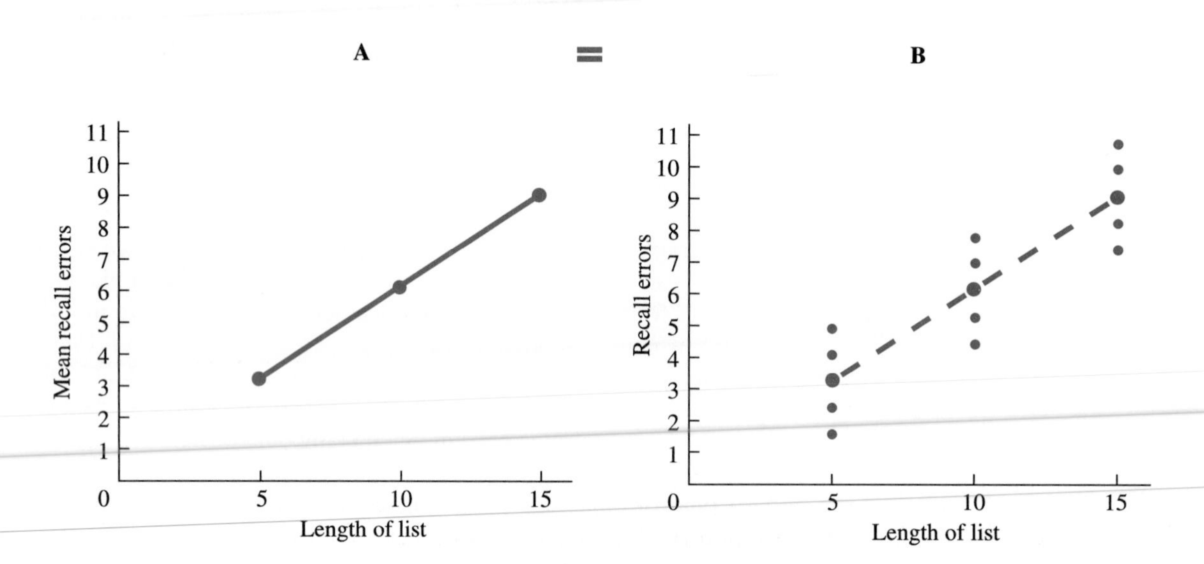

This graph conveys the same information as the sample means did in Table 4.2. There we saw that the different sample means indicate that for each condition there is a different distribution of scores, located at a different position on the dependent variable. In the graph we envision these distributions as shown on the right in Figure 4.12. Each mean implies a sample of scores and thus the corresponding data points that would occur *around*—above and below—the mean's data point. The different vertical locations of the means indicate that they have different values on the Y axis and therefore that there are different values of Y scores in each condition. Thus, because the vertical positions of the means change as the conditions change, we know that the raw scores also change, so there is a relationship here. Notice that you can easily spot such a relationship, because the differing positions of the means on the Y axis produce a line graph that is not horizontal. *On any graph, if the summary data points form a line that is not horizontal, the individual Y scores change as the X scores change and so a relationship is present.*

On the other hand, say that each condition had produced a mean of 5 errors. As shown on the left in Figure 4.13, plotting the same mean score of 5 for each condition results in a horizontal line. A horizontal (flat) line indicates that as list length changes, the mean error score stays the same. This implies that (as on the right) the individual scores stay the same, regardless of the condition. If the scores on the dependent variable stay the same as the conditions of the independent variable change, there is no relationship present. Thus, *on any graph, if the summary data points form a horizontal line, the individual Y scores do not change as the X scores change and so a relationship is not present.*

Bar graphs When the independent variable is a nominal or ordinal variable, we plot the results of the experiment by creating a bar graph. Notice that the rule here is the same as the rule for a frequency distribution discussed in Chapter

FIGURE 4.13 Line Graphs Showing (A) No Relationship for Mean Errors in Recall as a Function of List Length and (B) The Kind of Data Points We Envision Around Each Mean

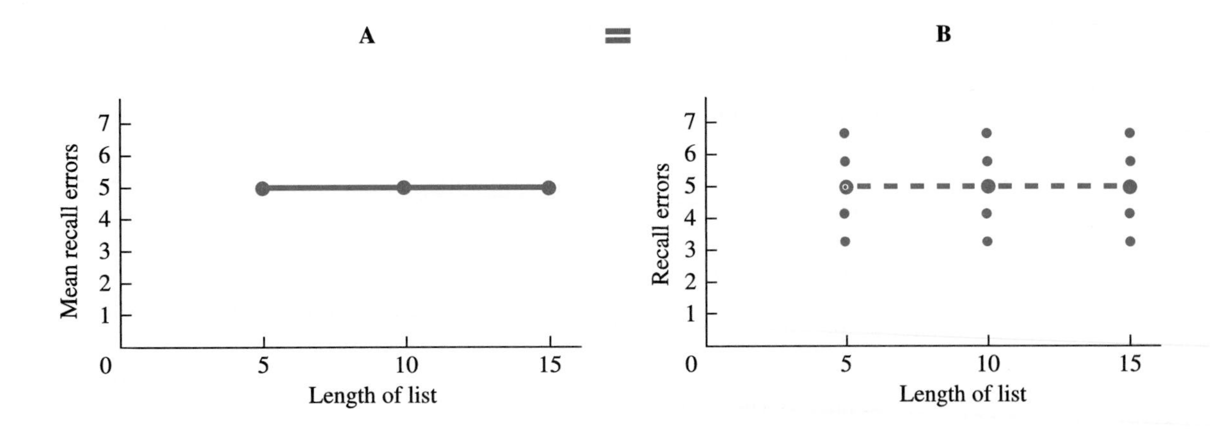

3: We create a bar graph whenever the scores plotted along the X axis are nominal or ordinal scores. Here each bar is centered over a condition on the X axis, and the height of the bar corresponds to the mean score for the condition.

For example, say that we conducted another recall experiment in which we compared the errors made by samples of psychology majors, English majors, and physics majors. Here, the independent variable of college major is a nominal, or categorical, variable, so we have the bar graph shown in Figure 4.14. The bars implicitly indicate that on the variable of college major, we arbitrarily assigned psychology a nominal score to the left of that of English, closer to zero. The bars also indicate that there is an unknown or undefined gap between categories. If, for example, we inserted an additional category of sociology

FIGURE 4.14 Bar Graph Showing Mean Errors in Recall as a Function of College Major

The height of each bar corresponds to the mean score for the condition.

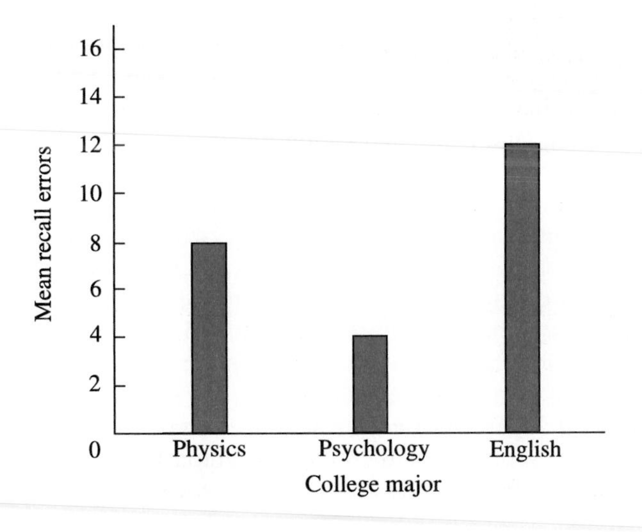

between psychology and English, we could not assume that the mean scores of sociology majors would fall on a line running between the means for psychology and English majors.

The bar graph in Figure 4.14 shows that a relationship is present, because the tops of the bars do not form a flat, or horizontal, line. This tells us that we have different means and thus different scores in each condition. We can again envision that we would see individual Y scores of around 8 for physics majors, around 4 for psychology majors, and around 12 for English majors. Thus, recall scores change as a function of changes in college major, so a relationship between these variables is present.

Notice that in a different experiment we might have measured a nominal or ordinal *dependent* variable. In that case, we would plot the modal or median scores for each condition along the Y axis. Then, again depending on the characteristics of our independent variable, we would create either a line or bar graph. However, we would still apply a similar interpretation, envisioning different values of Y scores as the vertical location of the measure of central tendency changes with each condition.

> ***STAT ALERT*** The type of dependent variable determines the measure of central tendency we calculate, but the type of independent variable determines the type of graph we create.

Inferring the Relationship in the Population

So far we have summarized the results of our recall experiment in terms of the sample data. But this is only part of the story. The big question remains: Do these data reflect a law of nature? Do longer lists produce more errors in recall for everyone in the population?

Recall that to make inferences about the population, we must first compute the appropriate inferential statistics in order to determine if the data are representative. For the moment we'll assume that the data from our original recall experiment passed the inferential test. Then we can conclude that each sample mean represents the population mean that would be found for that condition. The mean error score for the 5-item condition was 3, so we infer that if the population of subjects recalled a 5-item list, the mean of the population of error scores would be located at 3. In essence, we expect that everyone would have around 3 errors. Similarly, we infer that if the population recalled a 10-item list, μ would equal our sample mean of 6, and if the population recalled a 15-item list, μ would be 9.

We conceptualize the populations of error scores in the following way. Assuming that each sample mean provides a good estimate of the corresponding population mean, we know approximately where on the dependent variable each population of scores would be located. Further, assuming that recall errors are normally distributed in the population, we also have a good idea of the shape of each population distribution. Thus, based on our data and our assumptions, we can envision the population of recall errors we would expect to find for each list length as the simple frequency polygons shown in Figure 4.15. The population means in Figure 4.15 are located at the same points on the dependent variable as are the sample means in the line graph back in Figure 4.11. Because Figure 4.14 depicts frequency distributions, however, the scores on the dependent variable

FIGURE 4.15 Locations of Populations of Error Scores as a Function of List Length

Each distribution contains the recall scores we would expect to find if the population were tested under each condition.

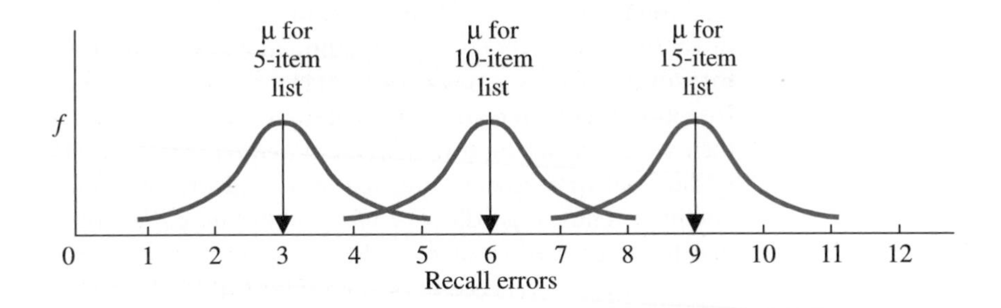

are on the X axis. The overlap among the distributions simply shows that some subjects in one condition make the same number of errors as other subjects in adjacent conditions.

Because these population distributions have different values of μ, we have described the relationship that we think exists in the population: as the conditions of the independent variable change, the scores on the dependent variable tend to change in a consistent fashion so that there is a different population of scores for each condition. In essence, it appears that for every 5 items in a list, everyone's score tends to increase by about 3 errors, and thus the distribution slides 3 units to the right for each condition.

Since a population of scores reflects the behavior of everyone of interest, a change in a population of scores indicates a change in everyone's behavior. If, as an independent variable changes, *everyone's* behavior changes, then we have learned about a law of nature involving that behavior. Above, we inferred a change in everyone's behavior in terms of recall errors as list length changes, so we have evidence of how a law of nature works: for every 5 items in a list, recall errors go up by about 3. That's about all there is to it: we have basically achieved the goal of our research.

If the process of arriving at the above conclusion sounds easy, it's because it is easy. In essence, statistical analysis of most experiments merely involves three steps: (1) computing each sample mean (and other descriptive statistics) so that we can summarize the scores and the relationship found in the experiment, (2) performing the appropriate inferential procedure to determine whether the data are representative of the population, and (3) describing the location of the population of scores that we expect would be found for each condition by describing the value of each μ. Once we have described the expected population for each condition, we are basically finished with statistical analysis.

FINALLY

As you may have noticed, psychological research and the statistical procedures therein rely on the mean as *the* measure of central tendency. To be literate in statistics, you should understand the mode and the median, but the truly important topics in this chapter involve the mean and its characteristics, especially when applied to the normal distribution. Become very familiar with these topics, because they form the basis of virtually everything else that we will discuss.

Recognize that we will eventually discuss inferential procedures and that they will tend to occupy all of your attention. But despite the emphasis that such procedures may receive, remember that the mean (or another central tendency measure) forms the basis for interpreting any study. We always want to say something like "the subjects scored around 3" for a particular situation, because then we are describing their typical *behavior* in that situation. Such a description is the goal of research, and using a measure of central tendency is *the* way to get there. Thus, regardless of what other fancy procedures we focus on, remember that to make sense out of your data, you must ultimately return to identifying *around* where the scores in each condition are located, as we did in this chapter.

CHAPTER SUMMARY

1. **Measures of central tendency** summarize the location of a distribution of scores on a variable. Each measure is a way of indicating around where the center of the distribution tends to be located. So that we accurately summarize the distribution, we base our choice of measure on (1) the scale used to measure the variable we are describing and (2) the shape of the distribution.

2. The **mode** is the most frequently occurring score or scores in a distribution. The mode is used primarily to summarize nominal data.

3. The **median,** symbolized by Mdn, is the score in a distribution located at the 50th percentile. The median is used primarily for describing ordinal data and for describing interval or ratio data that form a very skewed or nonsymmetrical distribution.

4. The **mean** is the score located at the exact mathematical center of a distribution, and it is computed in the same way as an average. The mean is used to describe interval or ratio data that form a symmetrical unimodal distribution such as the normal distribution. The symbol for a sample mean is $\overline{X}$, and the symbol for a population mean is μ.

5. Transforming raw scores by using a **constant,** K, results in a new value of the mean, median, or mode. The new value is equal to the one that would be obtained if the transformation were performed directly on the old value.

6. The amount a score **deviates** from the mean is computed as the score minus the mean, or $X - \overline{X}$. A **deviation** consists of a number and a sign ($+$ or $-$) and it indicates the location of the raw score relative to the mean and relative to the distribution. A positive deviation indicates that the score is above the mean and, when graphed, to the right of the mean. A negative deviation indicates that the score is below the mean and, when graphed, to the left of the mean. The size of the deviation, whether positive or negative, indicates the distance the score lies from the mean. In a normal distribution, scores

having larger deviations lie farther into the tail of the distribution. Thus, the larger the deviation, the less frequently the score and the deviation occur.

7. The **sum of the deviations around the mean,** or $\Sigma(X - \bar{X})$, always equals zero. When we do not know anything else about the scores, the mean is the best score to use when describing or predicting any individual's score. This is because the **total error** across all such estimates will be the sum of the deviations around the mean, which will always equal zero.

8. In graphing the results of an experiment, we plot the independent variable along the X axis and the dependent variable along the Y axis. When the independent variable is measured using a ratio or interval scale, we create a **line graph.** When the independent variable is measured using a nominal or ordinal scale, we create a **bar graph.**

9. On any graph, if the summary data points form a line that is not horizontal, then the individual Y scores change as a function of changes in the X scores and a relationship is present. If the data points form a flat, or horizontal, line, then the Y scores do not change as a function of changes in the X scores and a relationship is not present.

10. If the samples in an experiment are representative of the population, then we can infer that each condition of the experiment would produce a population of scores having a μ equal to the sample mean for that condition. If a relationship is present, there will be different values of μ for two or more conditions of the independent variable. Then we can conclude that the variables are related in nature, because we can infer that each condition of the independent variable would result in a different population of scores on the dependent variable.

PRACTICE PROBLEMS

(Answers for odd-numbered problems are provided in Appendix E.)

1. What does a measure of central tendency indicate?
2. What two aspects of the data determine which measure of central tendency to use?
3. What is the mode, and with what type of data is it most appropriate?
4. What is the median, and with what type of data is it most appropriate?
5. What is the mean, and with what type of data is it most appropriate?
6. Why is it best to use the mean as the measure of central tendency for a normal distribution?
7. Why is it inappropriate to use the mean as the measure of central tendency in a very skewed distribution?
8. Which measure of central tendency is used most often in psychological research? Why?

9. For the following data, compute (a) the mean and (b) the mode.

55 57 59 58 60 57 56 58 61 58 59

10. *a.* In problem 9, what is your best estimate of the median (without computing it)?

b. Explain why you think your answer is correct.

11. For the data below, compute the mean.

18 16 19 20 18 19 23 54 20 16
18 19 18 19 18 40 30 19 18 38

12. *a.* For the data in question 11, what is the mode?

b. By comparing the mean and the mode, determine the shape of the distribution. How do you know what it is?

13. A scientist collected the following sets of data. For each, indicate which measure of central tendency she should compute.

a. The following IQ scores:

60, 72, 63, 83, 68, 74, 90, 86, 74, 80

b. The following error scores:

10, 15, 18, 15, 14, 13, 42, 15, 12, 14, 42

c. The following blood types:

A−, A−, O, A+, AB−, A+, O, O, O, AB+

d. The following grades:

B, D, C, A, B, F, C, B, C, D, D

14. You misplaced two of the scores in a sample, but you have the data indicated below. What should you guess the value of the missing scores to be? Why?

100 120 130 140 110 140 150 130 120 130

15. What two pieces of information about the location of a score does a deviation score convey?

16. Why do we use the mean score in a sample to describe or predict any score that might be found in that sample?

17. On a normal distribution of scores, four subjects obtained the following deviation scores: −5, 0, +3, and +1.

a. Which subject obtained the lowest raw score? How do you know?

b. Which subject's raw score had the lowest frequency? How do you know?

c. Which subject's raw score had the highest frequency? How do you know?

d. Which subject obtained the highest raw score? How do you know?

18. In a normal distribution of scores, five subjects obtained the following deviation scores: +1, −2, +5, and −10.

a. Which subject obtained the highest raw score?

b. Which subject obtained the lowest raw score?

c. Rank-order the deviation scores in terms of their frequency, putting the score with the lowest frequency first.

19. Foofy says a deviation of +5 is always better than a deviation of −5. Why is she correct or incorrect?

20. What is μ, and how do we usually determine its value?

21. For the following experimental results, interpret specifically the relationship between the independent and dependent variables:

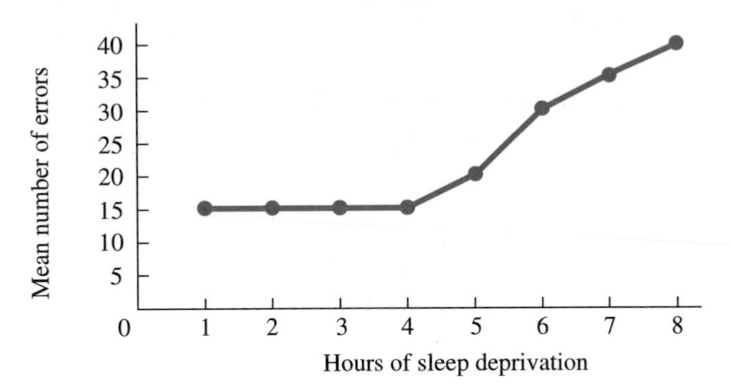

22. *a.* In problem 21, give a title to the graph, using "as a function of".

b. If you were a subject in the study in problem 21 and had been deprived of 5 hours of sleep, how many errors do you think you would make?

c. If we tested all of the subjects in the world after 5 hours of sleep deprivation, how many errors do you think each would make?

d. What symbol stands for your prediction in part c?

23. For each of the experiments listed below, determine (1) which variable should be plotted on the *Y* axis and which on the *X* axis, (2) whether the researcher should use a line graph or bar graph to present the data, and (3) how she should summarize scores on the dependent variable.

a. A study of income as a function of age

b. A study of politicians' positive votes on environmental issues as a function of the presence or absence of a wildlife refuge in their political district

c. A study of running speed as a function of carbohydrates consumed

d. A study of rates of alcohol abuse as a function of ethnic group

24. You hear that Dr. Grumpy's experimental data form a line graph of scores from the Grumpy Emotionality Test that slants downward as a function of increases in the amount of sunlight present on the day subjects were tested.

a. What does this tell you about the mean scores for the condition?

b. What does this tell you about the raw scores for each condition?

c. Assuming that the samples are representative, what does this tell you about the population μ's?

d. What do you conclude about whether there is a relationship between emotionality and sunlight in nature?

(Please turn the page for the Summary of Formulas.)

SUMMARY OF FORMULAS

1. *The formula for computing the sample mean is*

$$\bar{X} = \frac{\Sigma X}{N}$$

where ΣX stands for the sum of the scores and N is the number of scores.

5

Measures of Variability: Range, Variance, and Standard Deviation

To understand the upcoming chapter, from Chapter 4:

- Understand that the mean score is in the center of a distribution.
- Understand the difference between $\overline{X}$ and μ.
- Understand what a deviation score tells you.
- Understand why the sum of the deviations around the mean is zero.

Then your goals in this chapter are to learn:

- What is meant by variability.
- When the range and semi-interquartile range are used and how to interpret them.
- When the standard deviation and variance are used and how to interpret them.
- How to compute the sample variance and standard deviation, the estimated population variance and standard deviation, and the true population variance and standard deviation.
- How variance is used to measure errors in prediction, and what is meant by the proportion of variance accounted for.

In previous chapters you learned the sequence involved in applying descriptive procedures to any set of data: after considering the scale of measurement we've used, we examine the shape of the frequency distribution formed by the scores and then compute the appropriate measure of central tendency for them. This information simplifies the distribution and allows us to envision its general properties. In particular, it allow us to answer the question of whether the scores tend to be low or high.

But recall that, because of individual differences, all subjects will not behave in the same way. Therefore, all of the scores in a distribution will not be the same score, and there may be many, very different scores. Thus, to have a complete description of any set of data, we must also answer the question "Are there large differences between the scores, or are there small differences between the scores?" In this chapter we discuss procedures for describing and summarizing the differences between scores.

First, though, here are a couple of new symbols and terms that will appear later.

MORE STATISTICAL NOTATION

As part of our calculations, we often square each score in a sample. This operation is symbolized by X^2. Usually, we then add up the squared scores. This operation is symbolized by ΣX^2. The notation ΣX^2 indicates the **sum of the squared *X*'s:** it tells you to first square each X score and then find the sum of the squared X's. Thus, to find ΣX^2 for the scores 2, 2, and 3, add $2^2 + 2^2 + 3^2$, or $4 + 4 + 9$, which equals 17.

Learn right here to avoid confusing ΣX^2 with a similar looking, yet very different operation symbolized by $(\Sigma X)^2$. The symbol $(\Sigma X)^2$ stands for the **squared sum of *X*.** Recall that parentheses indicate to first find the quantity inside of the parentheses and then perform any other operations on that quantity. Because ΣX is inside the parentheses, first find the sum of the X scores and then square that sum. Thus, to find $(\Sigma X)^2$ for the scores 2, 2, and 3, you have $(2 + 2 + 3)^2$, which is $(7)^2$, or 49. Notice that for our example scores of 2, 2, and 3, ΣX^2 gives 17, while $(\Sigma X)^2$ gives the different answer of 49. Be careful when dealing with these terms.

With this chapter we will begin using subscripts. A *subscript* is a symbol placed below and to the right of a statistical symbol which identifies the scores used in computing the statistic. Pay attention to subscripts, because they are part of the symbols for certain statistics.

Finally, prepare yourself for the fact that many statistics have two different formulas, a definitional formula and a computational formula. A definitional formula defines a statistic. The reason you need to pay attention to this formula is so you understand where the answer comes from when you compute the statistic. However, definitional formulas tend to be very time-consuming to use, especially when N is large. Therefore, statisticians have reworked the formulas to produce computational formulas. As the name implies, we use the computational formula when actually computing a statistic. Trust me, computational formulas give exactly the same answers as definitional formulas, but they are much easier to use.

THE CONCEPT OF VARIABILITY

Recall that we most often have interval or ratio scores forming a normal distribution. In this situation, the mean score best describes the center of the distribution, and we use the mean as the typical score to represent or predict all of the scores in a distribution. However, although the mean may sound like a great invention, by itself it provides an incomplete description of any distribution. The mean tells us the central score and often where the most frequently occurring scores are, but it tells us little about scores that are not at the center of the distribution and/or that occur infrequently.

Consider the three samples of data shown in Table 5.1. As you can see, each sample has a mean of 6. If we did not look at the raw scores, we might believe that the three samples formed identical distributions. Obviously, they do not. Sample A contains relatively large differences between many of the scores, Sample B contains smaller differences between the scores, and Sample C contains no differences between the scores.

Thus, to accurately describe a set of data, we need to know not only the central tendency but also how much the individual scores differ from each other. The type of statistic we need is called a measure of variability. **Measures of variability** summarize and describe the extent to which scores in a distribution differ from each other. Thus, when we ask whether there are large differences between the scores or small differences between the scores, we are asking the statistical question "How much variability is there in the data?" When there are many relatively large differences between the scores, the data are said to be relatively *variable* or to contain a large amount of *variability.*

In Chapter 4 we saw that a score indicates a subject's location on the variable and that the difference between two scores is the distance that separates them. From this perspective, by telling us the extent to which scores in a distribution differ from each other, measures of variability tell us how *spread out* the scores are. For example, in Figure 5.1 we see a graphic representation of the distances separating the scores in our previous samples. There are relatively large differences between the scores in Sample A, so the distribution in Sample A is spread out. There are smaller differences between the scores in Sample B, so this distribution is not as spread out. There are no differences between the scores in Sample C, so there is no spread in this distribution. Thus, of the three samples, we can describe Sample A as having the largest differences between scores, or we can say that the scores in Sample A are spread out the most. In the language of statistics, we say that the scores in Sample A show the greatest variability.

We always measure the variability of a distribution, because it tells us about two important and related aspects of the data. First, the opposite of variability

TABLE 5.1 Three Different Distributions Having the Same Mean Score

Sample A	*Sample B*	*Sample C*
0	8	6
2	7	6
6	6	6
10	5	6
12	4	6
$\bar{X} = 6$	$\bar{X} = 6$	$\bar{X} = 6$

FIGURE 5.1 Distance Between the Locations of Scores in Three Distributions

An X over a score indicates a subject who obtained that score. Each arrow indicates how spread out the scores in the sample are.

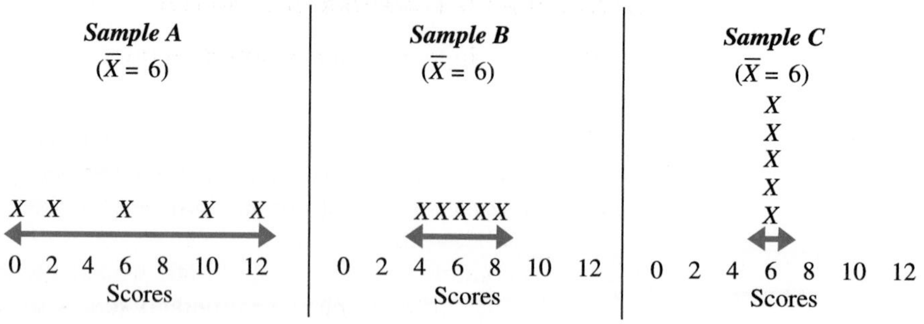

Sample A ($\overline{X} = 6$)

X X X X X

0 2 4 6 8 10 12
Scores

Sample B ($\overline{X} = 6$)

XXXXX

0 2 4 6 8 10 12
Scores

Sample C ($\overline{X} = 6$)

X X X X X

0 2 4 6 8 10 12
Scores

is how *consistent* the scores are. A small amount of variability indicates that there are not many large differences between the scores, so the scores must be rather similar and consistently close to the same value. Conversely, larger variability indicates that subjects were inconsistent in the scores they produced, each producing a score rather different from the next. Second, a measure of variability tells us how accurately our measure of central tendency describes the distribution: the greater the variability, the more the scores are spread out, and so the less accurately they are represented by one central score. Conversely, the smaller the variability, the closer the scores are to each other and to the one central score. Thus, by knowing the amount of variability in each of the above samples, we know Sample C contains consistent scores (and so 6 very accurately represents it), Sample B contains less consistent scores (and so 6 is not so accurate a summary), and Sample A contains very inconsistent scores (and so 6 is not even close to most scores).

STAT ALERT Measures of variability provide a number that indicates how spread out the scores are. While measures of central tendency indicate the *location* of most scores in a distribution, measures of variability indicate the *distance* between the scores in the distribution.

As with previous procedures, the specific measure of variability we compute depends on what it is we want to know and what type of data we have. In the following sections we will discuss three common measures of variability: the range, the variance, and the standard deviation.

THE RANGE

One way to describe the variability in scores is to determine how far the lowest score is from the highest score. In Figure 5.1, Sample A was spread out the most because the lowest and highest scores are farther apart than in the other two groups. The descriptive statistic that indicates the distance between the two most extreme scores in a set of data is called the **range.**

THE FORMULA FOR COMPUTING THE RANGE IS

$$\text{Range} = \text{highest score} - \text{lowest score}$$

In Sample A, the highest score is 12 and the lowest score is 0, so the range is 12 − 0, which equals 12. Sample B is less variable, because its range is 8 − 4 = 4. The range in Sample C is 6 − 6 = 0, so it has no variability.

Although each of these ranges does give an indication of the amount of spread in the sample, the range is a rather crude measure. Since it involves only the two most extreme scores, the range is based on the least typical and often least frequent scores. Therefore, we use the range as our sole measure of variability only with nominal or ordinal data or when interval or ratio scores form distributions that cannot be accurately described using other, better measures.

The Semi-interquartile Range

A special version of the range is the semi-interquartile range, which is used in conjunction with the median to describe highly skewed distributions of interval or ratio scores. There are four *quartiles,* each referring to a quarter of a distribution: the first quartile contains the lowest 25% of the distribution, the second quartile contains the next 25% (between the 25th percentile and the median), and so on. The **semi-interquartile range** is one-half of the distance between the scores at the 25th and 75th percentiles.

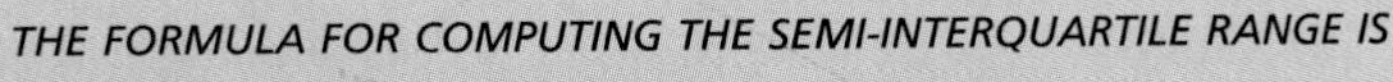

THE FORMULA FOR COMPUTING THE SEMI-INTERQUARTILE RANGE IS

$$\frac{\text{Score at 75th percentile} - \text{score at 25th percentile}}{2}$$

We determine the scores at the 25th and 75th percentiles using the formula in Chapter 3 for finding the score at a given percentile. Then we subtract the score at the 25th percentile from the score at the 75th percentile and divide by 2.

To see what the semi-interquartile range tells us, consider the distribution shown in Figure 5.2. We have determined that the score of 12 is at the 25th percentile and the score of 17 is at the 75th percentile. Using the above formula, we find that the semi-interquartile range is (17 − 12)/2, which is 5/2, or 2.5. Essentially we have calculated the average distance between the median and the scores at the 25th and 75th percentile. To see this, we can recalculate the semi-interquartile range by first finding that the range between the 25th percentile and the median (the 50th percentile) is 14 − 12, or 2. The range between the 50th

FIGURE 5.2 Semi-interquartile Range on a Positively Skewed Distribution

A total of 25% of all subjects scored between 12 and 14, and 25% scored between 14 and 17.

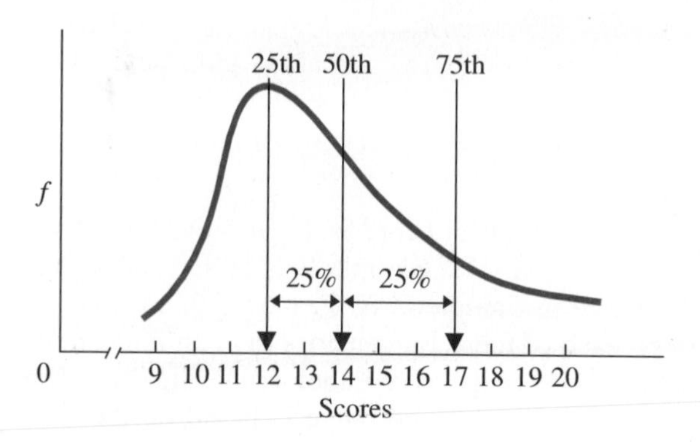

and 75th percentile is 17 − 14, or 3. The average of 2 and 3 is 5/2, or again 2.5. Thus, in both cases we have determined that the 25% of the distribution immediately below or above the median is, on average, within 2.5 points of the median. The semi-interquartile range describes the majority of the scores in the distribution more accurately than the overall range does, because the semi-interquartile range describes the range of scores more toward the center of the distribution and does not include the lopsided tails containing extreme scores.

UNDERSTANDING VARIANCE AND STANDARD DEVIATION

Most of the time in psychological research we have interval or ratio data that more or less fit a normal or at least a symmetrical distribution, so the mean is the best measure of central tendency. When the mean is the appropriate measure of central tendency, we use two measures of variability, called the variance and the standard deviation. Recognize from the outset that the variance and the standard deviation are two very similar ways to describe the variability in a distribution of scores. They are the best measures of variability because in computing them we consider every score in the distribution.

It is important to understand that although we use the variance and the standard deviation to describe and communicate how different the scores are from each other, they actually measure how much the scores differ from the mean. We use the mean as our reference point, because the mean is always the center of a distribution: when the scores in a distribution are spread out from each other, they are also spread out from the center of the distribution. Thus, if a high score is far from a low score, both scores are relatively distant from the mean. If all scores are close to each other, they are also close to the mean.

The variance and standard deviation are most appropriate with normal or other symmetrical distributions, because these distributions are balanced above and below the mean. Then the mean is the point *around* which a distribution is located, and the variance and standard deviation allow us to quantify "around." For example, if the grades in your statistics class form a normal distribution with a mean of 80, then you know that most people have a score around 80. But does this mean that most scores are between 79 and 81, or between 60 and 100, or what? By computing the variance and standard deviation, you can define "around."

The distance between a score and the mean is the numerical difference between the score and the mean. This difference is symbolized by the quantity $(X - \bar{X})$. Recall from Chapter 4 that $(X - \bar{X})$ defines the amount that a score deviates from the mean. Thus, a deviation indicates how much a score is spread out from the mean. Of course, some scores will deviate from the mean more than other scores. Because we want to summarize the variability of many scores that deviate by different amounts, it makes sense to determine the average amount that the scores deviate from the mean. We could call this amount the "average of the deviations." The larger the average of the deviations, the greater the spread, or variability, between all of the scores and the mean.

To compute an average, we sum the scores and divide by N. We *might* find the average of the deviations by first computing $(X - \bar{X})$ for each subject, then summing these deviations to find $\Sigma(X - \bar{X})$, and finally dividing by N, the number of deviations we summed. Altogether, the formula for the average of the deviations[1] would be

$$\text{Average of the deviations} = \frac{\Sigma(X - \bar{X})}{N}$$

We *might* compute the average of the deviations using this formula, except for a *big* problem. Recall from Chapter 4 that the sum of the deviations around the mean, $\Sigma(X - \bar{X})$, always equals zero, because the positive deviations cancel out the negative deviations. This means that, for any set of scores, the numerator of the average of the deviations will always be zero, so the average of the deviations will always be zero. So much for the average of the deviations!

But remember our purpose here: we want a statistic that tells us something *like* the average of the deviations, so that we know the average amount the scores are spread out around the mean. The trouble is, mathematically, the average of the deviations is always zero, so we will have to calculate this statistic in a more complicated way. Nonetheless, you should think of the variance and standard deviation as producing a number that, like an average, indicates the typical amount that the scores differ from the mean.

DESCRIBING THE SAMPLE VARIANCE

So how do we compute the average of the deviations? The problem is that the positive and negative deviations always cancel out to produce a sum of zero. The solution to this problem is to *square* the deviations. That is, we will find

[1]In advanced statistics there is a very real statistic called the average deviation. This isn't it.

the difference between each score and the mean and then square that difference. This removes all the negative deviations, so that the sum of the squared deviations is not necessarily zero and neither is the average of the squared deviations. (As we shall see, we also choose this solution because squaring the deviations results in statistics that have unique and very useful characteristics.)

In finding the average of the squared deviations, we are computing the variance. The **variance** is the average of the squared deviations of the scores around the mean. When we calculate this statistic for a sample of scores around the sample mean, we are computing the *sample variance.* The symbol for a sample variance is S_X^2. You must always include the squared sign (^{2}), because it is part of the symbol for variance. The capital S indicates that the quantity describes a sample, and the subscript X indicates that it is computed for a sample of X scores.

THE DEFINITIONAL FORMULA FOR THE SAMPLE VARIANCE IS

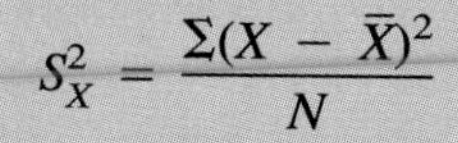

$$S_X^2 = \frac{\Sigma(X - \bar{X})^2}{N}$$

Finding the average of the squared deviations around the mean is simply one way of describing variability. In essence, S_X^2 stands for the answer we obtain when we use this formula to describe our sample, and this answer is called the sample variance. Recognize that we use the above formula *only* when we are describing a sample of data (as opposed to the population), simply to determine the variance of those scores in the sample.

STAT ALERT The symbol S_X^2 stands for the variance of the scores in a sample.

To see how this all works, say that as part of a very small study we measure the ages of our subjects and find they are 2, 3, 4, 5, 6, 7, and 8 years old, with a mean age of 5. To compute S_X^2 using the above formula, we arrange the data as shown in Table 5.2. First we compute each deviation $(X - \bar{X})$, by subtracting the mean from each score. Next, as shown in the far right column, we square

TABLE 5.2 Calculation of Variance Using the Definitional Formula

Subject	*Age Score*	−	$\bar{X}$	=	$(X - \bar{X})$	$(X - \bar{X})^2$
1	2	−	5	=	−3	9
2	3	−	5	=	−2	4
3	4	−	5	=	−1	1
4	5	−	5	=	0	0
5	6	−	5	=	1	1
6	7	−	5	=	2	4
7	8	−	5	=	3	9
	$N = 7$				$\Sigma(X - \bar{X})^2 =$	28

each of these deviations to get $(X - \bar{X})^2$. Then we add the squared deviations to find $\Sigma(X - \bar{X})^2$, which here is 28. The number of scores in the sample, N, is 7. Filling in our formula for S_X^2, we have

$$S_X^2 = \frac{\Sigma(X - \bar{X})^2}{N} = \frac{28}{7} = 4.0$$

Thus, in this sample, the variance, S_X^2, equals 4.0. In other words, the average squared deviation of the age scores around the mean is 4.0.

Computational Formula for the Sample Variance

To simplify the preceding definitional formula for variance, we replace the symbol for the mean with the formula for the mean. After reducing the formula, we have the computational formula. Like the definitional formula, this computational formula is used only when we wish to know the variance of the scores in a sample.

THE COMPUTATIONAL FORMULA FOR THE SAMPLE VARIANCE IS

$$S_X^2 = \frac{\Sigma X^2 - \dfrac{(\Sigma X)^2}{N}}{N}$$

In this formula, we first find the sum of the X's, ΣX, square that sum, and divide the squared sum by N. Then we subtract that amount from the sum of the squared X's, ΣX^2. Finally, we divide that quantity by N. The answer is the sample variance, S_X^2.

For example, using our previous age scores, we arrange the scores as shown in Table 5.3. The sum of the X's, ΣX, is 35, the sum of the squared X's, ΣX^2, is 203, and N is 7. Putting these quantities into the computational formula, we have

$$S_X^2 = \frac{\Sigma X^2 - \dfrac{(\Sigma X)^2}{N}}{N} = \frac{203 - \dfrac{(35)^2}{7}}{7}$$

The squared sum of X, $(\Sigma X)^2$, is 35^2, which is 1225, so we have

$$S_X^2 = \frac{203 - \dfrac{1225}{7}}{7}$$

Now, 1225 divided by 7 equals 175, so

$$S_X^2 = \frac{203 - 175}{7}$$

TABLE 5.3 Calculation of Variance Using the Computational Formula

X score	X^2
2	4
3	9
4	16
5	25
6	36
7	49
8	64
$\Sigma X = 35$	$\Sigma X^2 = 203$

Because 203 minus 175 equals 28, we have

$$S_X^2 = \frac{28}{7}$$

Finally, after dividing, we have

$$S_X^2 = 4.0$$

Thus, as we found using the definitional formula, the sample variance for these age scores is 4.0.

Do not read any further until you are sure that you understand how to work this formula!

Interpreting Variance

The good news is that the variance provides a direct way of measuring variability using all the scores in the data. Ideally, though, we want the average of the deviations, and the bad news is that the variance does not make much sense as an average deviation. There are two problems. First, because the variance is the average of the *squared* deviations—the squared distances between the scores and the mean—it is always an unrealistically large number. For the age scores of 2, 3, 4, 5, 6, 7, and 8, the $\overline{X}$ is 5 and S_X^2 is 4. But to say that the ages of these subjects differ from the mean age of 5 by an *average* of 4 years is plain silly! Not one age score actually deviates from the mean by as much as 4 years, so this is certainly not the average deviation. The second problem is that variance is rather bizarre because it measures in squared units: in the example above we measured the ages of our subjects, so the variance indicates that the scores deviate from the mean by 4 *squared* years (whatever that means!).

Thus, we cannot directly interpret the variance as telling us the "average" deviation in the data. Does this mean that computing the variance is a waste of time? No, because variance is used extensively in the statistics we will discuss later. Also, variance does communicate the relative variability of scores. If someone reports that one sample has $S_X^2 = 1$ and another sample has $S_X^2 = 3$, you know that the second sample is more variable, because it has a larger average squared deviation. This tells you that the scores are relatively less consistent and less accurately described by their mean score. Thus, think of variance as a number that communicates how generally variable the scores are: the larger the variance, the more the scores are spread out.

The measure of variability that more directly communicates the average deviation of the raw scores is the standard deviation.

DESCRIBING THE SAMPLE STANDARD DEVIATION

The variance is always an unrealistically large number because we square each deviation. We solve this problem by taking the square root of the variance. The answer is called the standard deviation. The **standard deviation** is the square root of the variance, or the square root of the average squared deviation of scores around the mean. To create the definitional formula for the standard deviation, we simply add the square root to our previous definitional formula for variance.

THE DEFINITIONAL FORMULA FOR THE SAMPLE STANDARD DEVIATION IS

$$S_X = \sqrt{\frac{\Sigma(X - \bar{X})^2}{N}}$$

Notice that the symbol for the sample standard deviation is also the square root of the symbol for the sample variance, because $\sqrt{S_X^2}$ is S_X. Conversely, if we square the standard deviation, we have the variance. (To remember all of this, think of the capital S in S_X as indicating the sample standard deviation and S_X^2 as the squared sample standard deviation, which we call the sample variance.)

> ***STAT ALERT*** The symbol S_X stands for the standard deviation of the scores in a sample.

To compute S_X using this formula, we first compute everything inside the square root sign, which gives us the variance: we square each score's deviation, sum the squared deviations, and then divide that sum by N. In our example with age scores above the variance (S_X^2) was 4. We then take the square root of this number to find the standard deviation. In this case,

$$S_X = \sqrt{4.0}$$

which gives us

$$S_X = 2.0$$

The standard deviation of the original age scores is 2.0.

Computational Formula for the Sample Standard Deviation

We create the computational formula for the standard deviation by merely adding the square root symbol to the computational formula for the variance.

THE COMPUTATIONAL FORMULA FOR THE SAMPLE STANDARD DEVIATION IS

$$S_X = \sqrt{\frac{\Sigma X^2 - \frac{(\Sigma X)^2}{N}}{N}}$$

This formula is used only when we are describing the standard deviation of a *sample* of scores.

To see how the formula works, we will again use our age scores. From Table 5.3 we know that ΣX is 35, ΣX^2 is 203, and N is 7. Putting these values in the formula gives

$$S_X = \sqrt{\frac{203 - \frac{(35)^2}{7}}{7}}$$

Completing the computations inside the square root symbol, we have the variance, which is 4.0. Thus, we have

$$S_X = \sqrt{4.0}$$

Taking the square root of 4.0, we once again find that the standard deviation is

$$S_X = 2.0$$

Interpreting the Standard Deviation

The standard deviation is as good as our measures of variability get: computing the standard deviation is as close as we come to computing the "average of the deviations." Thus, in our age scores, S_X equals 2.0, so first we interpret this as indicating that the scores differ, or deviate, from the mean by an "average" of about 2. Some scores will deviate by more and some by less, but overall the scores will deviate from the mean by something like an average of 2. Further, the standard deviation measures in the same units as the raw scores, so the scores differ from the mean age by an "average" of 2 *years*.

Second, because the average is influenced by the size of the deviation scores, if the value of S_X is relatively large, then we know that a relatively large proportion of scores are relatively far away from the mean and that few scores are close to it. If, however, S_X is relatively small, then most scores are close to the mean, and a relatively small proportion of scores are far from it. Thus the standard deviation allows us to gauge the extent to which the scores are relatively consistent or inconsistent, and correspondingly, the degree to which they are or are not accurately summarized by the mean score.

Finally, the standard deviation indicates how much the scores below the mean deviate from it and how much the scores above the mean deviate from it. Because of this we can further summarize a distribution by describing the scores that lie at "plus one standard deviation from the mean" ($+1S_X$) and "minus one standard deviation from the mean" ($-1S_X$). For example, our age scores of 2, 3, 4, 5, 6, 7, 8 produced a $\overline{X} = 5.0$ and a $S_X = 2.0$. The score that is $+1S_X$ from the mean is the score at 5 + 2, or 7. The score that is $-1S_X$ from the mean is the score at 5 − 2, or 3. As you can see, a good way to summarize these scores is to say that the mean score is 5 and the majority of the scores are between 3 and 7.

Applying the Standard Deviation to the Normal Curve

One reason we compute the standard deviation when the scores are normally distributed is because there is a precise mathematical relationship between the standard deviation and the normal curve. Therefore, describing a distribution in terms of the scores that are between $-1S_X$ and $+1S_X$ is especially useful. For example, earlier in this chapter we imagined a statistics class with a mean score of 80. Say that the S_X for the class is 5. The score at 80 − 5 is the score of 75, and the score at 80 + 5 is the score of 85. Figure 5.3 shows about where these scores are located on a normal distribution. First notice that there is an easy way to locate where the scores at $-1S_X$ and $+1S_X$ from the mean are located on any normal curve. At the scores that are close to the mean, the curve forms a downward convex shape (∩). As you travel away from the mean, at a certain point the curve changes its pattern to an upward convex shape (∪). The points at which the curve changes its shape are called "inflection points." Because of the mathematical relationship between a normal curve and the standard deviation, the scores at the inflection points are always the scores that are one standard deviation away from the mean.

Now, by looking at Figure 5.3, you can see how we summarize the distribution. First, by saying that the mean is 80, we imply that most scores are around 80. Then, by finding the scores at $-1S_X$ and $+1S_X$, we define "around": most of the scores are between 75 and 85 (from our parking lot perspective, this is where most of the subjects are standing). In fact, we can further define "around." Because of the relationship between the standard deviation and the normal distribution, approximately 34% of the scores in a perfect normal distribution are between

FIGURE 5.3 Normal Distribution Showing Scores at Plus or Minus One Standard Deviation

With $S_X = 5.0$, the score of 75 is at $-1S_X$ and the score of 85 is at $+1S_X$. The percentages are the approximate percentages of the scores falling in each portion of the distribution.

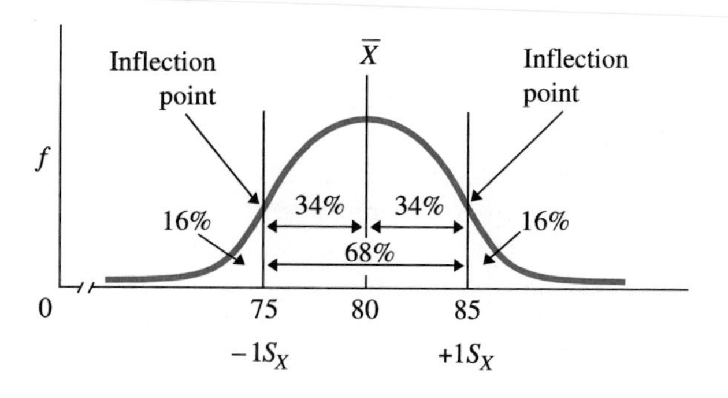

the mean and the score that is one standard deviation from the mean. Therefore, in total, approximately 68% of the scores are always between the scores at $+1S_X$ and $-1S_X$ from the mean. As illustrated in Figure 5.3, we see that about 34% of all students have scores between 75 and 80, and 34% have scores between 80 and 85. In total approximately 68% of all scores in the distribution fall between 75 and 85. Conversely, only about 32% of the scores are outside this range, with about 16% below 75 and 16% above 85. Thus, saying that most scores are between 75 and 85 is an accurate summary, because the majority of scores (68%) are here.

Of course it is very unlikely that the scores of a small statistics class would produce an ideal normal distribution. However, this example shows how we apply the normal curve model to real data. If we can assume that statistics grades are generally normally distributed, we can operate as if the class formed the ideal normal distribution. Then we *expect* about 68% of the scores in the class to fall between 75 and 85. The more that the distribution conforms to a perfect normal curve, the more that precisely 68% of the scores will be between 75 and 85.

Different data produce different normal distributions. By finding the scores at $-1S_X$ and $+1S_X$ from the mean, we can envision and communicate these differences. Figure 5.4 shows three variations of the normal curve for scores having a mean of 50. The larger the standard deviation (and variance), the wider the distribution, because the scores are more frequently spread out farther above and below 50. For example, in Distribution A, S_X is 4.0. This tells us that, as shown

FIGURE 5.4 Three Variations of the Normal Curve

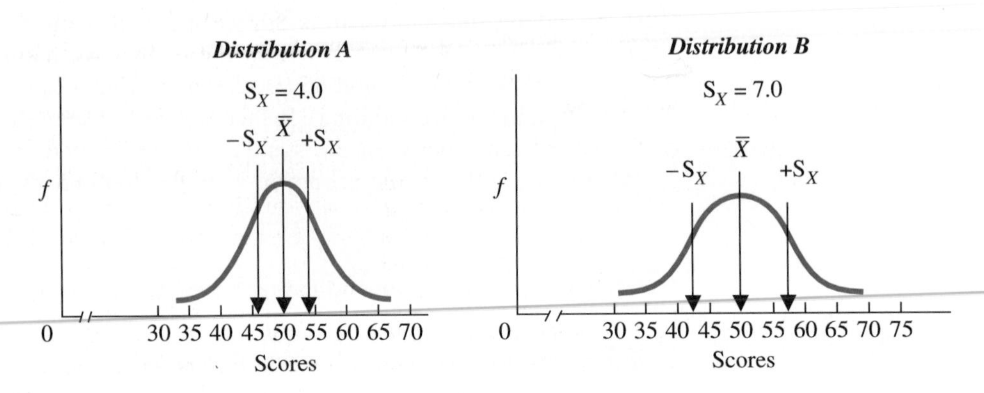

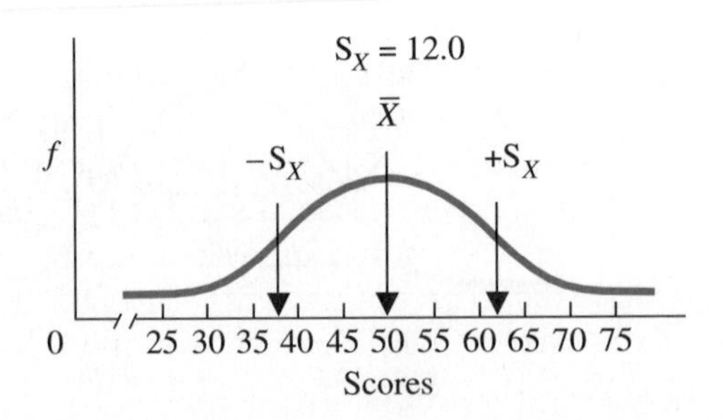

in the figure, the most frequent scores are between 46 and 54. Because the majority of the deviations are smaller than or equal to $+4$ and -4 and there are relatively few larger deviations, the "average" of these deviations is 4. Essentially, the most frequent scores are bunched relatively close to the mean and so most frequently there are small deviations and thus a small "average" or standard deviation. But look at Distribution B. Here S_X is 7.0, and this tells us that the relatively frequent scores are spread out over a wider range, between about 43 and 57. Now the majority of the deviations are larger, up to $+7$ and -7, producing an "average" deviation of about 7. Finally, in Distribution C, S_X is 12, telling us that the relatively frequent scores are between about 38 and 62, producing many deviations between $+12$ and -12. With so many large deviations, we have a large "average" deviation of about 12. Thus, a larger standard deviation (or variance) indicates that a distribution is wider, because in a wider distribution the relatively extreme scores occur more frequently, producing a larger "average" deviation.

STAT ALERT The larger the value of S_X (or S_X^2), the more the scores are spread out around the mean, and the wider the distribution.

Despite differences in shape, any normal distribution will have approximately 68% of the scores between the scores at $+1S_X$ and $-1S_X$ from the mean. This is because, mathematically, 68% of the *area* under the normal curve falls between $+1S_X$ and $-1S_X$. To understand this, visualize the above distributions as parking lots full of people. Say that each parking lot contains 1000 subjects: 68% of 1000 is 680. In Distribution A, the greater height of the curve indicates that more subjects are standing in line at those scores close to the mean. With so many small deviations, we will compute a small standard deviation. Correspondingly, because each line contains so many subjects, we will not travel very far above and below the mean before we accumulate 680 subjects (and cover 68% of the area). On the other hand, in Distribution C there are fewer subjects standing in line at each score near the mean. Here relatively large deviations are more frequent, producing a larger standard deviation. Correspondingly, because each line contains fewer subjects, we will travel farther above and below the mean before we accumulate 680 subjects (and cover 68% of the area). Thus, regardless of the size of S_X, approximately 68% of the area under the normal curve, and thus 68% of the scores, falls between the scores at $+1S_X$ and $-1S_X$ from the mean.

In summary, then, here is how the standard deviation (and variance) add to our description of a distribution. For the distributions in Figure 5.4 for example, if you know that they form normal distributions, you can envision their general shape. If you know that the mean is 50 in each, you know where the center of the distribution is and what the typical score is. And if you know that, for example, $S_X = 4$, you know that those subjects who did not score 50 missed it by an "average" of 4 points; the distribution is relatively "skinny," or narrow; and most scores (68% of them) are within 4 points of 50, or between 46 and 54. Conversely, if you know that the $S_X = 12$, you know that those subjects who did not score 50 missed it by an "average" of 12 points; the distribution is relatively "fat," or wide; and most scores (68% of them) are within 12 points of 50, or between 38 and 62.

Avoiding Some Common Errors in Computing S_X

Because the computations for S_X (and S_X^2) are somewhat involved, you should always examine your answers to be sure that they are correct. There are some answers you should never obtain.

First, remember that variability can never be a negative number. If you state that S_X is -5, for example, you are claiming that the distance between scores is less than zero! Also, all our formulas involve squared numbers, and squared numbers can never be negative numbers.

Second, watch for answers that simply do not make sense. If the raw scores range between 35 and 65, what should you conclude if you find that S_X is 20? My first guess is that you made a mistake! If these scores form anything like a normal distribution, the mean score will be around 50. If so, then the largest deviation for the extreme scores of 35 and 65 will be only 15 points, so the "average deviation" for all scores cannot be 20. Similarly, it is unlikely that the S_X for these scores is extremely small, such as .25. If there are only two extreme deviations of 15, imagine how many small deviations it would take to create an average of only .25. (Similarly, you can evaluate any variance you compute by first finding its square root, the standard deviation.)

Strange numbers for S_X and S_X^2 may be correct for strange distributions, but you should always be alert to whether statistical answers make sense. (They *do* make sense, you know.) Checking your calculations is the best way to ensure that you have the correct answer. However, a general rule of thumb for any roughly normal distribution is that the standard deviation equals about one-sixth of the overall range of scores in the data.

Mathematical Constants and the Standard Deviation

As we discussed in Chapter 4, we sometimes transform scores by adding, subtracting, multiplying, or dividing by a constant. What effect do such transformations have on the standard deviation and variance? The answer depends on whether we add (subtracting is adding a negative number) or multiply (dividing is multiplying by a fraction).

When we add a constant to the scores in a distribution, we merely shift the entire distribution to higher or lower scores. We do not alter the relative position of any score, so we do not alter the spread in the data. For example, take the scores 4, 5, 6, 7, and 8. The mean is 6. Now add the constant 10. The resulting scores of 14, 15, 16, 17, and 18 have a mean of 16. Before the transformation, the score of 4 was 2 points away from the mean of 6. In the transformed data, that score is now 14, but it is still 2 points away from the new mean of 16. In the same way, each score's distance from the mean is unchanged, so the standard deviation is unchanged. If the standard deviation is unchanged, then the variance is unchanged.

Adding or subtracting a constant does not change the value of the standard deviation or variance.

When we multiply by a constant, however, we alter the relative positions of scores, and therefore we change the variability. If we multiply the scores 4, 5,

6, 7, and 8 by the constant 10, they become 40, 50, 60, 70, and 80. The original raw scores that were 1 and 2 points from the mean of 6 are now 10 and 20 points from the new mean of 60. Each transformed score produces a deviation that is 10 times the original deviation, so the new standard deviation is also 10 times greater.

Multiplying or dividing the scores by a constant produces the same standard deviation as multiplying or dividing the old standard deviation by that constant.

Note that this rule does not apply to the variance. The new variance will equal the square of the new standard deviation.

THE POPULATION VARIANCE AND THE POPULATION STANDARD DEVIATION

In every chapter so far, we have seen how researchers and statisticians always discuss two things: samples and the populations they represent. The same is true when it comes to variance and standard deviations. We have described the variability of a sample, and now we will describe the variability of the population.

In Chapter 2 you learned that Greek letters symbolize numbers that describe a population (population parameters). The symbol for the true population standard deviation is σ_X. The σ is the lowercase Greek letter s, or sigma. Because the squared standard deviation is the variance, the symbol for the true population variance is σ_X^2. (In each case, the subscript X indicates that we have a population of X scores.)

The definitional formulas for σ_X and σ_X^2 are similar to those we saw previously for describing a sample:

Population Standard Deviation

$$\sigma_X = \sqrt{\frac{\Sigma(X - \mu)^2}{N}}$$

Population Variance

$$\sigma_X^2 = \frac{\Sigma(X - \mu)^2}{N}$$

The only difference is that, in describing the population, we determine how far each score deviates from the population mean, μ. Otherwise, the standard deviation and variance of the population tell us exactly the same things about the population that the sample standard deviation and variance tell us about the sample. Both are ways of measuring how much, "on average," the scores differ from μ, indicating how much the scores are spread out in the population.

STAT ALERT Anytime we are discussing the true population variability, we use the symbols σ_X^2 and σ_X.

Of course, we usually think of a population as being infinitely large. Since we usually cannot obtain all the scores in the population, we usually cannot use the above formulas to compute the true values of σ_X and σ_X^2. Instead, we make estimates, or inferences, about the population based on a random sample of scores. In Chapter 4 we saw that we can estimate the population mean, μ, by computing the sample mean, $\overline{X}$. We estimate that if we could measure the entire population, the population mean would equal our sample mean. Now we ask the question "What is our estimate of the variability of the scores around μ?"

Estimating the Population Variance and Population Standard Deviation

You might think that we would compute the sample variance as described previously and then use it to estimate the population variance (and, likewise, that we would use the sample standard deviation to estimate the population standard deviation). If, for example, the sample variance is 4, should we then guess that the population variance is also 4? Nope! The sample variance and standard deviation are used *only* to describe the variability of the scores in a sample. They are *not* the best way to estimate the corresponding population parameters.

To understand why this is true, say that we measure an entire population of scores and compute its true variance, σ_X^2. We then draw a series of random samples of scores from the population and compute the variance of each sample, S_X^2. Sometimes the sample variance will equal the actual population variance, but other times the sample will not be perfectly representative of the population. Then either the sample variance will be smaller than, or underestimate, the population variance, or the sample variance will be larger than, or overestimate, the population variance. Over many random samples, however, more often than not the sample variance will *underestimate* the population variance. The same thing happens if we perform the above operations using the standard deviation.

In statistical terminology, the formulas for S_X^2 and S_X are called the *biased estimators* of the population variance and standard deviation: they are biased toward underestimating the true population parameters. Using the biased estimators is a problem because, as we saw in Chapter 4, if we cannot be perfectly accurate in our estimates, we at least want our under- and overestimates to cancel out over the long run. (Remember the statisticians shooting targets?) With the biased estimators, the underestimates and overestimates will not cancel out. Thus, although the sample variance (S_X^2) and sample standard deviation (S_X) accurately describe the variability of the scores in a sample, they are too often too small to serve as accurate estimates of the true population variance (σ_X^2) and standard deviation (σ_X).

Why do S_X^2 and S_X produce biased estimates of the population variance and standard deviation? Because their formulas are intended to describe the variability in the sample and are not intended to estimate the population. Remember that to accurately estimate a population, we should have a random sample. We want the variability, or deviation, of each score to be random so that we can accurately estimate the variability of the scores in the population. Yet, when we measure the variability of a sample, we use the mean as our reference point. In doing so, we encounter the mathematical restriction imposed by the mean that the sum of the deviations, $\Sigma(X - \overline{X})$, must equal zero. Because of this restriction, not all of the variability in the sample reflects random variability.

For example, say that the mean of five scores is 6, and that four of the scores are 1, 5, 7, and 9. Their deviations are −5, −1, +1, and +3, so the sum of their deviations is −2. Without even looking at the final score, we know that it must be 8, because it must have a deviation of +2 so that the sum of all deviations is zero. Thus, given the sample mean and the deviations of the other scores, the deviation of the score of 8 is not random; rather, it is determined by those of the other scores. Therefore, only the deviations produced by the scores of 1, 5, 7, and 9 reflect the random variability found in the population. The same would be true for any four of the five scores. Thus, when N is 5, only four of the scores actually reflect the variability of the scores in the population. In general, out of the N scores in a sample, only $N - 1$ of them actually reflect the variability in the population.

The problem is that when we compute the variability using our previous formulas for the biased estimators, S_X and S_X^2, we want the "average deviation" reflected by the four scores, but these formulas require us to divide by an N of 5. Because we are dividing by too large a number, our answer tends to be too small, underestimating the actual variability in the population. Obviously, to accurately estimate the average variability based on four scores, we should divide by 4, or $N - 1$. By doing so, we compute the unbiased estimators of the population variance and standard deviation.

THE DEFINITIONAL FORMULAS FOR THE UNBIASED ESTIMATORS OF THE POPULATION VARIANCE AND STANDARD DEVIATION ARE

Estimated Population Variance

$$s_X^2 = \frac{\Sigma(X - \overline{X})^2}{N - 1}$$

Estimated Population Standard Deviation

$$s_X = \sqrt{\frac{\Sigma(X - \overline{X})^2}{N - 1}}$$

As you can see, we are still computing a number that is analogous to the average of the deviations in the sample, *but* (and this is the big but) here we divide by the number of scores in the sample minus one. We include all of the scores when we are computing the sum of the squared deviations in the numerator, but then we divide by $N - 1$.

Notice that the symbol for the unbiased estimator of the population standard deviation is s_X and the symbol for the unbiased estimator of the population variance is s_X^2. To keep all of your symbols straight, remember that the symbols for the sample variance and standard deviation involve the capital, or big, S, and in those formulas you divide by the entire, or big, value of N. The symbols for estimates of the population variance and standard deviation involve the lowercase, or small, s, and you divide by the smaller number, the quantity $N - 1$. Further, the small s is used to estimate the true population value, symbolized by the small Greek s, σ. Finally, think of s_X^2 and s_X as the inferential variance and the inferential standard deviation, because the *only* time you use them is to estimate, or infer, the variance or standard deviation of the population based on a sample. You can think of S_X^2 and S_X as the descriptive variance and standard deviation, because they are used to describe the sample.

> ***STAT ALERT*** Use the symbols S_X^2 and S_X and the corresponding formulas to describe the variability of scores in a sample of scores. Use the symbols s_X^2 and s_X and the corresponding formulas to estimate the variability of scores in the population.

For future reference, the quantity $N - 1$ has a special name; it is referred to as the degrees of freedom. The **degrees of freedom** is the number of scores in a sample that are free to vary so that they reflect the random variability in the population. The symbol for degrees of freedom is *df*. We shall use degrees of freedom frequently in later chapters, and they are not always equal to $N - 1$. Here, however, $df = N - 1$.

In the final analysis, you can think of degrees of freedom, or $N - 1$, as simply a correction factor. Because $N - 1$ is a smaller number than N, dividing by $N - 1$ produces a slightly larger answer than does dividing by N. Over the long run, this larger answer will prove to be a more accurate estimate of the population variability.

Computational formula for the estimated population variance

THE COMPUTATIONAL FORMULA FOR ESTIMATING THE POPULATION VARIANCE IS

$$s_X^2 = \frac{\Sigma X^2 - \frac{(\Sigma X)^2}{N}}{N - 1}$$

The only difference between this computational formula for the estimated population variance and the previous computational formula for the sample variance is that here the final division is by $N - 1$. Notice that in the numerator we still divide by N.

In our previous computational examples, the age scores we collected were 2, 3, 4, 5, 6, 7, and 8, with $N = 7$, $\Sigma X^2 = 203$, and $\Sigma X = 35$. Putting these quantities into the above formula gives

$$s_X^2 = \frac{\Sigma X^2 - \frac{(\Sigma X)^2}{N}}{N - 1} = \frac{203 - \frac{(35)^2}{7}}{6}$$

We work through this formula in exactly the same way we worked through the formula for the sample variance, except that here the final division involves $N - 1$, or 6. Since 35^2 is 1225, and 1225 divided by 7 equals 175, we have

$$s_X^2 = \frac{203 - 175}{6}$$

Now 203 minus 175 equals 28, so we have

$$s_X^2 = \frac{28}{6}$$

and the final answer is

$$s_X^2 = 4.67$$

Notice that this answer is slightly larger than the one we obtained when we computed the sample variance for these age scores. There S_X^2 was 4.0. Although 4.0 accurately describes the sample variance, it is likely to underestimate the actual variance of the population; 4.67 is more likely to be the population variance. In other words, if we could measure all the scores in the population from which this sample was drawn and then compute the true population variance, σ_X^2, we would expect our answer to be 4.67.

Computational formula for the estimated population standard deviation Creating the formula for the unbiased estimator of the population standard deviation involves merely adding the square root sign to the previous formula for the estimated population variance.

THE COMPUTATIONAL FORMULA FOR ESTIMATING THE POPULATION STANDARD DEVIATION IS

$$s_X = \sqrt{\frac{\Sigma X^2 - \frac{(\Sigma X)^2}{N}}{N - 1}}$$

In the previous section, we computed the estimated population variance from our sample of age scores to be $s_X^2 = 4.67$. Using the above formula, we find that the estimated population standard deviation, s_X, is $\sqrt{4.67}$, or 2.16. Thus, if we could compute the standard deviation using the entire population of scores, we would expect σ_X to be 2.16.

Interpreting the Estimated Population Variance and Standard Deviation

We interpret the estimated variance and standard deviation in the same way as we did S_X^2 and S_X, except that now we are describing how much we *expect* the scores to be spread out in the population, how consistent or inconsistent we *expect* the scores to be, and thus how accurately we *expect* the population to be represented by the mean score. Notice that, assuming our sample is representative of the population, this allows us to reach the ultimate goal of research: to describe an unknown population of scores. If we can assume that the distribution is normal, we have described its overall shape. Our sample mean, $\overline{X}$, provides a good estimate of the population mean, μ. The size of s_X^2 or s_X is our estimate of how spread out the population is—an estimate of the "average amount" that the scores in the population deviate from μ. Further, we expect that approximately 68% of the scores in the population lie between the scores at $+1s_X$ and $-1s_X$ from μ. Then, because scores reflect behaviors, we have a good idea of how most individuals in the population behave in a given situation (which is why we conduct research in the first place).

Recall that in addition to being able to describe an overall distribution of scores, another goal of research is to be able to predict anyone's *individual* score. As we will see in the next section, variability plays an important role in meeting this goal as well.

VARIANCE IS THE ERROR IN PREDICTIONS

In Chapter 4 we saw that the mean of a sample is the best single score to use to predict unknown scores. We employ measures of variability to determine how well we can predict these scores. To do this, we first determine how well we can predict the scores in a known sample: pretending that we don't know the scores, we predict them, and then we see how close we came to the actual scores.

For example, if a statistics class has a mean score of 80, then our best guess is that any student in the class has a grade of 80. Of course not every student will actually obtain a grade of 80. When we use the mean score to predict the actual scores in a sample, the amount by which we are wrong in a single prediction is the quantity $(X - \overline{X})$, the amount that the actual score differs, or deviates, from the predicted mean score. Since some predictions in a sample will contain more error than others, we summarize our error by finding the average amount that the actual scores deviate from the mean. As we have seen, our way of finding the "average" amount that scores deviate from the mean is by computing the variance and standard deviation. Because these statistics measure the difference between each score and the mean, they also measure the error in our predictions when we predict the mean score for all subjects in a sample. Thus, we have a slightly novel way of thinking about measures of variability. In this context, the larger the variability, the larger the differences between the mean and the scores, so the larger our error when we use the mean to predict scores.

Thus, if the mean score in the statistics class is 80 and the standard deviation is 5, then the actual scores differ from the mean by an "average" of 5 points. This indicates that if we predict a score of 80 for every member of the class, the actual scores will differ from our predicted score by an "average" of 5 points. Sometimes we will be wrong by more, sometimes by less, but overall we will be wrong by an amount equal to the sample standard deviation, S_X.

Similarly, the variance, S_X^2, indicates the average of the squared deviations from the mean, so the variance is the average of the "squared errors" we will have when we predict the mean score for each subject in a sample. Unfortunately, the concept of squared errors is rather strange. This is too bad, because in statistics the proper way to describe the amount of error in our predictions is to compute the variance. In fact, the variance is sometimes called *error:* it is our way of measuring the average error between the predicted mean score and the actual raw scores. Thus, for our purposes, simply remember that when we use the mean to predict scores, the larger the variance, the larger the error, and the smaller the variance, the smaller the error. To keep this concept in focus, think of the extreme case in which all the scores in a statistics class are the same score of 80. Then the mean is 80, and the variance, S_X^2, is zero. In this case, when we use the mean of 80 as the predicted score for each student, there is zero error: there is no difference between what we predict for students (the $\overline{X}$) and what they got (the X's), and that is exactly what $S_X^2 = 0$ indicates.

STAT ALERT The sample variance, S_X^2, is the "average" error we have when we use the sample's mean score as the predicted score for every subject in the sample.

Estimating the Error in Predictions in the Population

Of course, we can also estimate the amount of error we expect to have if we predict unknown scores not in our sample. The unknown scores we are predicting are the other scores in the population. Our best estimate of any score in the population is the population mean, μ, which we assume is equal to our sample mean (assuming that our data are representative of the population). Thus, for example, based on our statistics class mean of 80, we estimate that the population mean is 80. Then our best prediction is that any student in the population who takes this statistics class will receive a grade of 80.

To determine the error in our predictions for the population, we use the same logic that we used above for the sample. The amount of error in our predictions is the population variance, because it describes the differences between the population mean, which we predict for each subject, and the actual scores in the population. Since we usually cannot compute the true population variance, we instead compute the estimated population variance, s_X^2. Say that the s_X^2 for our statistics class is 5.75, so we expect that scores differ from the μ of the population by an "average" of about 5.75. Therefore, when we predict that other students taking this class will receive a grade equal to the μ of 80, the average amount we expect to be wrong is also 5.75.

STAT ALERT The estimated population variance, s_X^2, is the amount of error we expect when we assume that our sample mean equals the population mean and then use this score as the predicted score for every subject in the population.

Researchers can always measure a sample of scores, compute the mean, and use it to predict scores. Therefore, the value of S_X^2 is the largest error we are forced to accept when predicting the scores in a sample, and s_X^2 is the largest error we expect when predicting the scores in a population. As we will see in the next section, because the variance is the worst that we can do, it is our reference point. Anything that improves the accuracy of our predictions is measured relative to the variance.

Accounting for Variance

We use the mean to predict scores unless we have other information about them. As scientists, we look to relationships to provide additional information that will improve the accuracy of our predictions. Recall that when a relationship exists between two variables, a particular score on one variable tends to be consistently associated with a particular score on the other variable. Therefore, we can more accurately predict the scores on one variable if we know the corresponding scores on the other variable. For example, for children, the variable of height forms a consistent relationship with the variable of weight: the taller the child, the more he or she tends to weigh. Thus, if we know a child's height, we can use this relationship to more accurately predict his or her weight. An important use of variance is determining by how much a relationship improves the accuracy of our predictions.

With a relationship, our predictions will be more accurate, but "more accurate" than what? They will be more accurate than if we didn't use the relationship. That is, to describe how well we predict scores when using a relationship, we compare our errors in prediction when we use the relationship to the errors in prediction we'd have if we ignored the relationship. If we ignore the height-weight relationship, for example, we can still measure the weight of a sample of children and use the mean score as the predicted score. When we use the mean of the sample of weight scores to predict scores, our prediction error will be the sample variance. Then, if we don't ignore the relationship with height, our predictions will be more accurate, so our error will be less than the sample variance. Thus, to summarize the improvement in our predictions when using the relationship, we compare our error when using the relationship to the sample variance.

For example, the upper portion of Table 5.4 shows a sample of children's weight scores, ignoring their height. Here, our best guess is to predict the sample's

TABLE 5.4 Weight Scores, With and Without Considering Their Relationship to Height Scores

Sample of weight scores without considering the relationship to height.

80	85	90
81	86	91
82	87	92

mean weight = 86
average error = 4

Sample of weight scores as related to height scores: each column shows the weights of children at that height.

46 inches	47 inches	48 inches
80	85	90
81	86	91
82	87	92
predict 81	predict 86	predict 91

Average error = 1

overall mean weight of 86 for each child. Say that in round numbers our error then equals an S_X^2 of 4. However, as shown in the lower portion of Figure 5.4, by considering the relationship of height to weight we can pair certain weight scores with a particular height. Now we would predict a weight of around 81 pounds for children who are 46 inches tall, around 86 pounds for children who are 47 inches tall, and around 91 pounds for children who are 48 inches tall. These predictions are closer to each subject's actual weight score so we have less error. Later we'll see how to compute the errors in prediction here, but just by looking at the scores, it makes sense to say that in round numbers our average amount of error is now about 1.

Thus, when we do not use the height-weight relationship and instead predict the mean weight score from all subjects, we are off by an "average" of about 4. But when using the height-weight relationship, we are off by an "average" of only 1. Therefore, by using the relationship our prediction errors have gone from an average of 4 to an average of 1, so we have reduced our error by an average of 3, eliminating 3 points of error. As researchers, it is difficult to know whether eliminating 3 points of error should be considered a large amount. But recall that using the overall mean of a sample to predict scores produces the most error we must tolerate, so an average error of 4 is the worst we can do in this situation. Therefore, we interpret our improvement relative to this worst-case scenario: by eliminating 3 points of error, we have eliminated 3 of the 4 points of maximum error we could have in predicting these weight scores. In other words, because 3 out of 4 is .75, using the relationship eliminates .75 or 75% of the error we'd have if we did not use the relationship.

In statistical language, we have calculated the proportion of variance accounted for. The **proportion of variance accounted for** is that proportion of the error we have in our predictions when we use the mean to predict scores that is eliminated when we use the relationship with another variable to predict scores. In other words, the proportion of variance accounted for is the proportional improvement in our predictions achieved by using a relationship to predict scores, compared to if we did not use the relationship.

> ***STAT ALERT*** When we say that Variable A accounts for a certain proportion of the variance in Variable B, we mean that knowing subjects' scores on Variable A allows us to more accurately predict their scores on Variable B. The size of the proportion indicates how much our errors are reduced, relative to what they would be if we used only the overall mean of scores on Variable B to predict all scores on Variable B.

In later chapters we will discuss methods for computing the proportion of variance accounted for when using a relationship to predict scores in our sample, as well as methods for estimating the proportion of variance accounted for when using a relationship to predict scores in the population. Determining the proportion of variance accounted for is a very important procedure, because it is variance and variability that lead to scientific inquiry in the first place. When researchers ask, "Why does a person do this instead of that?" they are trying to predict and explain differences in scores. In other words, they are trying to account for variance. Ultimately, as behavioral scientists, we want to understand the laws of nature so that we know with 100% accuracy when a subject will get one score, reflecting one behavior, and when a subject will get a different score, reflecting a different behavior. In statistics, this translates into accounting for 100% of the variance in scores.

FINALLY

You can organize your thinking about measures of variability using the diagram in Figure 5.5. Remember that variability refers to the differences between scores. The variance and standard deviation are simply two related methods for describing variability, or differences. Constructing any formula for the standard deviation merely requires adding the square root to the corresponding formula for variance. For either the variance or the standard deviation, we can compute the descriptive sample version, which is a biased estimator of the population, or we can compute the inferential, or estimated, version, which is an unbiased estimator of the population. The difference is that inferential formulas require a final division by $N - 1$ instead of by N.

FIGURE 5.5 Organizational Chart of Descriptive and Inferential Measures of Variability

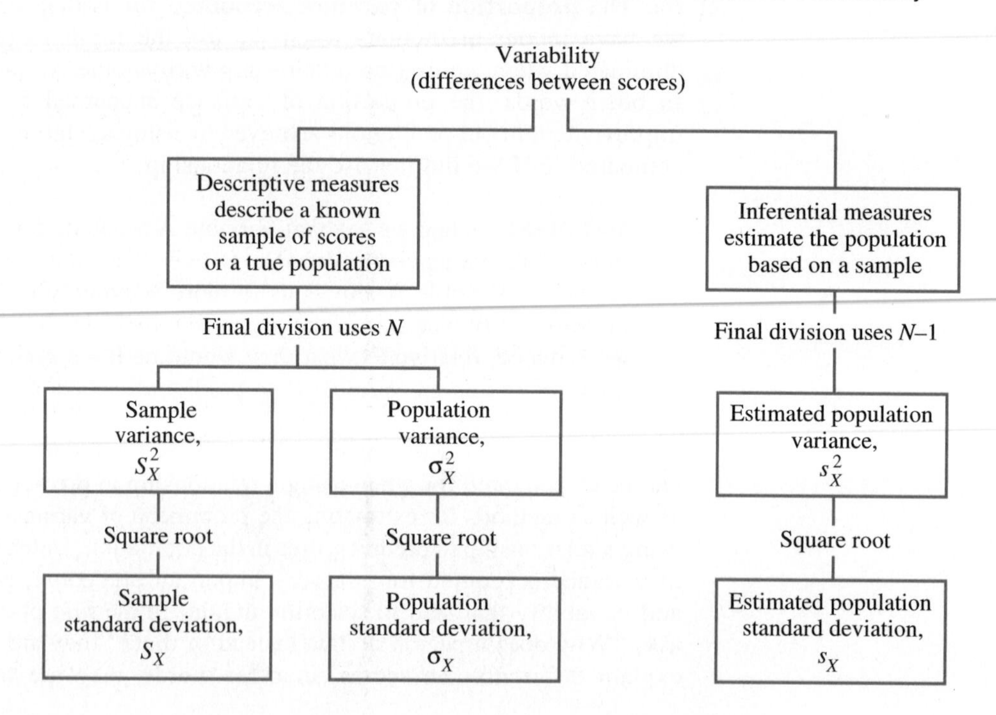

CHAPTER SUMMARY

1. To accurately describe any set of scores, we must know how different the scores are from each other. *Measures of variability* describe how much the scores differ from each other, or how much the distribution is spread out.

2. One measure of variability is the range. The *range* is the difference between the highest score and the lowest score.

3. The *semi-interquartile range* is used in conjunction with the median to describe skewed distributions. It is calculated as one-half of the distance between the scores at the 25th and 75th percentiles.

4. The *variance* is used in conjunction with the mean to describe symmetrical and normal distributions. It is the average of the squared deviations of the scores around the mean.

5. The *standard deviation* is also used in conjunction with the mean to describe symmetrical and normal distributions. It is computed as the square root of the variance. It can be thought of as indicating the "average" amount that the scores deviate from the mean.

6. If we transform scores by adding or subtracting a constant, we do not alter the standard deviation. If we transform scores by multiplying or dividing by a constant, we alter the standard deviation by the same amount as if we had multiplied or divided the original standard deviation by the constant.

7. There are three versions of the formula for variance. The version identified by S_X^2 is used to describe how far sample scores are spread out around the sample mean. The version identified by σ_X^2 describes how far the true population of scores is spread out around the population mean. The version identified by s_X^2 is computed using sample data, but it is used as the inferential, unbiased estimate of how far the scores in a population are spread out around a population mean.

8. There are three versions of the formula for the standard deviation. The version identified by S_X is used to describe how far the sample scores are spread out around the sample mean. The version identified by σ_X describes how far the true population of scores is spread out around the population mean. The version identified by s_X is computed using sample data, but it is used as the inferential, unbiased estimate of how far the scores in a population are spread out around a population mean.

9. The difference between the formulas for descriptive and inferential measures of variability is that the descriptive formulas (for S_X^2 and S_X) use N as the final denominator but the inferential formulas (for s_X^2 and s_X) use $N - 1$. The quantity $N - 1$ is known as the *degrees of freedom* in the sample.

10. When we predict the scores in a sample by predicting the mean score for each subject, the amount of error in our predictions is measured by the

sample variance, S_X^2. When we predict the scores in the population by predicting the population mean for each subject, we estimate the amount of error in our predictions using the estimated population variance, s_X^2.

11. When there is a relationship between two variables, knowing the scores on one variable allows us to more accurately predict the scores on the other variable. The improvement in prediction is described as the *proportion of variance accounted for.* It indicates the proportion of the error using the mean as the predicted score that is eliminated by using the relationship to predict scores.

PRACTICE PROBLEMS

(Answers for odd-numbered problems are provided in Appendix E.)

1. In any research study, what three characteristics of a distribution must the researcher know in order to accurately describe and summarize it?

2. What do measures of variability communicate about the following?
   a. the size of the differences between the scores in a distribution.
   b. how consistently the subjects scored.

3. *a.* What is the range?
   b. Why is the range not the most accurate summary of the variability of all scores in a distribution?

4. *a.* What do both the variance and standard deviation tell you about a distribution of scores?
   b. In most situations, which measure will you want to compute to describe a distribution? Why?

5. *a.* What is the mathematical definition of the variance?
   b. Mathematically, how is a sample's variance related to its standard deviation and vice versa?

6. *a.* What do S_X, s_X, and σ_X have in common in terms of what they communicate?
   b. How do they differ in terms of their use?

7. Why are your estimates of the population variance and standard deviation always larger than the corresponding values for describing a sample from that population?

8. You correctly compute the variance of a distribution to be $S_X^2 = 0$. What should you conclude about this distribution?

9. In a condition of a memory experiment, a researcher obtains the following error scores.

3	2	1	0	7	4	8	6	9	1
6	8	6	9	4	5	0	8	7	6

In terms of memory errors, interpret the variability of these data using the following:

a. the range

b. the variance

c. the standard deviation

10. If you could test the entire population in problem 9, what would you expect each of the following to be?

a. the typical error score

b. the variance

c. the standard deviation

d. the two scores between which we would see about 68% of all error scores in this situation

11. Say the sample in problem 9 above had an N of 1000. How many people would you expect to make fewer than about two errors? Why?

12. As part of studying the relationship between mental and physical health, you obtain the following heart rates.

73	72	67	74	78	84	79	71	76	78	76
79	81	75	80	78	76					

In terms of differences in heart rates, interpret these data using the following:

a. the range

b. the variance

c. the standard deviation

13. If you could test all possible subjects in problem 12, what would you expect each of the following to be?

a. the shape of the distribution

b. the typical heart rate

c. the variance

d. the standard deviation

e. the two scores between which about 68% of all heart rates fall

14. Foofy has a normal distribution of scores ranging from 2 to 9.

a. She has computed the variance of her data to be $-.06$. What should you conclude from this answer and why?

b. She recomputes the standard deviation to be 18. What should you conclude and why?

c. She recomputes the variance to be 1.36. What should you conclude and why?

15. From his statistics grades, Guchi has a $\overline{X}$ of 60 and $S_X = 20$. Pluto has a $\overline{X}$ of 60 and $S_X = 5$.

a. Who is the more consistent student and why?

b. Who is more accurately described as a 60 student and why?

16. *a.* For which student in problem 15 can you more accurately predict the next test score and why?

b. Who is more likely to do either extremely well or extremely poorly on the next exam?

17. On a final exam, the $\overline{X} = 65$ and $S_X = 6$. What score would you predict for each student, and if you're wrong, what do you expect will be the "average error" in your prediction?

18. The teacher who gave the test in problem 17 found a relationship between students' scores and the amount they studied. If she uses her knowledge of each student's study time to predict the corresponding exam grade, what will happen to her average error relative to the error described in problem 17?

19. If the teacher in problem 18 compares the error when using study times to predict exam scores to the error when using the mean exam score to predict exam scores, what statistical information is she computing?

20. Say that the teacher in question 19 finds that the relationship between study times and exam scores accounts for .40 of the variance in exam scores. What does this mean?

SUMMARY OF FORMULAS

1. *The formula for the range is*

$$\text{Range} = \text{highest score} - \text{lowest score}$$

2. *The formula for the semi-interquartile range is*

$$\frac{\text{Score at 75th percentile} - \text{score at 25th percentile}}{2}$$

3. *The computational formula for the sample variance is*

$$S_X^2 = \frac{\Sigma X^2 - \dfrac{(\Sigma X)^2}{N}}{N}$$

4. *The computational formula for the sample standard deviation is*

$$S_X = \sqrt{\frac{\Sigma X^2 - \frac{(\Sigma X)^2}{N}}{N}}$$

5. *The computational formula for estimating the population variance is*

$$s_X^2 = \frac{\Sigma X^2 - \frac{(\Sigma X)^2}{N}}{N - 1}$$

6. *The computational formula for estimating the population standard deviation is*

$$s_X = \sqrt{\frac{\Sigma X^2 - \frac{(\Sigma X)^2}{N}}{N - 1}}$$

6

z-Score Transformations and the Normal Curve Model

To understand the upcoming chapter:

- From Chapter 3, recall that relative frequency is the proportion of time certain scores occur, that it corresponds to the proportion of the total area under the normal curve, and that a score's percentile equals the percent of the total curve to the left of the score.
- From Chapter 4, recall that the mean score is the center of a distribution and that the larger a score's deviation from the mean, the lower the score's simple frequency and relative frequency.
- From Chapter 5, recall that S_X and σ_X indicate the "average" deviation of scores around $\overline{X}$ and μ, respectively.

Then your goals in this chapter are to learn:

- What a *z*-score is and what it tells you about a score's relative location.
- How the standard normal curve is used in conjunction with *z*-scores to determine expected relative frequency, simple frequency, and percentile.
- The characteristics of a sampling distribution of means and what the standard error of the mean is.
- How a sampling distribution of means is used with *z*-scores to determine the expected relative frequency of sample means.

The techniques discussed in the preceding chapters for graphing, measuring central tendency, and measuring variability comprise the descriptive procedures used in the vast majority of research in the behavioral sciences. If you understand these concepts, then congratulations, you are learning statistics. In this chapter we will use this knowledge to answer another question about our data: How does any one particular score compare to the other scores in a sample or population? We answer this question by performing what is known as the z-score transformation. Recall that we transform scores for two reasons: to compare scores on different variables and to make scores within the same distribution easier to work with and interpret. The z-score transformation is the Rolls-Royce of statistical transformations, because it allows us to interpret and compare scores from virtually any normal distribution of interval or ratio scores.

In the following sections we will first examine the logic of z-scores and discuss their computation. Then we will look at their various uses, both in describing individual scores and in describing sample means.

MORE STATISTICAL NOTATION

Statistics often involve negative and positive numbers. Sometimes, however, we want to consider only the size of a number, ignoring, for the moment, its sign. When we are interested in the size of a number, regardless of its sign, we are concerned with the *absolute value* of the number.

In this chapter you will encounter the symbol $\pm$, which means "plus or minus." It provides a shorthand method for describing two numbers or the range of numbers between them. If we talk about ± 1, we mean $+1$ and -1. If we describe the scores "between ± 1," we mean all possible scores from -1, through 0, up to and including $+1$.

UNDERSTANDING z-SCORES

Let's say that we conduct a study at Prunepit University in which we measure the attractiveness of a sample of male subjects. We train several judges to evaluate subjects on the variable of attractiveness, and each subject's score is the total number of points assigned by the judges. We want to analyze these attractiveness scores, especially those of three men: Slug, who scored 35; Binky, who scored 65; and Biff, who scored 90.

How do we interpret these scores? We might develop an absolute definition of attractiveness: if a subject scores above X, he is attractive. However, the problem is that we cannot easily justify such a definition. A frequent problem for behavioral scientists is that they usually do not know what a particular score, in and of itself, indicates about nature. Thus, it is difficult to know whether, in the grand scheme of things, a specific attractiveness score is high or low, good or bad, or what. Instead, like most variables in psychology, we interpret each score in *relative* terms: whether a score is good, bad, or indifferent is determined by comparing it to the other scores in the distribution. Essentially, we determine what a particular score indicates about a particular situation found in nature by comparing it to all other possible scores found in that situation. To illustrate, let's say that our sample of attractiveness scores forms the normal curve shown in Figure 6.1. By looking at the distribution and using the statistics we have learned, we can make several statements about each man's score.

What would we say to Slug? "Bad news, Slug. Your score is to the left of the mean, so you are below average in attractiveness in our sample. What's worse, you are below the mean by a large distance. Down in the tail of the distribution, the height of the curve above your score is not large, indicating a low f: not many men received this low score. The proportion of the total area under the curve at your score is also small, so the relative frequency—the proportion of all men who received your score—is small. (In terms of our parking lot approach to the normal curve, the proportion of all subjects standing at Slug's score is small.) Finally, Slug, your percentile rank is low: a small percentage of men scored below your score, while a large percentage scored above it. So, Slug, scores such as yours are relatively infrequent, and few scores are lower than yours. Relative to the others, you're fairly ugly!"

What would we tell Binky? "Binky, we have some good news and some bad news. The good news is that your score of 65 is above the mean of 60, which is also the median, or 50th percentile: you are better looking than more than 50% of the men in our study. The bad news is that your score is not *that* far above the mean. The area under the curve at your score is relatively large, and thus the relative frequency of your score is large: this means that the proportion of equally attractive men is large. What's worse, there is a relatively large part of the distribution with higher scores."

And then there is Biff. "Yes, Biff, as you expected, you are above average in attractiveness. In fact, as you have repeatedly told everyone, you are one of the most attractive men around. The area under the curve at your score, and thus the relative frequency of your score, is quite small, meaning that only a small proportion of the men are equally attractive. Also, the area under the curve to the left of your score is relatively large. This means that if we cared to figure it out,

FIGURE 6.1 Frequency Distribution of Attractiveness Scores at Prunepit U

Scores for three individual subjects are identified on the X axis.

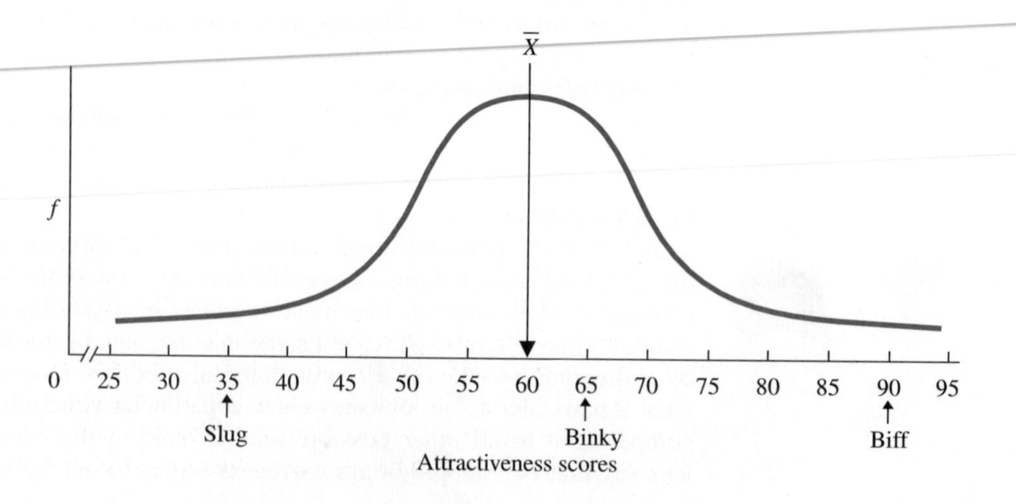

we'd find that you scored at a very high percentile rank: a large percentage of scores are below your score, while a small percentage are above your score."

Recognize that, although Biff is better looking than most men in our study, he may in fact have a face so ugly that it could stop a clock! Or Slug might be considered highly attractive at some other school. However, since we have defined attractive and ugly relative to this distribution, we have described each score's relative standing. The **relative standing** of a score reflects a systematic evaluation of the score relative to the characteristics of the sample or population in which the score occurs.

Notice that, in describing each man's relative standing above, we first determined his score's location relative to the mean. Based on this information, we then determined other measures of relative standing, such as the score's relative frequency and percentile. Thus, the essence of determining a score's relative standing is to determine how far the score is above or below the mean.

Describing a Score's Relative Location as a *z*-Score

In previous chapters we saw that the distance between a raw score and the mean is the amount that the score deviates from the mean, which equals $(X - \overline{X})$. To quantify a score's relative standing, we begin by computing the score's deviation. For example, Biff's raw score of 90 deviates from the mean of 60 by +30 (90 − 60 = +30). (The + sign indicates that he is above the mean.) A deviation of +30 *sounds* as if it might be a large deviation, but is it? As with a raw score, we do not necessarily know whether a particular deviation score should be considered large or small. Therefore we need a frame of reference, so we examine the relative standing of the deviation score. When we examine the entire distribution, we see that only a few scores deviate by as much as Biff's score, and that is what makes his score an impressively high score. Similarly, Slug's score of 35 deviates from the mean of 60 by −25 (35 − 60 = −25). This too is impressive, because only a few scores deviate below the mean by such an amount. Thus, a score is impressive if it is far from the mean, and "far" is determined by how frequently other scores deviate from the mean by that same distance.

To interpret a score's location, then, we need a way to compare its deviation to all deviations. For Biff, we need to quantify whether his deviation of +30 is impressive relative to all the deviations in the sample. To do this, we need a *standard* to compare to Biff's *deviation:* we need a standard deviation. As we saw in Chapter 5, calculating the standard deviation is our way of computing the average deviation of the scores around the mean. By comparing a score's deviation to the standard deviation, we can describe the location of an individual score in terms of the "average" deviation.

For example, say that for the data from Prunepit U., the sample standard deviation is 10 (10 attractiveness points). Biff's deviation of +30 attractiveness points is equivalent to 3 standard deviations: 30/10 = 3. Thus, another way to describe Biff's raw score is to say that it is located 3 standard deviations above the mean. We have simply described Biff's score in terms of its distance from the mean, measured in standard deviation units. We do the same type of thing when we convert inches to feet. In that case, the unit of measurement called a foot is defined as 12 inches, so a distance of 36 inches is equal to 3 of those units, or 3 feet. In our study, the unit of measurement called a standard deviation is defined as 10 attractiveness points. Since Biff's deviation from the mean is 30 attractiveness points, Biff's score is a distance of 3 of those units, or 3 standard deviations, from the mean.

By transforming Biff's deviation into standard deviation units, we have performed a *z*-score transformation and computed Biff's *z*-score. A ***z*-score** is the distance a raw score deviates from the mean when measured in standard deviations. Biff's *z*-score of +3.00 provides us with one number that summarizes his relative standing: Biff's raw score deviates from the mean by an amount that is three times the "average" amount that all scores in the sample deviate from the mean. Since his *z*-score is positive, he is 3 standard deviations above the mean.

Computing *z*-Scores

The symbol for a *z*-score is *z*. Above, we performed two mathematical steps in computing Biff's *z*. First we found the score's deviation by subtracting the mean from the raw score. Then we divided the score's deviation by the standard deviation.

THE FORMULA FOR TRANSFORMING A RAW SCORE IN A SAMPLE INTO A z-SCORE IS

$$z = \frac{X - \overline{X}}{S_X}$$

Technically, this is both the definitional and computational formula for *z*. If we are starting from scratch with a sample of raw scores, we first compute $\overline{X}$ and S_X and then substitute their values into the formula. Notice that we are computing a *z*-score from a known sample of scores, so we use the descriptive sample standard deviation, S_X (the formula in Chapter 5 that involves dividing by *N*, not $N - 1$).

To find Biff's *z*-score, we substitute into the above formula his raw score of 90, the $\overline{X}$ of 60, and the S_X of 10:

$$z = \frac{X - \overline{X}}{S_X} = \frac{90 - 60}{10}$$

We find the deviation in the numerator first, so we subtract 60 from 90 (always subtract $\overline{X}$ from *X*). Rewriting the formula gives

$$z = \frac{+30}{10}$$

We then divide, to find that

$$z = +3.00$$

Likewise, Binky's raw score of 65 produces a z-score of

$$z = \frac{X - \overline{X}}{S_X} = \frac{65 - 60}{10} = \frac{+5}{10} = +0.50$$

Binky's raw score is literally one-half of 1 standard deviation above the mean.

And finally, Slug's raw score is 35, so his z is

$$z = \frac{X - \overline{X}}{S_X} = \frac{35 - 60}{10} = \frac{-25}{10} = -2.50$$

Here 35 minus 60 results in a deviation of *minus* 25, which, when divided by 10, results in a z-score of -2.50. Slug's raw score is 2.5 standard deviations *below* the mean. In working with z-scores, always pay close attention to the positive or negative sign: it is part of the answer.

Usually our purpose is to describe the relative standing of a subject in a sample, so we use the previous formula. However, we can also compute a z-score for a score in a population, if we know the population mean, μ, and the true standard deviation of the population, σ_X.

THE FORMULA FOR TRANSFORMING A RAW SCORE IN A POPULATION INTO A z-SCORE IS

$$z = \frac{X - \mu}{\sigma_X}$$

This formula is identical to the previous formula except that now the answer indicates how far the raw score lies from the population mean, measured in units of the true population standard deviation. (Note that we do not compute z-scores using the estimated population standard deviation, s_X.)

Computing a Raw Score When z Is Known

Sometimes we know a z-score and want to find the corresponding raw score. For example, in our study at Prunepit U. (with $\overline{X} = 60$ and $S_X = 10$), say that another student, Bucky, scored $z = +1.0$. What is his raw score? His z-score indicates that he is 1 standard deviation above the mean, or, in other words, 10 points above 60, so he has a raw score of 70. What did we just do? We multiplied his z-score times the value of S_X and then added the mean.

THE FORMULA FOR TRANSFORMING A z-SCORE IN A SAMPLE INTO A RAW SCORE IS

$$X = (z)(S_X) + \overline{X}$$

Substituting values into the formula to transform Bucky's z-score of $+1.0$, we have

$$X = (+1.0)(10) + 60$$

so

$$X = +10 + 60$$

so

$$X = 70$$

To check this answer, compute the z-score for the raw score of 70. You should end up with the z-score you started with: $+1.0$.

Say that Fuzzy has a negative z-score of $z = -1.3$ (with $\overline{X} = 60$ and $S_X = 10$). Then

$$X = (-1.3)(10) + 60$$

so

$$X = -13 + 60$$

Adding a negative number is the same as subtracting its positive value, so

$$X = 47$$

Fuzzy has a raw score of 47.

After transforming a raw score or z-score, always determine whether your answer makes sense. At the very least, negative z-scores must correspond to raw scores smaller than the mean, and positive z-scores must correspond to raw scores larger than the mean. Further, as we shall see, we seldom obtain z-scores greater than ± 3.00 (plus or minus 3.00), although they are possible. Be very skeptical if you compute a z-score of greater than ± 3.00, and double-check your work.

How Variability Influences *z*-Scores

It is important to recognize that the size of a particular z-score depends on both the amount that the raw score deviates from the mean *and* on the variability (the standard deviation) in the distribution. For example, Biff's deviation of $+30$ produced a z-score of $+3.00$ when the standard deviation was 10. Such a standard deviation indicates that most scores were relatively close to the mean of 60, so a deviation of $+30$ is unusually large, and thus impressive. If, however, the data had produced a standard deviation of 30, then Biff's z-score would be $z = (90 - 60)/30 = +1.0$. Here the scores are more inconsistent relative to the mean, so Biff's deviation would equal the "average" deviation, indicating that his raw score is among the more frequent scores. In this case Biff's score would not be as impressive.

Thus, bear in mind that two factors produce a z-score having a large absolute value: (1) a large deviation from the mean and (2) a small standard deviation.

> ***STAT ALERT*** The magnitude of a z-score depends on the size of the raw score's deviation and the size of the standard deviation.

INTERPRETING *z*-SCORES: THE *z*-DISTRIBUTION

Creating a *z*-Distribution

The easiest way to interpret scores is to create a *z*-distribution. A **z-distribution** is the distribution produced by transforming all raw scores in a distribution into *z*-scores. By transforming all the attractiveness scores in our study into *z*-scores, we get the *z*-distribution shown in Figure 6.2.

Notice the three ways the *X*-axis is labeled. This shows that by creating a *z*-distribution, we have only transformed the way in which we identify each score. Saying that Biff has a *z* of +3 is merely another way to say that he has a deviation of +30, or a raw score of 90. And recognize this very important fact: because we are still looking at the same point on the distribution, Biff's *z*-score of +3.00 and his deviation of +30 have the same frequency, relative frequency, and percentile rank as his raw score of 90.

The advantage of looking at *z*-scores, however, is that they directly communicate each score's relative location in the distribution. The *z*-score of 0.0 here corresponds to the mean raw score of 60: a subject having the mean score is zero distance from the mean. For any other score, think of a *z*-score as having two parts: the number and the sign in front of it. The number indicates the absolute distance the score is from the mean, measured in standard deviations. The sign indicates the *direction* the score lies in relation to the mean. A + indicates that the score is above and graphed to the right of the mean. A − indicates that the score is below and graphed to the left of the mean. From this perspective,

FIGURE 6.2 *z*-Distribution of Attractiveness Scores at Prunepit U

The labels on the X axis show first the raw scores, then the deviations, and then the z-scores.

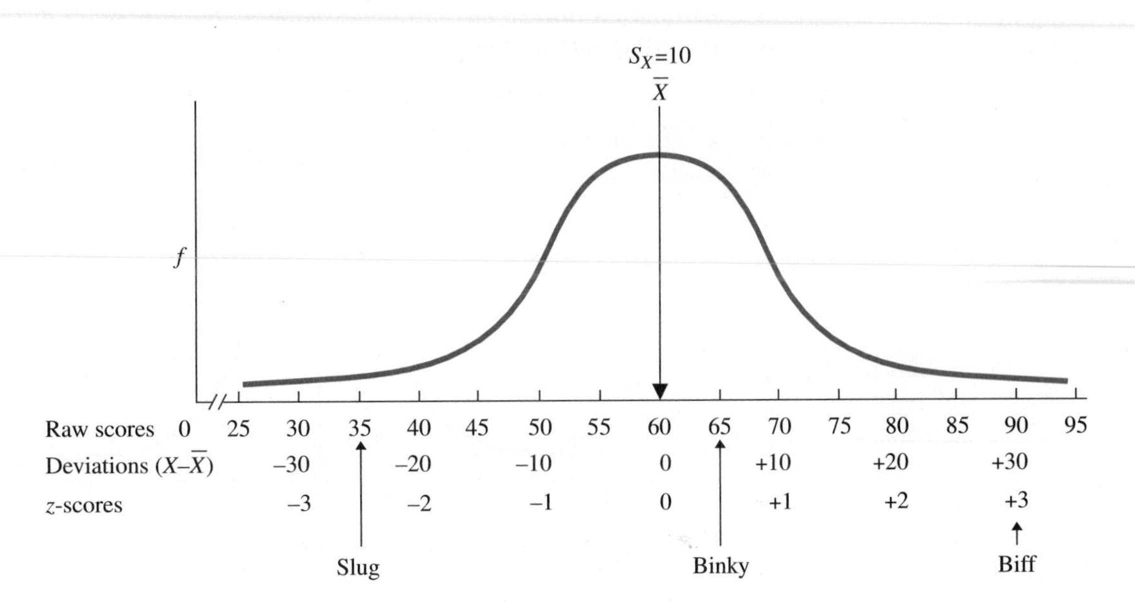

z-scores become increasingly larger numbers with a positive sign as we proceed to raw scores farther above and to the right of the mean. Conversely, z-scores become increasingly larger numbers with a negative sign as we proceed to raw scores farther below and to the left of the mean. However, remember that the farther below the mean the raw score is, the lower its value. Thus, for example, a z-score of -2.0 corresponds to a *lower* raw score than does a z-score of -1.0.

> ***STAT ALERT*** The farther a raw score is from the mean, the larger its corresponding z-score. On a normal distribution, the larger the z-score, whether positive or negative, the less frequently that z-score and the corresponding raw score occur.

It is important to recognize that a negative z-score is not automatically a bad score. How we interpret z-scores depends on the nature of the variable we have measured. For some variables, the goal is to have as low a raw score as possible (errors on a test, number of parking tickets, amount owed on a credit card bill). With these variables, negative z-scores are best. For example, say we measure the number of times that subjects are depressed during a one-month period, and we find that the $\bar{X}$ is 5 and the S_X is 1.0. Figure 6.3 shows the z-distribution for the sample. Ideally, we'd like never to be depressed, so the goal is to have the score that is the greatest distance *below* the mean. Thus, a z-score of -3.0 would be good because it corresponds to only 2 episodes of depression. Conversely, a z-score of $+3.0$ would not be good, because it corresponds to 8 episodes of depression.

Characteristics of the *z*-Distribution

The previous examples illustrate three important characteristics of any z-distribution.

1. *A z-distribution always has the same shape as the raw score distribution.* The preceding distributions form normal distributions only because the raw

FIGURE 6.3 z-Distribution of Number of Episodes of Depression and the Corresponding z-Scores

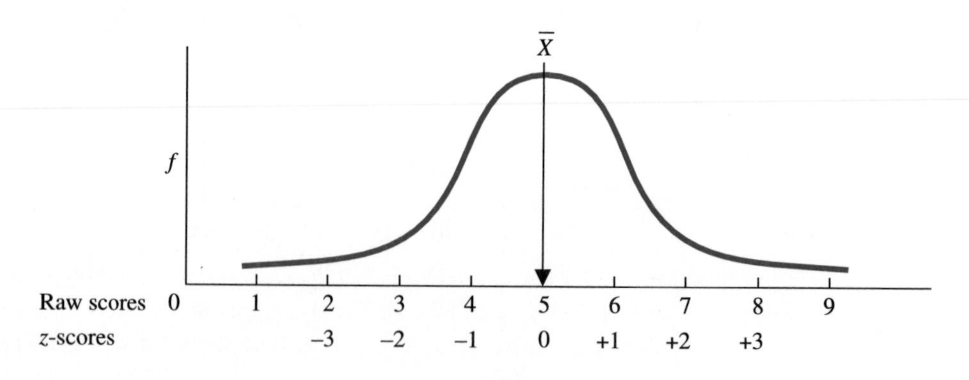

scores form normal distributions. Transforming a nonnormal distribution into z-scores will *not* "normalize" the distribution, making it form a normal curve.

2. *The mean of any z-distribution always equals 0.* The mean of the raw scores transforms into a z-score of 0. Or, if we computed the mean of all z-scores, the result would also be 0: the sum of the positive and negative z-scores is the sum of the deviations around the mean, and this sum always equals 0, so the mean z-score is 0.

3. *The standard deviation of any z-distribution always equals 1.* This is because 1 standard deviation unit for raw scores transforms into 1 z-score unit. Whether the standard deviation in the raw scores is 10.0 (as in Figure 6.2) or 1.0 (as in Figure 6.3), it is still 1 standard deviation, and 1 standard deviation is 1 z-score unit.

Because of these characteristics, all normal z-distributions are similar. It is very important that you recognize that any particular z-score will be at the same relative location on *any* normal distribution. Look again at Figures 6.2 and 6.3. Recall from Chapter 5 that the mathematical relationship between the standard deviation and the normal curve dictates that the raw scores that are $\pm 1\ S_X$ from the mean are always directly below the inflection points of a normal curve. Because z-scores of ± 1.0 are $\pm 1\ S_X$ from the mean, they are always below the inflection points too. Likewise, a z of ± 2 (2 S_X's from the mean) will always be about halfway to the tail of the distribution. And z of ± 3 will always be located in the extreme tail.

Now here's the nifty part: any raw score that produces a particular z-score will always have the same relative location within its raw score distribution. Thus, for example, a raw score on any normal distribution that produces a z of $+3.0$ will, like Biff's, be a relatively infrequent score, located at the extreme high end. Or, a raw score on any distribution that produces a z of $+.50$ will, like Binky's, be a relatively frequent score located close to the mean. In previous chapters we've seen how statistical procedures allow us to envision a distribution and the location of particular scores within the distribution. By transforming a raw score to a z-score and envisioning the z-distribution, we can identify the location of any raw score in *any* distribution.

STAT ALERT A z-score allows us to *locate* the corresponding raw score on any normal distribution.

Transforming z-Scores

Sometimes we must communicate the information in a z-distribution to people who are statistically unsophisticated, so we can transform the z-scores into numbers that are easier to understand. The advantage of such transformations is that they eliminate negative scores and reduce the number of decimal places.

One such transformation of z-scores occurs with college entrance exams such as the Scholastic Aptitude Test (SAT). Essentially, a grade on the SAT is a z-score that is then transformed using the formula

$$\text{SAT score} = z(100) + 500$$

In z-scores, the mean is 0, so in SAT scores, the mean is 500. Similarly, the standard deviation of z is 1, which becomes 100. You may have heard that the highest possible SAT score is 800. This is because a score of 800 corresponds to a z of $+3$. Since z-scores beyond $+3$ are so infrequent, for simplicity these scores are rounded to $+3.00$, or 800. Likewise, the lowest score is 200.

One general way of transforming z-scores is to produce T-scores.

THE FORMULA FOR A T-SCORE IS

$$T = z(10) + 50$$

For example, $z = -1.30$ becomes $T = -13.0 + 50$, which is 37. Multiplying z times 10 eliminates one decimal place, and adding 50 eliminates the negative sign of any z up to -5.00. The mean, which is 0 in z-scores, becomes a T of 50, and the standard deviation, which is 1 in z-scores, becomes 10. Thus, T-scores range between 0 and 100, with a mean of 50 and a standard deviation of 10. We interpret such transformed scores in the same way that we interpret their corresponding z-scores.

Regardless of whether it is transformed, the z-distribution provides us with a very useful statistical tool. As we shall see, we can use a z-distribution to:

1. Make comparisons between scores from different distributions.
2. Determine the relative frequency of raw scores within any distribution.
3. Define psychological attributes and categories.

USING THE *z*-DISTRIBUTION TO COMPARE DIFFERENT DISTRIBUTIONS

In research, comparing a score on one variable to a score on another variable may be a problem, because the two variables may not be comparable. For example, say that Althea received a grade of 38 on a statistics quiz and a grade of 45 on an English paper. If we say that she did better in English, we are comparing apples to oranges. These grades reflect scores on different kinds of variables, assigned by different instructors using different criteria. We can avoid this problem by transforming the raw scores from each class into z-scores. This gives us two z-distributions, each with a mean of 0, a standard deviation of 1, and a range of between about -3 and $+3$. Each z-score indicates an individual's relative standing in his or her respective class. Therefore, we can compare Althea's relative standing

FIGURE 6.4 Comparison of Two Distributions for Statistics and English Grades, Showing Raw Scores and z-Scores

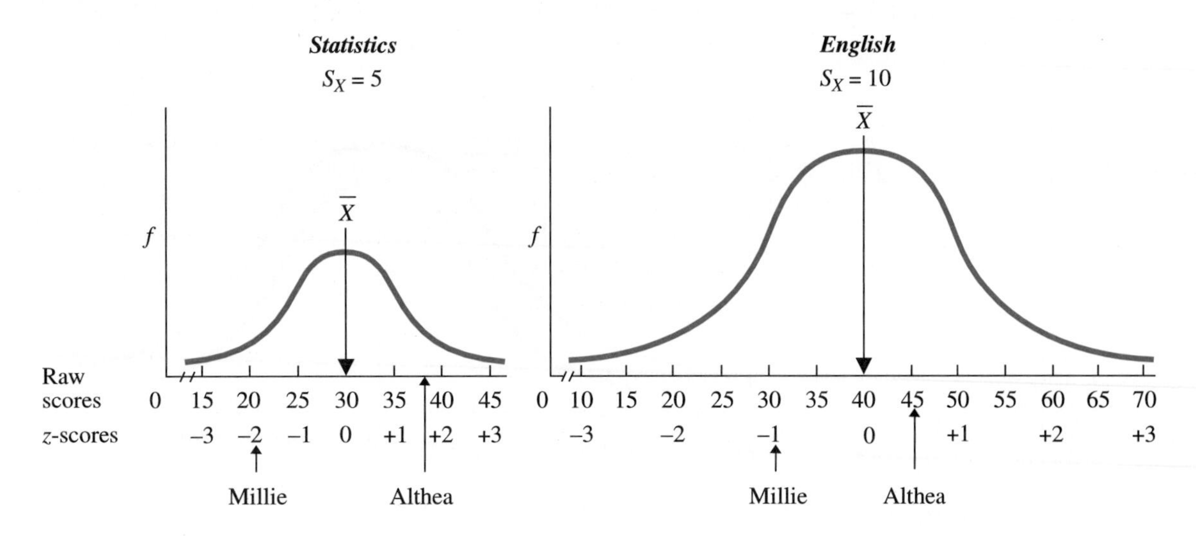

in English to her relative standing in statistics, and we are no longer comparing apples and oranges. Note: With the z-transformation, we equate, or standardize, the distributions. For this reason, z-scores are often referred to as **standard scores.**

Say that for the statistics quiz, the $\overline{X}$ was 30 and the S_X was 5. We transform all of the statistics grades to z-scores, including Althea's grade of 38, which becomes $z = +1.60$. For the English paper, the $\overline{X}$ was 40 and the S_X was 10, so Althea's grade of 45 becomes $z = +.50$. (Did you get the same answers?) Figure 6.4 shows the locations of Althea's z-scores on their respective z-distributions. A z-score of $+1.60$ is farther above the mean than a z-score of $+.50$. Thus, in terms of her relative standing in each class, Althea did better in statistics, because she is farther above the statistics mean than she is above the English mean.

Another student, Millie, obtained raw scores that produced $z = -2.00$ in statistics and $z = -1.00$ in English. In which class did Millie do better? Millie's z-score of -1 in English is better, because it is less distance below the mean.

Of course, it would be easier to compare these two distributions if we plotted them on the same set of axes, and z-scores enable us to do just that.

Plotting Different z-Distributions on the Same Graph

By transforming both the statistics and English scores into z-scores, we establish a common variable. Then, to see each student's relative location in each class, we can graph both of the previous distributions on one set of axes, as shown in Figure 6.5. As we've noted, all normal z-distributions are similar, so there are only two minor differences between the curves. First, the classes produced different standard deviations, so the raw scores for each class are spaced differently along the X-axis. (For example, going from the z-score of $+1$ to $+2$ corresponds to going from the raw scores of 35 to 40 in statistics, but from 50 to 60 in English.) Second, the greater height of the English distribution merely reflects a

FIGURE 6.5 Comparison of Distributions for Statistics and English Grades, Plotted on the Same Set of Axes

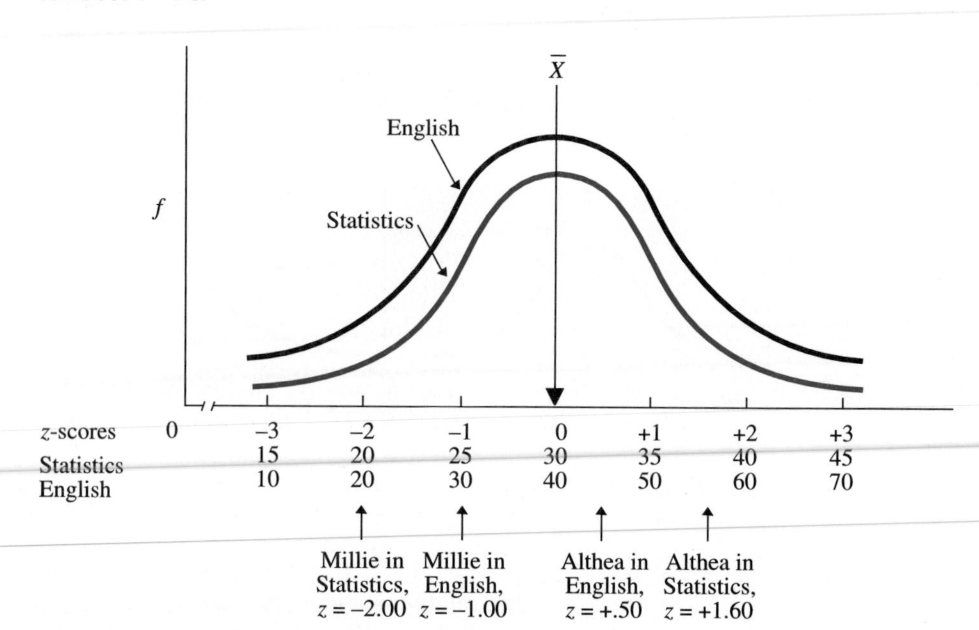

larger f for each score. Overall, the English class simply had a larger N. If the two raw score distributions had contained the same N, the curves would be identical.

By plotting the two distributions on the same set of axes, we can easily compare anyone's scores. Clearly, Althea scored better in statistics than in English, and Millie scored better in English than in statistics.

The fact that different z-distributions can be plotted on the same set of axes leads us to our second use of the z-distribution: determining the relative frequency of raw scores in any normal distribution.

USING THE *z*-DISTRIBUTION TO DETERMINE THE RELATIVE FREQUENCY OF RAW SCORES

The two z-distributions in Figure 6.5 illustrate a critical point:

> **Even though different normal distributions may not contain an equal *N*, the relative frequency of a particular *z*-score will be the same on all normal *z*-distributions.**

Recall that relative frequency is the proportion of time that a score occurs, and it can be computed as the proportion of the total area under the curve. For example, you know from previous chapters that 50% of all scores on a normal curve are to the left of the mean score. However, now you know that scores to the left of the mean produce negative z-scores, so in other words negative z-scores make up 50% of a distribution. Thus, regardless of their raw scores, the students in each class with the negative z-scores in Figure 6.5 constitute 50% of their respective distributions. Further, 50% of a distribution corresponds to a relative frequency of .50. On *any* normal z-distribution, the total relative frequency of the negative z-scores is .50.

Having determined the relative frequency of the z-scores, we can work backwards to find the relative frequency of the corresponding raw scores. In the statistics distribution in Figure 6.5, those students having negative z-scores have raw scores ranging between 15 and 30, so the total relative frequency of 15 to 30 is .50. Similarly, in the English distribution, those students having negative z-scores have raw scores ranging between 10 and 40, so the relative frequency of 10 to 40 is .50.

Similarly, recall from Chapter 5 that approximately 68% of all scores on a normal distribution fall between the score that is 1 standard deviation below the mean and the score that is 1 standard deviation above the mean. However, now you know that raw scores that are ± 1 standard deviation from the mean produce z-scores of ± 1.0, respectively. Thus, in Figure 6.5, students with z-scores between ± 1.0 constitute approximately 68% of their distributions. On *any* normal z-distribution, the relative frequency of z-scores between ± 1.0 is approximately .68. Having determined this, we can again work backwards to the raw scores. We see that statistics grades between 25 and 35 constitute approximately 68% of the statistics distribution, and English grades between 30 and 50 constitute approximately 68% of the English distribution.

In the same way, we can determine the relative frequencies within any set of scores once we envision it as a z-distribution. For example, in a normal distribution of IQ scores (whatever the $\overline{X}$ and S_X may be) we know that those scores producing negative z-scores have a *rel. f* of .50, and about 68% of all the IQ scores fall between the two scores corresponding to the z-scores of ± 1.0. The same will also hold true for a distribution of running speeds, or a distribution of personality test scores, or for *any* normal distribution.

We can also use z-scores to determine the relative frequency of scores in any other portion of a distribution. To do so, we employ the standard normal curve.

The Standard Normal Curve

Because the relative frequency of a given z-score is always the same on any normal z-distribution, we conceptualize all normal z-distributions as conforming to one standard curve. In fact, this curve is called the standard normal curve. The **standard normal curve** is a theoretical perfect normal curve, which serves as a model of the perfect normal z-distribution. (Since it is a z-distribution, the mean of the standard normal curve is 0, and the standard deviation is 1.0.)

We can use the standard normal curve to determine the relative frequency of any particular z-scores on a perfect normal curve. As we saw above, once we know the relative frequency of certain z-scores, we can work backwards to determine the relative frequency of the corresponding raw scores. Thus, the first step is to find the relative frequency of the z-scores. To do this, we look at the area under the standard normal curve. Statisticians have determined the proportion of the area under various parts of the normal curve. Look at Figure 6.6. Above the X axis, the numbers between the vertical lines indicate the proportions of the total area between z-scores. Below the X axis, the number on each arrow indicates the proportion of the total area between the mean and the indicated z-score.

The proportion of the total area under the curve is the same as relative frequency, so each proportion is the relative frequency of the z-scores located in that section of the ideal normal curve. Thus, on a perfect normal distribution, .3413 of the z-scores are located between $z = 0.00$ and $z = +1.00$. Since we can express a proportion as a percent by multiplying the proportion times 100, we can say that 34.13% of all z-scores fall between $z = 0.00$ and $z = +1.00$. Similarly, z-scores between $+1.00$ and $+2.00$ occur 13.59% of the time, and z-scores between $+2.00$ and $+3.00$ occur 2.15% of the time. Because the distribution is symmetrical, the same proportions occur between the mean and the corresponding negative z-scores.

To determine the relative frequency for larger areas, we add the above proportions. For example, .3413 of the distribution is located between $z = -1.00$ and the mean, and .3413 of the distribution is between the mean and $z = +1.00$. Thus, a total of .6826, or 68.26%, of the distribution is located between $z = -1.00$ and $z = +1.00$ (see, about 68% of the distribution really is between $\pm 1S_X$ from the mean). Likewise, we can add together nonadjacent portions of the curve. For example, .0228, or 2.28%, of the distribution is in the tail of the distribution beyond $z = -2.00$, and 2.28% is beyond $z = +2.00$. Thus, a total of 4.56% of all scores falls in the tails beyond $z = \pm 2.00$. (Don't worry: you won't need to memorize these proportions.)

Figure 6.6 shows why we seldom obtain z-scores greater than ± 3. This is a theoretical distribution representing an infinitely large distribution (the tails of

FIGURE 6.6 Proportions of Total Area Under the Standard Normal Curve

The curve is symmetrical: 50% of the scores fall below the mean, and 50% fall above the mean.

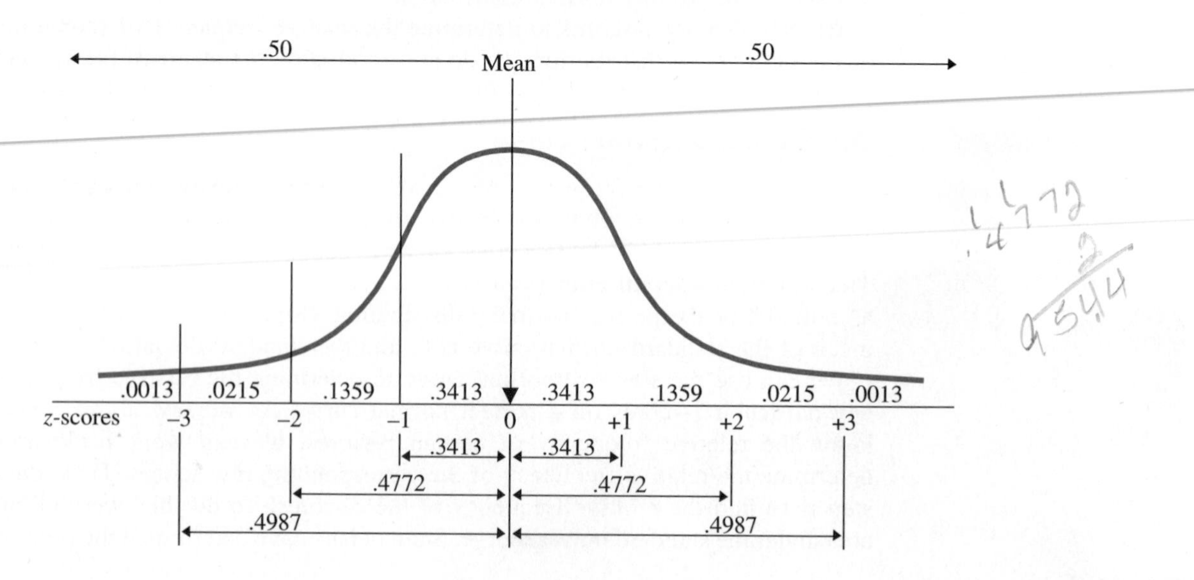

the polygon never touch the X axis). Even so, only .0013 of the scores are above $z = +3$ and only .0013 are below $z = -3.00$. In total, only .0026, or .26 of *1* percent, of the scores are beyond ± 3.00. This leaves 99.74% of all scores (100% − .26% = 99.74%) located between $z = \pm 3$. Thus, for all practical purposes, the range of z is between ± 3. Also, now you can see why, in Chapter 5, I said that for normally distributed scores, the value of S_X should be about one-sixth of the range of the raw scores. The range of the raw scores is approximately between $z = -3$ and $z = +3$, a distance of six times the standard deviation. If the range is six times the standard deviation, then the standard deviation is one-sixth of the range.

Applying the Standard Normal Curve Model

By using the standard normal curve, we can quickly determine the relative frequency of any score or scores in any normally distributed sample or population. This is true even though real data will not form a perfect normal curve. As we have seen in previous chapters, the normal curve model assumes that most populations of raw scores "more or less" form a normal distribution and that most samples represent such a distribution. When we conceptualize this model, however, we do not draw a "more or less" normal curve: we draw the ideal perfect normal curve. We then use this curve as a model of our distribution of raw scores, operating as if the scores form a perfect normal curve. If we operate as if the raw scores form a perfect normal curve, then when we transform the raw scores to z-scores, the z-distribution would also form a perfect normal curve. As we have seen, the perfect normal z-distribution is the standard normal curve.

> ***STAT ALERT*** The standard normal curve is our model for any roughly normal distribution when transformed to z-scores.

We can use this model to determine, for example, the relative frequency of scores between the mean and any raw score. We first transform the raw score into a z-score. Then, from the standard normal curve, we determine the proportion of the total area between the mean and this z. This proportion is the same as the relative frequency of the z-scores between the mean and this z on a perfect normal curve. The relative frequency of these z-scores is the same as the relative frequency of their corresponding raw scores in a perfect normal distribution. Thus, the relative frequency we obtain from the standard normal curve is the *expected* relative frequency of the raw scores in our data, if the data were to form a perfect normal distribution.

How accurately the expected relative frequency from the model describes our actual scores depends on three aspects of the data:

1. The closer the raw scores are to forming a normal distribution, the more accurately the model describes the data. Therefore, the standard normal curve model is appropriate *only* if we can assume that our data are at least approximately normally distributed.

2. The larger the sample N, the more closely the sample tends to fit the normal curve and therefore the more accurate the model will be. The model is most accurate when applied to very large samples or to populations.
3. The model is most appropriate if the raw scores are theoretically continuous scores (which can include decimals) measured using a ratio or interval scale (which has equal intervals between scores, and measures actual amounts).

Our original sample of attractiveness scores from Prunepit U. meets the above requirements, so we can apply the standard normal curve model to these data. Say that Cubby has a raw score of 80, which, with $\overline{X} = 60$ and $S_X = 10$, is a z of $+2.0$. We can show Cubby's location on the distribution as illustrated in Figure 6.7. We might first ask what proportion of scores are expected to fall between the mean and Cubby's score. From the standard normal curve we see that .4772 of the total area falls between the mean and $z = +2.00$. Because .4772 of all z-scores fall between the mean and a z of $+2$, we expect .4772, or 47.72%, of all attractiveness scores at Prunepit U. to fall between the mean score of 60 and Cubby's score of 80.

We might also ask how many people scored between the mean and Cubby's score. Then we would convert the above relative frequency to simple frequency by multiplying the N of the sample times the relative frequency. Say that the sample N at Prunepit was 1000. If we expect .4772 of all scores to fall between the mean and $z = +2$, then $(.4772)(1000) = 477.2$, so we expect about 477 scores to fall between the mean and Cubby's raw score of 80.

Finding percentile rank for a raw score Looking again at Figure 6.7, you'll see that we can also use the standard normal curve model to determine Cubby's

FIGURE 6.7 Location of Cubby's Score on the z-Distribution of Attractiveness Scores

Cubby is at approximately the 98th percentile.

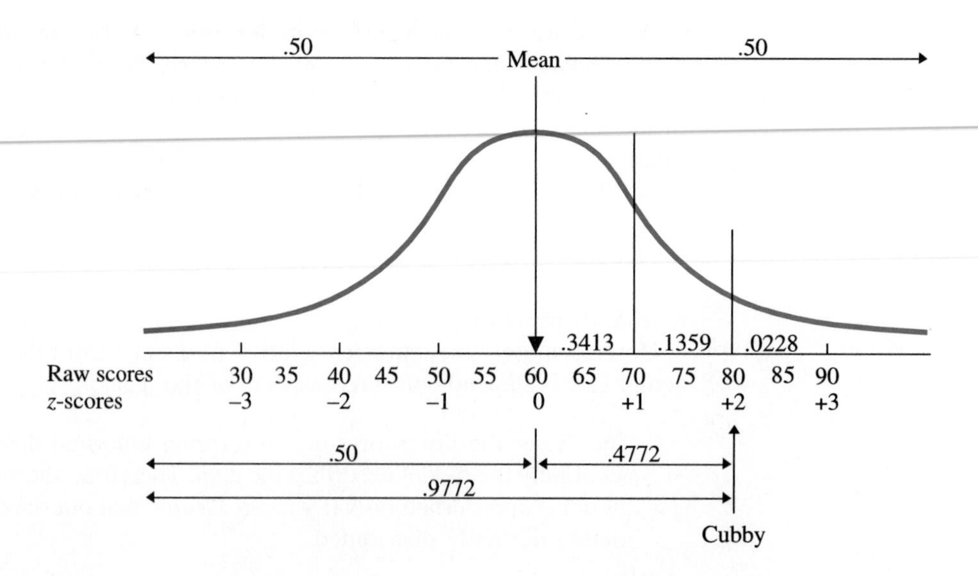

expected percentile. Recall that a percentile is the percent of all scores *below* (graphed to the left of) a score. On a normal distribution, the mean is the median (the 50th percentile). Any positive z-score is above the mean, so Cubby's z-score of $+2$ is above the 50th percentile. In addition, as Figure 6.7 shows, Cubby's score is above the 47.72% of the scores that fall between the mean and his z-score. Thus, we add the 50% of the scores below the mean to the 47.72% of the scores between the mean and his score. In total, 97.72% of all z-scores are below Cubby's z-score. We usually round off percentile to a whole number, so we conclude that Cubby's z-score is at the 98th percentile. Likewise, Cubby's raw score of 80 is expected to be at the 98th percentile. Conversely, if 97.72% of the curve is below $z = +2.00$, then, as shown in Figure 6.7, only 2.28% of the curve is above $z = +2.00$ (100% − 97.72% = 2.28%). Thus, anyone scoring above $z = +2.00$, or the raw score of 80 would be in about the top 2% of all scores.

On the other hand, say that Elvis obtained an attractiveness score of 40, producing a z-score of -2.00. We can find Elvis's percentile using Figure 6.8. Because .0215 of the distribution is between $z = -2$ and $z = -3$, and .0013 of the distribution is below $z = -3$, there is a total of .0228, or 2.28%, of the distribution below (to the left of) Elvis's score. With rounding, Elvis ranks at the 2nd percentile.

Finding a raw score at a given percentile rank We can also use the standard normal curve model to find a raw score at a particular relative frequency or percentile. Say that we want to find the raw score at the 16th percentile. Because the 16th percentile is below the 50th percentile, we are looking for a negative

FIGURE 6.8 Location of Elvis's Score on the z-Distribution of Attractiveness Scores

Elvis is at approximately the 2nd percentile.

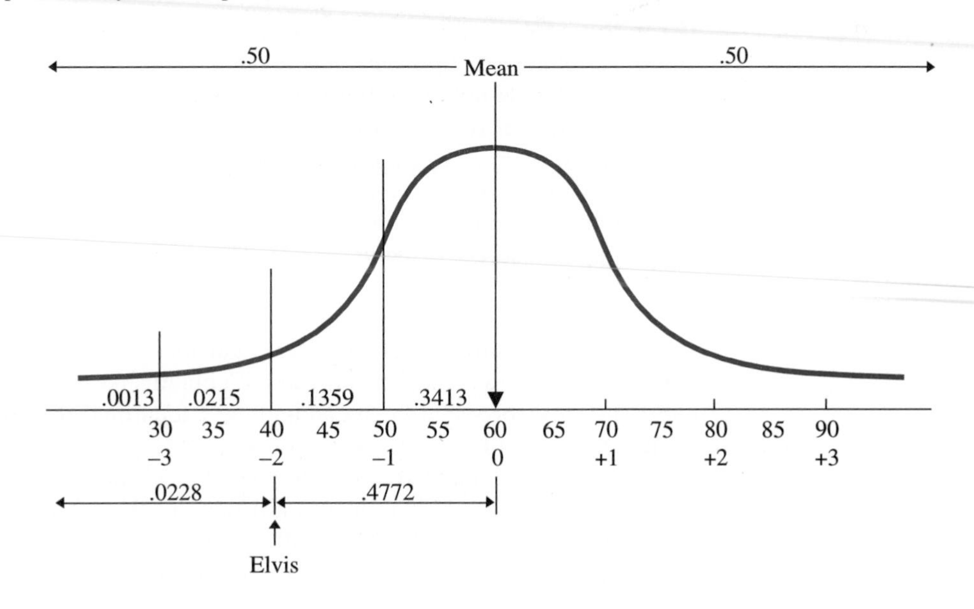

FIGURE 6.9 Proportions of the Standard Normal Curve at Approximately the 16th Percentile

The 16th percentile corresponds to a z-score of about −1.0.

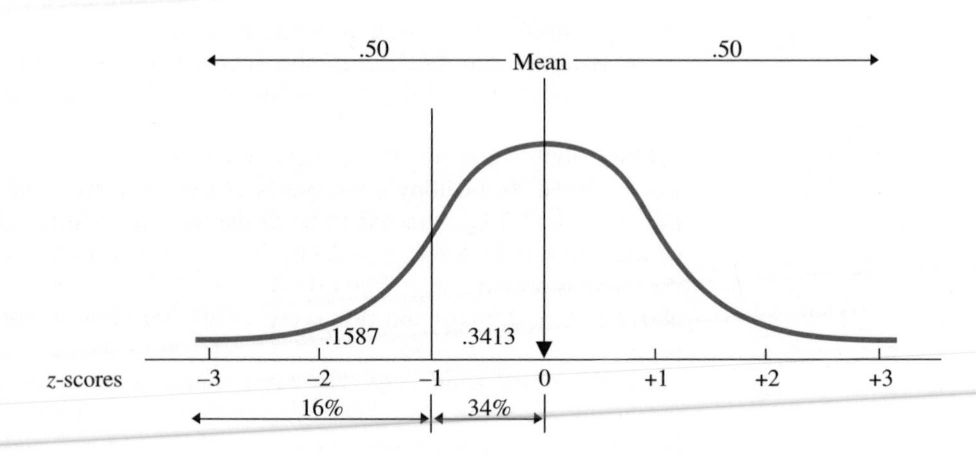

z-score. Consider Figure 6.9. If 16% of all scores are below the unknown score, then 34% of all scores are between it and the mean (50% − 16% = 34%). We saw that 34.13% of a normal z-distribution is between $z = -1$ and the mean, so that leaves 15.87% of the distribution to the left of a z of -1.0 (50% − 34.13% = 15.87%). Thus, with rounding, $z = -1.0$ is at approximately the 16th percentile.

We then use the formula $X = (z)(S_X) + \overline{X}$ to find the raw score at $z = -1.0$. With $\overline{X} = 60$ and $S_X = 10$ for this sample, we have $X = (-1.0)(10) + 60 = 50$. Thus, the raw score of 50 is expected to be at approximately the 16th percentile.

Using the z-table In the above examples, we rounded off the z-scores to keep things simple. With real data, however, we do not round z-scores containing decimals to whole numbers. Further, fractions of z-scores do *not* result in proportional divisions of the corresponding area. For example, even though $z = +.50$ is one-half of $z = +1.0$, the area between the mean and $z = +.50$ is *not* one-half of the area between the mean and $z = +1.0$. Instead, we find the proportion of the total area under the standard normal curve for any two-decimal z-score by looking in Table 1 of Appendix D. This table is called the *z-table.* A portion of the z-table is reproduced in Table 6.1.

Say that we seek the proportions corresponding to $z = +1.63$. First we locate $z = 1.63$ in column A, labeled "z," and then move to the right. Column B, labeled "Area between the mean and z," contains the proportion of the total area under the curve between the mean and the z identified in column A. Thus, .4484 of the curve, or 44.84% of all z-scores, is between the mean ($z = 0$) and $z = +1.63$. Column C, labeled "Area beyond z," contains the proportion of the total area under the curve that is in the tail beyond the z-score in column A. Thus, .0516 of the curve, or 5.16% of all z-scores, is in the tail of the distribution beyond $z = +1.63$. We can translate this information from the table as shown

TABLE 6.1 Sample Portion of the z-Table

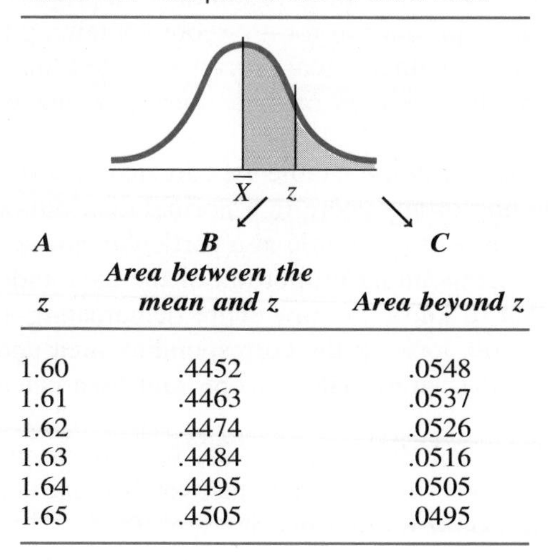

A	B	C
z	***Area between the mean and z***	***Area beyond z***
1.60	.4452	.0548
1.61	.4463	.0537
1.62	.4474	.0526
1.63	.4484	.0516
1.64	.4495	.0505
1.65	.4505	.0495

in Figure 6.10. (If you get confused when using the z-table, look at the normal distribution at the top of the table. The shaded portion and arrows indicate the part of the curve described in each column.)

We can also read the columns in the opposite order to find the z-score that corresponds to a particular proportion. First find the proportion in column B or C, depending on the area you seek, and then identify the corresponding z-score in column A. For example, say that we seek the z-score corresponding to 44.84% of the curve between the mean and z. In column B of the table, we find .4484, which corresponds to the z-score of 1.63.

Notice that the z-table contains no positive or negative signs. Because the normal distribution is symmetrical, only the proportions for one-half of the standard normal curve are given. *You* must decide whether z is positive or negative, based on the problem you are working on.

FIGURE 6.10 Distribution Showing the Area Under the Curve Above $z = +1.63$ and Between $z = +1.63$ and the Mean

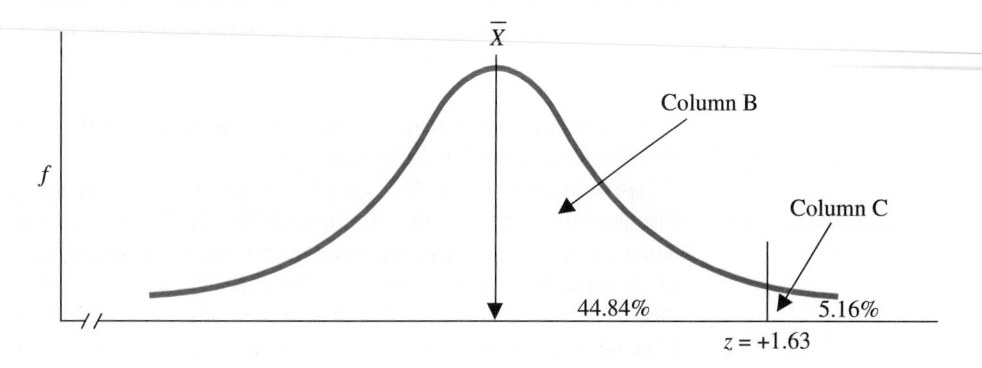

Sometimes you will not want to round off to a proportion given in the *z*-table, or you will need the proportion for a *z*-score containing three decimal places. In such cases, you use a mathematical procedure called linear interpolation. Instructions for interpolating *z*-scores or their corresponding proportions are given in Appendix A.

In summary, then, with the *z*-table we can answer virtually any question about the relative standing of any score in a normal distribution. If you want to know the relative frequency or percentile at a particular raw score, compute its *z*-score to then get the corresponding proportion of the area under the curve from the *z*-table. If you want to know the raw score demarcating a particular relative frequency or percentile, look up the corresponding area under the normal curve in the *z*-table to get the appropriate *z*-score, and then calculate the corresponding raw score.

In most of the remainder of this book, we will be working with *z*-scores or their equivalent, so it is *very* important that you become comfortable with interpreting their location on a normal distribution. To do so, always sketch a normal curve. Then, as we did in the previous examples, label the location of the mean score, identify the relevant portions under the curve, and indicate the corresponding *z*-scores and raw scores. By giving yourself a curve to look at, you'll greatly simplify the problem.

USING THE STANDARD NORMAL CURVE MODEL TO DEFINE PSYCHOLOGICAL ATTRIBUTES

Our third use of *z*-scores is in defining psychological categories or attributes. Recall that we often have difficulty interpreting what, in the grand scheme of things, a particular raw score indicates. Because of this, we also have difficulty deciding on the "cutoff" scores to use when classifying subjects based on their scores. What must someone do to be considered a genius? How do we define an "abnormal" personality? To answer such questions, psychologists often use a "statistical definition" based on relative standing. Essentially, this involves applying the normal curve model and defining an attribute in terms of a particular *z*-score. For example, we might statistically define the old-fashioned term "genius" as a person with a *z*-score of more than $+2.00$ on an intelligence test. Because a z greater than $+2.00$ falls in about the highest 2% of the distribution, we have defined a genius as anyone with a score in the top 2% of all scores on the intelligence test. Similarly, we might define as "abnormal" any person with a *z*-score beyond -1.5 on a personality test. Such scores are "abnormal" in a statistical sense, because they are very infrequent in the population and are among the most extreme low raw scores.

Instructors who "curve" grades generally use the normal curve and *z*-scores. They assume that grades are normally distributed, so they assign letter grades based on proportions of the area under the normal curve. If the instructor defines an A student as one who is in the top 2%, then students with *z*-scores greater than $+2$ receive A's. If the instructor defines B students as those in the next 13%, then students having *z*-scores between $+1$ and $+2$ receive B's, and so on.

USING THE STANDARD NORMAL CURVE MODEL TO DESCRIBE SAMPLES

So far we have discussed using the standard normal curve model to describe the relative standing of any single raw score. Now, using the same logic, we will determine the relative standing of an entire sample. This procedure is important, not only because it allows you to evaluate a sample, but also because it is the basis for inferential statistical procedures (and you will definitely be performing some of these in the very near future).

To see how the procedure works, say that we obtained the SAT scores of a random sample of 25 students at Prunepit U. Their mean score is 520. Nationally the mean of *individual* SAT scores is 500 (and σ_X is 100), so it appears that at least some Prunepit students scored relatively high, pulling the overall mean to 520. However, how do we interpret the performance of the sample as a whole? Is a sample mean of 520 impressively above average, or more mundane? By considering only our sample mean, we cannot answer this question. We have the same problem here that we had previously when examining an isolated individual raw score: without a frame of reference, we do not know whether, in the grand scheme of things, a particular sample mean is high, low, good, bad, or indifferent.

Our solution is to evaluate a sample mean in terms of its relative standing. Previously, we compared a particular raw score to all other possible scores in the distribution that we might have obtained. (Biff's score was impressive compared to any other score obtained.) Now, we will compare our sample mean to all other possible sample means that we might obtain. We envision all possible sample means that we might obtain as a distribution of sample means. Then by comparing our sample mean to this distribution, we can determine whether the Prunepit sample is relatively impressive.

Our first task is to envision the distribution of all possible sample means we might obtain, called the sampling distribution of means.

The Sampling Distribution of Means

By saying that nationally the average SAT score is 500, we are saying that in the population of SAT scores, the population mean (μ) is 500. Because we randomly selected a sample of 25 students and obtained their SAT scores, we essentially drew a random sample of 25 scores from this population of scores. To evaluate our sample mean, we will first create a distribution of all other possible means we might obtain when randomly selecting a sample of 25 scores from this population.

One way to do this would be to record everyone's raw score from the entire population of SAT scores on a slip of paper and deposit all the slips into a very large hat. We could then hire a statistician to sample this population an infinite number of times (she would be very bored, so the pay would have to be good). The statistician would randomly select from the hat a sample with the same size N as we used in our sample (25), compute the sample mean, replace the scores in the hat, draw another 25 scores, compute the mean, and so on. Because the scores selected in each sample would not be identical, all sample means would not be identical. By constructing a frequency distribution of the different values of sample means she obtained, the statistician would create a sampling distribution of means. The **sampling distribution of means** is the distribution of all possible values of random sample means when an infinite number of samples of the same size *N* are randomly selected from one raw score population. Thus, the sampling distribution of means is the population of all values of sample means that can occur when a raw score population is exhaustively sampled using a particular size *N*.

In subsequent chapters we will discuss the sampling distributions of many types of sample statistics. In general, a sampling distribution for any statistic has four important characteristics:

1. All the samples contain raw scores from the same population.

2. All the samples are randomly selected.

3. All the samples have the same size *N*, equal to that in our sample.

4. The sampling distribution reflects the infinite population of all possible values of the sample statistic we might obtain with our sample.

Of course, our bored statistician could not actually sit down and sample the population an infinite number of times. However, she could create a *theoretical* sampling distribution by applying the central limit theorem. The **central limit theorem** is a statistical principle that defines the mean, the standard deviation, and the shape of a theoretical sampling distribution. From the central limit theorem, we know that the sampling distribution of means will: (1) form an approximately normal distribution, (2) have a mean that equals the mean of the raw score population from which the sampling distribution was created, and (3) have a standard deviation that is mathematically related to the standard deviation of the raw scores.

Thus, the central limit theorem tells us that if the bored statistician sampled the population of SAT scores, she would create a sampling distribution of means that would look like the curve shown in Figure 6.11. You should conceptualize a sampling distribution of means in the same way that you conceptualize a distribution of raw scores. The only difference is that here each "score" along the *X* axis is a sample mean. (We can still think of the distribution as a parking lot full of people, except that now each person is the captain of a sample, having the sample's mean score and thus representing the sample.)

Understand that the different values of the sample means in the sampling distribution occur simply because of the luck of the draw of which scores are selected for each sample. The bored statistician would sometimes obtain a sample mean that is higher than 500 because, *by chance,* she randomly selected a sample of predominantly high scores. At other times she might select predominantly low scores, producing a mean below 500. Or she might, by luck, obtain a sample mean equal to 500. Thus the sampling distribution provides a picture of all of the different sample means—and their frequency—that occur due to chance when randomly sampling the SAT population. Because our Prunepit sample was a random sample from this population, this picture allows us to evaluate our sample mean relative to all other sample means we might have obtained.

FIGURE 6.11 Sampling Distribution of Random Sample Means of SAT Scores

The X axis is labeled to show the different values of $\overline{X}$ we obtain when we sample a population where the mean for SAT scores is 500.

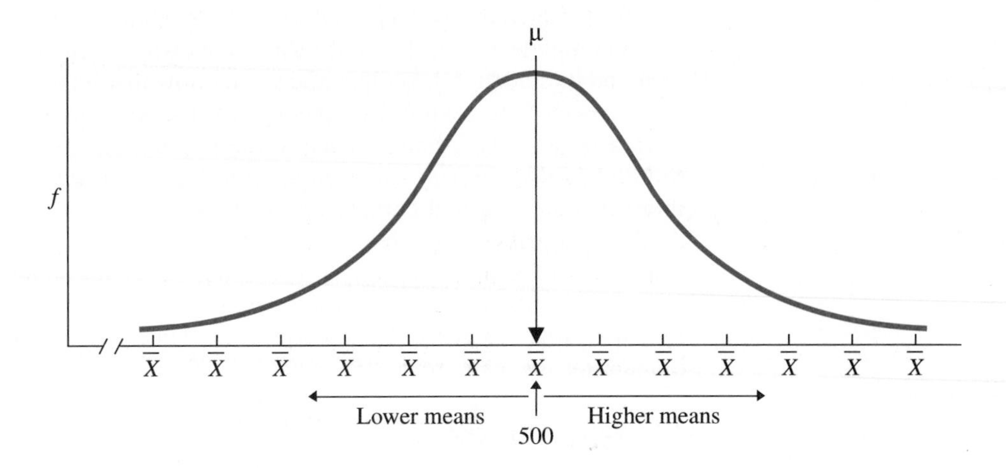

We know from the central limit theorem that the bored statistician would obtain such a sampling distribution, because, first, regardless of the shape of the raw score distribution, a sampling distribution is always an approximately normal distribution. This is because, over an infinite number of random samples, most often the sample mean will equal the population mean, μ. Sometimes, however, a sample will contain a few more high scores or low scores relative to the population, so the sample mean will be close to μ but slightly greater than or less than μ. Less frequently, random sampling will produce rather strange samples, with sample means farther above or below μ. Once in a great while, some very infrequent and unusual scores will be drawn, resulting in a sample mean that deviates greatly from μ. The larger the N in the samples used to create the sampling distribution, the more closely the sampling distribution will conform to the perfect normal curve.

Second, the mean of the sampling distribution is simply the average sample mean, so because the μ of individual SAT scores is 500, we know that the μ of the sampling distribution is 500. The mean of the sampling distribution always equals the mean of the raw score population, because the raw scores in a population are balanced around μ, so the sample means created from those scores will also be balanced around μ. Therefore, those means above and below μ will average out, and overall the mean of the sampling distribution will equal the mean of the raw score population.

Finally, although it's not shown in Figure 6.11, the sample means may be very different from one another and deviate greatly from the average sample mean (μ), or they may be very similar and deviate little from μ. As we'll see, the size of the "standard deviation" of the sampling distribution is related to the size of the standard deviation in the population of raw scores. If the raw scores are highly variable, then each time we sample the population we are likely to get a very different set of scores, and so our various sample means will differ greatly from one another. Conversely, if the raw scores are not variable, then different samples will tend to contain the same scores, and so the sample means, too, will be very similar to one another.

Notice that once we have computed its standard deviation, we will have a rather complete description of the sampling distribution of SAT means: we'll know its μ and its standard deviation, and we'll know that it is a normal distribution. The importance of the central limit theorem is that we obtain this information *without* having to actually infinitely sample the population of SAT scores. All we need to know is that the raw score population forms a normal distribution of ratio or interval scores (so that computing the mean is appropriate), that μ is 500 and σ_X is 100, and that we sampled using an N of 25. Further, because of the principles of the central limit theorem, we expect to produce the *same* sampling distribution with any population of raw scores having these characteristics, *regardless of what they measure.* Likewise, by knowing the characteristics of any other raw score population, we can create the corresponding sampling distribution of means.

> ***STAT ALERT*** The central limit theorem allows us to envision the sampling distribution of means that would be created by exhaustive random sampling of any raw score distribution.

Thus, Figure 6.11 provides a *model* of all the different sample means—and the frequency with which they occur—when sampling the SAT population. Now to evaluate our original SAT sample mean, we simply need to determine *where* a mean of 520 falls on the X-axis of the sampling distribution in Figure 6.11 and then interpret the curve accordingly. If 520 lies close to 500, then it is a rather frequent, common mean when sampling SAT scores (after all, the bored statistician would frequently obtain this result). But if 520 lies toward the upper tail of the distribution, far from 500, then it is a more infrequent and unusual sample mean (over an *infinite* number of samples, she would seldom encounter such a mean).

The sampling distribution is a normal distribution, and you already know how to determine the location of any "score" on a normal distribution: we use—you guessed it—z-scores. That is, we determine how far the sample mean deviates from the mean of the sampling distribution, measured using the "average" deviation in the distribution. With one number then, the z-score will tell us the sample mean's relative location within the sampling distribution, and thus indicate its relative standing among all possible means.

To calculate the z-score for our sample mean, we need one more piece of information: the "standard deviation" of the sampling distribution.

The Standard Error of the Mean

The standard deviation of the sampling distribution of means is called the **standard error of the mean.** Because the term "standard deviation" refers to the variability of raw scores, we use a different name to refer to the variability of sample means. However, as we've seen in previous chapters, the terms *error* and *deviation* are synonymous. Therefore, like a standard deviation, the standard error of the mean can be thought of as the "average" amount that the sample means deviate from the mean (μ) of the sampling distribution.

For the moment, we will discuss the *true* standard error of the mean, as if we had actually computed it using the entire sampling distribution. The symbol for the true standard error of the mean is $\sigma_{\overline{X}}$. Be careful, because the symbol for the true standard error of the sampling distribution is very similar to the symbol for the true standard deviation of the population of raw scores. In $\sigma_{\overline{X}}$, the σ indicates that we are describing the population, but the subscript $\overline{X}$ indicates that we are describing the population of sample means—what we call the sampling distribution of means. Also, be careful because we compute the true standard error using the true standard deviation. The central limit theorem tells us that $\sigma_{\overline{X}}$ can be found using the following formula:

THE FORMULA FOR THE TRUE STANDARD ERROR OF THE MEAN IS

$$\sigma_{\overline{X}} = \frac{\sigma_X}{\sqrt{N}}$$

This formula involves σ_X because, as we've seen, the variability of the raw score population influences the variability of the sampling distribution. When we know the true standard deviation of the raw scores, we will know the true "standard deviation" of the sample means.

STAT ALERT The true standard error of the means, $\sigma_{\overline{X}}$, is computed using the true standard deviation of the population of raw scores, σ_X.

The formula also involves N, the number of scores used to compute our and all other sample means in the sampling distribution. We factor in the size of N, because it influences how representative each sample is and thus determines how close each sample mean is to μ. With a very small N (say 2), each sample can easily contain scores that are very unrepresentative of the population, so that each sample mean differs greatly from the population μ, producing a large "average deviation" in the sampling distribution. However, with a larger N we should obtain more representative samples. At the furthest extreme, say that each sample contains virtually the entire population. Now each sample mean should equal the population mean, or be very close to it, so that the average deviation in the sampling distribution will be very small.

We can compute $\sigma_{\overline{X}}$ for the sampling distribution of SAT scores because, through recordkeeping, we *know* that the true standard deviation of the population of SAT scores is 100. Because our $N = 25$, the above formula says that the standard error of the mean for the sampling distribution is

$$\sigma_{\overline{X}} = \frac{100}{\sqrt{25}}$$

The square root of 25 is 5, so

$$\sigma_{\overline{X}} = \frac{100}{5}$$

and thus

$$\sigma_{\overline{X}} = 20$$

A value of $\sigma_{\overline{X}} = 20$ indicates that in the SAT sampling distribution, the individual sample means differ from the μ of 500 by an "average" of 20 SAT points when the N of each sample is 25.

Now, at last, we are ready to calculate a z-score for our sample mean.

Calculating a *z*-Score for a Sample Mean

Previously we saw that when we know the population mean, μ, and we know the true population standard deviation, σ_X, the formula for transforming a raw score into a z-score is

$$z = \frac{X - \mu}{\sigma_X}$$

Since the sampling distribution of means is the population of sample means and we know its μ and true "standard deviation," $\sigma_{\overline{X}}$, we can transform a sample mean into a z-score using a similar formula.

THE FORMULA FOR TRANSFORMING A SAMPLE MEAN INTO A z-SCORE IS

$$z = \frac{\overline{X} - \mu}{\sigma_{\overline{X}}}$$

Do not be confused by the minor difference in symbols between the preceding formulas. Conceptually we are doing the same thing in both: we are simply finding how far a score on a distribution falls from the mean of the distribution, measured in standard deviations of that distribution. With a sample mean, we find how far the sample mean is from the mean of the distribution, μ, measured in standard error units, or $\sigma_{\overline{X}}$.

STAT ALERT We use the above formula only when we know the true population standard deviation, σ_X, because only then can we compute $\sigma_{\overline{X}}$.

Using the above formula, we calculate the z-score for our sample from Prunepit U. With $\overline{X} = 520$, $\mu = 500$ and $\sigma_{\overline{X}} = 20$ when $N = 25$, we have

$$z = \frac{\overline{X} - \mu}{\sigma_{\overline{X}}} = \frac{520 - 500}{20} = \frac{+20}{20} = +1.0$$

Thus, a sample mean of 520 has a z-score of $+1.0$ on the sampling distribution of means that occurs when N is 25 and the sampling distribution is created from the SAT raw score population where $\mu = 500$ and $\sigma_X = 100$.

Using the Sampling Distribution to Determine Relative Frequency of Sample Means

Everything we said previously about a z-score for an individual raw score applies to a z-score for a sample mean. It makes no difference that the z-score now refers to the location of a sample mean: a z-score is a z-score! Thus, since our sample mean has a z-score of $+1.0$, we know that it is above the μ of the sampling distribution (the average sample mean) by an amount equal to the "average" amount that sample means deviate from μ. Therefore we know that, although they were not stellar, our Prunepit students did outperform a substantial proportion of comparable samples. Likewise, if another sample of 25 SAT scores (say from Podunk U.) produced a mean of 440, we'd know how poorly these students performed: here $z = (440 - 500)/20 = -3.0$, so this mean would be among the lowest SAT means we'd ever expect to obtain in this situation.

Notice that, as with an individual's z-score, the size of a sample mean's z-score depends both on the size of the deviation from μ *and* on the variability in the distribution. If, for example, $\sigma_{\bar{X}}$ above had been 5, then with $\bar{X} = 520$ we'd have $z = (520 - 500)/5 = +4$! Here, with all sample means deviating from 500 by an "average" of only 5, a mean that deviates by 20 would be *extremely* unusual (which is what a z of $+4$ indicates).

To obtain a more precise description of a sample mean, we can go one step further (and here's another nifty part): because the sampling distribution of means always forms at least an approximately normal distribution, if we transformed *all* of the sample means in the sampling distribution into z-scores, we would have a roughly normal z-distribution. Recall that the standard normal curve is our model of *any* normal z-distribution. This is true even if it is a z-distribution of random sample means! Thus, as we did previously for raw scores, we can use the standard normal curve model to determine the relative frequency of random sample means in any portion of a sampling distribution.

Figure 6.12 shows the standard normal curve applied to our SAT sampling distribution. This curve is the same curve, with the same proportions, that we saw when describing individual raw scores. Once again we see that the farther a score (here a $\bar{X}$) is from the mean of the distribution (here μ), the larger the absolute value of the z-score. The larger the z-score, the smaller the relative frequency and simple frequency of the z-score and of the corresponding sample mean. As we did with raw scores, we can now use the standard normal curve (and the z-table) to determine the proportion of the area under any part of the curve. This proportion is also the expected relative frequency of the corresponding sample means in that part of our sampling distribution.

For example, our sample from Prunepit U. has a z of $+1.0$. As we have seen, .3413, or 34.13%, of all scores fall between the mean and $z = +1.00$ on a normal distribution. With SAT scores, μ is 500 and a z of $+1$ produces a sample mean of 520. Therefore, 34.13% of all random SAT sample means are expected to fall between the μ of 500 and a sample mean of 520 (when N is 25). Further, by

FIGURE 6.12 Proportions of the Standard Normal Curve Applied to the Sampling Distribution of SAT Means

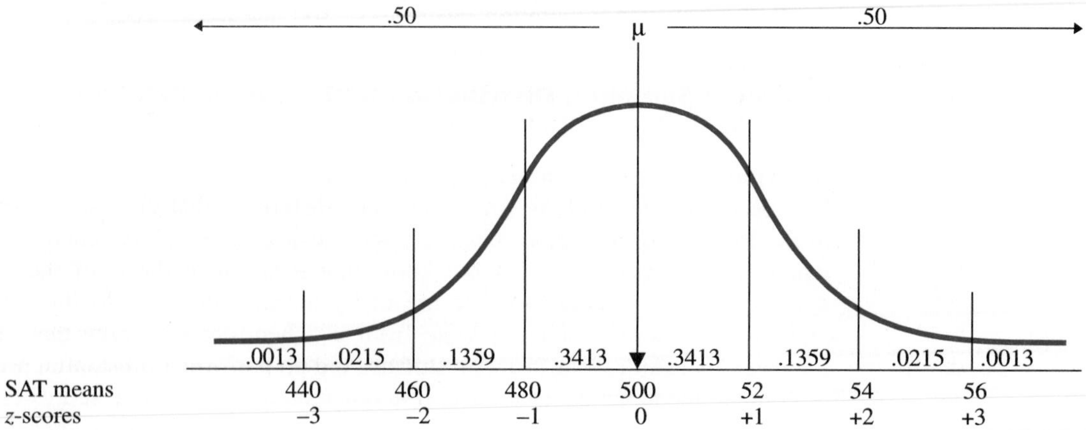

.50 μ .50

	.0013	.0215	.1359	.3413	.3413	.1359	.0215	.0013

SAT means	440	460	480	500	52	54	56
z-scores	−3	−2	−1	0	+1	+2	+3

adding in the 50% of the distribution below μ, we see that about 84% of all means fall below—are to the left of—a z of $+1$. Therefore, our sample mean of 520 ranks at about the 84th percentile among all such sample means. Similarly, a sample mean of 540 would have a z of $+2.00$ on this sampling distribution. Since we know that about 2% of all z-scores are above a z of $+2.00$, we expect that only about 2% of all such SAT sample means will be larger than 540.

We can use this same procedure to determine the relative frequency of any random sample means: to determine the relative frequency above or below a particular mean, compute its z-score and then look up the corresponding proportion of the area under the curve in the z-table. This is the expected relative frequency of the corresponding sample means in the sampling distribution of all possible sample means that we might obtain.

In fact, as we shall see in later chapters, we use a similar procedure to determine the relative frequency of any sample statistic. Why would we want to do this? Because eventually we will perform inferential statistical procedures. In the final analysis, all inferential statistics involve computing something like a z-score for a particular sample statistic so that we can determine its relative frequency in a particular sampling distribution. We will elaborate on these procedures in later chapters, but for now it is enough to understand that the basic approach is to compute z-scores and apply the standard normal curve model.

FINALLY

In this chapter we saw that we should apply the standard normal curve model to a distribution of raw scores only when we can make certain *assumptions* about the data. In subsequent chapters we will apply other statistical models and discuss their assumptions. Always bear in mind that a statistical model indicates what we can expect, assuming the rules and regulations of the model are being met.

The most important concept for you to understand is that any normal distribution of scores can be described using the standard normal curve model and z-scores. To paraphrase a famous saying, a normal distribution is a normal distribution is a normal distribution. Any normal distribution contains the same proportions of the total area under the curve between z-scores. Therefore, think z-scores! Picture a normal distribution and repeat after me: the larger the z-score, whether positive or negative, the farther the z-score and the corresponding raw score are from the mean of the distribution. The farther they are from the mean of the distribution, the lower the relative frequency of the z-score and of the corresponding raw score. This is true whether the raw score is an individual's score or a sample mean.

By the way, what was Biff's percentile?

CHAPTER SUMMARY

1. The *relative standing* of a score reflects a systematic evaluation of the score relative to the characteristics of a sample or population of scores. We determine a score's relative standing by computing its *z-score,* which indicates the distance the score is above or below the mean, measured in standard deviations.

2. The larger the value of a positive z-score, the farther the raw score is above the mean. The larger the absolute value of a negative z-score, the farther the raw score is below the mean. On a normal distribution, the larger the z-score, whether positive or negative, the less frequently it occurs, and thus the less frequently the corresponding raw score occurs.

3. By transforming an entire raw score distribution into z-scores, we create a *z-distribution.* The z-distribution will have the same shape as the raw score distribution, but the mean of a z-distribution is always 0 and the standard deviation is always 1. The z-distribution enables us to compare scores from different variables by transforming these scores to one common scale.

4. The *standard normal curve* is a mathematically perfect normal curve that can be used as a model of any z-distribution when (a) the distribution is at least roughly normally distributed and (b) the corresponding raw scores reflect an interval or ratio scale of measurement.

5. The *z-table* gives the proportion of the total area under the standard normal curve between the mean and any value of z. The proportion of the area under the curve is equal to the relative frequency of z-scores falling between the mean and z.

6. Using the relative frequency of z-scores in any portion of the standard normal curve, we can determine the *expected* relative frequency, simple frequency, and percentile rank of corresponding raw scores. These are the expected values, assuming the raw score distribution conforms to the perfect normal curve.

7. A *sampling distribution* is a theoretical distribution of the different values of a sample statistic obtained when samples of a particular size N are randomly selected from a particular raw score population an infinite number of times. The *sampling distribution of means* is the distribution of sample means obtained when samples of a particular size N are randomly selected from one raw score population. The different sample means in a sampling distribution of means occur solely because of chance—the luck of the draw when each sample is randomly selected.

8. The *central limit theorem* shows that (a) the sampling distribution of means will be an approximately normal distribution, (b) the mean of the sampling distribution of means will equal the mean of the raw score population, and (c) the variability of the sample means is related to the variability of the raw scores.

9. The true *standard error of the mean* is the standard deviation of the sampling distribution of means, and its symbol is $\sigma_{\bar{X}}$. The value of $\sigma_{\bar{X}}$ can be thought of as the average amount that the sample means deviate from the μ of the sampling distribution. We compute $\sigma_{\bar{X}}$ *only* when we know the true standard deviation of the raw score population, σ_X.

10. The location of a random sample mean on the sampling distribution of means can be described by calculating a z-score: the distance the sample mean score ($\bar{X}$) is from the mean of the distribution (μ), measured in standard error units ($\sigma_{\bar{X}}$).

11. We can apply the standard normal curve model to the sampling distribution of means. From the z-table, we determine the proportion of the area under the curve for any part of the distribution. This tells us the expected relative frequency of the sample means in that corresponding part of the sampling distribution of means.

12. Biff's percentile was 99.87.

PRACTICE PROBLEMS

(Answers for odd-numbered problems are provided in Appendix E.)

1. What does a z-score indicate?

2. On what factors does the size of a z-score depend?

3. What is a z-distribution?

4. What are the three uses of z-distributions in describing individual scores?

5. Why are z-scores referred to as standard scores?

6. In freshman English last semester, Foofy earned a 76 ($\bar{X} = 85$, $S_X = 10$) and her friend Bubbles, in a different class, earned a 60 ($\bar{X} = 50$, $S_X = 4$). Should Foofy be bragging about how much better she did in the course? Why?

7. Poindexter received a grade of 55 on a biology test ($\overline{X} = 50$) and a grade of 45 on a philosophy test ($\overline{X} = 50$). He is considering whether to ask his two professors to curve the grades using *z*-scores.
   a. What other information should he consider before making his request?
   b. Does he want the S_X to be large or small in biology? Why?
   c. Does he want the S_X to be large or small in philosophy? Why?
8. A student computes *z*-scores for a set of normally distributed exam scores. She obtains a *z*-score of -3.96 for 8 (out of 20) of the students. What does this mean?
9. *a.* What is the standard normal curve model?
   b. What is it used for?
   c. What criteria should be met for the model to give an accurate description of a sample?
10. Why are *z*-scores beyond ± 3.0 seldom obtained?
11. For the data 9 5 10 7 9 10 11 8 12 7 6 9
    a. Compute the *z*-score for the raw score of 10.
    b. Compute the *z*-score for the raw score of 6.
12. For the data in question 11, find the raw scores that correspond to the following.
    a. $z = +1.22$
    b. $z = -0.48$
13. Which *z*-score in each of the following pairs corresponds to the smaller raw score?
    a. $z = +1.0$, $z = +2.3$
    b. $z = -2.8$, $z = -1.7$
    c. $z = -.70$, $z = +.20$
    d. $z = 0.0$, $z = -2.0$
14. For each pair in question 13, which *z*-score has the higher frequency?
15. In a normal distribution of scores, what proportion of all scores would you expect to fall in each of the following areas?
    a. Between the mean and $z = +1.89$
    b. Below $z = -2.30$
    c. Between $z = -1.25$ and $z = +2.75$
    d. Above $z = +1.96$ and below -1.96
16. For a distribution in which $\overline{X} = 100$, $S_X = 16$, and $N = 500$, answer the following.
    a. What is the relative frequency of scores between 76 and the mean?
    b. How many subjects are expected to score between 76 and the mean?
    c. What is the percentile of someone scoring 76?
    d. How many subjects are expected to score above 76?

17. Poindexter may be classified as having a math dysfunction—and thus not have to take statistics—if he scores below the 25th percentile on a national diagnostic test. The μ of the test is 75 ($\sigma_X = 10$). Approximately what raw score is the cutoff score for him to avoid taking statistics?

18. Over the years, an IQ test has been given to so many people that we know the population $\mu = 100$ and the $\sigma_X = 16$. We are interested in creating the sampling distribution when $N = 64$.

a. What does that sampling distribution of means reflect?

b. What is the shape of the distribution of IQ means and the mean of the distribution?

c. Calculate $\sigma_{\overline{X}}$ for this distribution.

d. What is your answer in part *c* above called, and what does it indicate?

19. A recent graduate has two job offers and must decide which to accept. The job in City A pays $27,000. The average cost of living there is $50,000, with a standard deviation of $15,000. The job in City B pays $12,000. The average cost of living there is $14,000, with a standard deviation of $1,000. Assuming the data are normally distributed, which is the better job offer? Why?

20. A researcher obtained a sample mean of 68.4 when he gave a test to a random sample of 49 subjects. For everyone who has ever taken the test, the mean is 65 (and $\sigma_X = 10$). The researcher believes that his sample mean is rather unusual.

a. How often can he expect to obtain a sample mean that is higher than 68.4?

b. Why might the researcher obtain such an unusual mean?

21. If you took 1,000 random samples of 50 subjects each from a population where $\mu = 19.4$ and $\sigma_X = 6.0$, how many samples would you expect to produce a mean below 18?

22. Suppose you own shares of a company's stock, the price of which has risen so that, over the past ten trading days, its mean selling price is $14.89. Over the years, the mean price of the stock has been $10.43 ($\sigma_X = \$5.60$). You wonder if the mean selling price over the next ten days can be expected to go higher. Should you wait to sell, or should you sell now?

SUMMARY OF FORMULAS

1. *The formula for transforming a raw score in a sample into a z-score is*

$$z = \frac{X - \overline{X}}{S_X}$$

where X is the raw score, $\overline{X}$ is the sample mean, and S_X is the sample standard deviation.

2. *The formula for transforming a z-score in a sample into a raw score is*

$$X = (z)(S_X) + \bar{X}$$

3. *The formula for a T-score is*

$$T = z(10) + 50$$

4. *The formula for transforming a raw score in a population into a z-score is*

$$z = \frac{X - \mu}{\sigma_x}$$

5. *The formula for the true standard error of the mean is*

$$\sigma_{\bar{X}} = \frac{\sigma_X}{\sqrt{N}}$$

where σ_X is the true standard deviation of the raw score population.

6. *The formula for transforming a sample mean into a z-score on the sampling distribution of means is*

$$z = \frac{\bar{X} - \mu}{\sigma_{\bar{X}}}$$

where $\bar{X}$ is the sample mean, μ is the mean of the sampling distribution (which is also equal to the μ of the raw score distribution), and $\sigma_{\bar{X}}$ is the standard error of the mean of the sampling distribution of means.

PART 3

DESCRIBING RELATIONSHIPS

As you know by now, the purpose of research is to understand the laws of nature, and psychologists do this by describing relationships between variables. The final question that we answer with descriptive statistics is "What is the nature of the relationship we have found?" In the next two chapters we will discuss statistics that directly summarize and describe a relationship. In Chapter 7 we discuss the procedure known as correlation, and in Chapter 8 we discuss the procedure known as linear regression.

If it seems that the topics in the upcoming chapters are different from previous topics, it's because they *are* different. Although correlation and regression are major descriptive statistical procedures, their perspective is somewhat different from that of our previous techniques, which focused on an individual sample mean and standard deviation. Therefore, think of the upcoming chapters as somewhat of a detour. After we complete the detour, we will return to describing a sample using the mean and standard deviation. In particular, we will return to describing the location of a sample mean on a sampling distribution of sample means. Don't forget those procedures, because they form the basis for inferential statistics.

7

Describing Relationships Using Correlations

To understand the upcoming chapter:

- From Chapter 2, recall how we graph data points and identify when a relationship is present, and what is meant by its strength.
- From Chapter 4, understand how we summarize the relationship in an experiment by plotting the mean dependent score on the Y axis for subjects in each condition plotted on the X axis.
- From Chapter 5, understand that greater variability indicates that scores are not typically close to each other.

Then your goals in this chapter are to learn:

- The logic of correlational research and how it is interpreted.
- How to read and interpret a scatterplot and a regression line.
- How to identify the type and strength of a relationship.
- How to interpret a correlation coefficient.
- When to use the Pearson r, the Spearman r_s, and the point-biserial r_{pb}.
- The logic of inferring a population correlation based on a sample correlation.

Recall that in a relationship, as the scores on one variable change, there is a consistent pattern of change in the scores on the other variable (for example, the bigger they are, the harder they fall). In research, in addition to demonstrating a relationship, we also want to describe its characteristics: What is the nature of the relationship? How consistently do the scores change together? What direction do the scores change? In this chapter we discuss the descriptive statistical procedure used to answer such questions. This procedure is known as *correlation.*

In the following sections we will consider correlational research procedures, and then we will examine those characteristics of a relationship that are described by correlation. Finally, we will see how to calculate different types of correlation statistics. First, though, here are several new symbols that are used with correlation.

MORE STATISTICAL NOTATION

We perform correlational analysis when we have scores from two variables. We use X to stand for the scores on one variable and Y to stand for the scores on the other variable. Usually we obtain an X score and a Y score from the same subject, and then each subject's X is paired with the corresponding Y. If the X and Y scores are not from the same subject, there must be some other rational system for pairing the scores (for example, we might pair the scores of roommates).

In our computations, we use the same conventions for Y that we have previously used for X. Thus, ΣY is the sum of the Y scores, ΣY^2 is the sum of the squared Y scores, and $(\Sigma Y)^2$ is the squared sum of the Y scores. The mean of the Y scores is symbolized by $\overline{Y}$ and is equal to $\Sigma Y/N$. Similarly, the variance of a sample of Y scores is S_Y^2, and the standard deviation of Y is S_Y. To find S_Y^2 and S_Y, we use the same formulas that we used to find S_X^2 and S_X, except that now we plug in the Y scores instead of the X scores.

We will also encounter three new mathematical notations. We will see $(\Sigma Y)(\Sigma X)$, which tells us to first find the sum of the Xs and the sum of the Ys, and then multiply the two sums together. We will also see ΣXY, which is called the sum of the cross products. This tells us to first multiply each X score in a pair times its corresponding Y score and then sum all of the resulting products. Finally, we will see the symbol D. This stands for the numerical *difference* between the X and Y scores in a pair, which we find by subtracting one from the other.

Table 7.1 shows how we compute each of the above quantities. Notice that because we are using pairs of scores, there *must* be the same number of X and Y scores. Then, as the first two columns show, ΣX is 6 and ΣY is 8, so $(\Sigma X)(\Sigma Y)$ is (6)(8), which is 48. On the other hand, we obtain ΣXY by first multiplying each X times the corresponding Y and then summing, so $\Sigma XY = 18$. As shown

TABLE 7.1 Computing ΣXY, (ΣX) (ΣY), and D for Three Subjects' X and Y Scores

Subject	X	Y	XY	D
1	1	2	2	−1
2	2	2	4	0
3	3	4	12	−1
	$\Sigma X = 6$	$\Sigma Y = 8$	$\Sigma XY = 18$	

$(\Sigma X)(\Sigma Y) = (6)(8) = 48$

in the far right-hand column, we compute each D by subtracting each Y from its corresponding X.

Now, on to correlation.

UNDERSTANDING CORRELATIONAL RESEARCH

A statistical statement that captures the flavor of a relationship is "The scores on variables X and Y *covary.*" By this we mean that the scores on the two variables vary, or change, together. For example, it is commonly believed that as people drink more coffee, they become more nervous. Thus, if there is a relationship, amount of coffee consumed and nervousness will covary.

To demonstrate this relationship in an experiment, we would first actively manipulate the amount of coffee subjects consume: one condition might be given 1 cup, another would receive 2 cups, and a third would drink three cups. Then we would measure our subjects' nervousness (perhaps using some physiological device) to see if more coffee is related to more nervousness.

But recall that another way to demonstrate a relationship is through the correlational approach. In correlational research, we do *not* manipulate or select particular conditions of an independent variable, as in an experiment. Instead, we simply measure subjects' scores on two variables and describe the relationship that is present. In fact, the word *correlation* means *relationship.* (Think of a correlation as the shared, or "co," relation between two variables.) Thus, using this approach, we might simply ask each subject to report the amount of coffee he or she had consumed that day, and then measure how nervous each subject was. Likewise, in a different study, we might describe the relationship we find after asking subjects how many traffic accidents and how many speeding tickets they have had, or we might determine the relationship between subjects' scores on a personality test and their responses to an attitude survey.

As the name implies, in correlational research we typically use correlational statistics to summarize a relationship. On the one hand, computing a correlation does not automatically make research correlational: it is the absence of manipulation that creates this design. On the other hand, a correlation often implies correlational research. Because of the manner in which the data are collected in correlational research, we must be very careful when we interpret the results.

Correlation and Causality

Whenever we hear of a relationship between X and Y, we have a natural tendency to conclude that it is a *causal* relationship—that X causes Y. Thus, if we discovered that subjects who reported drinking more coffee also tended to be more nervous we might want to conclude that drinking more coffee causes people to be more nervous. In a correlational study, however, we only describe how the X and Y scores are paired, or match up, to form a relationship. *The fact that there is a relationship between two variables does not mean that changes in one variable cause the changes in the other variable.* A statistical relationship can exist even though one variable does not cause or influence the other. Therefore, correlational research *cannot* be used to infer a causal relationship between two variables.

There are two requirements for concluding that *X* causes *Y*. First, *X* must occur before *Y*. In correlational research, we have no way of knowing which factor occurred first. For example, in our study above, we simply measured the scores for amount of coffee and nervousness. Perhaps subjects who were first more nervous subsequently drank more coffee. If so, then it is greater nervousness that may actually cause greater coffee consumption. In any correlation it is possible that *Y* causes *X*. Second, *X* must be the only variable that can influence *Y*. But, in correlational research, we do not control or eliminate other variables that may potentially cause scores to change. For example, in our coffee study, some of the subjects may have had less sleep than other subjects the night before we tested them. Perhaps the lack of sleep caused those subjects to be more nervous *and* to drink more coffee. In any correlation some other variable may cause both *X* and *Y* to change.

Thus, a correlation by itself does not allow us to conclude causality. We must consider not only that a relationship exists, but also the research method used to demonstrate that relationship. In experiments we manipulate the independent variable *first* and we control other potential causal variables. Therefore, experiments provide the best evidence for drawing conclusions about the causes of a behavior. In a correlational study, however, the relationship we describe may be a *coincidence*. Sometimes the coincidental nature of the relationship is obvious. For example, there is a relationship between the number of toilets in a neighborhood and the number of crimes committed in the neighborhood: the more toilets, the more crime. Based on this correlation, should we conclude that indoor plumbing causes crime? Of course not! Crime tends to occur more frequently in large cities, especially in crowded neighborhoods. Coincidentally, there are more indoor toilets in crowded neighborhoods in large cities.

The problem is that it is easy to be trapped by more mysterious relationships involving variables that we do not understand. For example, there is a correlation between the amount of "adult literature" (pornography) sold in a state and the incidence of rape in the state: the more pornography sold, the more rape. Based solely on this correlation, can we conclude that pornography causes rape? Not unless we can also conclude that indoor plumbing causes crime! If we cannot use correlation to infer causality in one situation, we cannot use it in another. Pornography may actually cause rape, but, for all the reasons given above, the mere existence of this relationship is not evidence of causality.

STAT ALERT We never infer causality based solely on the existence of a correlation.

Distinguishing Characteristics of Correlation

There are four major differences between how we handle data in correlational analysis and how we handle data in an experiment. First, in our coffee experiment we would examine changes in the mean nervousness score (the *Y*-scores) as a function of changes in our conditions of amount of coffee consumed (the *X* scores). However, with correlational data, we typically have a rather large and unwieldy number of different X scores: our subjects would probably report a wide range of coffee consumption. Comparing the many corresponding mean nervousness scores would make it very difficult to summarize and describe the relationship. Therefore, in correlational procedures we use a different approach, not computing a mean Y score at each X. Instead, and this is an advantage of using correlation, we simultaneously summarize the *entire* relationship formed by all pairs of subjects' X-Y scores in our data.

A second difference is that, because we examine all pairs of X-Y scores, correlational procedures involve *one* sample containing all pairs of X and Y scores.

In correlational analysis, N stands for the number of pairs of scores in the data.

Third, in a correlational study, neither variable is called the independent or dependent variable, and either variable may be called the X or Y variable. How do we decide which variable is X and which is Y? As we have seen, we discuss any relationship in terms of changes in Y scores as a function of changes in X scores, and the X scores are the "given" scores. Thus, if we ask, "For a given amount of coffee, what are the nervousness scores?" then amount of coffee is the X variable (plotted on the X axis) and nervousness is the Y variable (plotted on the Y axis). Conversely, if we ask, "For a given nervousness score, what is the amount of coffee consumed?" then nervousness is the X variable and amount of coffee is the Y variable.

Finally, as we will see in the next section, the data are graphed differently in correlational research than in an experiment. We use the individual pairs of scores to create a scatterplot.

Plotting Correlational Data: The Scatterplot

A **scatterplot** is a graph that shows the location of each pair of X-Y scores in the data. The scatterplot in Figure 7.1 shows the data we might obtain if we actually conducted a study of coffee drinking and nervousness (for reference, the raw scores are also listed). We are interested in how nervousness scores change as a function of coffee consumption, so nervousness is the Y variable and coffee consumption is the X variable. Real research typically involves a larger N than we have here, and the data points may not form such a pretty pattern. Nonetheless, a scatterplot does summarize the data somewhat. In the table of raw scores on the left, we see that two subjects had scores of 1 on both coffee consumption and nervousness. On the scatterplot there is one data point for these subjects. (It is circled because some researchers do this to indicate that two data points are plotted on top of each other.) In real data, many subjects with a particular X score may score the same value of Y, so the number of data points may be considerably smaller than the number of pairs of raw scores.

Remember that the lower left-hand corner of the graph always has the low scores on the X and Y variables. The scores get larger as the Y variable goes up and as the X variable goes to the right. We ask, "As the X scores increase, what happens to the Y scores?" In Figure 7.1 we can see that the more coffee subjects drink, the more nervous they tend to be: subjects drinking 1 cup of coffee tend to have nervousness scores around 1, subjects drinking 2 cups of coffee tend to

FIGURE 7.1 Scatterplot Showing Nervousness as a Function of Coffee Consumption

Each data point is created using a subject's coffee consumption as the X score and the subject's nervousness as the Y score.

Cups of coffee: X	*Nervousness scores: Y*
1	1
1	1
1	2
2	2
2	3
3	4
3	5
4	5
4	6
5	8
5	9
6	9
6	10

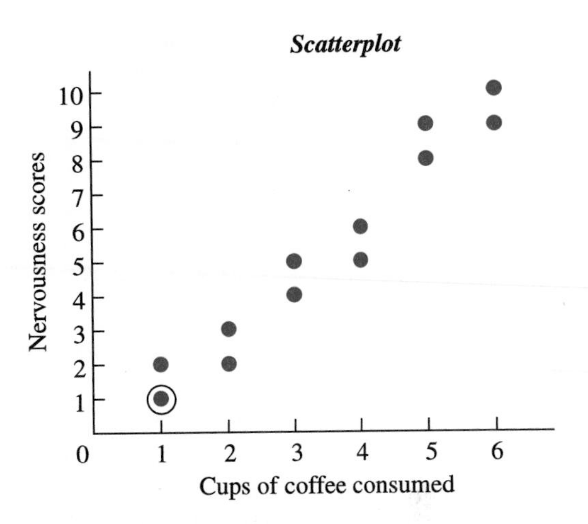

have nervousness scores around 2, and so on. Thus, the scatterplot reflects a relationship: one value or close to one value of *Y* tends to be paired with one value of *X*. As the *X* scores increase, the *Y* scores change so that a different value of *Y* tends to be paired with a different value of *X*.

> ***STAT ALERT*** When a relationship exists, a particular value of *Y* tends to be paired with one value of *X* and a different value of *Y* tends to be paired with a different value of *X*.

Always draw the scatterplot of a set of correlational data. A scatterplot allows you to see the nature of the relationship that is present and map out the best way to accurately summarize and describe it.

The shape of the scatterplot is an important indication of both the presence of a relationship and the nature of the relationship. We can visually summarize a scatterplot by drawing a line around its outer edges. As we shall see, a scatterplot may have any one of a number of shapes when a relationship is present. When no relationship is present, however, the scatterplot will be either circular or elliptical, oriented so that the ellipse is parallel to the *X* axis. The scatterplots in Figure 7.2 on the next page show what no relationship between coffee consumption and nervousness would look like. Here, as the *X* scores increase, the *Y* scores do not consistently change. No particular value of *Y* tends to be associated with a particular value of *X*. Instead, virtually every value of *Y* is associated with every value of *X*.

The scatterplots in Figure 7.2 are further summarized by the line drawn through them that seems to best fit the center of the scatterplot. This best-fitting summary

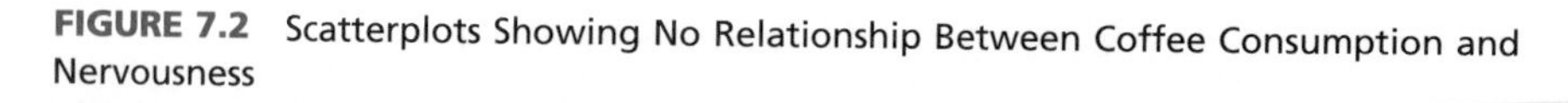

FIGURE 7.2 Scatterplots Showing No Relationship Between Coffee Consumption and Nervousness

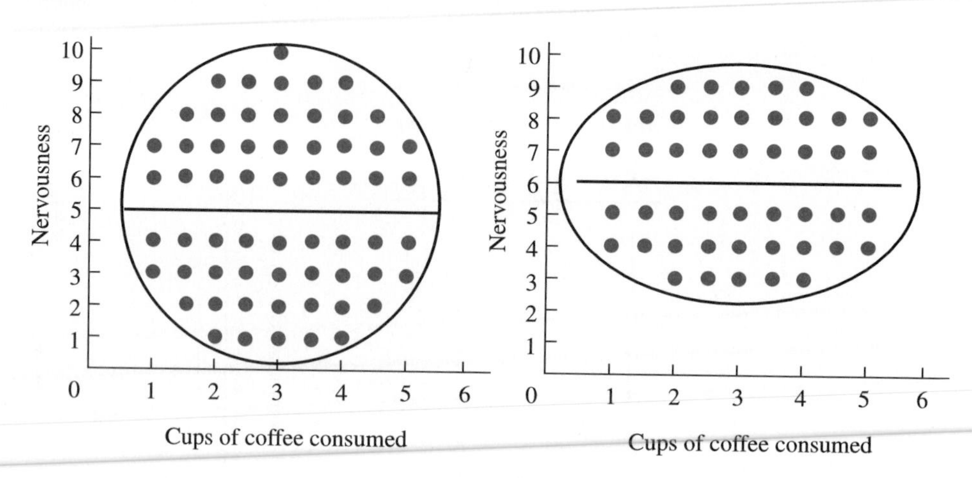

line is called the **regression line.** We will discuss the procedures for drawing a regression line in the next chapter. However, as you can see in Figure 7.2, the orientation of the regression line matches the orientation of the scatterplot. Thus, when no relationship is present, the regression line forms a horizontal straight line.

The Correlation Coefficient

When a relationship is present, the scatterplot will form some shape other than a circle or a horizontal ellipse, and the regression line will not be a horizontal straight line. The particular shape and orientation of a scatterplot reflect the characteristics of the relationship formed by the data. To summarize these characteristics, we compute the statistic known as the correlation coefficient. A **correlation coefficient** is a number that summarizes and describes the important characteristics of a relationship. There are two important characteristics of a relationship: the type of relationship and the strength of the relationship. In the following sections, we discuss these characteristics.

TYPES OF RELATIONSHIPS

The **type of relationship** that is present in a set of data is determined by the overall direction in which the Y scores change as the X scores change. There are two general types of relationships: linear and nonlinear relationships.

Linear Relationships

The term *linear* means "straight line," and a linear relationship has a summary regression line that is a straight line. The scatterplots in Figure 7.3 illustrate two

FIGURE 7.3 Scatterplots Showing Positive and Negative Linear Relationships

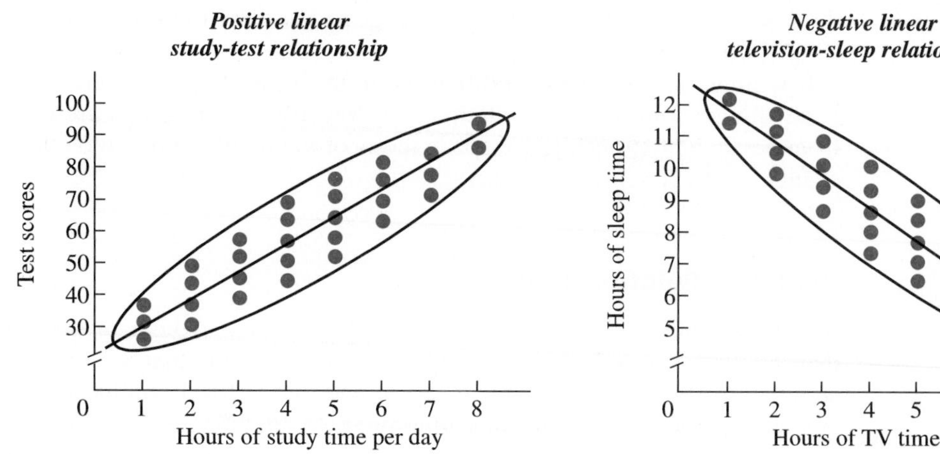

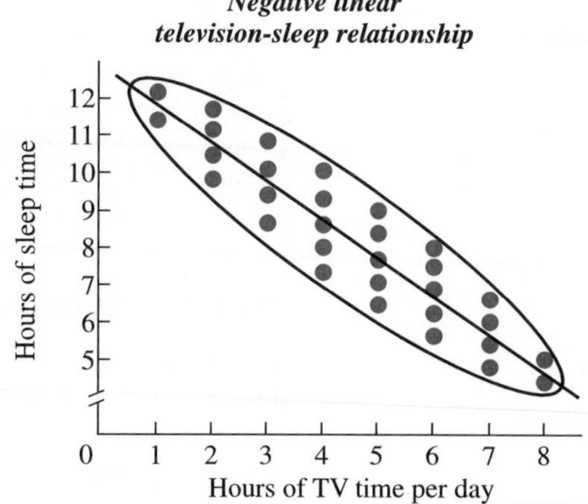

linear relationships. The scatterplot on the left illustrates the relationship between the amount of time students study and their test performance. The scatterplot on the right illustrates the relationship between the number of hours students spend watching television and the amount of time they spend sleeping.

Both scatterplots show a relationship because they do not form horizontal ellipses. Both indicate linear relationships because they are best summarized by a straight line. Any scatterplot that forms a *slanted ellipse* and has a *slanted, straight* regression line indicates that a linear relationship is present. In a **linear relationship,** as the X scores increase, the Y scores tend to change in only one direction. On the left, as students study longer, their grades tend only to increase. On the right, as students watch more television, their sleep tends only to decrease.

The above scatterplots reflect two different types of linear relationships. The type of linear relationship depends on the *direction* in which the Y scores change. The study-test relationship is an example of a positive relationship. In a **positive linear relationship,** as the scores on the X variable increase, the scores on the Y variable also tend to increase. Thus, low X scores are paired with low Y scores, and high X scores are paired with high Y scores. Any relationship that fits the general pattern "the more X, the more Y" is a positive linear relationship. Examples of positive linear relationships include the more you eat, the more you weigh; and the more long-distance phone calls you make, the more you pay. You can remember that such relationships are positive by remembering that as the X scores increase, the Y scores change in the direction away from zero, toward higher *positive* scores.

On the other hand, the television-sleep relationship is an example of a negative relationship. In a **negative linear relationship,** as the scores on the X variable increase, the scores on the Y variable tend to decrease. Low X scores are paired with high Y scores, and high X scores are paired with low Y scores. Any relationship that fits the general pattern "the more X, the less Y" is a negative linear relationship. Examples of negative linear relationships include the more you study statistics, the less difficult they are; and the more you eat, the less hungry you are. You can remember that such relationships are negative by remembering that as the X scores increase, the Y scores change in the direction toward zero, heading for *negative* scores.

It is important to understand that the term *negative* does not mean that there is anything wrong with the relationship. Negative relationships are no different from positive relationships *except* in terms of the direction in which the Y scores change as the X scores increase.

Nonlinear Relationships

If a relationship is not linear, then it is called nonlinear. *Nonlinear* does not mean that the data cannot be summarized by a line; it means that the data cannot be summarized by a *straight* line. Thus, another name for a nonlinear relationship is a curvilinear relationship. In a **nonlinear,** or **curvilinear, relationship,** as the X scores change, the Y scores do not tend to *only* increase or *only* decrease: the Y scores change their direction of change.

Nonlinear relationships come in many different shapes. Figure 7.4 shows two common nonlinear relationships. The scatterplot on the left illustrates the relationship between subjects' age and the amount of time they require to move from one place to another. Very young children locomote slowly, but as age increases, movement time decreases. Beyond a certain age, however, the time scores change direction so that as age continues to increase, movement time increases. Because of the shape of the scatterplot, such a relationship is called a *U-shaped function.* The scatterplot on the right illustrates the relationship between the number of alcoholic drinks subjects consume and their sense of feeling well. At first people tend to report feeling better as they drink, but beyond a certain point, drinking

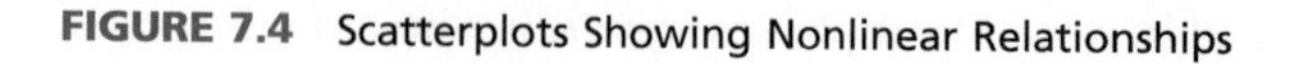

FIGURE 7.4 Scatterplots Showing Nonlinear Relationships

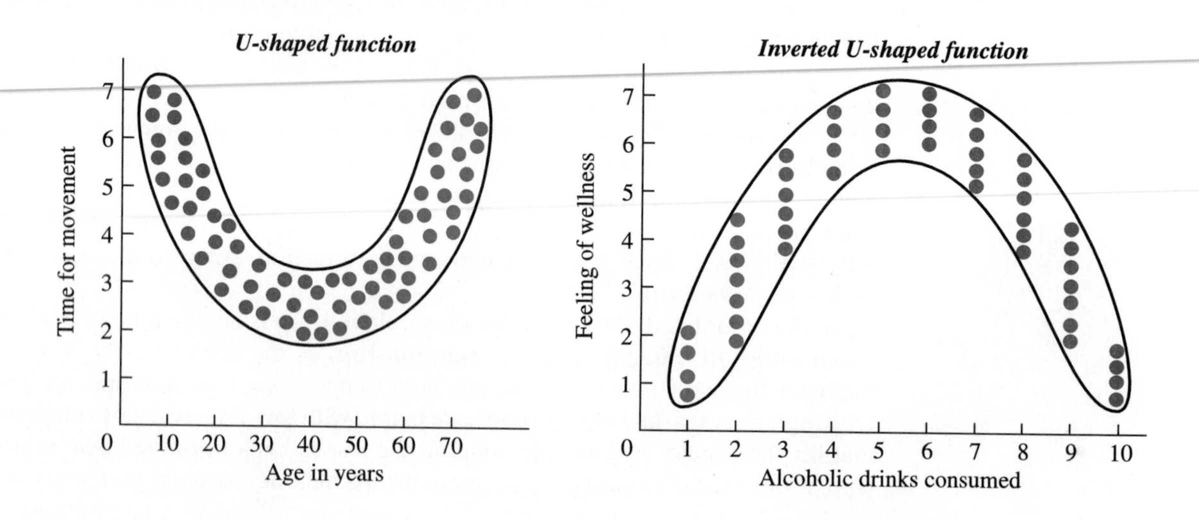

more makes them feel progressively worse. Such a relationship is called an *inverted U-shaped function.*

Curvilinear relationships may be even more complex than those shown in Figure 7.4, producing a wavy pattern that repeatedly changes direction. However, correlational research in psychology focuses almost entirely on linear relationships.

How the Correlation Coefficient Describes the Type of Relationship

Although there are correlation coefficients that describe nonlinear relationships, in this chapter we discuss only linear correlation. How do we know whether our data form a linear relationship? We make a scatterplot of the data. If the scatterplot seems to be best fit by a straight line, then linear correlation is appropriate. Sometimes, however, we may want to describe the extent to which a nonlinear relationship has a linear component and partially fits a straight line. In this case linear correlation is also appropriate. However, we do not summarize a nonlinear relationship by computing a linear correlation coefficient. Describing a nonlinear relationship with a straight line is like putting a round peg into a square hole: the relationship will not fit a straight line very well, and the correlation coefficient will not accurately describe the relationship.

With a correlation coefficient we communicate two things about the type of relationship we are describing. First, by virtue of the fact that we compute a linear correlation coefficient, we communicate that we are describing a linear relationship. Second, we communicate whether the linear relationship is positive or negative. If the coefficient we compute has a minus sign in front of it, we know that the linear relationship is negative. If the computed coefficient does not have a minus sign, then it is positive and we put a plus sign in front of it to indicate a positive linear relationship. Thus, a positive correlation coefficient indicates a positive linear relationship, and a negative correlation coefficient indicates a negative linear relationship.

The other characteristic of a relationship communicated by the correlation coefficient is the strength of the relationship.

STRENGTH OF THE RELATIONSHIP

Recall from Chapter 2 that, although in any relationship the Y scores tend to change consistently as the X scores change, different relationships will exhibit different degrees of consistency. The **strength of a relationship** is the extent to which one value of the Y variable is consistently paired with one and only one value of the X variable. The strength of a relationship is also referred to as the *degree of association* between the two variables. In stronger relationships, there is more consistently one value of Y or close to one value of Y associated with only one value of X. In weaker relationships, there is not one value of Y or close to one value of Y consistently associated with only one value of X.

The absolute value of the correlation coefficient (the size of the number we calculate) indicates the strength of the relationship. The largest value we can obtain is 1.0, and the smallest value we can obtain is 0.0. Thus, when we include the positive or negative sign, the correlation coefficient may be any value from -1.0 to 0.0 or from 0.0 to $+1.0$. The *larger* the absolute value of the coefficient, the *stronger* the relationship. In other words, the closer the coefficient is to ± 1.0, the more consistently one value of Y is paired with one and only one value of X.

STAT ALERT A correlation coefficient contains two components: the sign and the absolute value. The sign indicates either a positive or a negative linear relationship. The larger the absolute value, the stronger the relationship.

As we will eventually see, computing the correlation coefficient is not difficult. The difficulty comes in interpreting the strength of the relationship, because correlation coefficients do not directly measure units of "consistency." Thus, if one correlation coefficient is $+.40$ and another is $+.80$, we *cannot* conclude that $+.80$ describes a relationship that has twice as much consistency as the one with $+.40$. Instead, we evaluate any correlation coefficient by comparing it to the extreme values of 0.0 and ± 1.0. First we will discuss a coefficient of ± 1.0.

Perfect Association

A correlation coefficient of ± 1.0 describes a perfect linear relationship that is as strong as it can be. Figure 7.5 shows the data and resulting scatterplots for examples in which the correlation coefficient is $+1.0$ and -1.0.

There are four ways to think about what a correlation coefficient of ± 1.0 tells us about a relationship. First, it tells us that *every* subject who obtains a particular X score obtains one and only one value of Y. Every time X changes, the Y scores all change to one new value. Thus, ± 1.0 indicates a one-to-one correspondence, or perfectly consistent pairing, between the X and Y scores.

Second, a coefficient of ± 1.0 indicates that there are no differences among the Y scores associated with a particular X. In statistics, we describe differences between scores as variability, so ± 1.0 indicates that there is no variability, or spread, in the Y scores at each X.

Third, a coefficient of ± 1 ensures perfect predictability. Pretend that we do not know a particular subject's Y score. Since each Y score is associated with only one X score, if we know the subject's X score, then we know the subject's Y score. Thus, when the coefficient is ± 1.0, knowing subjects' X scores allows us to perfectly predict their corresponding Y scores. (We'll discuss how we predict the Y scores in the next chapter.)

Fourth, because it indicates that there is no spread in the Ys at each X, a coefficient of ± 1.0 tells us that the data points for the subjects at an X are all on top of one another. When we summarize this scatterplot with a regression line, all of the data points lie *on* the regression line.

FIGURE 7.5 Data and Scatterplots Reflecting Perfect Positive and Negative Correlations

Perfect positive coefficient = *+1.0*	
X	*Y*
1	2
1	2
1	2
3	5
3	5
3	5
5	8
5	8
5	8

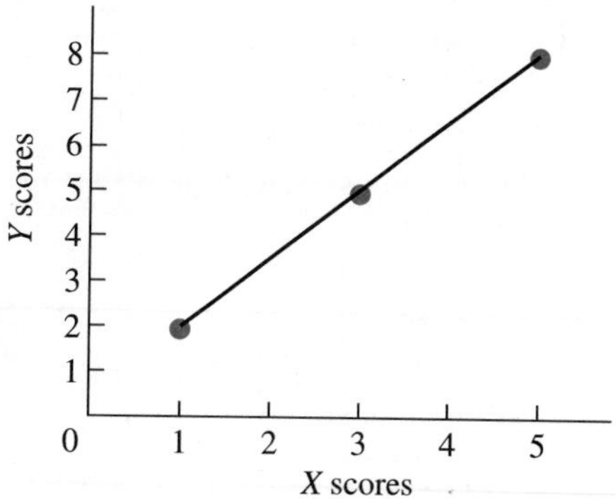

Perfect negative coefficient = *−1.0*	
X	*Y*
1	8
1	8
1	8
3	5
3	5
3	5
5	2
5	2
5	2

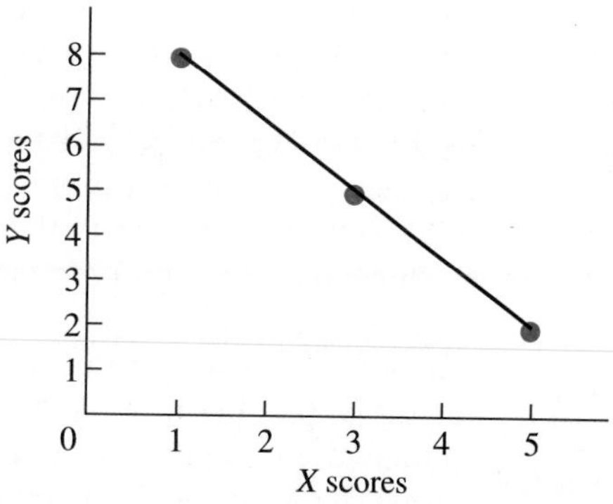

Intermediate Association

The essence of understanding any other value of the correlation coefficient is this: the strength of a linear relationship is the extent to which the data tend to form a perfect linear relationship. Thus, any correlation coefficient that is not equal to ± 1.0 indicates that only to some degree do the data form a perfect linear relationship.

> **STAT ALERT** The absolute value of a correlation coefficient indicates the extent to which the data conform to a perfect linear relationship.

The way to interpret any value of the correlation coefficient is to compare it to ± 1.0. For example, Figure 7.6 on the next page shows data and the resulting scatterplot that produce a correlation coefficient of $+.98$. Again we interpret the coefficient in four ways. First, a correlation coefficient with an absolute value less than 1.0 indicates that there is less than perfectly consistent association.

FIGURE 7.6 Data and Scatterplot Reflecting a Correlation Coefficient of +.98

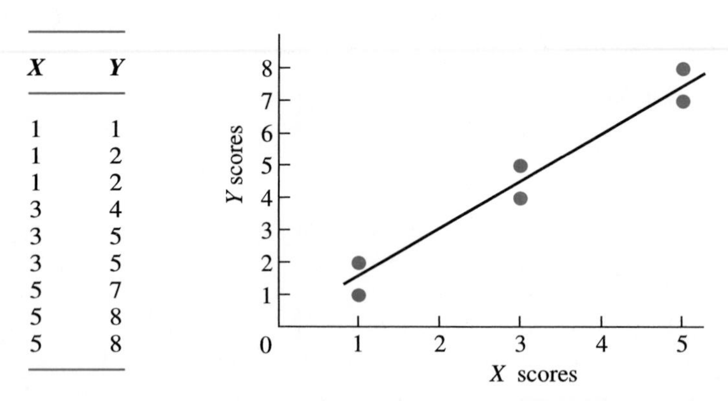

X	*Y*
1	1
1	2
1	2
3	4
3	5
3	5
5	7
5	8
5	8

Every subject obtaining a particular *X* did not obtain the exact same *Y*. However, here, since a coefficient of +.98 is close to a value of ±1.0, it tells us that there is "close" to perfect consistency between the *X* and *Y* scores. (That is, when subjects do produce different values of *Y* at the same *X*, all of their *Y* scores are relatively close to each other.)

Second, when the correlation coefficient is not ±1.0, *different Y* scores are associated with a single *X* score, so there is variability between the *Y* scores at each *X*. You can see this variability in the scatterplot in Figure 7.6 by looking at the vertical spread in the *Y* scores at each *X*. Subjects obtaining an *X* of 1 obtained a *Y* of 1 or 2, and subjects obtaining an *X* of 3 scored a *Y* of 4 or 5. However, since +.98 is close to +1.0, it tells us that the variability of the *Y* scores at each *X* is small *relative* to the overall variability of all *Y* scores in the data. Think of it this way: over the entire sample, *Y* scores are between 1 and 8, so the overall range is 8. At each *X* score, however, the *Y*s span a range of only 1. It is this small variability in *Y* at each *X* relative to the overall variability in all *Y* scores that produces a correlation coefficient close to +1.0.

Third, again pretend that we do not know subjects' *Y* scores. When the correlation coefficient is not ±1.0, knowing each subject's *X* score only allows us to predict *around* what his or her *Y* score will be. However, since a coefficient of +.98 is close to +1.0, we know that the predicted *Y* score will be close to the actual *Y* score the subject obtained.

Fourth, because there is now spread in the *Y*s at each *X*, all data points do not fall *on* the regression line: they fall above and below it. However, a coefficient of +.98 is close to +1.0, so we know that the *Y* scores are close to, or hug, the regression line. The spread between the lowest and highest *Y* scores at each *X* is small, resulting in a scatterplot that is a narrow, or skinny, ellipse. In fact, the absolute value of the correlation coefficient always tells us how skinny the scatterplot is. When the coefficient is ±1.0, the scatterplot forms a straight line, which is the skinniest ellipse possible. The closer the coefficient is to ±1.0, the skinnier the scatterplot, and vice versa.

FIGURE 7.7 Data and Scatterplot Reflecting a Correlation Coefficient of −.28

X	Y
1	9
1	6
1	3
3	8
3	6
3	3
5	7
5	5
5	1

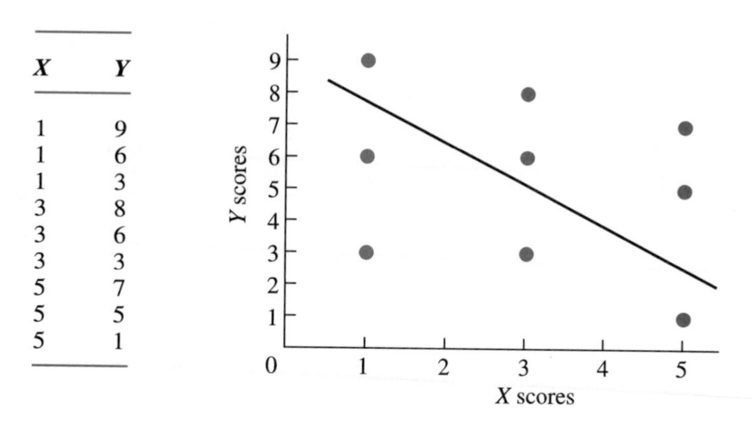

The key to understanding the strength of a relationship is this:

> **As the variability, or spread, in the *Y* scores at each *X* becomes relatively larger, the relationship becomes weaker.**

Therefore, as the variability in the *Y*s at each *X* becomes relatively larger, the value of the correlation coefficient approaches 0.0. In Figure 7.7 we see an example of data and the resulting scatterplot that produce a much smaller correlation coefficient of −.28. The fact that this is a negative relationship has nothing to do with its strength. Rather, as the scatterplot illustrates, the spread in the *Y* scores at each *X* is relatively large, so there tends to be great overlap between the *Y* scores at one *X* and those at another *X*. Therefore, instead of seeing one value of *Y* consistently paired with only one value of *X*, we tend to see the same value of *Y* paired with *different* values of *X*.

Thus, on a scale of 0.0 to ±1.0, a coefficient of −.28 is not very close to ±1.0, so we know that this relationship is not very close to forming a perfectly consistent linear relationship. In fact, only barely does one value of *Y* or even close to one value of *Y* tend to be associated with one value of *X*. In addition, such a small coefficient indicates that the variability in the *Y*s at each *X* is almost as large as the spread across all *Y* scores in the data. Further, this means that knowing a subject's *X* score will not allow us to come very close when predicting the actual *Y* score that he or she obtained. Finally, the coefficient tells us that there is a large vertical distance between the *Y*s at each *X*, so the scatterplot is fat and does not hug the regression line.

> ***STAT ALERT*** Greater variability in the *Y* scores at each *X* works against the strength of a relationship and decreases the size of the correlation coefficient.

Zero Association

The lowest possible value of the correlation coefficient is 0.0, indicating that no linear relationship is present. Figure 7.8 on the next page shows an example of data and the resulting scatterplot that produce a correlation coefficient of 0.0. A

FIGURE 7.8 Data and Scatterplot Reflecting a Correlation Coefficient of 0.0

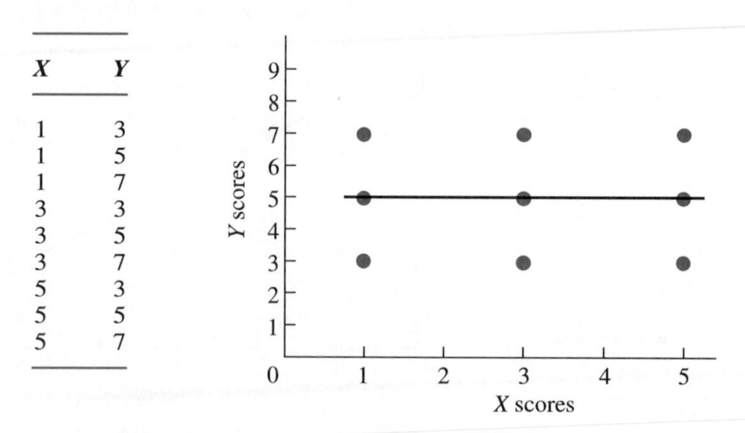

X	*Y*
1	3
1	5
1	7
3	3
3	5
3	7
5	3
5	5
5	7

scatterplot having this shape is as far from forming a slanted straight line as possible, and a correlation coefficient of 0.0 is as far from a correlation coefficient of ± 1.0 as possible. Therefore, we know that there is no value or values of *Y* consistently associated with one and only one value of *X*. Instead the *Y*s found at one *X* are the same as the *Y*s found at any other *X*. This also means that knowing a subject's *X* score will not in any way help us to predict the corresponding *Y* score. Finally, this indicates the spread in *Y* at any *X* equals the overall spread of *Y* in the data, producing a scatterplot that is a circle or horizontal ellipse that in no way hugs the regression line.

> ***STAT ALERT*** The larger the correlation coefficient, whether positive or negative, the stronger the relationship and the closer it is to forming a perfect linear relationship. The stronger the relationship, the less the *Y*s are spread out at each *X*, the more accurately we can predict a subject's *Y* score by knowing the subject's *X* score, and the closer the *Y* scores come to all falling on a straight line.

Although theoretically a correlation coefficient may be as large as ± 1.0, in real research such values do not occur. Remember that the *X* and *Y* scores reflect the behaviors of living organisms, and that living organisms, because of individual differences, do not show a great deal of consistency. Therefore, in psychological research, a correlation coefficient in the neighborhood of $\pm .50$ is considered to be quite respectable, and coefficients above $\pm .50$ are downright impressive. A correlation of ± 1.0 is so unlikely to occur with real data that, if you ever obtain one, you should assume you've made a computational error. (If you obtain a coefficient greater than ± 1.0, you have definitely made an error, because ± 1.0 indicates a perfect relationship and you cannot do better than that.)

COMPUTING THE CORRELATION COEFFICIENT

In the following sections we will discuss the three most common linear correlation coefficients: the *Pearson correlation coefficient,* the *Spearman rank-order correla-*

tion coefficient, and the *point-biserial correlation coefficient.* In each case a value of between 0 and ± 1.0 describes the degree of linear relationship in our sample data, and everything we have said previously about the interpretation of the correlation coefficient applies. The major difference among the three coefficients is that they are calculated differently. Each is designed for different types of variables, and—as when we are selecting other statistical procedures—the specific coefficient we compute in a particular situation depends on the type of scores we have measured.

The Pearson Correlation Coefficient

By far the most common correlation coefficient in psychological research is the Pearson correlation coefficient. The **Pearson correlation coefficient** is used to describe the linear relationship between two variables that are both interval or ratio variables. (Technically this statistic is the Pearson Product Moment Correlation Coefficient, but it is usually called the Pearson correlation coefficient. It was invented by Karl Pearson.) The symbol for the Pearson correlation coefficient is r. When you see r, think "relationship." (All of the examples in the previous section involved r.)

The statistical basis for r is that it compares how consistently each value of Y pairs with each value of X in a linear fashion. In Chapter 6 we saw that to compare scores from different variables, we first transform the scores from each variable into z-scores. Essentially, by calculating r, we transform each Y score into a z-score (call it z_Y), and we transform each X score into a z-score (call it z_X). We then determine the correspondence between each z_Y and its paired z_X. Usually, every pair will not exhibit the same degree of correspondence. Therefore, r indicates the "average" amount of correspondence between the z_Y's and the z_X's in the sample. The Pearson correlation coefficient is defined as

$$r = \frac{\Sigma(z_X z_Y)}{N}$$

Mathematically, when we multiply each z_X times the corresponding z_Y of the pair, sum the products, and then divide by N (the number of pairs), we have computed the average correspondence within all z_Y–z_X pairs.

Of course, there's an easier way to compute r.

Computing the Pearson correlation coefficient We can derive a computational formula from the above definitional formula. First we replace the symbols z_X and z_Y with their formulas. Then, in these formulas, we replace the symbols for the mean and standard deviation with their own computational formulas. As you can imagine, this produces a monster of a formula. When we reduce the formula to its simplest form, we have the smaller monster below.

THE COMPUTATIONAL FORMULA FOR THE PEARSON CORRELATION COEFFICIENT IS

$$r = \frac{N(\Sigma XY) - (\Sigma X)(\Sigma Y)}{\sqrt{[N(\Sigma X^2) - (\Sigma X)^2][N(\Sigma Y^2) - (\Sigma Y)^2]}}$$

As an example, say that we have collected scores from ten subjects on two variables: the number of times they visited a doctor in the last year and the number of glasses of orange juice they drink daily. We want to describe the linear relationship between juice-drinking and doctor visits, so, because we have ratio scores on both variables, we compute r. Table 7.2 shows a good way to set up the data. Then, from the computational formula for r we see that we need the values of ΣX, ΣX^2, $(\Sigma X)^2$, ΣY, ΣY^2, $(\Sigma Y)^2$, ΣXY, and N. First we find each XY by multiplying each X times its corresponding Y, as shown in the far right-hand column in Table 7.2. Then, summing the columns gives us ΣX, ΣX^2, ΣY, ΣY^2, and ΣXY. Squaring ΣX and ΣY gives us $(\Sigma X)^2$ and $(\Sigma Y)^2$.

Once we have obtained these quantities, we put them in the formula for r. Thus,

$$r = \frac{N(\Sigma XY) - (\Sigma X)(\Sigma Y)}{\sqrt{[N(\Sigma X^2) - (\Sigma X)^2][N(\Sigma Y^2) - (\Sigma Y)^2]}}$$

becomes

$$r = \frac{10(52) - (17)(47)}{\sqrt{[10(45) - 289][10(275) - 2209]}}$$

Let's compute the numerator first. We multiply 10 times 52, which is 520, and we multiply 17 times 47, which is 799. Rewriting the formula, we have

$$r = \frac{520 - 799}{\sqrt{[10(45) - 289][10(275) - 2209]}}$$

We complete the numerator by subtracting 799 *from* 520, which is -279. (Note the negative sign.)

Now compute the denominator. First we perform the operations within each bracket. In the left bracket, we multiply 10 times 45, which is 450, and from

TABLE 7.2 Sample Data for Computing the r Between Orange Juice Consumed (the X Variable) and Doctor Visits (the Y Variable)

	Glasses of juice per day		*Doctor visits per year*		
Subject	X	X^2	Y	Y^2	XY
1	0	0	8	64	0
2	0	0	7	49	0
3	1	1	7	49	7
4	1	1	6	36	6
5	1	1	5	25	5
6	2	4	4	16	8
7	2	4	4	16	8
8	3	9	4	16	12
9	3	9	2	4	6
10	4	16	0	0	0
$N = 10$	$\Sigma X = 17$	$\Sigma X^2 = 45$	$\Sigma Y = 47$	$\Sigma Y^2 = 275$	$\Sigma XY = 52$
	$(\Sigma X)^2 = 289$		$(\Sigma Y)^2 = 2209$		

that we subtract 289, obtaining 161. In the right bracket, we multiply 10 times 275, which is 2750, and from that we subtract 2209, obtaining 541. Rewriting one more time, we have

$$r = \frac{-279}{\sqrt{[161][541]}}$$

Now we multiply the quantities in the brackets together: 161 times 541 equals 87,101. After taking the square root of 87,101, we have

$$r = \frac{-279}{295.129}$$

We divide, and there you have it: $r = -.95$.

The value of r is not greater than ± 1.0, so our calculations may be correct. Also, we have computed a negative r, and in the raw scores we see that there is a negative relationship: as the orange juice scores increase, the number of doctor visits decreases. (If you have any doubt, make a scatterplot.) Had this been a positive relationship, the numerator of the formula would not contain a negative number and r would not be negative.

Thus, we conclude that there is a negative linear relationship between juice-drinking and doctor visits. On a scale of 0.0 to ± 1.0, where 0.0 is no relationship and ± 1.0 is a perfect linear relationship, this relationship is a $-.95$. Relatively speaking, this is a very strong linear relationship: each amount of orange juice is associated with one relatively small range of doctor visits, and as juice scores increase, doctor visits consistently decrease. (Of course, if the correlation were this large in real life, we'd all be drinking a lot more orange juice, thinking (incorrectly) that this would prevent doctor visits.)

The Spearman Rank-Order Correlation Coefficient

Sometimes our data involve ordinal or rank-order scores (first, second, third, etc.). The **Spearman rank-order correlation coefficient** describes the linear relationship between two variables measured using ranked scores. The symbol for the Spearman correlation coefficient is r_S. (The subscript s stands for Spearman; Charles Spearman invented this coefficient.)

In psychological research, ranked scores often arise in cases where variables are difficult to measure quantitatively. Instead, we evaluate each subject by making qualitative judgments, and then we use these judgments to rank-order the subjects. We use r_S to correlate the ranks on two such variables. Or, if we want to correlate one ranked variable with one interval or ratio variable, we transform the interval or ratio scores into ranked scores (we might rank the subject with the highest interval score as 1, the subject with the second highest score as 2, and so on). Either way that we obtain the ranks, r_S tells us the extent to which subjects' ranks on one variable consistently match their ranks on the other variable to form a linear relationship. If every subject has the same rank on both variables, r_S will equal $+1.0$. If every subject's rank on one variable is the opposite of his or her rank on the other variable, r_S will equal -1.0. If there is only some degree of consistent pairing of the ranks, r_s will be between 0.0 and ± 1.0, and if there is no consistent pairing, r_s will equal 0.0.

Because r_s describes the consistency with which ranks match, or agree, one use of r_s is to determine the extent to which two observers agree. For example, say that we ask two observers to judge how aggressively a sample of children behave while playing. Each observer assigns the rank of 1 to his or her choice for most aggressive child, 2 to the second-most-aggressive child, and so on. Figure 7.9 shows the sets of ranked scores and the resulting scatterplot the two observers might produce for 9 subjects. In creating the scatterplot and computing r_s, we treat each observer as a variable: the scores on one variable are the rankings assigned by one observer to the subjects, and the scores on the other variable are the rankings assigned by the other observer. Judging from the scatterplot, it appears that there is a positive relationship here. To describe this relationship, we compute r_s.

Computing the Spearman correlation coefficient

THE COMPUTATIONAL FORMULA FOR THE SPEARMAN RANK-ORDER CORRELATION COEFFICIENT IS

$$r_s = 1 - \frac{6(\Sigma D^2)}{N(N^2 - 1)}$$

N is the number of pairs of ranks, and D is the difference between the two ranks in each pair. (Notice that the formula always contains the 6 in the numerator.)

FIGURE 7.9 Sample Data for Computing r_s Between Rankings of Observer A and Rankings of Observer B

Subject	*Observer A:* *X*	*Observer B:* *Y*
1	4	3
2	1	2
3	9	8
4	8	6
5	3	5
6	5	4
7	6	7
8	2	1
9	7	9

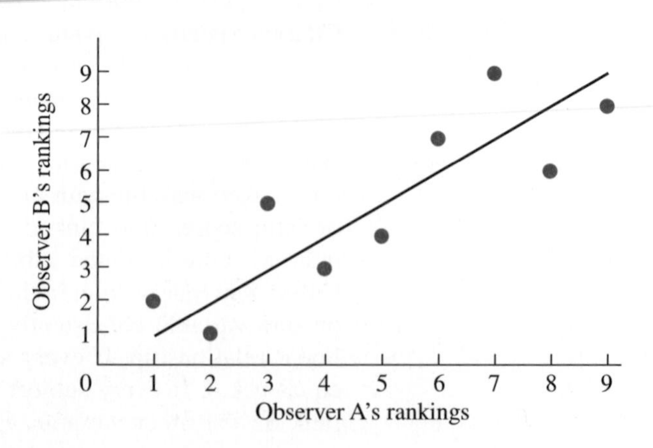

TABLE 7.3 Data Arrangement for Computing r_s

Subject	*Observer A:* *X*	*Observer B:* *Y*	*D*	D^2
1	4	3	1	1
2	1	2	−1	1
3	9	8	1	1
4	8	6	2	4
5	3	5	−2	4
6	5	4	1	1
7	6	7	−1	1
8	2	1	1	1
9	7	9	−2	4
				$\Sigma D^2 = 18$

When using this formula, first arrange the data as shown in Table 7.3. The column labeled *D* contains the difference between the rankings in each pair. Here I have subtracted each *Y* from the corresponding *X*, but you can subtract *X* from *Y*. In the right-hand column, after finding the *D*s, compute D^2 by squaring the difference in each pair. Finally, determine the sum of the squared differences, ΣD^2 (here ΣD^2 is 18). To compute r_s, we also need *N*, the number of *X*-*Y* pairs (here $N = 9$), and N^2 ($9^2 = 81$). Placing these quantities in the formula, we have

$$r_s = 1 - \frac{6(\Sigma D^2)}{N(N^2 - 1)} = 1 - \frac{6(18)}{9(81 - 1)}$$

In the numerator, 6 times 18 equals 108. In the denominator, $81 - 1$ is 80, and 9 times 80 is 720. Now we have

$$r_s = 1 - \frac{108}{720}$$

After dividing, we have

$$r_s = 1 - .15$$

Subtracting yields

$$r_s = +.85$$

Thus, on a scale of 0 to ± 1, the rankings form a linear relationship to the extent that $r_s = +.85$. This tells us that a child receiving a particular ranking from one observer tended to receive close to the same ranking from the other observer.

Notice that a negative r_s would indicate that the observers did not agree, because low ranks by one observer would tend to be paired with high ranks by the other observer, and vice versa.

Tied ranks A **tied rank** occurs when two subjects receive the same rank-order score on the *same* variable. A problem arises with tied ranks because the computed

TABLE 7.4 Sample Data Containing Tied Ranks

Runner	*Race X*	*Race Y*		*To resolve ties*		*New Y*
				Tie uses up ranks		
A	4	1 }	· · · →	1 and 2,	· · · →	{ 1.5
B	3	1 }		becomes 1.5		{ 1.5
C	2	2}	· · · →	Becomes 3rd	· · · →	{3
D	1	3}	· · · →	Becomes 4th	· · · →	{4
·	·	·				·
·	·	·				·
·	·	·				·

value of r_s will be incorrect. Therefore we must first resolve—correct—these tied ranks before computing r_s. As an example, say that we wish to correlate the finishing positions of the runners in two races. Table 7.4 shows such data: In race *Y*, runners A and B were tied for first place.

We resolve tied ranks using the following logic: if runners A and B had not tied for first place, then one of them would have been first and one would have been second. *To resolve any tie, we assign each tied subject the mean of the ranks that would have been used had there not been a tie.* The mean of 1 and 2 is 1.5, so, as in Table 7.4, we assign Runners A and B each a new *Y* score of 1.5. Now, in a sense, we have used up both first and second place (1 and 2), so we assign runner C a new *Y* of 3. (After all, he was the third person to cross the finish line.) Likewise, we assign runner D the rank of 4. If there were additional runners, we would assign each of them their new ranks based on what we have done above: e.g., if runners E, F, and G were originally tied for fourth place in race *Y*, they would now be tied for fifth. We would then resolve this tie using the mean of 5, 6, and 7, and runner H would be ranked 8.

Once we have resolved all ties in the *X* and *Y* ranks, we compute r_s using the new ranks and the above formula.

The Point-Biserial Correlation Coefficient

Sometimes we want to correlate the scores from a continuous interval, or ratio variable with the scores from a dichotomous variable (which has only two categories). The **point-biserial correlation coefficient** describes the linear relationship between the scores from one continuous variable and one dichotomous variable. The symbol for the point-biserial correlation coefficient is r_{pb} (the pb stands for point-biserial, and no, no one named Point and Biserial invented this one).

Computing the point-biserial correlation coefficient Say that we wish to correlate the dichotomous variable of gender (male/female) with the interval scores from a personality test. We cannot quantify "male" and "female," so we first arbitrarily assign numbers to represent these categories. We can assign any numbers, but an easy system is to use 1 to indicate male and 2 to indicate female. Think of each number as indicating whether a person scored "male" or "female." Then r_{pb} describes how consistently certain personality test scores are paired with one and only one gender score.

THE COMPUTATIONAL FORMULA FOR THE POINT-BISERIAL CORRELATION COEFFICIENT IS

$$r_{\text{pb}} = \left(\frac{\overline{Y}_2 - \overline{Y}_1}{S_Y}\right)(\sqrt{pq})$$

We always call the dichotomous variable the X variable and the interval or ratio variable the Y variable. Then $\overline{Y}_1$ stands for the mean of the Y scores for one of the two groups of the dichotomous variable. (In our example, $\overline{Y}_1$ will be the mean personality score for males.) The symbol $\overline{Y}_2$ stands for the mean of the Y scores for the other group. ($\overline{Y}_2$ will be the mean personality score for females.) The S_Y is the standard deviation of *all* the Y scores. The p stands for the proportion of all subjects that is in one of the groups of the dichotomous variable, and q stands for the proportion of all subjects in the other group. Each proportion is equal to the number of subjects in that group divided by the total N of the study.

Say that we tested 10 subjects and obtained the scores and scatterplot shown in Figure 7.10. The first thing we must do is compute the standard deviation of Y, S_Y. Below is the formula for the sample standard deviation, written using Ys. We substitute into the formula the data in Figure 7.10:

$$S_Y = \sqrt{\frac{\Sigma Y^2 - \frac{(\Sigma Y)^2}{N}}{N}} = \sqrt{\frac{26019 - \frac{(503)^2}{10}}{10}} = 8.474$$

To compute r_{pb}, we see that the first four subjects scored "male," and their mean test score, $\overline{Y}_1$, is 45.50. The remaining six subjects scored "female," and their mean test score, $\overline{Y}_2$, is 53.50. The S_Y for all Y scores is 8.474. We will call

FIGURE 7.10 Example Data for Computing r_{pb}

Subject	*Gender: X*	*Test: Y*	
	Males		
1	1	50	
2	1	38	$\overline{Y}_1 = 45.50$
3	1	41	
4	1	53	
	Females		
5	2	60	
6	2	50	
7	2	44	$\overline{Y}_2 = 53.50$
8	2	68	
9	2	53	
10	2	46	
$N = 10$		$\Sigma Y = 503$	

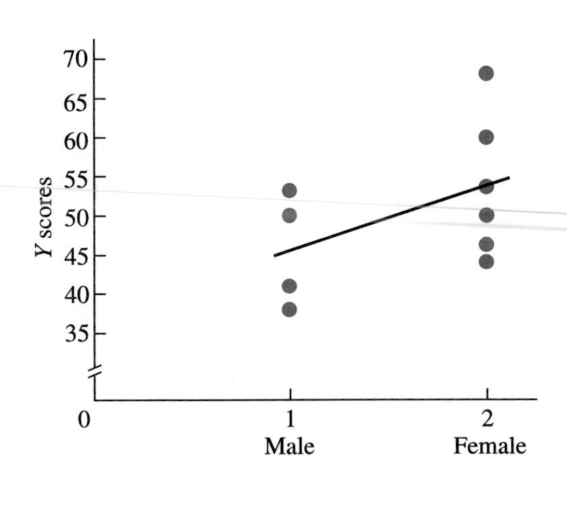

p the proportion of subjects who scored "male," so p is 4/10, or .40. We will call q the proportion of subjects scoring "female," which is 6/10, or .60.

Now we put these values into the formula for r_{pb}.

$$r_{pb} = \left(\frac{\overline{Y}_2 - \overline{Y}_1}{S_Y}\right)(\sqrt{pq}) = \left(\frac{53.50 - 45.50}{8.474}\right)(\sqrt{(.40)(.60)})$$

When we subtract 45.50 from 53.50, we obtain 8.00, so we have

$$r_{pb} = \left(\frac{8.00}{8.474}\right)(\sqrt{(.40)(.60)})$$

Dividing 8.00 by 8.474 gives .944. Also, .40 times .60 is .24, and the square root of .24 is .489. Thus, we have

$$r_{pb} = .944(.489)$$

Multiplying, we find that

$$r_{pb} = +.462$$

Thus, with rounding, our r_{pb} is $+.46$. We interpret this statistic as indicating that, on a scale of 0.0 to ± 1.0, we have a medium-strength relationship: test scores close to one value tend to be associated with one gender, and test scores close to a different value tend to be associated with the other gender.

It is important to recognize that in this example the dichotomous variable is a qualitative variable, so the scores of 1 and 2 do not reflect an amount of the gender variable. Therefore, our r_{pb} is positive only because we arbitrarily assigned a 1 to males and a 2 to females. Had we assigned females a 1 and males a 2, their locations on the X axis of the scatterplot would be reversed, and we would have a negative relationship. Likewise, the formula is $\overline{Y}_2 - \overline{Y}_1$, so above we had $53.50 - 45.50$. Had we chosen to call the female mean $\overline{Y}_1$ and the male mean $\overline{Y}_2$, we would have had $45.50 - 53.50$, which would have resulted in a negative r_{pb} of $-.46$. Thus, for any qualitative variable, the absolute value of r_{pb} will accurately describe the strength of the relationship, but whether it is positive or negative depends on how we have arbitrarily arranged the data.

The Restriction of Range Problem

In collecting scores for a correlation, it is important to avoid the restriction of range problem. **Restriction of range** arises when the range between the lowest and highest scores on one or both variables is small, or restricted.

Recall that the correlation coefficient reflects the spread in Y at each X *relative* to the overall spread in all Ys. Look at Figure 7.11. When we have the full range of X scores, the spread in the Y scores at each X turns out to be small relative to the overall variability in Y, and the data form a narrow ellipse that hugs the regression line. Therefore, r will be relatively large, and we will conclude that there is a strong relationship between these variables in nature.

If, however, we restrict the range of the X scores by collecting scores only between score A and score B in Figure 7.11, we will have just the data in the shaded part of the scatterplot. Now the spread in Ys at each X will be large

FIGURE 7.11 Scatterplot Showing Restriction of Range in X Scores

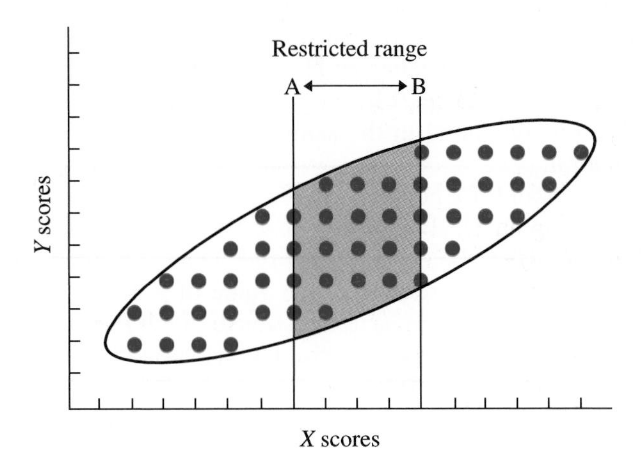

relative to the overall spread in all Ys in the shaded area. Further, the vertical width of the scatterplot is now close to equaling its horizontal width, producing a more circular scatterplot. Therefore, if we calculate a correlation coefficient using only the data in the shaded portion, it will be relatively small and we will conclude that there is a weak relationship in nature. However, this conclusion will be erroneous in that, if we had not restricted the range, we would have found a much stronger relationship. Thus, restriction of range leads to an erroneous *underestimate* of the degree of association between two variables. (Since either variable can be called the X or Y variable, restricting the range of Y has the same effect.)

How do we avoid restricting the range of our data? Generally, restriction of range occurs when researchers are too selective in obtaining scores or subjects. Thus, if we are interested in the relationship between subjects' high school grade averages and their subsequent salaries, we should not restrict the range of grades by including only honor students: we should measure all students to get the entire range of grades. Or, if we are correlating personality types with degree of emotional problems, we should not restrict our study to only college students. People with severe emotional problems tend not to be admitted to or remain in college, so using a college sample eliminates their scores and thus restricts the range. Instead, we should include the full range of subjects from the general population. In all cases, the goal is to allow a wide range of scores to occur on both variables, so that we have a complete description of the relationship.

CORRELATIONS IN THE POPULATION

As we have seen in previous chapters, ultimately we want to use our sample to describe the population we would find if we could measure it. Now we want to use our sample correlation coefficient to estimate the correlation coefficient we would obtain if we could correlate the *X* and *Y* scores of everyone in the population. However, remember that before we can estimate any population parameter with confidence, we *must* perform inferential statistics to determine whether the sample is representative. (In Chapter 11 we discuss the inferential procedures used with correlation coefficients.) If our data pass the inferential test, we can assume that the correlation found in the sample is about the same as we would find in the population.

The symbol for a population correlation coefficient is ρ. This is the Greek letter rho (the Greek r). Technically ρ is the symbol for the population correlation coefficient when the Pearson *r* is used. Thus, we compute *r* for a random sample, which gives us an estimate of the value of ρ. If our data involve ranked scores, then we compute r_s and have an estimate of the population coefficient symbolized by ρ_s. If our data involve one continuous variable and one dichotomous variable, we compute r_{pb} to estimate ρ_{pb}.

We interpret a population correlation coefficient in the same way we interpret the sample correlation coefficient. Thus, ρ stands for a number between 0.0 and ± 1.0, indicating either a positive or a negative linear relationship in the population. The larger the absolute value of ρ, the stronger the relationship: the greater the extent to which one and only one value of *Y* is associated with each *X* and the more closely the scatterplot for the population hugs the regression line. We interpret ρ_s and ρ_{pb} in the same manner.

FINALLY

It should be obvious why we compute a correlation coefficient whenever we wish to summarize a relationship. As in previous chapters, this descriptive statistic provides us with one number that allows us to envision and summarize the important information in a distribution. For example, consider again our study in which we collect data on nervousness and the amount of coffee consumed by a sample of subjects. Say that I tell you that the *r* in the study equals $+.50$. *Without even seeing the data,* you know there is a positive linear relationship such that as coffee consumption scores increase, nervousness scores also tend to increase. Further, because an absolute value of .50 indicates a reasonably strong, consistent relationship, you know that, given a subject's coffee score, you can achieve some accuracy in predicting his or her nervousness score. And you know that the data form a relatively narrow, elliptical scatterplot. There is no other type of statistic that so directly summarizes a relationship. Therefore, as we'll see in later chapters, even when you conduct an experiment, always think "correlation coefficient" to describe the strength and type of relationship you have observed.

Also be aware that correlation coefficients are tools that researchers use to check out various aspects of a research design, especially the reliability of their measurements. The term *reliability* means "consistency." Because a correlation coefficient indicates the degree to which *X* and *Y* scores are consistently paired, we demonstrate reliability by demonstrating a relatively large correlation. For example, we used the Spearman r_s to determine how reliably our observers ranked the aggressiveness of children (that is, to determine what is called "interrater reliability"). Likewise, a test is reliable if the subjects' scores we measured yesterday correlate positively with the same subjects' scores we obtain today—if low scores today are consistently paired with low scores from yesterday, and high scores today are consistently paired with high scores from yesterday. Although an in-depth discussion of reliability is beyond the scope of this text, remember that when researchers describe a measurement as reliable, they mean that the measurement is consistent over time and, typically, that they have examined the scores in a way that produced a high positive correlation.

CHAPTER SUMMARY

1. A *relationship* exists when the scores on one variable change consistently as the scores on another variable change, so that one value of Y tends to be associated with one value of X. No relationship is present when, as the X scores change, the Y scores do not form a consistent pattern of change.
2. In a correlational study, the researcher measures but does not manipulate the variables. The existence of a relationship does not necessarily mean that there is a causal relationship between the two variables. Therefore, we never infer causality based on correlational research.
3. There are three types of relationships. In a *positive linear relationship,* as the X scores increase, the Y scores tend to increase. In a *negative linear relationship,* as the X scores increase, the Y scores tend to decrease. In a *nonlinear,* or *curvilinear, relationship,* as the X scores increase, the Y scores do not only increase or only decrease.
4. A *scatterplot* is a graph that shows the location of each pair of X-Y scores in the data.
5. We summarize a scatterplot by drawing a best-fitting line called the *regression line.*
6. Circular or elliptical scatterplots that produce horizontal regression lines indicate no relationship. Sloped elliptical scatterplots with regression lines oriented so that as X increases, Y increases indicate a positive linear relationship. Sloped elliptical scatterplots with regression lines oriented so that as X increases, Y decreases indicate a negative linear relationship. Scatterplots producing curved regression lines indicate curvilinear relationships.
7. A *correlation coefficient* is a statistic that describes the *type* of relationship and the *strength* of the relationship.
8. A *positive correlation coefficient* indicates that a positive linear relationship is present. A *negative correlation coefficient* indicates that a negative relationship is present.
9. The absolute value of the coefficient may be as small as 0.0 (indicating no linear relationship) or as large as 1.0 (indicating a perfect linear relationship). The absolute value of the correlation coefficient indicates the *strength of the relationship,* which is the extent to which one value of Y is consistently

paired with one and only one value of X. When the coefficient equals ± 1.0, one value of Y is paired with one and only one value of X, and we can perfectly predict a subject's Y score if we know the subject's X score. The smaller the absolute value of the coefficient, the greater the variability, or spread, in the Y's associated with a particular value of X, and the less accurately we can predict Y scores.

10. A large absolute value of the coefficient also indicates a relatively narrow scatterplot, with most data points relatively close to the regression line. A small value indicates a relatively wide scatterplot, with many data points more distant from the regression line.

11. The *Pearson correlation coefficient, r,* describes the linear relationship between two interval or ratio variables.

12. The *Spearman rank-order correlation coefficient,* r_s, describes the linear relationship between two variables that have been measured using ranked scores.

13. The *point-biserial correlation coefficient,* r_{pb}, describes the linear relationship between scores from one continuous interval or ratio variable and one dichotomous variable.

14. The *restriction of range problem* occurs when the range of scores collected on one or both variables is too restricted. The resulting correlation coefficient erroneously underestimates the strength of the relationship that would be found if the range were not restricted.

15. Ultimately, after performing the appropriate inferential statistical procedures, we use a sample correlation coefficient to estimate the corresponding population correlation coefficient, symbolized by the Greek letter rho. We use r to estimate ρ, r_s to estimate ρ_s, and r_{pb} to estimate ρ_{pb}.

PRACTICE PROBLEMS

(Answers for odd-numbered problems are provided in Appendix E.)

1. What is the difference between an experiment and a correlational study in terms of how the researcher typically

   a. collects the data?

   b. examines the relationship?

2. What is the advantage of computing a correlation coefficient?
3. What are the two reasons why we can't conclude that we have demonstrated a causal relationship based on correlational research?

4. A researcher has just completed a correlational study. She measured the number of boxes of tissue purchased per week and the number of vitamin tablets consumed per week for each subject.

   a. Which is the independent and the dependent variable in this study?

   b. Which variable is X? Which is Y?

5. *a.* What is a scatterplot?

   b. What is a regression line?

6. What two characteristics of a linear relationship are described by a correlation coefficient?

7. *a.* Define a positive linear relationship.

   b. Define a negative linear relationship.

   c. Define a curvilinear relationship.

8. As the value of r approaches ± 1.0, what does it indicate about the following?

   a. The shape of the scatterplot

   b. The variability of the Y scores at each X

   c. The closeness of Y scores to the regression line

   d. The accuracy with which we can predict Y if X is known

9. What does a correlation coefficient equal to 0.0 indicate about the four characteristics in problem 8?

10. For each of the following, indicate whether it is a positive linear, negative linear, or nonlinear relationship.

    a. Quality of performance (Y) increases with increased arousal (X) up to an optimal arousal level; then quality of performance decreases with increased arousal.

    b. Heavier jockeys (X) tend to win fewer horse races (Y).

    c. As number of minutes of exercise increases each week (X), dieting individuals lose more pounds (Y).

    d. The number of bears in an area (Y) decreases as the area becomes increasingly populated by humans (X).

11. Poindexter sees the data in problem 10*d* and concludes, "We should stop people from moving into bear country so that we can preserve our bear population." What is the problem with Poindexter's conclusion?

12. For each of the following pairs of variables, give the symbol for the correlation coefficient you should compute.

    a. SAT scores and IQ scores

    b. Taste rankings of tea by experts with those by novices

    c. Presence or absence of a head injury and scores on a vocabulary test

    d. Finishing position in a race and amount of liquid consumed during the race

13. Poindexter finds that the correlation between the X variable of number of hours studied and the Y variable of number of errors on a statistics test is

$-.73$. He also finds that the correlation between the X variable of time spent taking the statistics test and the Y variable of number of errors on the test is $+.36$. He concludes that the time spent taking a test forms a stronger relationship with the number of errors than does the amount of study time.

a. Describe the relative shapes of the two scatterplots.

b. Describe the relative amount of variability in Y scores at each X in each study.

c. Describe the relative closeness of Y scores to the regression line in each study.

d. Is Poindexter correct in his conclusion? If not, what's his mistake?

14. In the correlation between orange juice consumed and number of doctor visits discussed in this chapter, does drinking more orange juice cause people to be more healthy so that they don't have to go to the doctor?

15. *a.* What is the restriction of range problem?

b. What produces a restricted range?

c. How is it avoided?

16. Foofy and Poindexter are investigating the relationship between IQ score and high school grade average. They have examined data from a large sample of students from PEST (the Program for Exceptionally Smart Teenagers) and computed $r = +.03$. They conclude that there is virtually no relationship between IQ and grade average. Should you agree or disagree with this conclusion? (Is there a problem with their study?)

17. *a.* What does ρ stand for?

b. How is the value of ρ determined?

c. What does ρ tell you?

18. We obtain a correlation of $+.20$ after measuring a sample of subjects' creativity test scores as a function of their intelligence test scores.

a. Can we conclude that this is similar to the relationship we'd find between all similar subjects found in nature?

b. Once we have performed the necessary procedures, what is the expected relationship between IQ and creativity?

c. Describe this relationship in terms of its consistency, its scatterplot, and whether it could be used to accurately predict creativity scores.

19. Why can't one obtain a correlation coefficient greater than ± 1.0?

20. A researcher measures the following scores for a group of subjects: the X variable is the number of errors on a math test, and the Y variable is the subjects' level of satisfaction with their performance.

a. With such ratio scores, what should the researcher conclude about this relationship?

b. How well will he be able to predict satisfaction scores using this relationship?

Subject	*Errors* *X*	*Satisfaction* *Y*
1	9	3
2	8	2
3	4	8
4	6	5
5	7	4
6	10	2
7	5	7

21. The data below reflect whether or not a subject is a college graduate (Y or N) and the score he or she obtained on a self-esteem test. To what extent is there a positive or negative linear relationship here?

Subject	*College Graduate* *X*	*Self-Esteem* *Y*
1	Y	8
2	Y	7
3	Y	12
4	Y	6
5	Y	10
6	N	2
7	N	8
8	N	6
9	N	1
10	N	9

22. In the data below, the X scores reflect subjects' rankings in a freshman class, and the Y scores reflect their rankings in a sophomore class. To what extent do these data form a linear relationship? (Caution: think before you calculate.)

Subject	*Fresh.* *X*	*Soph.* *Y*
1	2	3
2	9	7
3	1	2
4	5	7
5	3	1
6	7	8
7	4	4
8	6	5
9	8	6

23. You want to know if a nurse's absences from work in one month (Y) can be predicted by knowing her score on a test of psychological "burnout" (X). What do you conclude from the ratio data on the next page?

Subject	*Burnout* *X*	*Absences* *Y*
1	2	4
2	1	7
3	2	6
4	3	9
5	4	6
6	4	8
7	7	7
8	7	10
9	8	11

24. You hypothesize that students who sit toward the front of a classroom (those with a 0 on the X variable) perform better than those who sit toward the back of the classroom (1 on X) when given a brief quiz (the Y scores). Do these data support your hypothesis? (Call the group with 0s group 2.)

Subject	*Location* *X*	*Quiz* *Y*
1	0	4
2	0	6
3	0	11
4	0	5
5	1	8
6	1	5
7	1	8
8	1	11
9	1	7
10	1	4

25. A researcher observes the behavior of a group of monkeys in the jungle. He determines each monkey's relative position in the dominance hierarchy of the group (with an X of 1 being most dominant), and also notes each monkey's relative weight (with a Y of 1 being the lightest). What is the relationship between dominance and weight?

Subject	*Dominance* *X*	*Weight* *Y*
1	1	10
2	2	8
3	5	6
4	4	7
5	9	5
6	7	3
7	3	9
8	6	4
9	8	1
10	10	2

SUMMARY OF FORMULAS

1. *The computational formula for the Pearson correlation coefficient is*

$$r = \frac{N(\Sigma XY) - (\Sigma X)(\Sigma Y)}{\sqrt{[N(\Sigma X^2) - (\Sigma X)^2][N(\Sigma Y^2) - (\Sigma Y)^2]}}$$

where X and Y stand for the scores on the X and Y variables and N is the number of pairs in the sample.

2. *The computational formula for the Spearman rank-order correlation coefficient is*

$$r_s = 1 - \frac{6(\Sigma D^2)}{N(N^2 - 1)}$$

where N is the number of pairs of ranks and D is the difference between the two ranks in each pair.

3. *The computational formula for the point-biserial correlation coefficient is*

$$r_{pb} = \left(\frac{\overline{Y}_2 - \overline{Y}_1}{S_Y}\right)(\sqrt{pq})$$

where

$\overline{Y}_1$ is the mean of the scores on the continuous variable for one group of the dichotomous variable,

$\overline{Y}_2$ is the mean of the scores on the continuous variable for the other group of the dichotomous variable,

S_Y is the standard deviation of all the continuous Y scores,

p is the proportion of all subjects in the sample in one dichotomous group and q is the proportion of all subjects in the sample in the other dichotomous group. Each is found by dividing the number of subjects in the group by N, the total number of X-Y pairs in the study.

8

Linear Regression and Predicting Variability

To understand the upcoming chapter:

- From Chapter 5, understand that the variance indicates the "average difference" between scores and the mean, so when predicting scores, the variance indicates the "average error" in our predictions; and that using a relationship to predict subjects' Y scores allows us to reduce this error, or to "account for variance."
- From Chapter 7, understand that the summary line drawn through a scatterplot is called the regression line, and that the stronger the relationship, the less the Y scores are vertically spread out at each X score, so the more the Y scores hug the regression line.

Then your goals in this chapter are to learn:

- How a regression line, composed of values of Y', summarizes a scatterplot.
- How the regression equation is calculated and how it is used to predict Y for a given X.
- How the correlation coefficient implies the accuracy with which Y scores can be predicted.
- Which statistics we use to measure the errors in prediction when using regression.
- What the proportion of variance accounted for tells us about the accuracy of predictions when using a relationship to predict Y scores.

If the laws of nature produce a relationship, then certain Y scores are naturally paired with certain X scores. Therefore, if we know an individual's X score and the relationship between X and Y, we can predict the individual's Y score. As we shall see in this chapter, it is through the linear regression line that we predict Y scores from the corresponding X scores. In the following sections we will first examine the logic behind the regression line and then discuss how we use it to predict scores. Finally, we will look at ways of measuring the amount of error in our predictions.

MORE STATISTICAL NOTATION

In this chapter we will discuss the variance and standard deviation of the Y scores, or S_Y^2 and S_Y. Be sure that you understand that we compute the variance and standard deviation of the Y scores in the same ways that we compute these values for the X scores. Also recognize that the difference between two Y scores is reflected by their different locations along the Y-axis of a graph, and that the larger the values of S_Y^2 or S_Y, the more the Y scores are vertically spread out in the scatterplot.

UNDERSTANDING LINEAR REGRESSION

Linear regression is the procedure for describing the best-fitting straight line that summarizes a linear relationship. When do we use this procedure? We use it in conjunction with the Pearson correlation. The r is the *statistic* that summarizes the entire linear relationship all at once, and the regression line is the *line* that summarizes the entire linear relationship in the scatterplot all at once. However, we do not always create the regression line. We first compute r to determine whether a relationship exists. If $r = 0.0$, then there is no linear relationship present and regression techniques are unnecessary. All that the regression line would tell us is "Yup, there really is no relationship here." If the correlation coefficient is not 0.0, then we produce the linear regression line to further summarize the relationship.

Summarizing the Scatterplot Using the Regression Line

What does a regression line tell us? An easy way to understand a regression line is to compare it to a line graph. In Chapter 4 we created a line graph by computing the mean of the Y scores at each X. After graphing each X-$\overline{Y}$ data point, we connected the mean scores using straight lines. The left-hand scatterplot in Figure 8.1 shows the line graph of a study in which subjects produced Y scores at one of four values of X. We summarize the data by drawing a straight line from the $\overline{Y}$ at X_1 to the $\overline{Y}$ at X_2, another line from the $\overline{Y}$ at X_2 to the $\overline{Y}$ at X_3, and a third line from the $\overline{Y}$ at X_3 to the $\overline{Y}$ at X_4. To read the graph, we travel vertically from any X until we intercept the line, and then we travel horizontally until we intercept the Y axis. For example, as the arrows indicate, the mean of Y at X_3 is 3.0. Because this mean is the central score, it is the one score that best represents these Y scores. In essence, those subjects scoring at X_3 also scored around a Y of 3.0, so a score of 3.0 is our best single description of their scores. Further, if we know that a subject scored at X_3, then the mean Y score for that group (3.0) is our best prediction of that subject's Y score.

FIGURE 8.1 Comparison of a Line Graph and a Regression Line

Each data point is formed by a subject's X-Y pair. Each asterisk (*) indicates the mean Y score at an X.

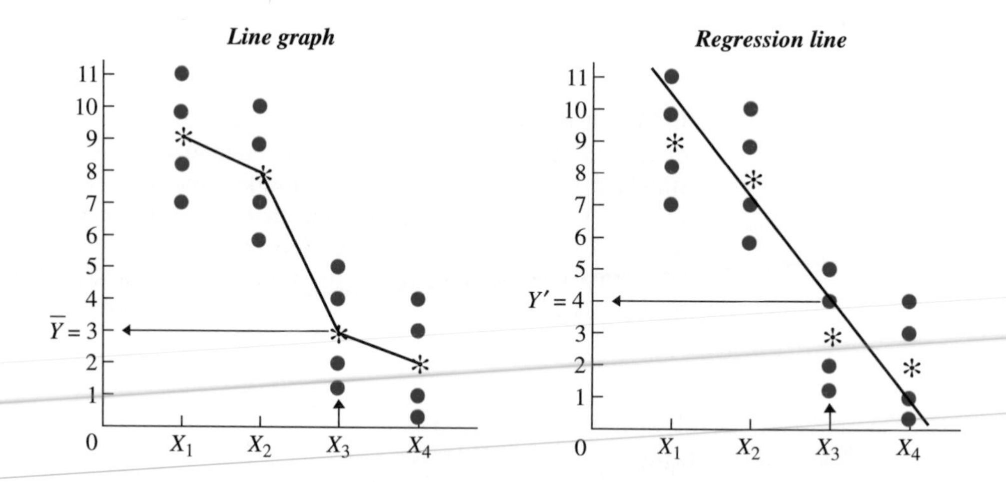

However, it is difficult to see the *linear* (straight-line) relationship in these data, because all of the means do not fall in a perfectly straight line. Therefore, as in the right-hand graph in Figure 8.1, we summarize the linear relationship hidden in the data by drawing a regression line. Think of the regression line as a straightened-out version of the line graph: it is drawn so that it comes as close as possible to connecting the mean of Y at each X while still producing one straight line. Although all means are not on the line, the distance that some means are above the line will average out with the distance that other means are below the line. Thus, the regression line is the *best-fitting* line, because "on average" it passes through the center of the various Y means. Because each Y mean is located in the center of the corresponding Y scores, by passing through the center of the Y means, the regression line passes through the center of the Y scores. Thus the **linear regression line** may be defined as the straight line that summarizes the linear relationship in a scatterplot by, on average, passing through the center of the Y scores at each X. As usual, this is another descriptive procedure that allows us to summarize and envision a distribution. We can think of the regression line as the line that would be formed *if* the data formed a perfect linear relationship, with all data points falling on the line. Because the actual Y scores fall above and below the line, the data only more or less fit this line. But we have no system for drawing a more or less linear relationship, so the regression line is how we envision what a perfect version of the linear relationship hidden in the data would look like.

We read the regression line by traveling vertically from any X until we intercept the regression line. Then we travel horizontally until we intercept the Y axis. As the arrows on the right-hand graph of Figure 8.1 indicate, the value of Y at X_3 is now 4. The symbol for this value is Y', pronounced "Y prime." Like the means of a line graph, each Y' is a summary of the Y scores for that X. However, the advantage of the regression line is that each Y' summarizes the scores based on the linear relationship across *all* X-Y pairs in the data. Thus, considering the entire linear relationship in Figure 8.1, we find that subjects at X_3 scored around 4.0, so 4.0 is our best single description of their scores. Further, if we know that a subject scored at X_3, then the Y' of 4.0 is our best prediction of that subject's Y score. Thus, the symbol Y' stands for a **predicted Y score.** Each Y' is our best description and prediction of the Y scores at a particular X, based on the linear relationship that is summarized by the regression line.

It is important to recognize that the Y' for the subjects who score at any value of X is the value of Y falling *on* the regression line. A line is made up of an infinite series of adjacent dots, so in essence, the regression line is created by plotting all of the data points formed by pairing all possible values of Y' with the corresponding possible values of X. If we think of the line as reflecting a perfect version of the linear relationship hidden in the data, then each value of Y' is the Y score everyone would have at a particular X if a perfect version of this relationship were present.

Predicting Scores Using the Regression Line

Why are we interested in predicting Y scores? After all, we have the Y score of every subject in the sample right in front of us. The answer is that regression techniques are a primary statistical device for predicting *unknown* scores based on a linear relationship. To do this, we first establish the relationship in a sample by computing the correlation. Then, by performing the inferential procedures for r (described in Chapter 11), we determine whether the sample is representative of a relationship in the population. If it is, we use the regression line to determine the Y' for each X. Then we can measure the X scores of other subjects who are not in our sample, and Y' is our best prediction of their Y scores. Thus, the importance of linear regression is that it allows us to predict subjects' unknown Y scores if we know their X scores on a correlated variable.

For example, the reason students must take the Scholastic Aptitude Test (SAT) to be admitted to some colleges is that researchers have established that SAT scores are positively correlated with college grades: to some extent, the higher the SAT score, the higher the college grades. Therefore, by using the SAT scores of students who are about to enter college, we can predict their future college performance. Applying regression techniques, we use a student's X score on the SAT to obtain a predicted Y' of the student's college grade average. If the predicted grades are too low, then the student is not admitted to the college.

The emphasis on prediction in correlation and regression brings two important terms into play. In the discussions of regression procedures in this text, we will use the X variable to predict scores on the Y variable. (There are procedures out there for predicting X scores from Y.) In statistical lingo, when the X variable is used to predict unknown scores, then X is called the **predictor variable** (because X does the predicting). When the unknown scores being predicted are on the Y variable, then Y is called the **criterion variable.** In our SAT example, SAT scores are the predictor variable, and college grade average is the criterion variable. (To remember the word "criterion," remember that your predicted grades must pass a certain criterion for you to be admitted to the college.)

Because we derive our predictions of Y based on the regression line, after we determine that there is a correlation in the sample, the next step is to determine the regression line for the sample. To do this, we compute the linear regression equation.

THE LINEAR REGRESSION EQUATION

To draw the regression line that summarizes a particular relationship, we don't simply eyeball the scatterplot and sketch in something that looks good. Instead, we use the linear regression equation. This is a mathematical equation for computing the value of Y' at each X. When we plot the data points formed by the X-Y' pairs, they all fall in a perfectly straight line. Then when we draw a line connecting the data points, we have the regression line that summarizes the scatterplot and the underlying relationship. Thus, the **linear regression equation** is an equation that, by describing the value of Y' at each X, defines the straight line that summarizes a linear relationship. The regression equation describes two characteristics of the regression line: its slope and its Y-intercept.

The **slope of a line** is a number that indicates how slanted the line is and the direction in which it slants. In algebra, the slope is defined as the ratio of the average change in Y scores to the average change in X scores. Figure 8.2 shows examples of regression lines having different slopes. When there is no relationship, the regression line is horizontal, such as line A. Then the slope of the line is zero. A positive linear relationship yields a regression line such as line B or C; each of these has a slope that is a positive number. Because line C is steeper, its slope is a larger positive number. A negative linear relationship yields a regression line such as line D, with a slope that is a negative number.

The ***Y*-intercept of a line** is the value of Y at the point where the regression line intercepts, or crosses, the Y axis. In other words, at the Y axis, the value of X is 0, so the Y-intercept is the value of Y' when X equals 0. In Figure 8.2, line

FIGURE 8.2 Regression Lines Having Different Slopes and Y-Intercepts

Line A indicates no relationship, lines B and C indicate positive relationships having different slopes and Y-intercepts, and line D indicates a negative relationship.

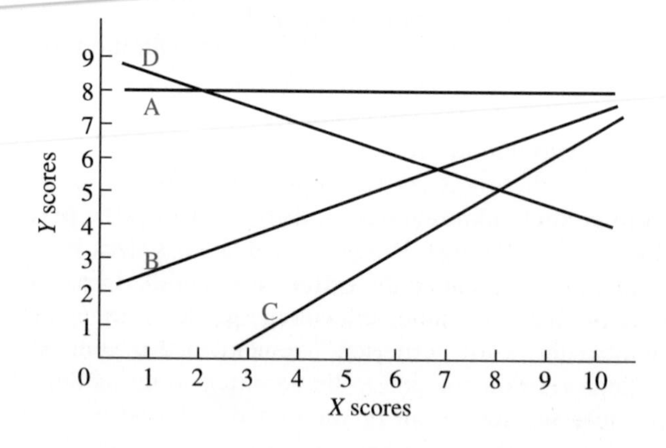

B intercepts the Y axis at $+2$, so its Y-intercept is $+2$. If we extended line C, it would intercept the Y axis at a point below the X axis, so its Y-intercept is a negative Y score. Because line D reflects a negative relationship, its Y-intercept is the relatively high Y score of 9. Finally, line A exhibits no relationship, and its Y-intercept equals $+8$. Notice that here the value of Y' for any value of X is always $+8$. *When there is no relationship, the slope of the regression line is zero, and every Y' equals the Y-intercept.*

The regression equation works in the following way: the slope summarizes the *direction* in which Y scores change as X increases, and also the *rate* at which they change. Thus, in line C in Figure 8.2, the Y scores increase more drastically as the X's increase than they do in line B. The Y intercept summarizes the starting point from which the Y scores begin to change as the X scores increase. Thus, together, the slope and intercept summarize how, starting at a particular value, the Y scores change with changes in X. Then, for each X, the summary of the Y scores is Y'.

The symbol for the slope of the regression line is b. The symbol for the Y-intercept is a.

THE LINEAR REGRESSION EQUATION IS

$$Y' = bX + a$$

This formula says that to find the value of Y' for a given X, multiply the slope (b) times X and then add the Y-intercept (a).

Suppose that a researcher has developed a paper-and-pencil test to identify people who will be productive widget-makers. To find out whether test scores help to predict "widgetability," the researcher first determines whether subjects' test scores are correlated with their widget-making ability. The researcher gives the test to an unrealistically small N of 11 subjects and then measures the number of widgets each makes in an hour. Figure 8.3 on the next page shows the resulting scatterplot, with test scores as the predictor (X) variable. The raw scores are also listed, arranged for computing r.

The first step is to find r:

$$r = \frac{N(\Sigma XY) - (\Sigma X)(\Sigma Y)}{\sqrt{[N(\Sigma X^2) - (\Sigma X)^2][N(\Sigma Y^2) - (\Sigma Y)^2]}}$$

$$r = \frac{11(171) - (29)(58)}{\sqrt{[11(89) - 841][11(354) - 3364]}}$$

The result is $r = +.736$, which rounds to $r = +.74$. (What did you get?) Thus, we have a positive linear relationship, which, with $r = +.74$, is quite strong.

To predict widget-making scores, we must compute the linear regression equation. To do that, we compute the slope and the Y-intercept. (In statistical language, the following formulas produce a regression line using the *least-squares method,* a term I'll explain in a later section.)

Compute the slope first.

FIGURE 8.3 Scatterplot and Data for Widgetability Study

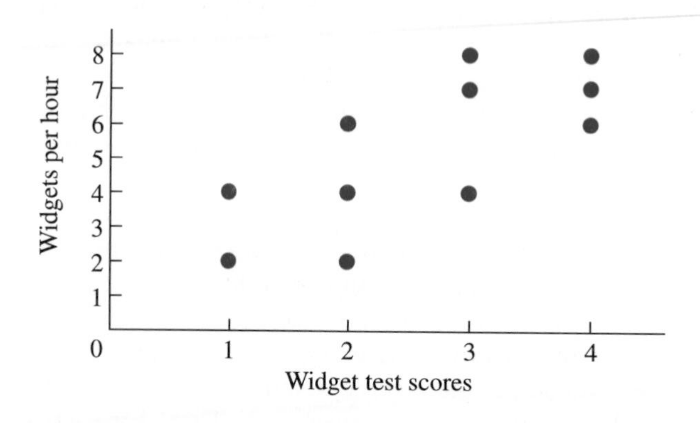

Subject	*Widget test score:* X	*Widgets per hour:* Y	XY
1	1	2	2
2	1	4	4
3	2	4	8
4	2	6	12
5	2	2	4
6	3	4	12
7	3	7	21
8	3	8	24
9	4	6	24
10	4	8	32
11	4	7	28
$N = 11$	$\Sigma X = 29$	$\Sigma Y = 58$	$\Sigma XY = 171$
	$\Sigma X^2 = 89$	$\Sigma Y^2 = 354$	
	$(\Sigma X)^2 = 841$	$(\Sigma Y)^2 = 3364$	
	$\bar{X} = 29/11 = 2.64$	$\bar{Y} = 58/11 = 5.27$	

Computing the Slope

THE FORMULA FOR THE SLOPE OF THE LINEAR REGRESSION LINE IS

$$b = \frac{N(\Sigma XY) - (\Sigma X)(\Sigma Y)}{N(\Sigma X^2) - (\Sigma X)^2}$$

N is the number of pairs of scores in the sample, and X and Y are the scores in the sample. This is not a difficult formula, because we typically compute the Pearson r first. The numerator of the formula for b is the same as the numerator of the formula for r, and the denominator of the formula for b is the left-hand quantity in the denominator of the formula for r. [If your calculator directly calculates r, an alternative formula for the slope is b = (r) (S_X / S_Y).]

For our widget study, we substitute the appropriate values from our computations of r into the formula for b, and we have

$$b = \frac{N(\Sigma XY) - (\Sigma X)(\Sigma Y)}{N(\Sigma X^2) - (\Sigma X)^2} = \frac{11(171) - (29)(58)}{11(89) - 841}$$

After multiplying and subtracting in the numerator, we have

$$b = \frac{199}{11(89) - 841}$$

After completing the denominator, we have

$$b = \frac{199}{138} = +1.44$$

Thus, the slope of the regression line for our widgetability data is $b = +1.44$. This positive slope indicates a positive relationship. (As a double check, note that the data do form a positive relationship, with $r = +.74$.) Had the relationship been negative, the formula would have produced a negative number for the slope.

We're not finished yet! Now we compute a, the Y-intercept.

Computing the *Y*-Intercept

THE FORMULA FOR THE Y-INTERCEPT OF THE LINEAR REGRESSION LINE IS

$$a = \overline{Y} - (b)(\overline{X})$$

First we multiply the mean of all X scores, $\overline{X}$, times the slope of the regression line, b. (See why we compute b first?) Then we subtract that quantity from the mean of all Y scores, $\overline{Y}$.

For our widgetability data, b is $+ 1.44$; in Figure 8.3, $\overline{Y}$ is 5.27 and $\overline{X}$ is 2.64. Filling in the formula for a, we have

$$a = 5.27 - (+1.44)(2.64)$$

After multiplying, we have

$$a = 5.27 - (+3.80) = +1.47$$

Thus, the Y-intercept of the regression line for our widgetability study is $a = +1.47$: when X equals 0, Y' equals $+1.47$.

We're still not finished!

Describing the Linear Regression Equation

Once we have computed the Y-intercept and the slope, we rewrite the regression equation, substituting our computed values for a and b. Thus,

$$Y' = +1.44X + 1.47$$

TABLE 8.1 Summary of Computations for the Linear Regression Equation

1. Compute r.
2. Compute the slope, b, where $b = \frac{N(\Sigma XY) - (\Sigma X)(\Sigma Y)}{N(\Sigma X^2) - (\Sigma X)^2}$.
3. Compute the Y-intercept, a, where $a = \overline{Y} - (b)(\overline{X})$.
4. Substitute the values of a and b into the formula for the regression equation

$$Y' = (b)(X) + a.$$

This is the finished regression equation that describes the linear regression line for the relationship between widget test scores and widgets-per-hour scores.

Putting all of this together, the preceding computations are summarized in Table 8.1.

Our final step is to plot the regression line.

Plotting the Regression Line

To plot the regression line, we must have some pairs of X and Y' scores to use as data points. Therefore, we choose some values of X, insert each in the finished regression equation, and calculate the value of Y' for that X. Actually, we need only two data points to draw a straight line: an X-Y' pair where X is low and an X-Y' pair where X is high. (An easy low X to use is $X = 0$: when X equals 0, Y' equals the Y-intercept, a.)

To see how the calculations work, we will compute Y' for all of the X scores from our widget study. We begin with the finished regression equation:

$$Y' = +1.44X + 1.47$$

First we find Y' for $X = 1$. Substituting 1 for X, we have

$$Y' = +1.44(1) + 1.47$$

Multiplying 1 times +1.44 and adding 1.47 yields a Y' of 2.91. Thus, we predict that anyone scoring 1 on the widget test will make 2.91 widgets per hour. Using the same procedure, we obtain the values of Y' for the remaining X scores of 2, 3, and 4. These are shown in Figure 8.4.

To graph the regression line, we simply plot the data points for the X-Y' pairs and draw the line. (Note that, as shown, we typically do not include the scatterplot, nor do we draw the regression line through the Y-intercept.)

Now we're finished. (Really.)

Using the Regression Equation to Predict *Y* Scores

When we compute Y' for a particular X, we are computing the predicted Y score for all subjects who have that X score. Above, for example, we found that those subjects scoring an X of 1 had a predicted Y' of 2.91. Therefore, in essence, we predict that anyone else not in our sample who scores an X of 1 will also have a Y' of around 2.91. Further, we can compute Y' for any value of X that falls

FIGURE 8.4 Regression Line for Widgetability Study

Widget test scores: X	*Predicted widgets per hour:* Y
1	2.91
2	4.35
3	5.79
4	7.23

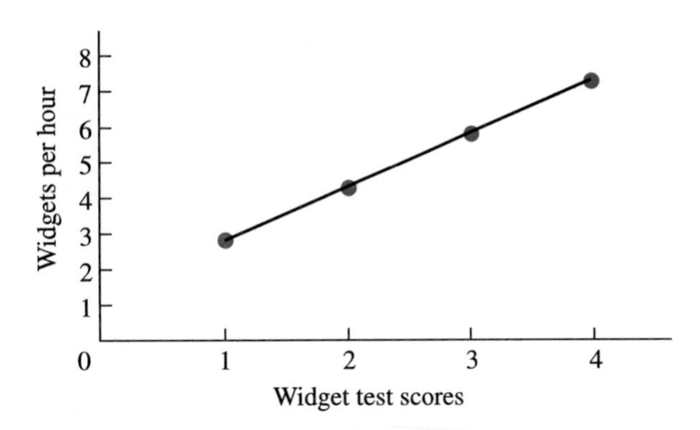

within the range observed in our sample, even if it is one that was not obtained by our original subjects. None of our subjects obtained a widget test score of 1.5. Yet, inserting an X of 1.5 into the regression equation yields a Y' of 3.63. (Notice that we'd obtain the same result from the regression line in Figure 8.4 if we could see decimal scores on the graph: if we traveled vertically from an X of 1.5 to the regression line and then traveled horizontally to the Y axis, we would find Y' is 3.63.) Thus, we predict that any individual who scores an X of 1.5 on our test will have a Y score of 3.63 widgets per hour.

Of course, any prediction we make may be wrong. Therefore, as we will see in the following sections, whenever we use the regression equation to predict Y scores, we also want to describe the amount of error in our predictions. This topic is important for two reasons. First, a complete description of our data includes the descriptive statistics that summarize the amount of error we have when we use a linear regression equation to predict Y scores. Second, understanding how we describe the amount of error in predictions when using the regression equation sets the stage for understanding how we describe the amount of error in predictions when using other statistical procedures. (In other words, the following topics will not go away.)

ERROR WHEN THE LINEAR REGRESSION EQUATION IS USED TO PREDICT SCORES

We describe the amount of error in our predictions by describing how well we can predict the actual Y scores in our sample: we pretend we do not know our subjects' scores, we predict them, and then we compare the predicted Y' scores to the subjects' actual Y scores. However, in this imperfect world, our predictions for some subjects will be close to their actual Y scores while our predictions for other subjects may contain considerably more error. Therefore, to summarize the error across the entire relationship, we compute something like the average error in our predictions.

The amount of error in any single prediction is the amount that a subject's actual score differs, or deviates, from the corresponding Y': in symbols this is $(Y - Y')$. To find the average error, we will find something like the average of the deviations (yes, here we go again). The first step is to find the deviations for all subjects. We compute Y' for each subject in our data and then subtract each Y' from its corresponding Y score. However, we cannot merely sum these deviations and find the average. Like with a mean, the Y' scores are in the center of the Y scores, so the actual Y's are equally spread out above and below the Y' scores. Therefore, like a mean, the sum of the deviations of the Y scores from their Y' scores is always zero, because the positive and negative deviations cancel out. Thus, on the one hand, as with other procedures we have seen, we like to use the regression equation to predict scores because, over the long run, our overestimates and underestimates cancel each other out. On the other hand, the problem is that if $\Sigma(Y - Y')$ is always zero, then the average error between the Y and Y' scores is always zero also.

To solve this problem, we *square* each deviation. The sum of the squared deviations of $Y - Y'$ is not necessarily zero, so neither is the average squared deviation. (Does this sound familiar?) When we find the average of the squared deviations between the Y and corresponding Y' scores, the answer we obtain is a type of variance. With this variance, we are describing the "average" spread of the actual Y scores around—above and below—their predicted Y' scores.

Computing the Variance of the *Y* Scores Around *Y'*

The symbol for the variance of a sample of Y scores around Y' is $S^2_{Y'}$. The S^2 indicates that we are computing the sample variance or error, and the subscript Y' indicates that it is the error associated with using Y' to predict Y scores.

THE DEFINITIONAL FORMULA FOR THE VARIANCE OF THE Y SCORES AROUND THEIR CORRESPONDING Y' SCORES IS

$$S^2_{Y'} = \frac{\Sigma(Y - Y')^2}{N}$$

This formula tells us to subtract each Y' we predict for a subject from his or her corresponding Y score, square each deviation, sum the squared deviations, and finally divide by N. The answer we obtain gives us the average squared difference between each Y and the corresponding Y' in our sample. It is one way to measure the amount of error we have when we use Y' to predict Y scores.

Remember our widgetability study? Table 8.2 shows the X and Y scores we obtained, as well as the Y' scores we computed for each X using our regression equation. In the column labeled $Y - Y'$, we subtract each Y' from the corresponding Y. In the column labeled $(Y - Y')^2$, we square each difference. Then we sum the squared differences to find $\Sigma(Y - Y')^2$.

TABLE 8.2 Widgetability Data with Computed Y' Scores

Subject	*Widget test score:* X	*Widgets per hour:* Y	*Predicted widgets:* Y'	$Y - Y'$	$(Y - Y')^2$
1	1	2	2.91	−.91	.83
2	1	4	2.91	1.09	1.19
3	2	4	4.35	−.35	.12
4	2	6	4.35	1.65	2.72
5	2	2	4.35	−2.35	5.52
6	3	4	5.79	−1.79	3.20
7	3	7	5.79	1.21	1.46
8	3	8	5.79	2.21	4.88
9	4	6	7.23	−1.23	1.51
10	4	8	7.23	.77	.59
11	4	7	7.23	−.23	.05
$N = 11$		$\Sigma Y = 58$ $\Sigma Y^2 = 354$ $(\Sigma Y)^2 = 3364$			$\Sigma(Y - Y')^2 = 22.07$

Filling in the formula for $S^2_{Y'}$, we have

$$S^2_{Y'} = \frac{\Sigma(Y - Y')}{N} = \frac{22.07}{11}$$

After dividing, we have

$$S^2_{Y'} = 2.006$$

With rounding,

$$S^2_{Y'} = 2.01$$

Thus, in these data the average squared difference between the actual Y scores and their corresponding values of Y' is 2.01. This indicates that, when using variance to measure our error, we are "off" by an average of 2.01 when we predict subjects' widgets-per-hour (Y) score based on their widget test (X) scores.

> **STAT ALERT** $S^2_{Y'}$ is a way to describe the average error we have when we predict Y scores using the corresponding Y' scores.

Now we can explain the term *least-squares regression method.* We saw that using Y' to predict the Y scores results in a sum of deviations, $\Sigma(Y - Y')$, of zero. Obviously, zero is the minimum that the sum can be. Because the sum of deviations is the minimum we can obtain, the sum of the squared deviations is also the minimum we can obtain. We shorten the term *squared deviations* to *squares.* Then, the least-squares method is the name for a way of computing the regression equation so that the sum of the squares between Y and Y' is the least that it can be. Any other method leads to greater error in our predictions, resulting in a larger sum of squared deviations between Y and Y' and a larger value of $S^2_{Y'}$.

Using the definitional formula for $S^2_{Y'}$ is very time-consuming. Luckily, we compute r before we compute $S^2_{Y'}$, and there is a mathematical relationship between the value of $S^2_{Y'}$ in a set of data and the value of r. From this relationship we derive a quick computational formula.

THE COMPUTATIONAL FORMULA FOR THE VARIANCE OF Y SCORES AROUND Y' IS

$$S^2_{Y'} = S^2_Y(1 - r^2)$$

The formula says to find the variance of the Y scores in the data and to square r. Subtract r^2 from 1 and then multiply the result times S^2_Y. The answer is $S^2_{Y'}$.

In our widget study, r was $+.736$. Using the data from Table 8.2, we compute S^2_Y to be 4.38. Placing these numbers in the above formula, we have

$$S^2_{Y'} = 4.38(1 - .736^2)$$

After squaring $+.736$ and subtracting the result from 1, we have

$$S^2_{Y'} = 4.38(.458)$$

so

$$S^2_{Y'} = 2.01$$

Again, the average squared difference between the actual Y scores and their corresponding values of Y' is 2.01.

There is a problem, however, in interpreting $S^2_{Y'}$. By squaring the difference between each Y and Y', we obtain an unrealistically large number. Also, our error is measured in squared units. (This *must* sound familiar.) To solve this problem, we take the square root of the variance, and the result is a type of standard deviation. To distinguish the standard deviation found in regression from other standard deviations, we call this standard deviation the *standard error of the estimate.*

Computing the Standard Error of the Estimate

The **standard error of the estimate** indicates the amount that the actual Y scores in a sample differ from, or are spread out around, their corresponding Y' scores. It is the clearest way of describing the average amount of error we have when we use Y' to predict Y scores. The symbol for the standard error of the estimate is $S_{Y'}$. (Remember, S measures the *error* in the sample, and Y' is our *estimate* of a subject's Y score.)

THE DEFINITIONAL FORMULA FOR THE STANDARD ERROR OF THE ESTIMATE IS

$$S_{Y'} = \sqrt{\frac{\Sigma(Y - Y')^2}{N}}$$

This is the same formula we used previously for the variance of Y scores around Y', except that here we have added the square root sign. Thus, to compute $S_{Y'}$, we first compute the variance of the scores around Y', or $S^2_{Y'}$, and then we find its square root.

In our widget study, we computed the variance of the Y scores around their corresponding Y' scores as $S^2_{Y'} = 2.01$. Taking the square root of this variance produces

$$S_{Y'} = 1.42$$

Previously we saw that the shortcut way to compute the variance of Y around Y' was to use the formula

$$S^2_{Y'} = S^2_Y(1 - r^2)$$

By taking the square root of each component, we construct the computational formula.

THE COMPUTATIONAL FORMULA FOR THE STANDARD ERROR OF THE ESTIMATE IS

$$S_{Y'} = S_Y\sqrt{1 - r^2}$$

This formula says to first find the standard deviation, S_Y, of the actual Y scores in the data. Then, after finding the quantity $1 - r^2$, take the square root and multiply it times S_Y. The answer is $S_{Y'}$.

For our widget study, S^2_Y was 4.38 so S_Y is 2.093 and r was $+.736$. Placing these numbers in the above formula, we have

$$S_{Y'} = 2.093\sqrt{1 - .736^2}$$

Squaring $+.736$ yields .542, which when subtracted from 1 gives .458. The square root of .458 is .677. Thus, we have

$$S_{Y'} = 2.093(.677)$$

so again

$$S_{Y'} = 1.42$$

The standard error of the estimate is 1.42. Because our Y scores measure the variable of widgets per hour, the standard error of the estimate is 1.42 widgets per hour. Therefore, we conclude that when we use the regression equation to predict widgets-per-hour scores based on widget test scores, we will be wrong by an "average" of about 1.42 widgets per hour.

> ***STAT ALERT*** The standard error of the estimate describes the amount of error in our predictions of Y scores when we use Y' from the regression equation as the predicted score.

Assumptions of Linear Regression

As we have seen, $S_{Y'}$ measures the differences between Y and Y' and thus indicates the amount the Y scores are spread out around the Y' scores. Likewise, because all values of Y' fall *on* the regression line, $S_{Y'}$ also describes the amount the actual Y scores are spread out around the regression line. In order for $S_{Y'}$ to accurately describe this spread, we must be able to make two assumptions about how the Y scores are distributed.

First, we assume that the Y scores are equally spread out around the regression line throughout the relationship. This assumption goes by the funny little name of homoscedasticity. **Homoscedasticity** occurs when the Y scores at each X are spread out to the same degree as they are spread out at every other X. The left-hand scatterplot in Figure 8.5 shows what homoscedastic data from our widget study would look like. Because the vertical distance separating the Y scores is the same at each X, the spread of the Y scores around the regression line and around their Y' is the same at each X. Therefore, the computed value of $S_{Y'}$ will be reasonably accurate in describing this spread and in describing the error in predicting Y scores for any X score. Conversely, the right-hand scatterplot shows an example of heteroscedastic data. **Heteroscedasticity** occurs when the spread in Y is not equal throughout the relationship. In such cases, our computed value of $S_{Y'}$ will not accurately describe the "average" error throughout the entire

FIGURE 8.5 Illustrations of Homoscedastic and Heteroscedastic Data

The vertical width of the scatterplot above an X indicates how spread out the corresponding Y scores are. On the left, the Y's have the same spread at each X.

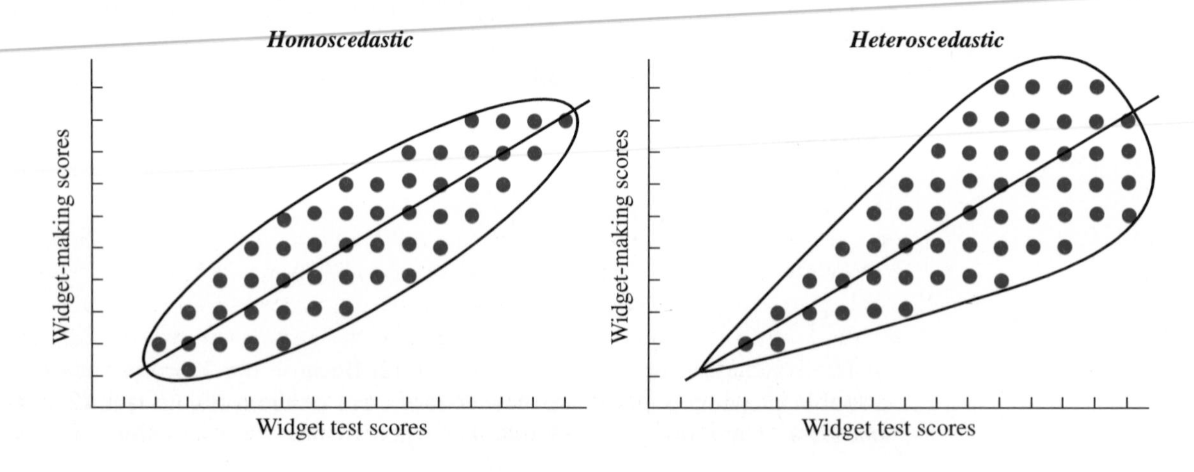

FIGURE 8.6 Scatterplot Showing Normal Distribution of *Y* Scores at each *X*

At each X, there is a normal distribution of different Y scores centered around Y′.

relationship. In Figure 8.5, for example, our $S_{Y'}$ will be much greater than the actual average error in predicting widget scores associated with low test scores, and much less than the average error in predicting widget scores associated with high test scores.

Our second assumption is that the sample of *Y* scores at each *X* represents an approximately normal distribution. That is, if we constructed a frequency polygon of the *Y* scores at each *X*, we would expect to have a normal distribution centered around Y'. Figure 8.6 illustrates this assumption for our widgetability study. Recall that if a distribution is normal, we expect approximately 68% of all scores to fall between ± 1 standard deviation from the mean. In the same way, if the *Y* scores are normally distributed around each Y', we expect approximately 68% of all *Y* scores to be between $\pm 1S_{Y'}$ from the regression line. In our widget study, the standard error of the estimate is 1.42, so we expect approximately 68% of the actual *Y* scores to be between ± 1.42 from each value of Y'.

In summary, $S_{Y'}$ and $S^2_{Y'}$ tell us how much the *Y* scores are spread out around the Y' scores and thus indicate the amount of error we have when we use Y' to predict the *Y* scores. However, $S_{Y'}$ and $S^2_{Y'}$ are not the only statistics that indicate how spread out the *Y* scores are. Recall that the correlation coefficient also reflects how much the scatterplot is spread out around the regression line. For this reason, *r* indirectly tells us the amount of error in our predictions.

Strength of the Relationship and Amount of Prediction Error

As you know, the larger the absolute size of *r*, the stronger the relationship: the more consistently one value of *Y* is associated with one and only one value of *X*. In turn, the more consistent the association, the smaller the spread in *Y* scores at each *X* and so the smaller the amount of error we have when we use Y' scores to predict *Y* scores.

Our minimum error occurs when *r* is ± 1.0, because there is no spread in the *Y* scores at each *X* and the scatterplot *is* the regression line. The left-hand graph

FIGURE 8.7 Scatterplots and Regression Lines When $r = +1.0$ and when $r = 0.0$

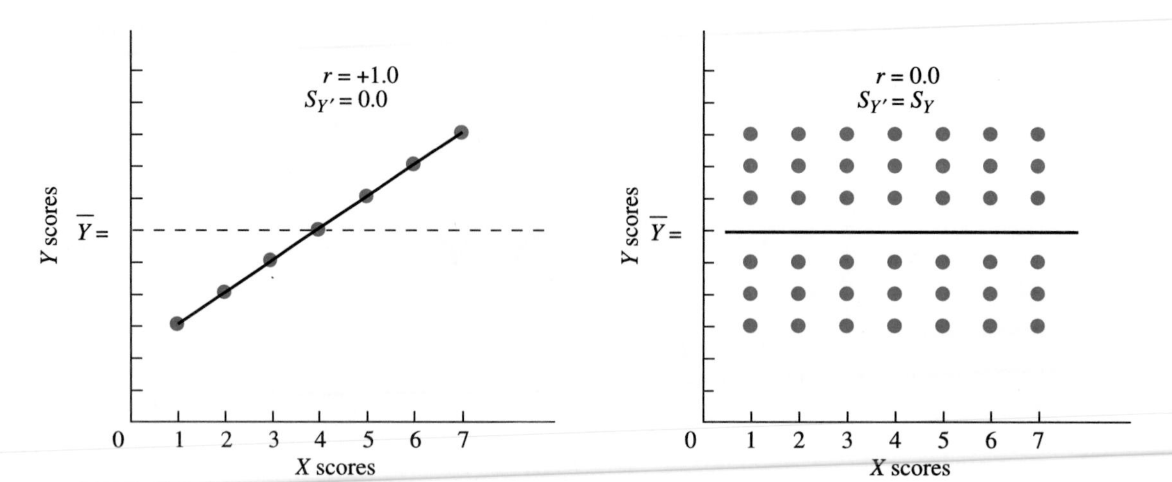

in Figure 8.7 shows such a perfect relationship. The Y' for each X equals the one Y score every subject obtained for that X. Therefore, the difference between each Y and corresponding Y' is zero, so the error in our predictions, $S_{Y'}$, equals zero (as does $S^2_{Y'}$).

At the other extreme, r equals 0.0 in the right-hand graph of Figure 8.7. Here the regression line is horizontal, and all values of Y' equal the Y-intercept. In order to understand the error in prediction here, recognize that the Y-intercept is the overall mean of all Y scores, or $\overline{Y}$. This is because when there is no relationship, the Y scores at each X are virtually the same as the Y scores at any other X. Thus, the regression line passes through the center of all Y scores in the sample, and the central score in a sample is the mean. *When $r = 0.0$, the Y-intercept is equal to the mean of all Y scores in the sample, and the predicted Y' for all subjects is the overall mean of Y.* As we saw in previous chapters, when we have no additional information about the scores, we should predict the overall mean for each score. When $r = 0.0$, there is no relationship to provide additional information about Y scores, so, even when using the regression equation, we end up predicting $\overline{Y}$ for each Y score.

When $r = 0.0$, the standard error of the estimate is at its maximum value, and that value is equal to S_Y, the standard deviation of all the actual Y scores in the sample. Why? Because when no relationship exists, the value of Y' for every subject is always the overall mean of the Y scores, $\overline{Y}$. We can call our prediction Y' or we can call it $\overline{Y}$, but it is the same score. Therefore, we can replace the symbol Y' with the symbol $\overline{Y}$ in the formula for the standard error of the estimate, as shown here:

$$S_{Y'} = \sqrt{\frac{\Sigma(Y - Y')^2}{N}} = \sqrt{\frac{\Sigma(Y - \overline{Y})^2}{N}} = S_Y$$

The resulting formula on the right is the formula for the standard deviation of all Y scores, S_Y. Thus, when $r = 0.0$, the standard error of the estimate, $S_{Y'}$, is

FIGURE 8.8 Scatterplots of Strong and Weak Relationships

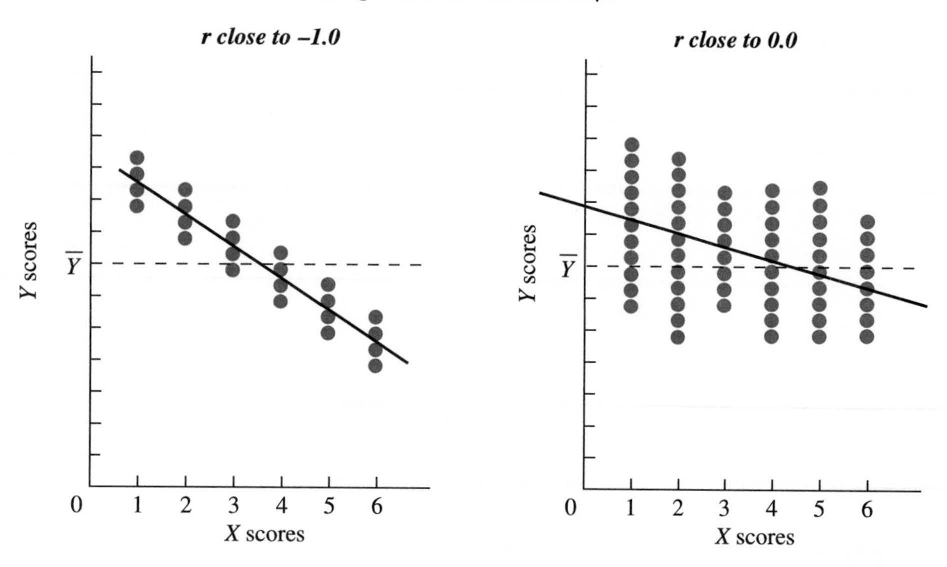

equal to the total variability, or spread, in the Y scores, S_Y. (Likewise, $S^2_{Y'}$ is equal to S^2_Y.)

Thus, the absolute size of r is *inversely* related to the size of $S_{Y'}$. When r is ± 1.0, we have the minimum amount of error in our predictions, which is zero, yielding a value of zero for $S_{Y'}$. When r is 0.0, we have the maximum amount of error in our predictions, with the value of $S_{Y'}$ equal to S_Y. When r is a value between 0.0 and ± 1.0, the value of $S_{Y'}$ is less than S_Y but greater than zero. Figure 8.8 shows two such intermediate relationships. The scatterplot on the left shows a relationship with an r close to -1.0. The Y scores at each X are not spread out much around the corresponding Y' on the regression line, so the actual Y scores are relatively close to the predicted Y'. Therefore, our error ($S_{Y'}$ and $S^2_{Y'}$) will be relatively close to zero. Conversely, the scatterplot on the right shows a relationship with an r much closer to 0.0. The Y scores at each X are spread out around the corresponding Y' on the regression line, meaning that many of the actual Y scores are relatively far from the predicted Y'. Therefore, there will be considerable error when Y' is used to predict subjects' Y scores, and $S_{Y'}$ (and $S^2_{Y'}$) will be relatively large.

Now we arrive at the essence of understanding how correlation and regression work together. Return to Figures 8.7 and 8.8, and notice that when r is close to ± 1.0, the actual Y scores are considerably closer to the corresponding values of Y' than they are to the overall mean of Y. However, when r is close to 0.0, the actual Y scores are not that much closer to the corresponding values of Y' than they are to the value of $\overline{Y}$. Thus, r indicates how much better our predictions are if we use subjects' X scores and the regression equation to predict Y scores, rather than merely predicting $\overline{Y}$ for every subject.

When r is 0.0, the X scores are no help whatsoever in predicting Y scores. If we use the X scores and the regression equation to compute Y', each predicted

Y' score is the mean of Y. If we do not use the X scores and the regression equation, we still predict that each Y score is the mean of Y. Thus, when there is no relationship, our error when we use the X scores, $S_{Y'}$, equals our error when we do not use the X scores, S_Y (and $S^2_{Y'}$ equals S^2_Y).

When r is not 0.0, however, using the X scores reduces our error in prediction. Because there is a relationship, the actual Y scores are closer to the corresponding values of Y' than to the mean of Y. Therefore, our error when we use the regression equation to predict Y scores, $S_{Y'}$, is less than our error when we use the overall mean of Y to predict Y scores, S_Y (and $S^2_{Y'}$ is less than S^2_Y). As the strength of the relationship increases and r approaches ± 1.0, the actual Y scores are closer to the corresponding values of Y', producing even less error (and $S_{Y'}$ is even smaller relative to S_Y).

> ***STAT ALERT*** The larger the absolute value of r, the smaller our error in prediction if we use the relationship $S_{Y'}$ compared to our error if we do not use the relationship S_Y.

Obviously, if we decrease the error in our predictions as r approaches ± 1.0, we increase the accuracy of our predictions. Although the value of r implicitly communicates the accuracy of our predictions, to directly compute the accuracy of our predictions, we compute the proportion of variance accounted for by the relationship.

PREDICTING VARIABILITY: THE PROPORTION OF VARIANCE ACCOUNTED FOR

The term "proportion of variance accounted for" is a short way of saying "the proportion of total variance in the Y scores that is accounted for by the relationship with X." As we saw in Chapter 5, this proportion is determined by comparing how well we predict scores when using a relationship to how well we predict scores if we ignore the relationship. The variance we account for is S^2_Y, the variance of the Y scores around the overall $\overline{Y}$. We use this frame of reference because, if we ignore the relationship, we can still compute $\overline{Y}$ and use it to predict scores, so S^2_Y is the greatest error we can have in our predictions. The total variance in Y is simply a way to measure the *differences* between all of the scores in the study. When using $\overline{Y}$ to predict every score, we will not predict any of the different scores that subjects obtain, so any differences among Y scores will result in error. When we use the relationship, by knowing subjects' X scores we are closer to knowing when they will have one value of Y and when they will have a *different* value of Y. Therefore, using the relationship helps us to reduce our error: we predict, or "account for," some of these differences, so we say that we account for some of the variance in Y.

> ***STAT ALERT*** Accounting for the variance in Y means that we can predict differences in Y scores.

Computations for the proportion of variance accounted for can be difficult to conceptualize. Therefore, we will first calculate it for only one subject, and then we will see how it is computed for an entire sample.

Understanding the Variance Accounted For

Say that in a study where we have discovered a relationship, we want to predict the Y score of one subject, Dorcas. The overall mean of the Y scores in the sample is 9, but Dorcas obtained an actual Y score of 4. Using the mean as her predicted score, we predict a score of 9 for her, and we are wrong by 5 points ($9 - 4 = 5$). Since our error in prediction when we use the overall mean of Y is the worst that we can do, think of this amount as the *total error.*

Say that, using the relationship and the regression equation, we predict a Y' of 6 for Dorcas. We are now off by 2 points ($6 - 4 = 2$). Thus, of the 5 points of error we have without using the relationship, we are still off by 2 of those points when we use the relationship. Think of this amount as the *error remaining.*

By reducing our error from 5 points of total error to 2 points of error remaining, we are now 3 points closer to Dorcas's actual Y score than we were when we predicted her score as the overall $\overline{Y}$. Think of this improvement as the *error eliminated.* We calculate the error eliminated by subtracting the error remaining from the total error. Thus,

$$\begin{matrix}\text{Total error} \\ \text{without using the} \\ \text{relationship}\end{matrix} - \begin{matrix}\text{error remaining} \\ \text{when using the} \\ \text{relationship}\end{matrix} = \begin{matrix}\text{error eliminated} \\ \text{when using the} \\ \text{relationship}\end{matrix}$$

As we've seen in previous discussions, it is difficult to judge whether a quantity like 3 points of error should be considered a large amount in the grand scheme of things in nature. To make our results easier to interpret, we need a frame of reference. Therefore, we describe the above components in terms of the proportion each represents of the total error we have when not using the relationship. We end up with

$$\frac{\text{Total error}}{\text{Total error}} - \frac{\text{error remaining}}{\text{total error}} = \frac{\text{error eliminated}}{\text{total error}}$$

On the left, the total error divided by the total error equals 1.0, or 100%, of the total error. On the right, the error eliminated divided by the total error is the proportion of total error that is eliminated by using the relationship. Therefore, we can state the above formula as

$$1.0 - \left(\frac{\text{error remaining}}{\text{total error}}\right) = \text{proportion of total error eliminated}$$

For Dorcas, the error remaining when using the relationship (2) divided by the total error without using the relationship (5) is 2/5, or .40. Thus, when we use the relationship, we still have .40, or 40%, of the error we have when we do not use the relationship. Since .40 of the total error remains when we use the relationship, the error eliminated is $1.0 - .40$, or .60. (The 3 points of error eliminated is .60 of 5.) Thus, .60, or 60%, of the total error we have when we do not use the relationship is eliminated when we use the relationship to predict her score. We are .60, or 60%, closer to Dorcas's score if we use the regression equation to predict her Y score than if we use the overall mean of the Y scores as her predicted score.

Of course, we never examine only one subject. Instead, we compute the proportion of variance accounted for in the entire sample.

Computing the Proportion of Variance Accounted For

To describe the entire sample, we use the above logic, except that we compute the appropriate measures of variance.

When we ignore the relationship and predict the overall mean Y score for every subject, we do not accurately predict the different scores. In that case, based on $\Sigma(Y - \overline{Y})^2$, the variance in Y, S^2_Y, measures the "average" difference between all Y scores and $\overline{Y}$, so it indicates the "average" error we have when we use the mean score to predict Y scores. Thus, S^2_Y is analogous to the average total error we have without using the relationship.

When we use the relationship and predict a Y' for each subject, we may not perfectly predict each Y score. Based on $\Sigma(Y - Y')^2$, the variance of the Y scores around Y', $S^2_{Y'}$, measures the "average" difference between Y and Y', so it indicates the "average" error we have when we use Y' to predict Y scores. Thus, $S^2_{Y'}$ is analogous to the average error remaining when we use the relationship.

From here, we follow the same procedure as we did with Dorcas. There we saw that

$$\text{Proportion of total error eliminated} = 1.0 - \left(\frac{\text{error remaining}}{\text{total error}}\right)$$

Because we measure the total error using the variance in Y scores, another name for the proportion of total error eliminated is the proportion of variance accounted for. Substituting the symbols $S^2_{Y'}$ and S^2_Y, in the above formula gives us the following definitional formula.

THE DEFINITIONAL FORMULA FOR THE PROPORTION OF VARIANCE IN Y THAT IS ACCOUNTED FOR BY A LINEAR RELATIONSHIP WITH X IS

$$\text{Proportion of variance accounted for} = 1 - \left(\frac{S^2_{Y'}}{S^2_Y}\right)$$

For example, using this formula, we can return to our widgetability study and determine the proportion of variance in subjects' widget-making scores that is accounted for by the relationship with their test scores. From the data back in

Table 8.2, we computed that the variance in the Y scores, S_Y^2, was 4.38, and $S_{Y'}^2$ was 2.01. Then, the proportion of variance accounted for is

$$1 - \frac{2.01}{4.38} = 1.00 - .46 = .54$$

By dividing 2.01 (the average error remaining) by 4.38 (the average total error), we find that .46 of the total error remains, even when we use the relationship. But, if only .46 of the error remains, then using the relationship eliminates 1 − .46, or .54, of the average error we have when not using the relationship. Thus, we say that this relationship accounts for .54 of the variance in Y scores: if we know subjects' scores on the widget test, we are, "on average," .54, or 54%, more accurate at predicting their different widgets-per-hour scores than we would be if we did not use this relationship. In other words, when we use the relationship to compute Y', our predictions are, on average, 54% closer to the actual Y scores than when we do not use the relationship and instead use $\overline{Y}$ as the predicted Y score for each subject.

> ***STAT ALERT*** The proportion of variance accounted for is the proportion by which we improve our accuracy in predicting scores on variable Y by using the relationship with X.

Using *r* to Compute the Proportion of Variance Accounted For

Computing the proportion of variance accounted for with the above definitional formula is rather time-consuming. However, we've seen that the size of r is related to the amount of error in our predictions, $S_{Y'}^2$, by the formula

$$S_{Y'}^2 = S_Y^2(1 - r^2)$$

In fact, this formula contains all the components of the previous definitional formula, so solving for the proportion of variance accounted for, we have

$$r^2 = 1 - \frac{S_{Y'}^2}{S_Y^2}$$

The squared correlation coefficient equals 1 minus the ratio $S_{Y'}^2/S_Y^2$. Since 1 minus this ratio is the definitional formula for the proportion of variance accounted for, we have the following computational formula.

THE COMPUTATIONAL FORMULA FOR THE PROPORTION OF VARIANCE IN Y THAT IS ACCOUNTED FOR BY A LINEAR RELATIONSHIP WITH X IS

$$\text{Proportion of variance accounted for} = r^2$$

Not too tough! All we do is compute r (which we would do anyway) and square it. Then we have computed the proportion of variance in Y scores that is accounted for by the relationship with *X*. (Yes, it has taken a long time to get to such a simple method, but to understand r^2 you must understand $1 - S^2_{Y'}/S^2_{Y}$.)

Above we computed that in our widgetability study, the relationship accounted for .54 of the variance in *Y* scores. Since *r* for this study was +.736, we can also compute the proportion of variance accounted for as $(.736)^2$, which again is .54.

In statistical language, r^2 is called the *coefficient of determination,* which is merely another name for the proportion of variance accounted for. The proportion of variance *not* accounted for is called the *coefficient of alienation.*

THE COMPUTATIONAL FORMULA FOR THE PROPORTION OF VARIANCE NOT ACCOUNTED FOR IS

$$\text{Proportion of variance not accounted for} = 1 - r^2$$

This value equals the ratio $S^2_{Y'}/S^2_{Y}$, so it is the proportion of total error remaining. In our widget study, the coefficient of determination is .54, so we still cannot account for 1 − .54, or .46, of the variance in the *Y* scores.

Note that r^2 describes the proportion of *sample* variance that is accounted for by the relationship. If the *r* passes the inferential statistical test, we can conclude that this relationship holds for the population. Then the value of r^2 is a *rough* estimate of the proportion of variance in *Y* scores that is accounted for by the relationship in the population. Thus, we expect to be roughly 54% more accurate if we use the relationship and our widget test scores to predict any other, unknown widget-making scores.

Using the Variance Accounted For

The logic of r^2 can be applied to any relationship. For example, in the previous chapter we discussed the correlation coefficients r_s and r_{pb}. Squaring these coefficients also indicates the proportion of variance accounted for. (It is as if we performed the appropriate regression analysis, computed $S^2_{Y'}$ and S^2_{Y}, and so on.) Likewise, as we shall see in later chapters, we can determine the proportion of variance accounted for in experiments. We describe the proportion of variance in the dependent variable (the *Y* scores) that is accounted for by the relationship with the independent variable (the *X* scores). This proportion tells us how much more accurately we can predict subjects' scores on the dependent variable if we take into account the condition of the independent variable under which the subjects were tested, rather than merely using the overall mean of the dependent scores as the predicted score.

Although theoretically a relationship may allow us to account for any proportion of the variance, in the behavioral sciences we get very excited if we find a relationship that accounts for around 25% of the variance. Remember, this is an *r* of ±.50, which is pretty good. Given the complexity of the behaviors of living organisms, we are unlikely to find an *r* that is very close to ±1.0, so we are also unlikely to find that a relationship accounts for close to 100% of the variance.

(If we ever found a relationship that accounted for 100% of the variance, we'd get ready for the Nobel prize!)

The reason we make such a big deal out of the proportion of variance accounted for is that it is *the* statistical measure of how "important" a particular relationship is. Remember, scores reflect behavior. When we use a relationship with X to predict different Y scores, we are actually identifying and predicting differences in behavior Y. The goal of psychological research is to understand differences in behavior. Therefore, the greater the proportion of variance accounted for by the relationship, the more accurately we can identify and predict differences in behavior, and thus the more scientifically important and informative the relationship is.

Thus, whenever we wish to compare two correlation coefficients to see which relationship is more informative, we first square each r. Say that we find a relationship between the length of a person's hair and his or her creativity, but r is only $+.02$. Yes, this r indicates a relationship, but such a weak relationship is virtually useless. The fact that $r^2 = .0004$ indicates that knowing a subject's hair length improves predictions about creativity by only four-hundredths of *one* percent! However, say that we also find a relationship between a subject's age and his or her creativity, and here r is $-.40$. This relationship is more important, at least in a statistical sense, because $r^2 = .16$. If we want to understand differences in creativity, age is the more important variable because knowing subjects' ages gets us 16% closer to accurately predicting their creativity. Knowing their hair length gets us only .04% closer to accurately predicting their creativity.

STAT ALERT The proportion of variance accounted for is our basis for evaluating the scientific importance or usefulness of a relationship.

A WORD ABOUT MULTIPLE CORRELATION AND REGRESSION

Sometimes we discover several X variables that help us to more accurately predict a Y variable. For example, there is a positive correlation between a person's height and his or her ability to shoot baskets in basketball: the taller people are, the more baskets they tend to make. There is also a positive correlation between how much people practice basketball and their ability to shoot baskets: the more they practice, the more baskets they tend to make. Obviously, to be as accurate as possible in predicting how well people shoot baskets, we want to consider both how tall they are and how much they practice. In this example there are two predictor variables (height and practice) that predict the criterion variable (basket shooting). In other words, there are multiple predictor variables. When we wish to simultaneously consider multiple predictor variables for one criterion variable, we use the statistical procedures known as *multiple correlation* and *multiple regression.*

Although the computations involved in multiple correlation and multiple regression are beyond the scope of this text, the logic is the same as the logic of the procedures we have discussed here. The correlation coefficient, called the *multiple r*, indicates the strength of the relationship between the multiple predictors taken together and the criterion variable. The multiple regression equation allows us to predict a subject's Y score by simultaneously considering the subject's scores on all X variables. The squared multiple r is the proportion of variance in the Y variable accounted for (the proportion of error eliminated) by using the relationship with the X variables to predict Y scores.

FINALLY

This chapter and the previous one have introduced many new symbols and concepts. However, they boil down to three major topics.

1. *Correlation.* When there is a relationship between the X and Y variables, a particular value or values of Y tends to be paired with one value of X. The stronger the relationship, the more consistently one and only one value of Y is paired with one and only one value of X. The type and strength of a relationship are described by r.
2. *Regression.* In a relationship, the X and Y scores are paired in a certain way. If we know the relationship, we can use a particular X score to help us predict the corresponding Y score. We predict Y scores by calculating Y' scores using the linear regression equation, and we graph these predictions as the linear regression line.
3. *Error in prediction.* The proportion of variance in Y that is accounted for by X is the amount by which our errors in predicting Y scores are reduced when we use the relationship, compared to what they would be if we did not use the relationship. This proportion equals r^2.

CHAPTER SUMMARY

1. *Linear regression* is the procedure for describing the best-fitting straight line that summarizes a linear relationship. The line is called the *linear regression line,* and it runs through the center of the Y scores.
2. The *linear regression equation* describes the regression line. The equation includes the *slope,* which indicates how much and in what direction the line slants, and the *Y-intercept,* which indicates the value of Y at the point where the line crosses the Y axis.
3. Using the regression equation, we compute for each X the summary Y score that best represents the Y scores at that X. This summary score is Y', the predicted Y score at each X. The regression line is formed by drawing a line that connects all X-Y' data points.
4. The *error in our predictions* is the difference between an actual Y score at an X and the predicted Y' at that X. The error in predictions across the entire relationship may be summarized by the *standard error of the estimate,* symbolized by $S_{Y'}$. The $S_{Y'}$ is a type of standard deviation indicating the

spread in the Y scores around the Y' scores, and thus the spread in the Y scores above and below the regression line. The differences (and error) between Y and Y' also may be summarized by the *variance of the Y scores around* Y', which is symbolized by $S^2_{Y'}$.

5. We assume that the Y scores are *homoscedastic*—that the spread in the Y scores around any Y' is the same as the spread in the Y scores around any other Y'. We also assume that the Y scores at each X are normally distributed around the corresponding value of Y'.

6. The stronger the relationship and the larger the absolute value of r, the smaller the values of $S_{Y'}$ and $S^2_{Y'}$. This is because the stronger the relationship, the closer the Y scores are to Y' and thus the smaller the difference, or error, between Y and Y'.

7. When r equals ± 1.0, every Y score equals its corresponding Y', and thus there is zero error in predictions and both $S_{Y'}$ and $S^2_{Y'}$ equal zero. When r equals 0.0, $S_{Y'}$ equals S_Y and $S^2_{Y'}$ equals S^2_Y. This is because when no relationship exists, the overall mean of Y is the predicted score for all subjects, and thus the differences between the actual Y scores and the predicted scores equal S_Y (or S^2_Y). When r is between 0.0 and ± 1.0, the value of $S_{Y'}$ is between the value of S_Y and 0.0.

8. The *proportion of variance in Y that is accounted for by X* describes the proportional improvement in our predictions that is achieved when we use the relationship to predict Y scores, rather than predicting Y scores without using the relationship. It indicates the proportion of the total prediction error that is eliminated when we use Y' as the predicted score instead of $\overline{Y}$. The proportion of variance accounted for is computed by squaring the correlation coefficient.

9. The proportion of variance not accounted for is computed as $1 - r^2$. This is the proportion of the total prediction error that is not eliminated when we use Y' as the predicted score instead of $\overline{Y}$.

10. The proportion of variance accounted for indicates the statistical importance of a relationship.

11. *Multiple correlation* and *multiple regression* are procedures for describing the relationship when multiple predictor variables (several X variables) are simultaneously used to predict scores on one criterion variable (Y variable).

PRACTICE PROBLEMS

(Answers for odd-numbered problems are provided in Appendix E.)

1. What is the linear regression line?
2. What is the linear regression procedure used for?
3. What is Y', and how do we compute it?

4. What is the general form of the linear regression equation? Identify its component symbols.
5. a. What does the Y-intercept of the regression line indicate?
   b. What does the slope of the regression line indicate?
6. Distinguish between the predictor variable and the criterion variable in linear regression.
7. a. What is the name for $S_{Y'}$?
   b. What does $S_{Y'}$ tell you about the spread in the Y scores?
   c. What does $S_{Y'}$ tell you about your errors in prediction?
8. a. What two assumptions must you be able to make about the data in order for the standard error of the estimate to be accurate, and what does each mean?
   b. How does heteroscedasticity lead to an inaccurate description of the data?
9. a. How is the value of $S_{Y'}$ related to r?
   b. When is $S_{Y'}$ at its maximum value? Why?
   c. When is $S_{Y'}$ at its minimum value? Why?
10. When are multiple regression procedures used?
11. a. What are the two statistical names for r^2 ?
    b. How do you interpret r^2?
12. A researcher determined that the correlation between statistics grades and scores on an admissions test to graduate school in psychology is $r = +.41$, $S_y = 3.90$.
    a. Compute the standard error of the estimate for these data.
    b. If the researcher predicts the overall mean score on the admissions test for each student, using variance, on average how much error can she expect?
    c. If the researcher predicts admissions test scores based on the regression equation and the statistics grades, using variance, on average how much error can she expect?
    d. What proportion of the error in part *b* remains even after using the regression equation?
    e. What proportion of the error in part *b* is eliminated by using the regression equation?
    f. What is your answer to part *e* called?
13. Bubbles has a statistics grade of 70, and Foofy has a grade of 98.
    a. Based on the data in problem 12, who is predicted to have a higher grade on the admissions test? Why?
    b. Subsequently, Bubbles received the higher test score. How can this be explained?

14. Poindexter conducted a correlational study measuring subjects' ability to concentrate and their ability to remember, finding $r = +.30$. He also correlated subjects' ability to visualize information and their memory ability, obtaining an $r = +.60$. He concludes that there is twice as consistent, and therefore twice as informative a relationship between subjects' visualization ability and memory ability as there is between their concentration ability and memory ability. Why do you agree or disagree?

15. *a.* In problem 14, what two statistical procedures can Poindexter employ to improve his predictions about memory ability even more?

b. Say that the resulting correlation coefficient is .67. Using the proportion of variance accounted for, explain what this means.

16. A researcher finds that variable A accounts for 25% of the variance in variable B. Another researcher finds that variable C accounts for 50% of the variance in variable B. Why, by accounting for greater variance, does variable C produce a relationship that is scientifically more important?

17. In a study, you measure how much subjects are initially attracted to a person of the opposite sex and how anxious they become during their first meeting with him or her. For the following ratio data, answer the questions below.

Subject	***X***	***Y***
1	2	8
2	6	14
3	1	5
4	3	8
5	6	10
6	9	15
7	6	8
8	6	8
9	4	7
10	2	6

a. Compute the statistic that describes the nature of the relationship formed by the data.

b. Compute the linear regression equation.

c. What anxiety score do you predict for any subject who produces an attraction score of 9?

d. When using this data, what is the "average" amount of error you should expect in your predictions?

18. *a.* For the relationship in problem 17, what is the proportion of variance in Y that is accounted for by X?

b. What is the proportion of variance not accounted for?

c. Why is or is not this a valuable relationship?

19. A researcher computes a Spearman r_S of $+.20$ when correlating the rankings of students in their statistics class (X) with their rankings in terms of how studious they are (Y). Another researcher computes a point-biserial correlation of $-.20$ when correlating a subject's gender (X) with his or her studiousness. Using the proportion of variance accounted for, interpret each result.

20. Using two brief questionnaires, a researcher measures how positive a subject's mood is (X) and how creative he or she is (Y), obtaining the following interval scores.

Subject	*X*	*Y*
1	10	7
2	8	6
3	9	11
4	6	4
5	5	5
6	3	7
7	7	4
8	2	5
9	4	6
10	1	4

a. Compute the statistic that summarizes this relationship.

b. What is the predicted creativity score for any subject scoring 3 on the mood questionnaire?

c. Assuming that your prediction is in error, what is the amount of error you expect to have?

d. How much smaller will your error be if you use the regression equation than if you merely use the overall mean creativity score as the predicted score for all subjects?

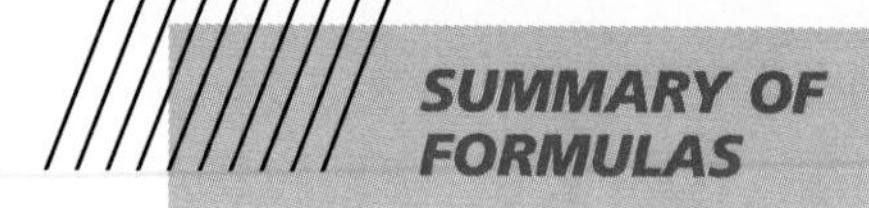

SUMMARY OF FORMULAS

1. *The formula for the linear regression equation is*

$$Y' = bX + a$$

where b stands for the slope of the line, X stands for an X score, and a stands for the Y-intercept.

2. *The formula for the slope of the linear regression line is*

$$b = \frac{N(\Sigma XY) - (\Sigma X)(\Sigma Y)}{N(\Sigma X^2) - (\Sigma X)^2}$$

where N is the number of pairs of scores in the sample and X and Y are the scores in the sample.

3. *The formula for the Y-intercept of the linear regression line is*

$$a = \overline{Y} - (b)(\overline{X})$$

where $\overline{Y}$ is the mean of all Y scores, b is the slope of the regression line, and $\overline{X}$ is the mean of all X scores.

4. *The definitional formula for the standard error of the estimate is*

$$S_{Y'} = \sqrt{\frac{\Sigma(Y - Y')^2}{N}}$$

where Y is each score in the sample, Y' is the corresponding predicted Y score, and N is the number of pairs in the sample.

5. *The computational formula for the variance of Y scores and Y′ is*

$$S^2_{Y'} = S^2_Y (1 - r^2)$$

6. *The computational formula for the standard error of the estimate is*

$$S_{Y'} = S_Y \sqrt{1 - r^2}$$

where S_Y is the standard deviation of the Y scores in the sample.

7. *The definitional formula for the proportion of variance in Y that is accounted for by a linear relationship with X is*

$$\text{Proportion of variance accounted for} = 1 - \frac{S^2_{Y'}}{S^2_Y}$$

where $S^2_{Y'}$ is the squared standard error of the estimate and S^2_Y is the variance of all Y scores in the sample.

8. *The computational formula for the proportion of variance in Y that is accounted for by a linear relationship with X is*

$$\text{Proportion of variance accounted for} = r^2$$

9. *The computational formula for the proportion of variance not accounted for is*

$$\text{Proportion of variance not accounted for} = 1 - r^2$$

PART 4

INFERENTIAL STATISTICS

Now that we have covered the various methods for describing scores and relationships, we are finally ready to discuss inferential statistical procedures.

Throughout our previous discussions, we've seen that descriptive statistics allow us to summarize and envision the distributions and relationships formed by our data. Descriptive statistics actually form the basis for *all* statistical interpretations, so now we will use these techniques as we perform inferential statistical procedures. We need inferential procedures because, as researchers, we want to answer these questions: Based on our sample, what would we expect to find if we could perform this study on the entire population? Would we find that the population has approximately the same mean as our sample? If we observe different samples that produce different means, would we find approximately the same difference between the population means? Would we find that the correlation coefficient in the population is about the same as the correlation coefficient in our sample?

Our problem is that we cannot *know* what the population contains, so the best we can do is to place an intelligent bet. In essence, inferential statistical procedures are ways to make decisions about the population that have a high probability of being correct. The first step in understanding these procedures is to understand probability. In the next chapter we will discuss the concept of probability and see how it is used to make statistical decisions. Subsequent chapters will then deal with the various inferential procedures we use, depending on the design of a particular study and the way we measure the variables. As you read each chapter and become familiar with the names and symbols for the different procedures, remember to keep one eye on the big picture: all inferential procedures involve making decisions about the scores and relationship we would find in the population, if we could observe it.

Probability: Making Decisions About Chance Events

To understand the upcoming chapter:

- From Chapter 6, understand the *z*-score formulas for describing the relative location of a score in a distribution of raw scores or the relative location of a sample mean in a sampling distribution of means; understand that by using *z*-scores we can determine the proportion of the area under the normal curve for any part of the curve; and understand that this proportion is the relative frequency we expect for the corresponding raw scores or sample means in that part of their distribution.

Then your goals in this chapter are to learn:

- What probability communicates.
- How probability is computed based on an event's relative frequency in the population.
- How the probability of raw scores is computed using *z*-scores and the standard normal curve.
- How the probability of sample means is computed using *z*-scores and the standard normal curve.
- How to set up and use a sampling distribution of means to determine whether a sample mean is likely to represent a particular population.

This chapter introduces you to the wonderful world of probability. As we shall see, psychologists combine their knowledge of probability with the standard normal curve model to make decisions about their data. We will keep the discussion simple, because psychologists do not need to be experts in probability. However, they do need to understand the basic logic of chance.

MORE STATISTICAL NOTATION

In daily conversation we use the words *chances, odds,* and *probability* interchangeably. In statistics, however, there are subtle differences among these terms. Odds are expressed as fractions or ratios ("The odds of winning are 1 in 2"). Chance is expressed as a percentage ("There is a 50% chance of winning"). Probability is expressed as a decimal ("The probability of winning is .50"). You should express the answers you compute as probabilities.

The symbol we use to represent probability is the lowercase letter p. When we describe the probability of a particular event such as event A, we write it as $p(\mathrm{A})$, which is pronounced "p of A" or "the probability of A."

THE LOGIC OF PROBABILITY

Probability is used to describe random, chance events. Such events occur when nature is being fair—when there is no bias toward one event over another (no rigged roulette wheels or loaded dice). Thus, a chance event occurs or does not occur merely because of the luck of the draw. In statistical work, luck is a very important concept, and probability is our way of mathematically describing how luck operates to produce an event.

But hold on! How can we describe an event that may happen only by luck? Well, we begin by paying attention to how often the event does occur when only luck is operating. The probability of any event is based on how often the event occurs *over the long run.* Intuitively, we use this logic all the time: if event A happens frequently over the long run, then we tend to think that A is likely to happen at any moment, and we say that it has a high probability. If event B happens infrequently, then we tend to think that B is unlikely to happen, and we say that it has a low probability.

When we decide that event A happens frequently, we are making a relative judgment. Compared to anything else that might happen in this situation, event A happens frequently. Intuitively, we determine the *relative frequency* of event A: the proportion of time that event A occurs, out of all possible events that might occur in this situation. In statistical terminology, we call all possible events that can occur in a given situation the *population* of events. Thus, the probability of an event is equal to the event's relative frequency in the population of all possible events that can occur.

> ***STAT ALERT*** The relative frequency of an event in the population equals the probability of the event.

If a population is thought of as all possible events that might occur, then the event or events that do occur make up a sample from that population. Thus, when we ask if a particular event will occur, we are actually asking what is the probability that the sample will contain that event when we randomly sample from a particular population. You already know that a score's relative frequency in the population is also the score's expected relative frequency in any sample.

We use probability to express a particular event's expected relative frequency in any single random sample. For example, I am a rotten typist, and while typing the manuscript for this book, say I randomly made typos 80% of the time. This means that in the population of my typing, typos occur with a relative frequency of .80. We expect the relative frequency of typos to continue at a rate of .80 in any random sample of my typing. We express this expected relative frequency as a probability, so the probability is .80 that I will make a typo when I type the next woid.

As the above example illustrates, a **probability** is a mathematical statement indicating the likelihood of an event when we randomly sample a particular population. It is our way of expressing what we expect to occur when only chance is operating, and it indicates our confidence in any particular event. For example, if event A has a relative frequency of zero in a particular situation, then the probability of event A is zero. This means that we do not expect A to occur in this situation, because it never does. But if event A has a relative frequency of .10 in this situation, then A has a probability of .10. This means that we have some—but not much—confidence that the event will occur in any particular sample. Because it occurs only 10% of the time in the population, we expect it to occur in only 10% of our samples. On the other hand, if event A has a probability of .95, we are confident that A will occur. Because it occurs 95% of the time in the population, we expect it to occur in 95% of our samples. At its most extreme, event A's relative frequency can be 1.0. If event A is 100% of the population, its probability is 1.0. Then we are positive it will occur in this situation because it always does.

Notice that an event cannot happen less than 0% of the time nor more than 100% of the time, so a probability can *never* be less than 0 or greater than 1.0. Notice also that all events in a population together constitute 100% of the time, or 1.0 of the population. This means that the relative frequencies of all events must add up to 1.0, so the probabilities of all events in the population must also add up to 1.0. Thus, if the probability of my making a typo at any moment is .80, then because $1.0 - .80 = .20$, the probability is .20 that any word I type will be error free.

It is important to remember that except when p equals either 0 or 1, we are never absolutely certain that an event will or will not occur in a particular situation. The probability of an event is its relative frequency over the long run, or in the infinite population. It is up to luck whether a particular sample contains the event. For example, even though I make typos at a rate of .80 over the long run, I may go for quite a while without making a typo. That 20% of the time I make no typos has to occur sometime. Thus, although the probability is .80 that I will make a typo in each word, it is only over the long run that we truly expect to see precisely 80% typos.

COMPUTING PROBABILITY

Computing the probability of an event is rather simple: we need only determine its relative frequency in the population. When we know the relative frequency of every possible event in a population, we have a probability distribution. A **probability distribution** tells us the probability of every possible event in a population.

Creating Probability Distributions

There are two ways to create a probability distribution. One way is to actually measure the relative frequency of every event in the population. If we "empirically" determine the relative frequencies of all events in a population, we create an *empirical probability distribution.* Typically, however, we cannot measure all of the events in a population, so we create an empirical probability distribution from samples of those events. We assume that the relative frequency of an event in a random sample represents the relative frequency of the event in the population.

For example, say that Dr. Fraud is sometimes a very cranky individual, and as near as we can tell, his crankiness is random. We observe him on 18 days and determine that he is cranky on 6 of these days. Relative frequency equals the frequency of an event divided by N, so the relative frequency of Dr. Fraud's crankiness is 6/18, or .33. Assuming that this sample represents the population of how Dr. Fraud always behaves, we expect that he will continue to be cranky on 33% of all days. Thus, the probability that he will be cranky today is $p = .33$. Conversely, he is not cranky on 12 of the 18 days we observe him, so the relative frequency of his not being cranky is 12/18, or .67. Thus, the probability that he will not be cranky today is $p = .67$. Because his cranky days plus his noncranky days constitute all possible events, we have an empirical probability distribution for his crankiness.

Statistical procedures usually rely on the other way to create a probability distribution, which is to create a theoretical probability distribution. A **theoretical probability distribution** is a theoretical model of the relative frequencies of events in a population. As with previous models, we devise theoretical probability distributions based on how we assume nature distributes events in the population. From such a model, we determine the relative frequency of each event in the population. This relative frequency is then the probability of the event in any random sample.

Let's first look at common random events such as tossing a coin and drawing playing cards. (As we shall see, determining probability in such situations is analogous to determining probability in research.) With coin tossing, we assume that we are discussing random flips of a fair coin. When we toss a coin in the air, there are two possible outcomes that can occur: a head or a tail. We assume that nature has no bias toward heads or tails, so that over the long run we will see 50% heads and 50% tails. In other words, we expect the relative frequency of heads to be .50 and the relative frequency of tails to be .50. Since relative frequency in the population *is* probability, we have devised our theoretical probability distribution. The probability of a head on any toss of the coin is $p = .50$, and the probability of a tail is $p = .50$.

When we draw a playing card, we select 1 of the 52 cards in the deck. We assume that there is no bias favoring one card over another, so each has the same relative frequency in the population. Over the long run, we expect each card to occur at a rate of once out of every 52 draws, so each card has a relative frequency of 1/52, or .0192. Therefore, the probability of drawing any specific card on a single random draw is $p = .0192$.

And that is the logic of probability. We either theoretically or empirically devise a model of how events are assumed to be distributed in the population. This model gives us the expected relative frequency of each event in the population. From this, we derive the expected relative frequencies of events in any sample, expressed as p. By computing the probability for all possible events in a particular situation, we create the probability distribution for that situation.

General Formula for Computing Probability

Hidden in the above examples is the method for directly computing the probability of a specific event whenever all events in the population are equally likely.

THE FORMULA FOR COMPUTING PROBABILITY WHEN EVENTS ARE EQUALLY LIKELY IS

$$p(\text{event}) = \frac{\text{number of outcomes that satisfy event}}{\text{total number of possible outcomes}}$$

In the numerator, we place the number of possible outcomes that can satisfy the requirements of the event we are describing. In the case of flipping a coin, for example, there is one outcome that satisfies the condition of showing a head. In the denominator we place the total number of possible outcomes that can occur. There are two possible outcomes that can occur when we flip a coin: either head or tail. Therefore, the probability of getting a head is 1/2, which equals .5.

Likewise, we might define the event as randomly drawing a king from a deck of cards. There are four kings in a deck, and any one of these would satisfy the event. With a total of 52 possible outcomes that can occur, the probability of randomly drawing a king on any single draw is 4/52, or .0769.

We can even apply this formula to real life! For example, raffle tickets are sometimes sold one for a dollar and sometimes sold three for a dollar. Many people think they have a better chance of winning when everyone gets three tickets for a dollar. Boy, are they wrong! By applying the above formula, we can see why. Say that 100 people each spend a dollar and get one ticket. Then each person's probability of winning is equal to 1/100, or .01. However, if 100 people each buy 3 tickets for a dollar, then each person's probability of winning equals 3/300. The kicker is that 3/300 equals 1/100, so each person still has a probability of winning equal to .01.

The above formula can be used to find the probability of any event when all possible events are equally likely. However, using this formula is rather tedious when the event can be satisfied by complex sequences or alternatives, such as when we describe the event using the word "and" or "or." Therefore, shortcut formulas for three of the most common complex events are provided in Appendix B.

Factors Affecting the Probability of an Event

All random events are not the same. In computing the probability of an event, we must consider two important factors. First, events may be either independent or dependent. Two events are **independent events** when the probability of one event is *not* influenced by the occurrence of the other event. For example, contrary to popular belief, washing your car does not make it rain. These are independent events, so the probability of rain does not increase when you wash your car. On the other hand, two events are **dependent events** when the probability of one event is influenced by the occurrence of the other event. For example, whether you pass an exam usually depends on whether you study: the probability of passing increases or decreases depending on whether studying occurs.

Second, the probability of an event is affected by the type of sampling we perform. **Sampling with replacement** is the procedure in which we replace any previously selected samples back into the population before drawing any additional samples. For example, say we select two cards from a deck of playing cards. Sampling with replacement occurs if, after drawing the first card, we return it to the deck before drawing the second card. The probabilities on the first and second draw are each based on 52 possible outcomes, and the probability of any particular card's being selected either time is constant. On the other hand, **sampling without replacement** is the procedure in which we do *not* replace any previously selected samples in the population before selecting again. Thus, sampling without replacement occurs if we discard the first playing card after drawing it. In this case, the probability of a card's being selected on the first draw is based on 52 possible outcomes, but the probability of a card's being selected on the second draw is based on only 51 possible outcomes. With fewer possible outcomes, the probability of drawing a particular card is slightly larger on the second draw than on the first draw.

In statistics we usually assume that the events we are considering are independent and sampled with replacement.

The reason we discuss probability is not that we have an uncontrollable urge to flip coins and draw cards. Researchers use probability to make decisions about scores and samples of scores. To do that, they use the standard normal curve.

OBTAINING PROBABILITY FROM THE STANDARD NORMAL CURVE

As we've seen, once we know the distribution of events in the population, we can determine their relative frequencies and thus their probabilities. In statistics we usually assume that the events—the scores—in the population form a normal distribution. For this reason, our most common theoretical probability distributions are based on the standard normal curve. Here is how they work. In Chapter 6 we saw that by using z-scores, we can find the proportion of the total area under the normal curve in any part of a distribution. This proportion corresponds to the relative frequency of the scores in that part of the distribution. The relative frequency of scores in the population equals the probability of those scores. Thus,

FIGURE 9.1 *z*-Distribution Showing the Area for Scores Below the Mean

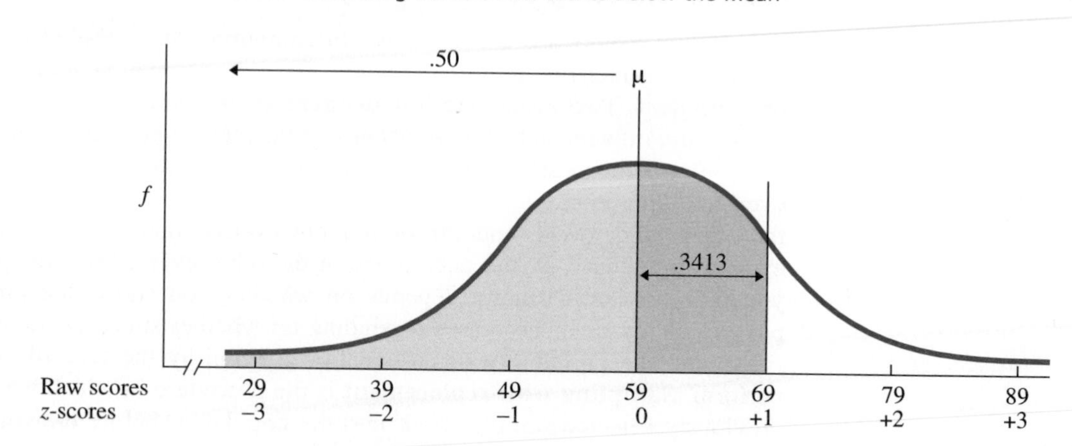

The proportion of the area under the standard normal curve for scores in any part of the distribution equals the probability of those scores!

For example, as shown in Figure 9.1, we know that 50% of the area and thus 50% of all scores fall below the mean, producing negative *z*-scores. Therefore, because negative *z*-scores occur .50 of the time in the population, the probability of randomly selecting a negative *z*-score—or a subject who produces a negative *z*-score—is .50. In other words, just as we saw when flipping coins, we can obtain either a positive or a negative *z*-score, and they both occur with equal frequency. Therefore, because we expect to run into negative *z*-scores 50% of the time, the probability that the next score we draw will be a negative *z*-score is .50.

We obtain the same answer by using our previous formula for probability. We know that out of a total of 100% of the possible scores in the population, 50% of them satisfy the event "negative *z*-score." So, 50%/100% equals $p = .50$.

Once we know the probability of selecting any range of *z*-scores, we know the probability of selecting the corresponding raw scores. In Figure 9.1, we have applied our model to a distribution of raw scores where the mean is 59. Raw scores below the mean of 59 produce negative *z*-scores. Thus, the probability is also .50 that any individual we select will have a raw score below 59.

We can obtain the proportion under the curve, and thus the probability for any part of a normal distribution, by using the *z*-tables in Appendix D. For example, from the *z*-tables we know that *z*-scores between the mean and a *z* of $+1.0$ occur .3413 of the time in the population. Thus, the probability of randomly selecting any one of these *z*-scores is .3413. Further, looking again at the distribution in Figure 9.1, we see that the mean raw score is 59 and a score of 69 produces a *z*-score of $+1.0$. Therefore, the probability of randomly selecting a raw score between 59 and 69 in this distribution is also .3413.

Likewise, we can determine the probability of randomly selecting any score greater than a certain *z*-score. For example, what is the probability of selecting a *z*-score larger than $+2.0$ (From the right hand shaded area in Figure 9.2?) From the *z*-tables we see that all of the *z*-scores above $z = +2.0$ comprise .0228 of

FIGURE 9.2 Area Under the Curve Beyond $z = \pm 2.0$

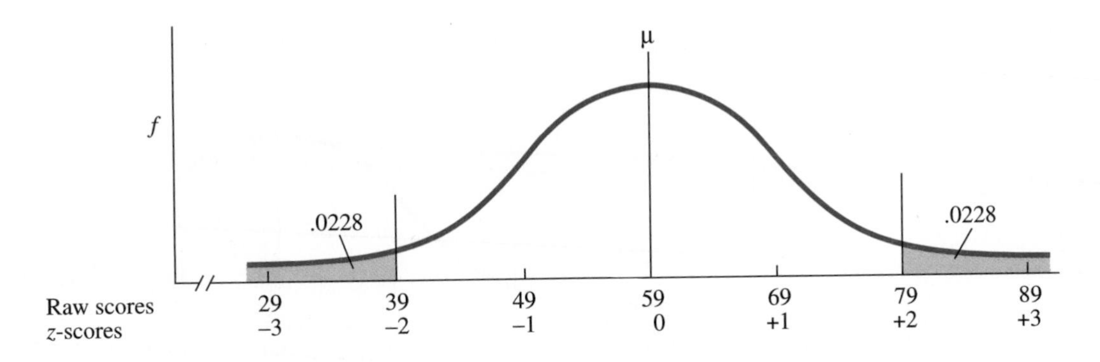

the population and thus occur .0228 of the time. Therefore, the probability of randomly selecting a score above $z = +2.0$ is $p = .0228$ (which in this case is also the probability of selecting a raw score above 79).

What is the probability of selecting a score that is beyond a z of ± 2.0? The phrase "beyond ± 2.0" means that we will be satisfied by any score above $+2.0$ or below -2.0, so we must determine the probability of drawing a score from either of the shaded areas shown in Figure 9.2. All scores beyond $z = +2.0$ constitute .0228 of the curve, and all scores beyond $z = -2.0$ constitute .0228 of the curve. In total, .0228 + .0228, or .0456, of the curve contains scores that will satisfy us. Because z-scores beyond ± 2.0 occur .0456 of the time, the probability of randomly selecting a z beyond ± 2.0 is $p = .0456$. On the distribution, a raw score of 39 corresponds to a z of -2.0, and a raw score of 79 corresponds to a z of $+2.0$. Therefore, the probability is also .0456 that we will randomly select a raw score below 39 or above 79.

Using the above procedure, we can determine the probability of randomly selecting scores from any portion of any normal distribution. In fact, we can even use this procedure to find the probability of randomly selecting "scores" that are sample means.

Determining the Probability of Sample Means

In Chapter 6 we conceptualized the sampling distribution of means as the frequency distribution of all possible sample means that would result if our bored statistician randomly sampled a raw score population an infinite number of times using a particular sample N. The sampling distribution is graphed like any other normally distributed population except that sample means, instead of raw scores, are plotted along the X axis. Because a sampling distribution shows how sample means are distributed in the population, it forms a theoretical probability distribution.

In Chapter 6 we also found a sample mean's location on a sampling distribution by first computing the standard error of the mean (the standard deviation of the sampling distribution) using the formula $\sigma_{\overline{X}} = \sigma_X/\sqrt{N}$. Then we computed a z-score using the formula $z = (\overline{X} - \mu)/\sigma_{\overline{X}}$. Then, by applying the standard normal

FIGURE 9.3 Sampling Distribution of SAT Means When $N = 25$

curve model, we determined the relative frequency of sample means falling above or below that z-score. Now, in the same way that we just determined the probability of randomly selecting certain raw scores, we can determine the probability of randomly selecting certain sample means.

Let's look again at the sampling distribution of SAT means discussed in Chapter 6, which is shown in Figure 9.3. The population mean, μ, of this distribution is 500; and when N is 25, the standard error of the mean is 20. First let's examine the portion of the sampling distribution containing means with z-scores between 0 and $+1.0$. The relative frequency of such z-scores is .3413, so the relative frequency of sample means that produce these z-scores is also .3413. Because an event's relative frequency in the population equals its probability, the probability of randomly selecting a sample mean with a z-score between 0 and $+1.0$ from this population is $p = .3413$. On this SAT sampling distribution, sample means between 500 and 520 produce z-scores between 0 and $+1.0$. Thus, when N is 25, the probability is .3413 that we will randomly obtain a sample mean between 500 and 520 from this SAT raw score population.

Think about this: randomly selecting a sample mean is the same as randomly selecting a sample of raw scores and then computing their mean. Likewise, randomly selecting a sample of raw scores is the same as randomly selecting a sample of subjects and then measuring their raw scores. Therefore, because the area under a portion of the curve provides the probability of selecting certain sample means, it also provides the probability of selecting the corresponding samples. Thus, we can rephrase our above finding: when we randomly select 25 subjects from this SAT population, the probability of selecting a sample that produces a mean between 500 and 520 is .3413.

Likewise, we saw in the previous section that z-scores beyond ± 2.0 have a probability of .0456. Thus, the probability that we will select a random sample that produces a sample mean having a z-score beyond ± 2.0 is also .0456. On our SAT sampling distribution, z-scores of ± 2.0 correspond to means of 540 and 460, respectively. Therefore, when we sample this SAT population, the probability is .0456 that we will randomly select a sample of 25 scores producing a mean below 460 or above 540.

In the same way, we can determine the probability of selecting any range of sample means from any normally distributed raw score population. From a different distribution of, for example, personality test scores, say that we compute that sample means of 88 and 78 have z-scores of ± 2.0, respectively. Then the probability of selecting a random sample of subjects who produce a mean personality test score above 88 or below 78 is also .0456.

Computing a z-score for sample data and then determining its probability on a sampling distribution is the basis for all inferential statistical procedures. Therefore, think of a sampling distribution as a "picture of chance." It shows us how often, over the long run, we can expect random chance to produce samples having certain sample means when we randomly sample a particular raw score population.

STAT ALERT A sampling distribution of means provides us with a theoretical probability distribution that describes the probability of obtaining any sample mean when we randomly select a sample of a particular N from a particular raw score population.

Recall that the reason we are discussing probability is so that ultimately we can make decisions about our data. Now that you understand how to compute probability, the next step is to understand how to make decisions using probability.

MAKING DECISIONS BASED ON PROBABILITY

The probability that I will make a typo at any moment is .80. Should you decide that I will make a typo on the next word? Yes, it's a good bet. Over the long run, I make typos 80% of the time, so on any word you are likely to win the bet. What is the probability that you made the correct decision? The probability of a correct decision equals the probability that chance will produce the event you are predicting. To figure this probability, assume that you make the same decision every time, over the long run. If 100% of the time you bet that I will make a typo and 80% of the time I make a typo, then you will be correct 80% of the time. Thus, the relative frequency of your winning the bet is .80, so the probability of winning any single bet is $p = .80$. Remember, though, that I do not make typos 20% of the time, so out of 100%, you will lose the bet 20% of the time. Because the relative frequency of your losing is .20, the probability of losing any single bet is $p = .20$. Conversely, if you bet that I will not make a typo, $p = .20$ of winning the bet, and $p = .80$ of losing the bet.

As this illustrates, we bet in favor of high-probability events because by doing so we maximize the probability of winning the bet. Conversely, we do not bet in favor of low-probability events, because if we did, the probability of winning the bet would be low. In statistics, we also make bets. As we shall see, the bet we make involves deciding whether random chance produced a particular sample from a particular population.

Deciding Whether Chance Produced a Particular Event

It is essential to remember that any probability implies "over the long run." For example, say we flip a coin 7 times and obtain 7 heads in a row. This will not concern us, assuming that we have a fair coin. If we see 7 heads (or 70 heads) in a row, we still assume that over the long run, the relative frequencies of heads and tails will each be .50. We understand that any single sample of coin tosses may not be *representative* of the population of coin tosses, because a head's relative frequency in the sample may not equal its relative frequency in the population. We have merely shown up at a time when luck produced an unrepresentative sample.

People who fail to understand this principle fall victim to the "gambler's fallacy." If we have obtained 7 heads in a row, the fallacy is that a head is now less likely to occur, because it has already occurred too frequently (as if the coin said, "Hold it. That's enough; no more heads for a while!"). If it is a fair coin, the probability of a head on the next toss is still .50. The mistake of the gambler's fallacy is failing to recognize that each coin toss is an independent event and we have merely observed an unrepresentative sample of coin tosses.

But what if we are not certain whether the coin is a fair coin? If you are betting money on the flips of a coin and another player obtains 7 heads in a row, you should stop playing: it's a good bet that the coin is rigged. How do we reach this decision? We compute the probability of obtaining such an outcome if the coin is *not* rigged.

To compute a probability, we must determine the total number of possible outcomes. Therefore, we determine all possible series of heads and tails we could obtain by flipping a coin 7 times. In addition to the one way of obtaining all 7 heads, there are many ways we can obtain 1 tail, such as

tail, head, head, head, head, head, head

or

head, tail, head, head, head, head, head

or

head, head, tail, head, head, head, head

and so on. Next we determine all the ways we can obtain 2 tails in 7 tosses, 3 tails in 7 tosses, and so on. We will find a total 128 different arrangements of heads and tails that can be obtained when a coin is tossed 7 times in a row. Only one of these outcomes will satisfy us, and that is obtaining all 7 heads. Thus, the probability of obtaining 7 heads in 7 coin tosses is 1/128, or about .008. (A shortcut formula for computing such probabilities, called the binomial expansion, is presented in Appendix B.)

Now we can decide about the fairness of the coin. If this coin is fair, then 7 heads in a row will hardly ever happen by chance. We expect its relative frequency to be 8 out of every 1000 times, or .8% of the time. Because it is difficult to believe a player would be that lucky, we reject the idea that random chance has produced such an unlikely outcome, and we leave the game.

Notice that we made a definitive, all-or-nothing decision: the probability suggests that the coin is crooked, so we decide that it is definitely crooked. In statistics, whenever we use probability to make a decision, our decision is a definite yes or no, period.

However, our decision could be wrong!

The Probability of Incorrectly Rejecting Chance

Even though 7 heads in a row occurs relatively infrequently, we cannot be certain that the coin is rigged. Unlikely events are just that: they are unlikely, not impossible. Random chance does produce unusual samples every once in a while. In fact, we computed that we would expect chance to produce 7 heads in a row .8% of the time. Perhaps this was one of those times. Thus, regardless of the probability, we can never *know* whether random chance was operating or not. The best we can do is determine the probability that we mistakenly rejected the explanation that chance produced the event.

We determine the probability that we incorrectly rejected the chance explanation in the same way that we determine the probability of losing any bet. When chance is operating (and the coin is fair), it produces 7 heads in a row .8% of the time. Because this is such an infrequent event, 100% of the time that we see seven heads in a row, we will bet that chance did not produce the result. If we make the bet 100% of the time and chance produces 7 heads on .8% of those times, then .8% of the time we will lose the bet. Since the relative frequency of our losing the bet is .008, the probability of losing any single bet is $p = .008$.

> ***STAT ALERT*** Anytime we reject the idea that chance produced an event, the probability that we are incorrect is equal to the relative frequency with which chance does produce the event.

Be sure that you feel comfortable with the above logic of deciding we had a crooked coin and then determining the probability that we were wrong. You need to understand this logic, because we apply it when making statistical decisions.

USING PROBABILITY TO MAKE DECISIONS ABOUT THE REPRESENTATIVENESS OF A SAMPLE

Recall that in research, we use a random sample of data to draw conclusions about the behavior of the population of our subjects. In essence, we want to say that the way in which the sample behaves indicates the way in which the population would behave if we could observe it. However, there is an insurmountable problem: we can never be certain how that population would behave, because there is no guarantee that our sample accurately reflects the population. In other words, we are never certain that a sample of scores is *representative* of a particular population of scores.

Back in Chapter 2 we said that a representative sample is a mini-version of the population, having the same characteristics as the population. However, representativeness is not all or nothing. A sample can be more or less representative, having more or less the same characteristics as the population. This is because how representative a sample is depends on random *chance*—the luck of the draw by which scores were selected. By chance the sample may be somewhat different from the population from which it is selected, and thereby represent that population somewhat poorly. The problem is that when a sample is different from the population it actually represents, it has the characteristics of some other population and thus appears to represent that other population. The end result is that although a sample always represents some population, we are never sure *which* population it represents: the sample may poorly represent one population, or it may represent another population.

This was the problem we faced with the crooked coin: we were uncertain whether the sample of coin tosses was representative of the population of tosses with a fair coin or the population of tosses with a crooked coin. We decided whether our sample represented the population of fair coin tosses by determining the probability of obtaining our sample from that population. Implicitly, we used this logic: if we flip a coin several times and obtain 51% heads and 49% tails, this sample still appears to represent the population of fair coin tosses containing 50% heads and 50% tails. Even 60% heads and 40% tails is reasonably representative of fair coin tosses. In these situations, the fact that we do not have a 50-50 split between heads and tails is written off as being due to chance: chance produced a less than perfectly representative sample.

Beyond a certain point, however, we begin to doubt that random chance is producing our results. For example, with 70% heads and 30% tails, we grow suspicious that mere luck produced such an unrepresentative sample of the population of fair coin tosses. And when we saw 7 heads in a row, we seriously doubted the honesty of the other player: it was *too unlikely* that we would obtain such a sample by chance if the sample actually represented the population of fair coin tosses. Therefore, we rejected the idea that luck merely produced a less than perfectly representative sample from that population. The only alternative was that the sample represented a population of tosses from a crooked coin.

To summarize our decision making above, we used a theoretical probability distribution based on one population (that of fair coin tosses). We determined that if our sample represents this population, the probability of obtaining a sample of 7 heads by chance is only $p = .008$. Because the probability of obtaining this sample from this population is so small, we concluded that our sample does not represent this population. Instead, we concluded that our sample represents some other population.

Here is another example. You obtain a sample paragraph of someone's typing, but you do not know whose. Is it mine? Does this sample represent the population of my typing? Say that you find zero typos in the paragraph. You cannot know for certain whether I typed this paragraph, but there is a good chance that I did not. Since I type errorless words only 20% of the time, the probability that I would type an entire errorless paragraph is very small. Thus, the probability of obtaining such a sample paragraph is very small if the sample represents the population of my typing. Because chance is *too unlikely* to produce such a sample when you randomly sample the population of my typing, you decide against this low-probability event. You reject the idea that chance produced this highly unrepresentative sample from the population of my typing and instead conclude that the sample represents some other population, such as that of a competent typist.

On the other hand, say that there are typos in 79% of the words in the paragraph. Since I make typos 80% of the time, this paragraph is reasonably consistent with what you would expect if the sample represents my typing. Although you expect 80% typos from me over the long run, you do not expect precisely 80% typos in every sample. Rather, a sample with 79% errors seems likely to occur simply by chance when the population of my typing is sampled. Thus, you can accept that this paragraph is more or less representative of my typing, but because of random chance, there are slightly fewer typos in the sample than in the population.

As we shall see, we use this same logic in research to decide whether a sample of scores is representative of a particular population of scores. This is the essence of all inferential statistical procedures: based on the probability of obtaining a particular sample from a particular population, we decide whether the sample represents that population. If the sample is likely to occur when the population is sampled, then we decide that it may represent that population. If the sample is unlikely to occur when that population is sampled, then we decide that the sample does not represent that population and instead represents some other population.

> ***STAT ALERT*** The essence of all inferential statistical procedures is to decide whether a sample of scores is likely or unlikely to represent a particular population of scores.

The next chapter puts all of this into a research context. In the following sections of this chapter, we will consider the basics of deciding whether a sample of scores represents a particular population of scores.

Making Decisions About Samples of Scores

Say that we have returned to Prunepit U. and obtained a random sample of subjects' SAT scores that has the surprising mean of 550! This is surprising because we think that the students at Prunepit U. are terminally average. Because the everyday, national population of SAT scores has a μ of 500, we should have obtained a sample mean of 500 if our sample was perfectly representative of this population. How do we explain having a sample mean of 550? On the one hand, the simplest explanation is that we obtained a sample of relatively high SAT scores merely because of random chance—the luck of the draw that determined who was selected to be in the sample. It is possible that chance produced a less than perfectly representative sample of the population where μ is 500. On the other hand, perhaps our sample does not come from or represent the everyday, national population of SAT scores: after all, these *are* Prunepit students, so they may belong to a very different population of students, having some other μ.

Here is the logic of our procedure: To decide whether our sample represents the population of SAT scores where μ is 500, we will determine the probability of randomly selecting a sample mean of 550 from this population. As we saw previously, to determine the probability of obtaining a particular sample mean from a particular population, we create the sampling distribution of means for that population. For our Prunepit problem, we envision the sampling distribution as showing the frequencies of all the different means that our bored statistician would obtain if, using our *N,* she randomly sampled the SAT population an infinite number of times. Any differences between sample means *are* due to the luck of the draw—which scores she happened to select for each sample. Thus, the sampling distribution provides us with a model of how often *random chance* will produce any particular sample mean when the samples *do* represent a population in which μ is 500.

Then, using *z*-scores, we determine the location of our sample mean on the sampling distribution and thereby determine the probability of obtaining that mean. If it is likely that random chance would produce a sample mean of 550 from the SAT population in which μ is 500, we will accept that our Prunepit sample may represent this population. Conversely, if it is too unlikely that chance would produce a sample mean of 550 from this population, we will reject the idea that it was merely the luck of the draw that produced such a mean. Instead, we conclude that our sample represents some other population of scores, having some other μ.

Be sure you understand the above logic before proceeding. If you do, you're ready to perform the mechanics of the procedure. Notice that we have two tasks to perform: we must determine the probability of obtaining our sample from this population, and we must decide whether our sample is too unlikely for it to represent this population. In statistics, we perform both of these tasks simultaneously. To do so, we first set up the sampling distribution of means.

Identifying the Region of Rejection

Earlier we decided that we did not have a fair coin because the occurrence of 7 heads in a row was too unlikely for us to believe that chance produced it. Likewise, we must decide what it will take to convince us that a particular sample mean is too unlikely for it to represent a particular population. To do so, we define the *criterion* for making our decision. (In Chapter 8 you saw that the *Y* variable is called the criterion variable. Here we use the term "criterion" differently, to refer to a probability.) The **criterion** is the probability that we use to decide whether a sample is too unlikely to occur by chance for it to represent a particular population. When the probability of obtaining a sample from a particular population is *less* than our criterion, then we decide that it is too unlikely that the sample represents that population. We do not accept the idea that the sample poorly represents that population because of chance; instead, we decide that the sample must represent some other population.

Psychologists usually use the probability of .05 as their criterion. Because we must make a yes or no decision, we will reject the idea that chance produced *any* of the sample means that, together, have only a .05 probability of occurring in a particular population.

> ***STAT ALERT*** The criterion probability that defines samples as too unlikely is usually $p = .05$.

FIGURE 9.4 Setup of Sampling Distribution of Means Showing the Region of Rejection

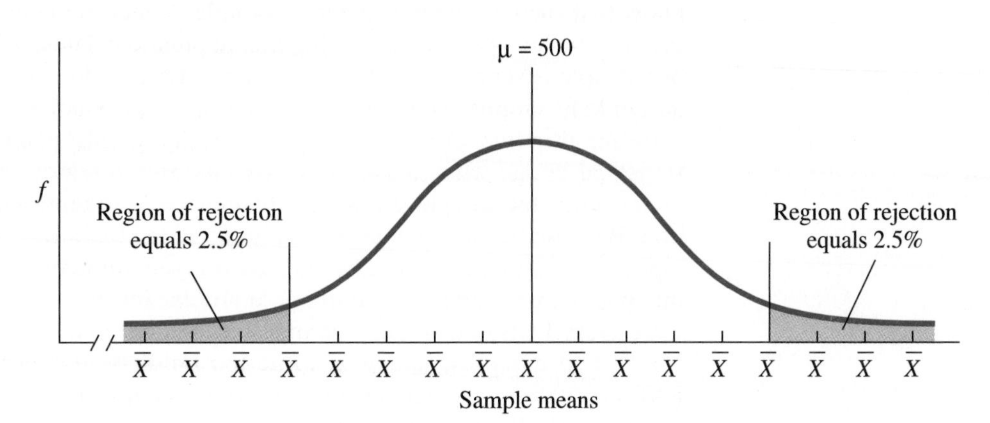

As you know, on a normal distribution, the scores having a low frequency and thus a low probability lie in the tails of the distribution. Because probability is the same as the proportion of the total area under the curve, those sample means having a total probability of .05 make up the extreme 5% of the curve. Thus, all samples falling in the extreme 5% of the sampling distribution are treated the same and considered too unlikely to accept as representing the particular population.

Because there are two tails, we can divide this 5% in half and demarcate the extreme 2.5% in each tail of the sampling distribution. Figure 9.4 shows how we will do this for our SAT example. Quite literally, we have drawn a line in each tail of the distribution that defines "too unlikely." In statistical terms, the shaded areas beyond the lines make up the region of rejection. The **region of rejection** is that portion of a sampling distribution containing values that are considered to be too unlikely to occur by chance. The size of the area under the curve that comprises the region of rejection is always equal to our criterion. Usually our criterion is .05, so the total region of rejection usually comprises .05 of the curve.

It is crucial that you understand what the region of rejection represents. Because the μ of the population is 500, sample means of 500 are perfectly representative. The more the means in Figure 9.4 differ from 500, the more they are both unrepresentative of the population μ and unlikely to occur by the luck of the draw when selecting a sample. Therefore, very infrequently are samples *so* poor at representing the population that they have means lying within the region of rejection. With our criterion, we have simply defined the most unrepresentative means as the most unlikely, extreme 5%. Thus, the region of rejection contains the worst, most unrepresentative sample means we could obtain when sampling from this population.

But remember, we already have a sample mean. The question is, what population does it represent? The sampling distribution shows all of the means our statistician obtained when she definitely *was* representing the SAT population. Therefore, if our Prunepit sample mean lies within the region of rejection, then either by chance our sample is extremely unrepresentative of this population, or it represents some other population. By its being in the region of rejection, however, we also know that such an unrepresentative sample is very unlikely to occur, if we—like the statistician—were representing this population. Thus, it is not a good bet that our sample *is* representing this population. Essentially, we "shouldn't" get such an unlikely sample mean when representing this population. If we did get such a mean, then we probably aren't representing this population. Therefore, in statistical terms, when a sample mean falls in the region of rejection, we *reject* the idea that the sample is one that, by chance, is unrepresentative of the population. Because the statistician was so unlikely to obtain such means when she was representing this population, we decide it is too unlikely that we are representing this population whenever we obtain a similar mean.

Conversely, if our Prunepit sample mean is not in the region of rejection, then it is not so unrepresentative as to be too unlikely to occur when sampling this SAT population. In fact, by our criterion, sample means not in the region of rejection are likely to occur by chance when this population is sampled. In statistical terms, in this case we *retain* the idea that chance may have produced our particular sample mean and that the sample may represent, although poorly, this population of SAT scores.

Identifying the Critical Value

We locate our sample mean on the sampling distribution by computing a z-score for it. Because the absolute values of z-scores get larger as we go farther into the tail of the distribution, if the absolute value of our z-score is large enough, then our sample mean falls in the region of rejection. How large must the z-score be? With a criterion of .05, we set up the region of rejection so that in each tail is half that amount, or .025 of the total area under the curve. From the z-table we see that the extreme .025 of the curve lies beyond the z-score of ± 1.96, and thus, the region of rejection lies beyond these values. Therefore, as shown in Figure 9.5, for a sample mean to lie in the region of rejection, its z-score must lie *beyond* ± 1.96. In this example, ± 1.96 is the critical value of z. A **critical value** marks the edge of the region of rejection and thus defines the value required for a sample to fall in the region of rejection.

Now, deciding whether our sample represents the SAT population in which μ is 500 boils down to comparing the sample's z-score to ± 1.96. Any sample mean producing an absolute value of z that is larger than the critical value of 1.96 lies in the region of rejection: Therefore, we will reject the idea that chance produced such an unrepresentative sample mean from this population, and thus reject the idea that the sample represents this SAT population.

Conversely, any sample mean producing an absolute value of z that is smaller than or equal to the critical value is *not* in the region of rejection. Therefore, we will retain the idea that chance may have produced an unrepresentative sample mean from this population and retain the idea that our sample may represent this SAT population.

> ***STAT ALERT*** When the z-score for our sample lies beyond the critical value, we *reject* the idea that the sample represents the particular raw score population reflected by the sampling distribution. When the z-

FIGURE 9.5 Setup of Sampling Distribution of SAT Means Showing Region of Rejection and Critical Values

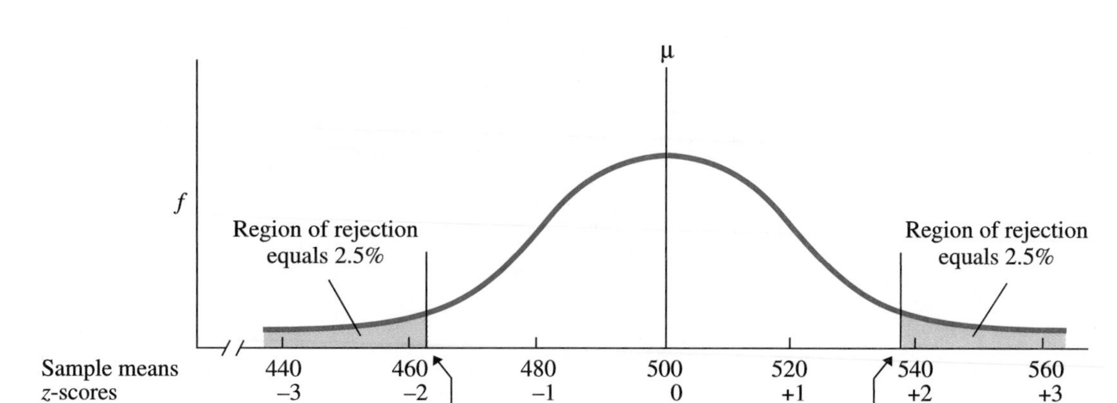

score does not lie beyond the critical value, we *retain* the idea that our sample may represent that raw score population.

Deciding If a Sample Is Representative

Now, at long last, we can evaluate our sample mean of 550 from Prunepit U. On the sampling distribution created from the population of SAT scores for which μ is 500, σ_X is 100, and N is 25, the standard error of the mean is $100/\sqrt{25}$, which is 20. Then our sample mean of 550 has a z-score of $(550 - 500)/20$, which is $+2.5$. Think about this z-score. If the sample represents this SAT raw score population, it seems to be doing a very poor job of it. With a population mean of 500, a perfectly representative sample would also have a mean of 500 and thus have a z-score of 0. Good old Prunepit produced a z-score of $+2.5$!

To confirm our suspicions, we examine our sampling distribution shown in Figure 9.6. Our sample's z-score of $+2.5$ is beyond the critical value of ± 1.96, so it is in the region of rejection. Because the z-score is in the region of rejection, the corresponding sample mean of 550 is also in the region of rejection. This tells us that our sample mean of 550 is among those means that we consider to be too unlikely to occur by chance when sampling the population of SAT scores where μ is 500. It is not a good bet that, through random chance, we merely obtained an unrepresentative sample from that population, because samples that are this unrepresentative hardly ever happen. After all, our bored statistician hardly ever got such a mean, and she sampled SAT scores an *infinite* number of times. Thus, essentially, we've determined that if we conclude that our sample represents this population, then we must also conclude that we had a tremendous amount of luck in getting our sample mean of 550. But! We don't believe we're that lucky! Therefore, we reject that chance produced this sample from this population and that our sample represents that population of SAT raw scores having a μ of 500.

FIGURE 9.6 Sampling Distribution of SAT Means Showing Location of the Prunepit U. Sample Relative to the Critical Value

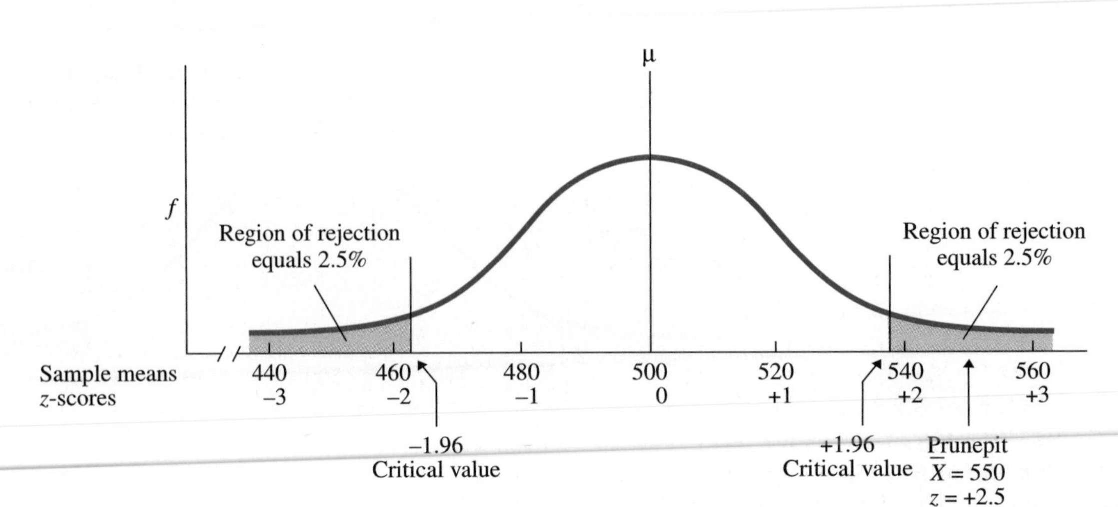

Notice that, as with the crooked coin, we make a yes or no decision. Because our sample would be unlikely to occur if it represented the SAT raw score population where μ is 500, we decide that no, it definitely does not represent that population.

Once we reject the idea that our sample represents the raw score population reflected by the sampling distribution, it is only logical that the sample must represent some other population of scores, where such a sample mean is more likely. For example, a high sample mean of 550 is more likely if students lie about their SAT scores. Perhaps our Prunepit subjects obtained a mean of 550 because they lied about their scores, so they may represent the population of students who lie about the SAT. Or, because Prunepit U. is located on a toxic waste dump, perhaps toxic wastes produce higher SAT scores. Our sample may represent the population of SAT scores of students who live on toxic waste dumps.

In fact, recall that we can use the value of a sample mean as our estimate of the value of the corresponding population μ. And if you stop and think about it, a population having a μ of 550 is most likely to produce a sample having a mean of 550. Therefore, because we have rejected that our sample represents the population where μ is 500, our best guess is that, for whatever reason, the SAT population represented by our sample has a μ of 550.

Thus, in sum, as we did when betting on my typos, we decide against the low-probability event that the sample represents the SAT population where μ is 500, and we decide in favor of the high-probability event that the sample represents a population where μ is 550.

On the other hand, say that our sample mean had been 474, resulting in a z-score of $(474 - 500)/20$, which is -1.30. Because -1.30 does not lie beyond our critical value of ± 1.96, our sample mean is not in the region of rejection. Looking at the sampling distribution, we see that when our bored statistician sampled this population, such a z-score and corresponding sample mean were relatively frequent and thus likely. Because of this, we can accept that random chance produced a less than perfectly representative sample for us but that our sample probably represents this SAT population.

Notice that, in essence, we have described a system for evaluating the *difference* between our sample mean and what we would expect to obtain if the sample represented a particular population. We expect a representative sample mean to be "close" to the population mean. We cannot clearly recognize whether our sample is representative of the population, because we don't know which sample means should be considered to be close to the population mean of 500. Because a z-score describes a sample mean's relative standing, it allows us to determine whether or not the sample mean is relatively close to the population mean. From this perspective, the critical value of ± 1.96 defines "close." If our sample has a z-score lying beyond the critical value, the sample mean is not close to the population mean. Then, the difference between our sample mean and the population mean is so large that we do not consider the sample to be representing that population.

Other Ways to Set Up the Sampling Distribution

In the above example, we placed the region of rejection in both tails of the distribution because we wanted to make a decision about any sample mean having an extreme positive or negative z-score. However, we can also set up the distribution to examine only negative z-scores or only positive z-scores. (In the next chapter we discuss why we might want to do this.)

Say that we are interested only in sample means less than 500, having negative z-scores. Our criterion is still .05, but now we place the entire region of rejection in the lower left-hand tail of the sampling distribution, as shown in Figure 9.7. Notice that this produces a different critical value of -1.645. From the z-table (and using the interpolation procedures described in Appendix A), we see that the extreme lower 5% of a distribution lies beyond a z-score of -1.645. Therefore,

FIGURE 9.7 Setup of SAT Sampling Distribution to Test Negative z-Scores

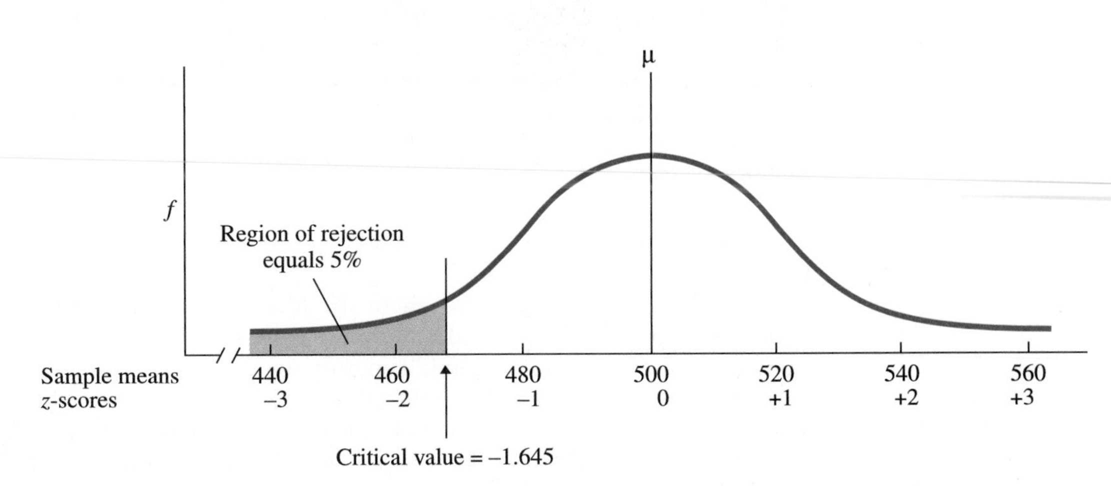

FIGURE 9.8 Setup of SAT Sampling Distribution to Test Positive *z*-Scores

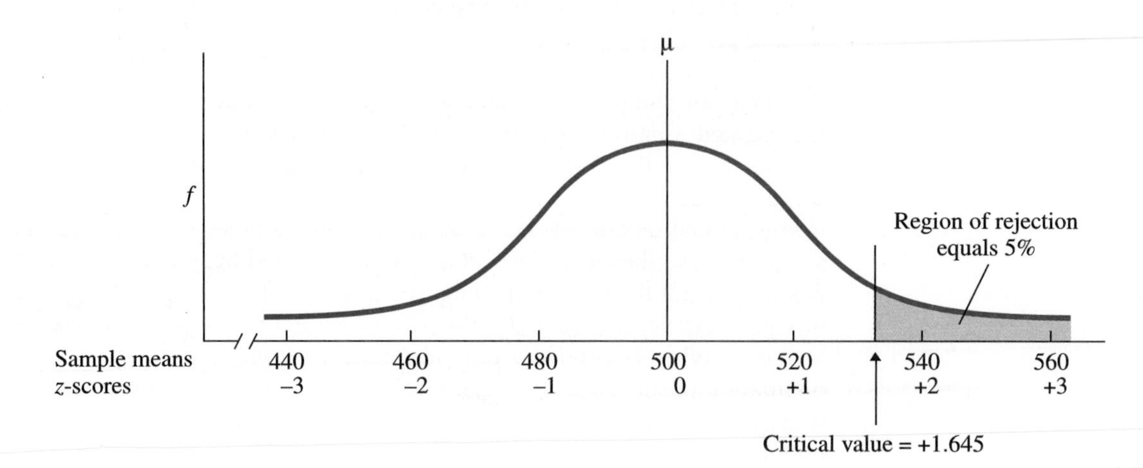

the z-score for our sample must lie beyond -1.645 for our sample mean to be in the region of rejection. Then, as usual, we conclude that such a sample is too unlikely to occur when sampling the SAT raw score population where $\mu = 500$, so we reject the idea that our sample represents this population.

On the other hand, say that we are interested only in sample means greater than 500, having positive z-scores. Here we place the entire region of rejection in the upper, right-hand tail of the sampling distribution, as shown in Figure 9.8. However, the critical value that defines the beginning of the upper extreme 5% of a distribution is a z-score of *plus* 1.645. If the z-score for our sample is beyond $+1.645$, our sample mean lies in the region of rejection. Then we again reject the idea that our sample represents the underlying raw score population.

On Being Wrong When We Decide About a Sample

When we rejected the idea that we were playing with a fair coin, we might have been wrong: maybe the player obtained 7 heads in a row with a fair coin by chance. Seven heads in a row do occur sometimes, and maybe we happened to observe one of those times. In the same way, we might be wrong when we reject the idea that the sample from Prunepit U. represents the SAT population where μ is 500. The sampling distribution clearly shows that sample means in the region of rejection do occur sometimes when the sample does represent the SAT population where μ is 500. Maybe our Prunepit mean was one of those means. In the same way, anytime we reject the idea that a sample represents a particular population, we may be wrong.

We can also be wrong if we retain the idea that our sample represents the typical SAT population. Consider the most extreme case, in which we obtain a sample mean of 500! This sample certainly appears to represent the population of scores where μ is 500. Using the above procedures, we would compute a z-score of 0., so we would retain this idea. But, even with such an apparently perfectly representative sample, it is possible that this sample actually represents some other population. Perhaps, for example, the sample is actually a very unrepresentative sample from the population where μ is 550! That is, maybe, simply by the luck of the draw, our sample contains too many low scores, so that, coincidentally, it appears to represent the population where μ is 500. In the same way, anytime we retain the idea that a sample represents a particular population, we may be wrong.

Thus, regardless of which population we decide our sample represents, we have not proven anything, and there is always the possibility that we have made an incorrect decision. In the next chapter we shall discuss these errors and their probabilities. For now, recognize that even though such errors are possible, they are not likely. By incorporating probability into our decision making, we can be confident that we have made the correct decision. Thus, in the final analysis, inferential statistical procedures help us to be more confident that a sample represents a particular population.

FINALLY

The decision-making process discussed in this chapter is the essence of all inferential statistics. As we shall see in the remaining chapters, the basic question we address is "Did chance produce an unrepresentative sample from a particular raw score population?" To answer this question, we create a sampling distribution based on that particular population. (This will be *much* easier if you sketch the distribution and label the region of rejection, as I did in this chapter.) Then we compute a statistic, such as a *z*-score, to describe the results of our study. If the *z*-score lies beyond the critical value, then the *z*-score and the corresponding sample data fall in the region of rejection. Any sample in the region of rejection is, by definition, too unlikely to occur when sampling the raw score population reflected in the sampling distribution. Therefore, we reject the idea that chance produced an unrepresentative sample from this raw score population, and we conclude that the sample represents some other population that is more likely to produce such data.

CHAPTER SUMMARY

1. *Probability,* or *p*, is a mathematical statement indicating the likelihood of an event when random chance is operating. The probability of an event is its relative frequency in the population and thus is the expected relative frequency of the event in any random sample.
2. When computing the probability of equally likely events, we determine the number of outcomes that can satisfy the event out of the total number of possible outcomes that can occur.
3. The probability that we are correct when we predict a chance event is equal to the probability that the event will occur. The probability that we are incorrect when we predict a chance event is equal to the probability that the event will not occur.

4. Events are *independent events* if the probability of one event is not influenced by the occurrence of the other events. Events are *dependent events* if the probability of one event is influenced by the occurrence of the other events. *Sampling with replacement* involves replacing a sample in the population before another sample is selected. *Sampling without replacement* involves *not* replacing one sample in the population before another sample is selected.

5. A *theoretical probability distribution* is a theoretical model of the relative frequencies of all possible events in a population when random chance is operating.

6. The standard normal curve model is a theoretical probability distribution that can be applied to any normal raw score distribution. Raw scores can be transformed into z-scores, and then the proportion of the area under the curve is the probability of randomly selecting those z-scores. This is also the probability of selecting the corresponding raw scores from the underlying raw score distribution.

7. The standard normal curve model can also be applied to a sampling distribution. Sample means can be transformed into z-scores, and then the proportion of the area under the curve is the probability of randomly selecting those z-scores. This is also the probability of selecting the corresponding sample means from the underlying raw score population.

8. The probability of randomly selecting a particular sample mean is the same as the probability of randomly selecting a sample of subjects whose scores produce that sample mean.

9. All inferential statistical procedures involve deciding whether a sample represents a particular population by determining the probability of obtaining the sample from that population. If the sample is likely to occur when the particular population is sampled, then we decide that it may represent that population. If the sample is unlikely to occur when the particular population is sampled, then we decide that it does not represent that population and that it instead represents some other population.

10. The *region of rejection* is located in the extreme tail or tails of a sampling distribution. Any z-score and corresponding sample mean that falls in the region of rejection is, by definition, too unlikely for us to accept as having occurred by chance when we sample from the raw score population reflected by the sampling distribution.

11. The edge of the region of rejection closest to the mean of the sampling distribution is at the *critical value.* In order for a sample to fall in the region of rejection, the sample's z-score must fall beyond the critical value.

12. The size of the region of rejection is determined by our criterion. The *criterion* is the probability that defines a sample as too unlikely. We usually use the criterion of .05. This produces a region of rejection that constitutes the extreme .05 of the sampling distribution.

PRACTICE PROBLEMS

(Answers for odd-numbered problems are provided in Appendix E.)

1. *a.* What does a probability convey about a random event in a sample?

b. What is the probability of a random event based on?

2. *a.* What is the difference between an empirical probability distribution and a theoretical probability distribution?

b. Why is the proportion of the area under the normal curve equal to probability?

3. What is the probability of:

a. Getting a six when rolling a die?

b. Selecting a diamond when cutting a deck of cards?

c. Randomly guessing the correct answer to a multiple choice question with four choices?

d. Selecting the ace of diamonds twice in a row when sampling a deck of cards without replacement?

4. *a.* When are events independent?

b. When are they dependent?

c. How does sampling without replacement affect the probability of events, compared to sampling with replacement?

5. A couple with eight children, all girls, decides to have one more baby, because the next one is bound to be a boy! Is this reasoning accurate?

6. Foofy read in the newspaper that there is a .05% chance of swallowing a spider while you sleep. She subsequently developed insomnia.

a. What is the probability of swallowing a spider?

b. Why isn't her insomnia justified on the basis of this probability?

c. Why is her insomnia justified on the basis of this probability?

7. Poindexter's uncle is planning to build a house on a piece of coastal property that has been devastated by hurricanes 160 times in the past 200 years. Because there hasn't been a major storm there in 13 years, his uncle is certain that this investment is a safe one. His nephew argues that he is wrong, and that there definitely will be a hurricane in the next year or so. What are the fallacies in the reasoning of both the uncle and the nephew?

8. Four airplanes from different airlines have crashed in the past two weeks. Bubbles must travel on a plane, but now she is terrified that it will crash. Her travel agent claims that the probability of a plane crash is minuscule. Who is correctly interpreting the situation? Why?

9. For each of the following, indicate whether the first event is dependent on or is independent of the second event:
   a. Playing golf; the weather.
   b. Buying new shoes; buying a new car.
   c. Losing weight; eating fewer calories.
   d. Winning the lottery; playing the same numbers each time.
10. What is the probability of randomly selecting a subject who scores the following?
    a. $z = +2.03$ or above
    b. $z = -2.8$ or above
    c. z between -1.5 and $+1.5$
    d. z beyond ± 1.72
11. For a distribution in which $X = 43$ and $S_X = 8$, what is the probability of randomly selecting the following?
    a. A score of 27 or below
    b. A score of 51 or above
    c. A score between 42 and 44
    d. A score below 33 or above 49
12. You are shopping for a used car. The salesperson says that over the life of the car you are thinking of buying, the probability of engine trouble is only .65.
    a. If you conclude that the engine will malfunction, what is the probability that you are correct? What is the probability that you are incorrect?
    b. If you conclude that the engine will not malfunction, what is the probability that you are correct? What is the probability that you are incorrect?
    c. Should you purchase this car? Why?
13. The mean of a population of raw scores is 18 ($\sigma_X = 12$). What is the probability of randomly selecting a sample of 30 scores having a mean above 24?
14. The mean of a population of raw scores is 50 ($\sigma_X = 18$). What is the probability of randomly selecting a sample of 40 scores having a mean below 46?
15. A sample produces a mean that is different from the μ of the population that we think the sample may represent. What are the two possible reasons for this difference?
16. When testing the representativeness of a sample mean,
    a. What is the criterion?
    b. What is the region of rejection?
    c. What is the critical value of z, and what is it used for?
17. Suppose that for the data in problem 13 you obtained a sample mean of 24. Using the .05 criterion, with the region of rejection in both tails of the

sampling distribution, should you consider the sample to be representative of the population in which $\mu = 18$? Why?

18. Suppose that for the data in problem 14 you obtained a sample mean of 46. Using the .05 criterion, with the region of rejection in both tails of the distribution, should you consider the sample to be representative of the population in which $\mu = 50$? Why?

19. In a study, you use a questionnaire and obtain the following data representing the aggressive tendencies of some football players:

    40, 30, 39, 40, 41, 39, 31, 28, 33

    a. Researchers have found that in the population of nonfootball players, μ is 30 ($\sigma_X = 5$). Using both tails of the sampling distribution, determine whether your football players represent a different population.

    b. What do you conclude about the population of football players and its μ?

20. On a standard test of motor coordination, a sports psychologist found that the population of average bowlers had a mean score of 24, with a standard deviation of 6. She tested a random sample of 30 bowlers at Fred's Bowling Alley and found that the sample had a mean of 26. A second random sample of 30 bowlers at Ethel's Bowling Alley had a mean of 18. Using the criterion of $p = .05$ and both tails of the sampling distribution, what should she conclude about each sample's representativeness of the population of average bowlers?

21. a. In problem 20, if each sample did not represent the population of average bowlers, what would be your best estimate of the μ of the population it does represent?

    b. Explain the logic behind this conclusion.

22. Foofy computes the $\overline{X}$ from data that her professor says is a random sample drawn from population Q. She determines that this sample mean has a z-score of $+41$ on the sampling distribution for population Q (and she computed it correctly!). Foofy claims she has proven that this could not be a random sample from population Q. Do you agree or disagree? Why?

SUMMARY OF FORMULAS

1. *The formula for computing the probability for equally likely events is*

$$p(\text{event}) = \frac{\text{number of outcomes that satisfy event}}{\text{total number of possible outcomes}}$$

2. *The formula for the true standard error of the mean is*

$$\sigma_{\overline{X}} = \frac{\sigma_X}{\sqrt{N}}$$

where σ_X is the true standard deviation of the raw score population.

3. *The formula for transforming a sample mean into a z-score on the sampling distribution of means is*

$$z = \frac{\overline{X} - \mu}{\sigma_{\overline{X}}}$$

where $\overline{X}$ is the sample mean, μ is the mean of the sampling distribution (which is also equal to the μ of the raw score distribution), and $\sigma_{\overline{X}}$ is the standard error of the mean of the sampling distribution of means.

10

Overview of Statistical Hypothesis Testing: The *z*-Test

To understand the upcoming chapter:

- From Chapter 4, recall that we envision a relationship in the population by noting the different means of the conditions of our independent variable, each mean representing a distribution of scores at a different μ and thus at a different location on the dependent variable.
- From Chapter 9, recall that when a sample's *z*-score falls in the sampling distribution's region of rejection, the sample's mean occurs so infrequently when sampling the underlying raw score population that we reject the idea that our sample represents this population. Instead, we assume that the sample represents some other raw score population.

Then your goals in this chapter are to learn:

- Why the possibility of sampling error leads a researcher to perform inferential statistical procedures.
- When your experimental hypotheses lead to either one-tailed or two-tailed statistical tests.
- How to set up a sampling distribution of means to test the null hypothesis for one- and two-tailed tests.
- How to interpret significant and nonsignificant results.
- How Type I errors, Type II errors, and power impact on the interpretations a researcher makes.

In Chapter 9 we discussed the essential basics of all inferential statistics. In this chapter we'll simply put these procedures in a research context and present the statistical language and symbols used to describe them.

MORE STATISTICAL NOTATION

Five new symbols will be used in stating mathematical relationships.

1. The symbol for *greater than* is >. We read from left to right, so, for example, $A > B$ means that A is greater than B. (The large opening in the symbol > is always on the side of the larger quantity, and the symbol points toward the smaller quantity.)
2. The symbol for *less than* is <. The notation $B < A$ means that B is less than A. (Again the symbol points toward the smaller quantity.)
3. The symbol for *greater than or equal to* is ≥. The notation $B \geq A$ indicates that B is greater than or equal to A.
4. The symbol for *less than or equal to* is ≤. The notation $B \leq A$ indicates that B is less than or equal to A.
5. The symbol for *not equal to* is ≠. The notation $A \neq B$ means that A is different from B.

THE ROLE OF INFERENTIAL STATISTICS IN RESEARCH

As we saw in the previous chapter, a random sample may be more or less representative of a population. Just by the luck of the draw, the sample may contain too many high scores or too many low scores relative to the population. Because the sample is not perfectly representative of the population from which it is selected, the sample mean does not equal the population mean.

We have a shorthand term for describing when random chance produces an unrepresentative sample. We say that the sample reflects sampling error. **Sampling error** results when random chance produces a sample statistic (such as $\overline{X}$) that is not equal to the population parameter it represents (such as μ). Because of the luck of the draw, the sample is in error to some degree in representing the population.

STAT ALERT Sampling error is the result of chance factors that produce a sample statistic that is different from the population parameter it represents.

Sampling error is the reason researchers perform inferential statistics. There is always the possibility of sampling error in any sample. Whenever a sample mean is different from a particular population mean, it may be because (1) due to sampling error, the sample poorly represents that population of scores, or (2) the difference is not due to sampling error, and instead the sample represents some other population. This creates a dilemma for a researcher who is trying to infer that a relationship exists in nature. Recall that in an experiment, we change the conditions of the independent variable in hopes of seeing scores on the dependent variable change in a consistent fashion. We want to infer that we would find this relationship in the population: if we measured the entire population, we would find a different population of scores located around a different value of μ under each condition. But here is where sampling error raises its ugly little head. Even though we obtain a different sample mean for each condition, perhaps we are being misled by sampling error. Maybe the samples are all more or less unrepresentative of the *same* population, so that if we tested everyone in the population under each condition, we would find the same population of scores, having the same μ, in each condition (and then there would not be a relationship).

Thus, because of sampling error, we never know if the relationship in our sample data represents a *real* relationship in the population. By a "real" relationship, I mean one in which there is some underlying aspect of nature that actively ties certain Y scores to certain X scores so that a relationship is produced. When the differences between the means of our conditions are actually due to sampling error, then our sample relationship does not reflect a real relationship: instead, by chance, the data coincidentally form the pattern of a particular relationship, in the same way that if you drop a handful of coins, by chance they may land so that they form a particular pattern. Thus, even though our sample data shows a relationship, it is still possible that there is no real relationship in the population. Or there may be a real relationship in the population, but because of sampling error, it is different from the relationship presented by our sample data.

To deal with the possibility of sampling error, we apply inferential statistics. **Inferential statistics** are the procedures we use to decide whether our sample data represent a particular relationship in the population. Using the process we discussed in the previous chapter, we decide whether our samples are likely to represent populations that form a particular relationship or whether they are likely to represent populations that do not form the relationship.

The specific inferential procedure we employ in a given research situation depends upon our research design and on the scale of measurement we used when measuring the *dependent variable.* There are two general categories of inferential statistics: parametric and nonparametric. **Parametric statistics** are procedures that require certain assumptions about the parameters of the raw score populations represented by the sample data. Recall that parameters describe the characteristics of a population, so parametric procedures are used when we can assume the population has certain characteristics. In essence, the assumptions of a procedure are the rules for using it, so think of them as a checklist for selecting the procedure to use in a particular study. There are specific assumptions for each procedure, but two assumptions are common to all parametric procedures: (1) we assume the population of raw scores forms a normal distribution, and (2) we assume the data are interval or ratio scores. Thus, parametric procedures are used when it is appropriate to summarize the scores by calculating their mean. This chapter and Chapters 11 through 14 discuss the most common parametric procedures.

On the other hand, as the name implies, **nonparametric statistics** are inferential procedures that do not require such stringent assumptions about the parameters of the populations represented by the sample data. Usually these procedures are employed when we have nominal or ordinal scores or when we have a skewed interval or ratio distribution (when the data are most appropriately described by the median or mode.) Chapter 15 presents the most common nonparametric procedures.

We choose nonparametric procedures if our data clearly violate the assumptions of parametric procedures. However, we can use a parametric procedure if our data come close to meeting its assumptions. This is because parametric procedures are robust. With a **robust procedure,** if we do not meet the assumptions of the procedure perfectly, we will have only a negligible amount of error in the inferences we draw. So, for example, if our data represent a population that is approximately normally distributed, we can still use a parametric procedure.

In the remainder of this book, we will become immersed in the details of inferential statistics. However, do not lose sight of the fact that in every procedure we ultimately decide whether our data represent a "real" relationship in the population—in nature. A study is *never* finished until we have performed the appropriate inferential procedure and made that decision.

In this chapter we will discuss one parametric procedure so that you can see the general format and terminology used in all inferential procedures.

SETTING UP INFERENTIAL STATISTICAL PROCEDURES

It is not appropriate to think about our statistical analysis only after we have collected the data. There are several decisions that we should make beforehand, for two reasons. First, the fact that we are performing a study indicates that we have some idea about how nature operates. To fairly test this idea, we must state it clearly before collecting the data. Second, we want to check that the data we plan on collecting will be "analyzable." It is possible to collect data and then find out that there are no appropriate statistical procedures we can apply.

In setting up a study, we do four things before collecting any data. First, we create our experimental hypotheses, which describe the relationship that the experiment either will or will not demonstrate. Then we design the experiment to test the experimental hypotheses. Then we translate the experimental hypotheses into statistical hypotheses, which can be tested using inferential statistics. Finally, we select and set up the appropriate statistical procedure. Once we have accomplished these steps, we collect the data, perform the analysis, and draw our conclusions about the study.

Creating the Experimental Hypotheses

Whenever we conduct research, we have some idea of how it should turn out. We formally express this expectation in the form of two experimental hypotheses. **Experimental hypotheses** describe the predicted relationship we may or may not find in our experiment. One experimental hypothesis states that we will demonstrate the predicted relationship (our manipulation of the independent variable will work as expected), and the other experimental hypothesis states that we will not demonstrate the predicted relationship (our manipulation of the independent variable will not work as expected). These two hypotheses are important because they help us to precisely define the purpose of a study, they force us to decide how we will recognize whether or not the study works, and they form the basis for setting up our statistical analysis.

For statistical purposes, our hypotheses describe the predicted relationship in one of two ways. The simplest prediction is that there is some kind of relationship, but we are not sure whether scores will increase or decrease as we change the independent variable. The other, more complicated prediction not only states that there will be a relationship, but also indicates the direction in which the scores will change: we may predict that as we change the independent variable, the dependent scores will only increase, or we may predict that they will only decrease.

In the following discussion we will first examine a study in which we merely predict some kind of a relationship. Say that we are researchers interested in intelligence as measured by the intelligence quotient, or IQ. We have developed a new IQ pill that we believe will affect IQ, but we are not sure whether the pill will make people smarter or whether it will make them dumber. Therefore, we hypothesize that there is a relationship such that the more of the pill a subject consumes, the more his or her IQ will change. The amount of the pill is our independent variable, and the subject's IQ is our dependent variable. Here are our two experimental hypotheses:

1. We will demonstrate that the pill works by either increasing or decreasing IQ scores.
2. We will not demonstrate that the pill works, because IQ scores will not change.

Once we understand the nature of the relationship our study is supposed to demonstrate, we design the study.

Designing a Single-Sample Experiment

Although there are many ways we could design our IQ pill study, we will choose the simplest approach: a single-sample experiment. We will randomly select one sample of subjects and give each subject one pill. After waiting for the pill to work, we will give the subjects a standard IQ test. Our sample will represent the population of IQ scores of all subjects when they have taken one of our pills, and the sample $\bar{X}$ will represent that population's μ.

To demonstrate a relationship, we must demonstrate that *different* amounts of the pill produce *different* populations of IQ scores, having different μ's. Therefore, we must compare the population represented by our sample to some other population receiving some other amount of the pill. *To perform any single-sample experiment, we must already know the population mean under some other condition of the independent variable.* One amount of the pill is zero, or no amount. The IQ test we are using has been given over the years to many subjects who have not taken our IQ pill, and this population of IQ scores has a μ of 100. We will compare this population without the pill to the population with the pill represented by our sample. If we find that the population represented by our sample has a different μ than the population without the pill, then we will have demonstrated a relationship in the population.

Creating the Statistical Hypotheses

Statistical hypotheses are translations of experimental hypotheses. They allow us to rephrase our experimental hypotheses in such a way that we can apply statistical procedures to them. **Statistical hypotheses** are statements that describe the population parameters our sample data will represent if the predicted relationship exists or does not exist. There are always two statistical hypotheses, the alternative hypothesis and the null hypothesis.

The alternative hypothesis Although we can create the two statistical hypotheses in either order, it is often easier to first create the alternative hypothesis, because it corresponds to the experimental hypothesis that the experiment *does* work as predicted. The **alternative hypothesis** describes the population parameters that the sample data will represent if the predicted relationship exists. The alternative hypothesis is always the hypothesis of a difference; it says that changing the independent variable produces the predicted difference in the population of scores. Therefore, it is our hypothesis that the sample data represent a "real" relationship in nature.

If we could test the entire population and if our IQ pill works as predicted, then we would find one of two outcomes. Figure 10.1 shows the change in the population if the pill *increases* IQ scores. This shows a relationship because, by changing the conditions (the amount of pill) for everyone in the population, IQ is increased so that the distribution moves to the right, over the higher IQ scores. To summarize such a relationship, we use μ. We do not know how much IQ

FIGURE 10.1 Relationship in the Population If the IQ Pill Increases IQ Scores

As the amount of the pill is changed from 0 pills to 1 pill, the IQ scores in the population tend to increase in a consistent fashion.

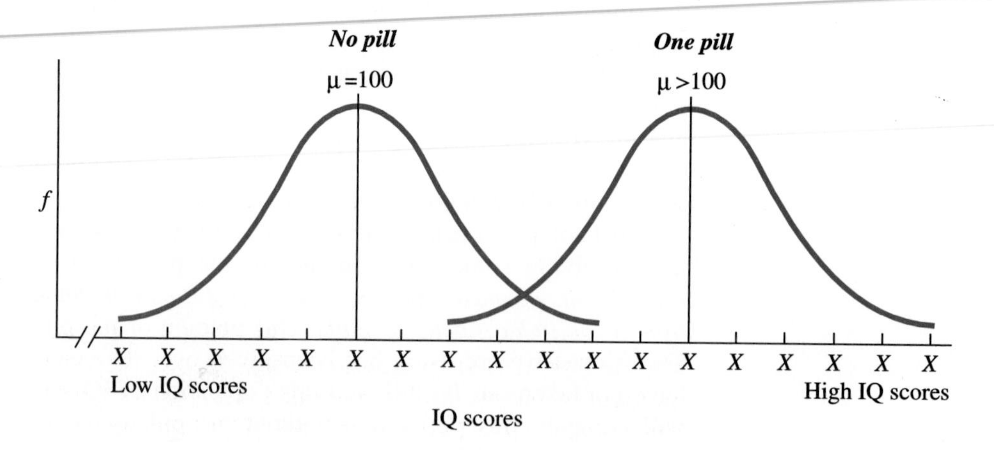

FIGURE 10.2 Relationship in the Population If the IQ Pill Decreases IQ Scores

As the amount of the pill is changed from 0 pills to 1 pill, the IQ scores in the population tend to decrease in a consistent fashion.

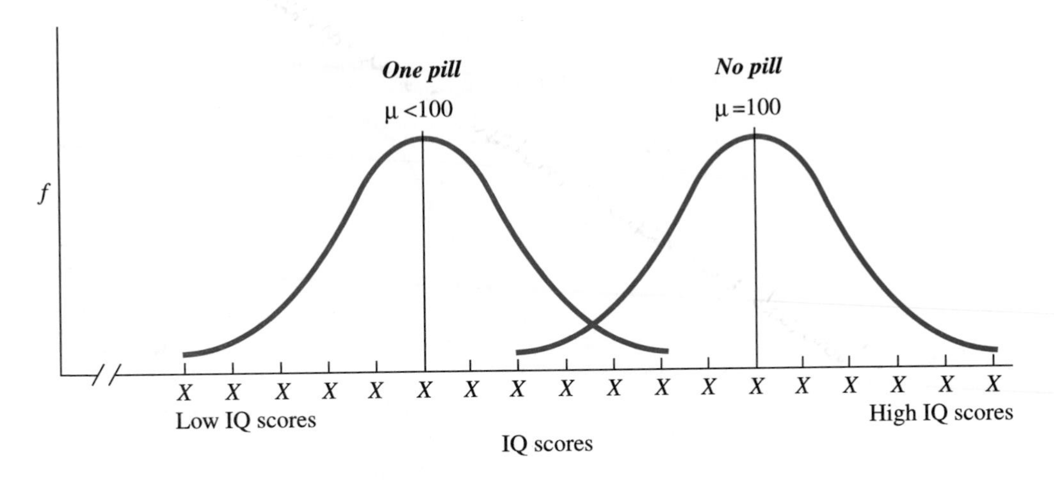

scores will increase, so we do not know the value of μ with the pill. But we do know that if the pill increases IQ, then the μ of the population with the pill will be *greater* than 100, the μ of the population without the pill.

On the other hand, Figure 10.2 shows the change in the population if the pill *decreases* IQ scores. This also shows a relationship because by changing the condition for everyone in the population, IQ decreases so that the distribution is moved to the left, over the lower IQ scores. Therefore, the μ of the population with the pill is *less than* 100, the μ of the population without the pill.

The alternative hypothesis provides a shorthand way of communicating all of the above. If the pill works as predicted, then the population with the pill will have a μ that is either greater than or less than 100. In other words, the population mean with the pill will *not equal* 100.

The symbol for the alternative hypothesis is H_a. (The H stands for hypothesis, and the a stands for alternative.) For our IQ pill experiment, we would write the alternative hypothesis as

$$H_a\colon \mu \neq 100$$

H_a implies that our sample mean represents a population mean not equal to 100. If, with the pill, μ is not 100, then this indicates a real relationship in the population. Thus, we can interpret H_a as stating that our independent variable "really" works as predicted in nature.

The null hypothesis The statistical hypothesis corresponding to the experimental hypothesis that the independent variable does *not* work as predicted is called the null hypothesis. The **null hypothesis** describes the population parameters that the sample data will represent if the predicted relationship does *not* exist. The null hypothesis is the hypothesis of "no difference": it says that changing the

FIGURE 10.3 Population of Scores If the IQ Pill Does Not Affect IQ Scores

Here there is no relationship.

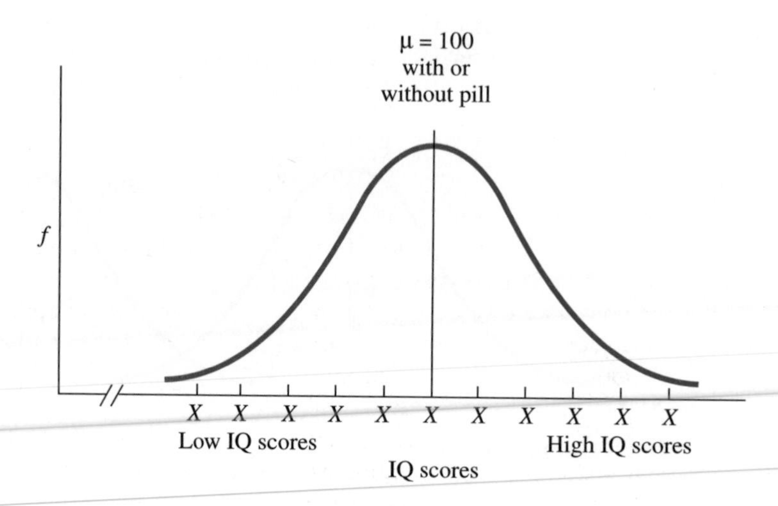

independent variable does *not* produce the predicted difference in the population of scores. Thus, it is also our hypothesis that even though the sample data may form a relationship, this results from sampling error so the data do *not* represent a "real" relationship in nature.

If our IQ pill does not work, then it would be as if the pill were not present. The population of IQ scores without the pill has a μ of 100. Therefore, if the pill does not work, then after everyone has taken the pill, the population of IQ scores will be unchanged and μ will still be 100. Thus, if we measured the population with and without the pill, we would have one population of scores, located at the μ of 100, as shown in Figure 10.3.

The null hypothesis is a shorthand way of communicating the above. The symbol for the null hypothesis is H_0. (The subscript is 0 because *null* means zero, as in zero relationship.) We write the null hypothesis for our IQ pill study as

$$H_0\text{: } \mu = 100$$

H_0 implies that our sample mean represents a population mean equal to 100. If μ is still 100 with the pill, then we have not demonstrated the predicted relationship. Thus, we interpret H_0 as implying that our independent variable really does not work as predicted.

The final step prior to collecting data is to select and set up the appropriate statistical procedure. However, we will violate the order of things and go directly to our data so that you can understand what it is we are setting up.

The Logic of Statistical Hypothesis Testing

The statistical hypotheses for our IQ pill study are

$$H_0\text{: } \mu = 100$$
$$H_a\text{: } \mu \neq 100$$

Notice that, together, H_0 and H_a always include all possibilities, so one or the other of them must be true for a given sample: our sample mean represents either a μ equal to 100 or a μ not equal to 100. We use inferential procedures to test (choose between) these two statistical hypotheses. To see why we need statistical hypothesis testing, say that we randomly selected a sample of 36 subjects, gave them the IQ pill, and then measured their IQ. We found that the mean of our sample of IQ scores was 105. Can we conclude that the IQ pill works?

We would like to say this: People who have not taken this IQ pill have a mean IQ of 100. If the pill did not work, then our sample mean should have been 100. However, our sample mean was 105. This suggests that the pill does work, raising IQ scores about 5 points. If the pill does this for the sample, we assume that it would do this for the population. Therefore, we expect that the population that received the pill would have a μ of 105. Our results appear to be consistent with our alternative hypothesis, H_a: $\mu \neq 100$, which states that our sample represents a population mean not equal to 100. Thus, it seems that if we were to measure everyone in the population with and without the pill, we would have the distributions shown previously in Figure 10.1, with the population that received the pill located at a μ of 105. Conclusion: we have demonstrated that our IQ pill works. In nature there is a relationship such that increasing the amount of the pill from 0 to 1 is associated with increasing IQ scores.

But hold on! Not so fast! Remember sampling error? We just assumed that our sample is *perfectly* representative of the population it represents. We said that if the pill did not work we *should* have obtained a sample mean of 100. What we should have said is "If the pill did not work *and there was no sampling error,* the sample mean should have been 100." But what if there *was* sampling error? Maybe we obtained a mean of 105 not because the pill works, but because we inaccurately represented the situation where the pill does *not* work. Maybe the pill does nothing, but we happened to select a smarter-than-average sample. Maybe the null hypothesis is the correct hypothesis: even though it does not look like it, maybe our sample represents the population where μ is still 100. Maybe we have not demonstrated that the pill works.

Strange, isn't it? We conducted this research because we did not know whether the pill works. Now, after we have conducted the research, we *still* don't know whether the pill works! The truth is that we can never *know* whether our IQ pill works based on the results of one study. Whether our sample mean is 105, 1050, or 105,000, it is still possible that the null hypothesis is true: the pill doesn't work, the sample actually represents the population where μ is 100, and the sample mean is different from 100 because of sampling error.

> **STAT ALERT** Regardless of our results, the null hypothesis steadfastly maintains that the sample data reflect sampling error and do not really demonstrate the predicted relationship.

To know for certain whether our pill actually works, we would have to give it to the entire population and see whether μ was 100 or 105. We cannot do that. Instead, the best we can do is to use our study to increase our *confidence* that the pill works. Before we tested the pill, our confidence that the pill works was based solely on our confidence that we had derived the correct formula for the magic ingredients in the pill. We conducted empirical research to increase our confidence that the pill works. But now we see that the null hypothesis says that sampling error gave the *impression* that the pill works. If we can reject the null hypothesis, then we can be even more confident that the pill works. From this perspective, statistical hypothesis testing is merely a slight detour we take—a tool we use—to increase our confidence that our study demonstrates the predicted relationship.

So first we must make a decision about that pesky sampling error. In statistical hypothesis testing, the hypothesis that we actually test is the null hypothesis: here we will test the idea that we have sampling error in representing the population having a μ of 100 and so the pill does not really work. If sampling error can reasonably explain our results, then we will *not* accept that the pill works. After all, it makes no sense to believe that the pill works if our results can be easily explained as sampling error from the population of IQ scores *without* the pill. As scientists trying to describe nature, we want to be very careful not to conclude that we have demonstrated a relationship when actually our results are due to chance factors. Therefore, we won't buy that the pill works unless we are convinced that the results are *not* due to sampling error. Only if the null hypothesis fails our test do we accept the alternative hypothesis (that the pill works).

Although we cannot determine whether the null hypothesis is true, we can determine how *likely* it is that a sample mean of 105 will occur because of sampling error when the sample actually represents the population where μ is 100. If such a mean is too unlikely, then we reject H_0, rejecting the idea that our sample poorly represents the population where μ is 100. If this sounds familiar, it's because this is the procedure we discussed in the previous chapter. In fact, all parametric and nonparametric inferential statistics involve the same logic: we test our null hypothesis by determining the probability of obtaining our sample data from the population described by the null hypothesis.

Which particular inferential procedure we use depends on the design of our experiment and the characteristics of our dependent variable. In other words, our study must meet the assumptions of the procedure. Our IQ pill study meets the assumptions of the parametric inferential test known as the z-test.

TESTING STATISTICAL HYPOTHESES WHEN σ_X IS KNOWN: THE z-TEST

You already know how to perform the z-test. The **z-test** is the procedure for computing a z-score for a sample mean on the sampling distribution of means discussed in previous chapters. The formula for the z-test is the formula we used in Chapter 6 and again in Chapter 9 (and we will see it again in a moment). The assumptions of the z-test are as follows:

1. We have randomly selected one sample.
2. The dependent variable is at least approximately normally distributed in the population, it involves an interval or ratio scale, and the mean is the appropriate measure of central tendency.

3. We know the mean of the population of raw scores under some other condition of the independent variable.
4. We know the true standard deviation (σ_X) of the population described by the null hypothesis. (It is *not* estimated using the sample.)

We can use the *z*-test for our IQ pill study because we have one sample of IQ scores, such scores are from an interval variable, the population of IQ scores is normally distributed, and we know σ_X for the population. Say that from our reading of the literature on IQ testing, we know that in the population where μ is 100, the standard deviation, σ_X, is 15.

STAT ALERT The *z*-test is used only if the raw score population's σ_X is *known*.

Now that we have chosen our statistical procedure, we can set up our sampling distribution.

Setting Up the Sampling Distribution of Means for a Two-Tailed Test

In our IQ pill study, H_0 is that the pill does not work, so our sample represents the population where μ is 100 (and σ_X is 15). To test this H_0, we will examine the sampling distribution of means created from the raw score population where μ is 100 and σ_X is 15. To create the sampling distribution, assume we have again hired our bored statistician (she's getting very bored, but she's a good sport). Using our *N* of 36, she infinitely samples the raw score population of IQ scores where μ is 100. The central limit theorem tells us that the distribution of her sample means will be normally distributed and that the μ of the distribution will equal the μ of the raw score population, 100. Recall that this is also the value of μ in our null hypothesis, H_0: $\mu = 100$. Our H_0 says that we are sampling from the raw score population where μ is 100, so our average sample mean should also be 100.

STAT ALERT The mean of the sampling distribution always equals the value described by H_0.

We can call this sampling distribution the null, or H_0, sampling distribution, because it describes the situation *when null is true:* it describes random samples from a population where μ *is* 100 and any sample mean not equal to 100 occurs *solely* because of sampling error—the luck of the draw that determined who our statistician selected for that particular sample. (Always add the phrase "when null is true" to any information you obtain from a sampling distribution.)

Recall that those means that we consider to be "too unlikely" are located in the extreme tail of the sampling distribution, the area we call the region of rejection. Once we envision the sampling distribution, we set up our statistical test by identifying the size and location of the region of rejection.

Choosing alpha: the size of the region of rejection Recall from Chapter 9 that our *criterion* is the probability we use to define sample means that are "too unlikely" to represent the underlying raw score population. The criterion is also the theoretical size of the region of rejection. (Previously, when our criterion was .05, the region of rejection comprised the extreme 5% of the curve.) Here is a new symbol: the symbol for our criterion and the theoretical size of our rejection region is α, the Greek letter **alpha.** Psychologists usually set their criterion at .05, so in code they say $\alpha = .05$.

Locating the region of rejection in the two-tailed test Recall that the region of rejection can be located either in both tails of the distribution or in only one tail. Which arrangement we use depends entirely on our experimental hypotheses and thus on our statistical hypotheses—in particular, our alternative hypothesis. We predicted that the pill would make people either smarter or dumber, resulting in the alternative hypothesis that our sample mean represents either a μ larger than or smaller than 100. For us to be correct, our sample mean must be either larger than 100 or smaller than 100 and, in either case, too unlikely to represent the population for which $\mu = 100$. Sample means that are either larger or smaller than 100 and also too unlikely to represent that population are those in either the upper or lower tails of the distribution. Therefore, as shown in Figure 10.4, we place part of the region of rejection in the tail above $\mu = 100$ and part of it in the tail below $\mu = 100$.

Because we have placed the region of rejection in the two tails of the distribution, we are testing our null hypothesis with a two-tailed test. A **two-tailed test** is used to test statistical hypotheses in which we predict that there is some kind of relationship, but we do not specifically predict that scores will only increase or only decrease.

FIGURE 10.4 H_0 Sampling Distribution of IQ Means for a Two-Tailed Test

There is a region of rejection in each tail of the distribution, marked by the critical values of ±1.96.

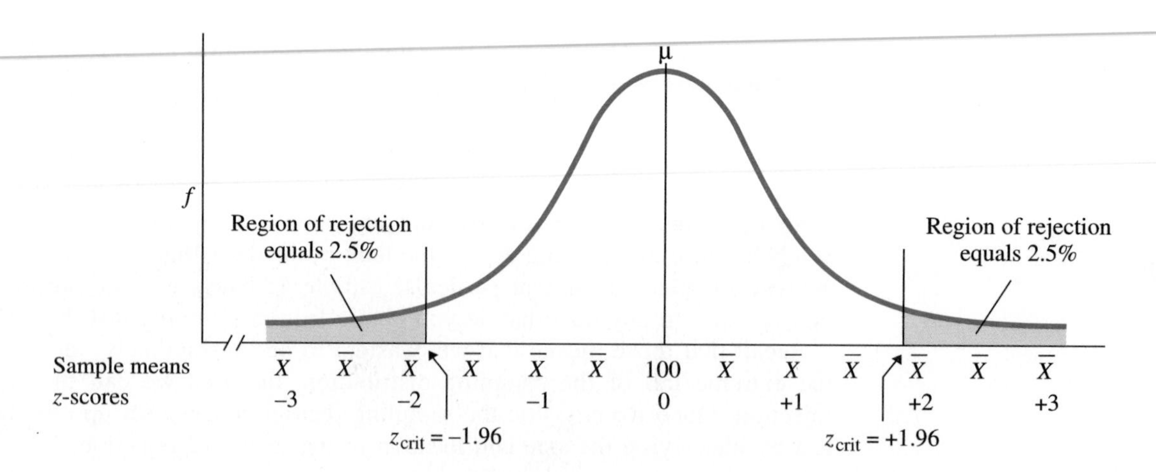

STAT ALERT A two-tailed test is used whenever we do *not* predict the direction in which scores will change.

You have a two-tailed test whenever your null hypothesis simply states that the population parameter equals some value and your alternative hypothesis states that the population parameter does not equal that value.

Determining the critical value Once we have decided that we have a two-tailed test, we identify the *critical value of z*. In code, we abbreviate the critical value of z as z_{crit}. Therefore, as shown in Figure 10.4, z_{crit} demarcates the region of rejection. With $\alpha = .05$, the region of rejection in each tail contains the extreme 2.5% of the distribution. From the z-tables we see that a z-score of 1.96 demarcates the extreme 2.5% of the curve. Because either $+1.96$ or -1.96 demarcates a part of our region of rejection, our z_{crit} is ± 1.96.

Now our test of H_0 boils down to this: if the z-score for our sample lies beyond ± 1.96, then our sample mean lies in the region of rejection. Thus, the final step is to compute the z-score for our sample.

PERFORMING THE *z*-TEST

Here is some more code. Because the z-score we compute is the z-score we obtain from our sample data, we identify this z-score as *z obtained,* which we abbreviate as z_{obt}.

As we saw in previous chapters,

THE COMPUTATIONAL FORMULA FOR THE z-TEST IS

$$z_{obt} = \frac{\overline{X} - \mu}{\sigma_{\overline{X}}}$$

$\overline{X}$ is the value of our sample mean. μ is the mean of the sampling distribution when H_0 is true: it is the μ of the raw score population that H_0 says the sample represents. $\sigma_{\overline{X}}$ is the standard error of the mean, which is computed as

$$\sigma_{\overline{X}} = \frac{\sigma_X}{\sqrt{N}}$$

where N is the N of our sample and σ_X is the true population standard deviation.

We first find the standard error of the mean, $\sigma_{\overline{X}}$. For our IQ pill study, I said the population standard deviation, σ_X, is 15, and the sample N is 36. Putting these values into the formula for $\sigma_{\overline{X}}$ gives

$$\sigma_{\overline{X}} = \frac{\sigma_X}{\sqrt{N}} = \frac{15}{\sqrt{36}} = \frac{15}{6} = 2.5$$

Thus, our sampling distribution of means has a $\sigma_{\bar{X}}$ of 2.5.

Now we compute the z-score for our sample mean of 105 by putting the appropriate values into the formula for z_{obt}:

$$z_{obt} = \frac{\bar{X} - \mu}{\sigma_{\bar{X}}} = \frac{105 - 100}{2.5} = \frac{+5}{2.5} = +2.0$$

On the sampling distribution from the population where $\mu = 100$ (and $\sigma_{\bar{X}} = 2.5$), our sample mean of 105 has a z-score of $+2.0$.

If H_0 is correct, then ideally we'd expect a sample mean of 100 and thus a z_{obt} of 0. Even allowing for some sampling error, we'd expect a mean close to 100 and thus a z_{obt} close to 0. However, our sample actually produced a z_{obt} of $+2.0$! To confirm our suspicions, we compare z_{obt} to z_{crit}.

Interpreting z_{obt}: Rejecting H_0

Remember that our H_0 implies that the pill does not work and that, through sampling error, our sample actually represents the raw score population where μ is 100. If we are to believe H_0, then a mean of 105 should be likely to occur when sampling from this population. The H_0 sampling distribution shows how frequently—and thus how likely—sampling error will produce any given sample mean if we, like our bored statistician, *were* sampling from the raw score population where μ is 100. Therefore, if we are to believe H_0, the sampling distribution should show that a mean of 105 occurs relatively frequently and is thus likely in this situation. However, looking at the sampling distribution in Figure 10.5, we see just the opposite.

FIGURE 10.5 Sampling Distribution of IQ Means

The sample mean of 105 is located at $z_{obt} = +2.0$.

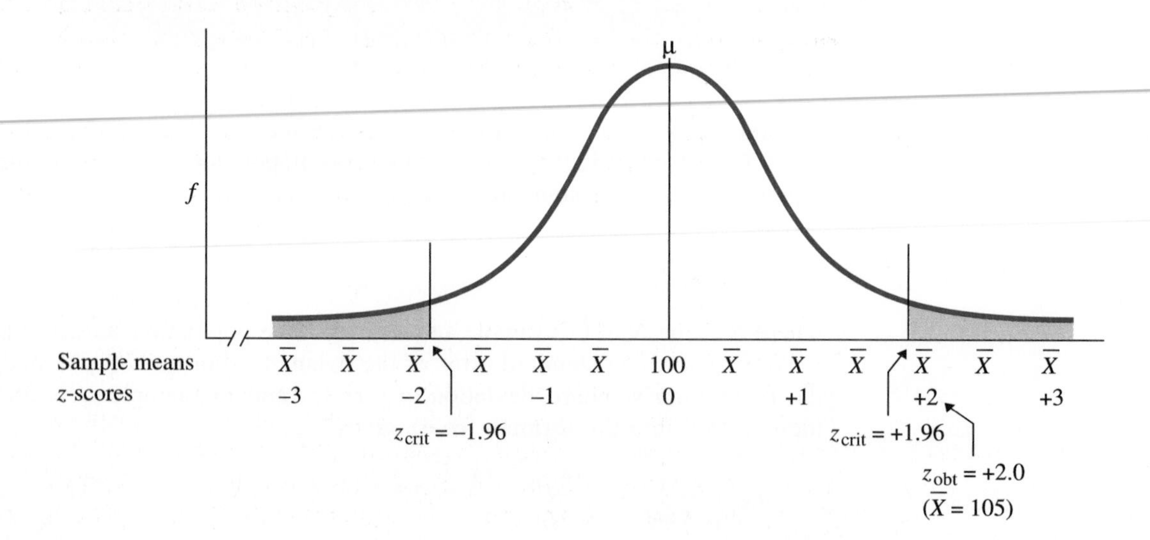

The location of our z_{obt} of $+2.0$ tells us that the bored statistician would hardly ever obtain a sample mean of 105 when she *was* drawing samples that represent the population where μ is 100. This makes it difficult to believe that we obtained our sample mean by drawing from the population where μ is 100. In fact, because our z_{obt} of $+2.0$ is beyond the z_{crit} of ± 1.96, our sample is in the region of rejection. Therefore, we conclude that our mean of 105 is "too unlikely" to accept as representing the population where $\mu = 100$. That is, the idea that our sample is merely a poor representation of the population where $\mu = 100$ is not reasonable: samples are seldom *that* poor at representing this population. Therefore, because we are too unlikely to obtain such a sample when representing the population where $\mu = 100$, we reject the idea that our sample represents this population.

We have just rejected the null hypothesis. H_0 states that our sample represents the population where μ is 100, and we have found that this is not a reasonable hypothesis. Whenever our sample mean falls in the region of rejection, we say that we "reject H_0." If we reject H_0, then we are left with H_a. Therefore, when we reject H_0, we also say that we "accept H_a." Here H_a is $\mu \neq 100$, so we accept that our sample represents a population where μ is not 100. Thus, in sum, we have determined that our sample is unlikely to represent the population where μ is 100, so we conclude that it is likely to represent a population where μ is not 100.

> ***STAT ALERT*** When a sample statistic falls beyond the critical value, the sample statistic lies in the region of rejection on the H_0 sampling distribution. This indicates that the statistic is too unlikely to accept as representing the parameter described by H_0, so we reject H_0 and accept H_a.

Reporting significant results The shorthand way of communicating that we have rejected H_0 and accepted H_a is to use the term *significant.* (Statistical hypothesis testing is sometimes called significance testing.) *Significant* does *not* mean important or impressive. When we say that our results are **significant,** we mean we have decided that our results are too unlikely to occur if the predicted relationship does not occur in the population. Therefore, we imply that the relationship found in the experiment is "believable," representing a "real" relationship found in nature and not a chance pattern resulting from sampling error.

> ***STAT ALERT*** Whenever we use the term *significant,* we mean that we have rejected the null hypothesis and believe our data reflect a real relationship.

We use the term *significant* in several ways. In our IQ pill study, we might say that our pill produced a "significant difference" in IQ scores. This indicates that the difference between our sample mean and the μ found without the pill is too large to accept as occurring by chance if our data represent that μ. Or we might say that we obtained a "significant z": our sample mean has too large a z-score to accept as occurring by chance if the sample represents the μ described by H_0. Or we can say that there is a "significant effect of our IQ pill": we have decided that the change in IQ scores reflected by our sample mean is not caused by chance sampling error, so presumably it is an effect caused by changing the conditions of the independent variable.

It is very important to remember that we decide either yes, we reject H_0, or no, we do not reject H_0. All z-scores in the region of rejection are treated the same, so one z_{obt} cannot be "more significant" than another. Likewise, there is no such thing as "very significant" or "highly significant." (That's like saying "very yes" or "highly yes.") If z_{obt} is beyond z_{crit}, regardless of how far it is beyond, we completely and fully reject H_0 and our results are simply significant, period!

Recognize that whether we have a significant result depends solely on how we define "too unlikely." *Significant* implicitly means that *given our* α and therefore the size of our region of rejection, we have decided that our data are too unlikely to represent the situation described by our null hypothesis. Because our decision depends on our criterion, anytime we use the word *significant,* we also report our value of α. In reporting the results of any significant statistical test, we indicate the statistic we computed, the obtained value, and the α we used. Thus, to report our significant z_{obt} of $+2.0$, we would write

$$z_{obt} = +2.0, \quad p < .05$$

Notice that instead of indicating that α equals .05, we indicate that the probability, p, is less than .05, or $p < .05$. We'll discuss the reason for this shortly.

Interpreting significant results In accepting H_a, we also accept the corresponding experimental hypothesis that the independent variable works as predicted: apparently, our IQ pill study has demonstrated that the pill works. However, there are three very important restrictions on how far we can go in claiming that the pill works.

First, we did not *prove* that H_0 is false. Statistics don't prove anything! All we have "proven" is that a sample of 36 scores is unlikely to produce a mean of 105 if the scores represent the population where $\mu = 100$ and σ_X is 15. However, as the sampling distribution plainly shows, unrepresentative means of 105 *do* occur once in a while when we *are* representing this population. Maybe the pill did not work, and our sample was simply very unrepresentative. There is *always* that possibility.

Second, by accepting H_a, we are only accepting the hypothesis that our sample mean represents a μ not equal to 100. But we have not proven that the *pill* produces these scores. We have simply obtained data consistent with the idea that the pill works. By saying we have demonstrated a "real" relationship, we are saying that our sample reflects *some* variable in nature that is associated with higher IQ scores. But that variable may *not* be the pill. The higher IQ scores in our sample might actually have occurred because our subjects cheated on the IQ test, or because there was something in the air that made them smarter or because there were sunspots, or who-knows-what! If we have performed a good experiment and can eliminate such factors, then we can *argue* that it is our pill that produced higher IQ scores.

Finally, assuming that the pill increased IQ scores and produced the mean of 105, then it is logical to assume that if we gave the pill to everyone in the population, the resulting μ would be 105. However, even if the pill works, the μ might not be *exactly* 105. Our sample may reflect (you guessed it) sampling error! That is, our sample may accurately indicate that the pill influences IQ, but it may not be perfectly representative of *how* the pill influences scores. If we gave the pill to the population, we might find a μ of 104, or 106, or *any* other value. However, since a sample mean of 105 is most likely when the population μ is around 105, we would conclude that the μ resulting from our pill is probably *around* 105.

Bearing these qualifications in mind, we can return to our sample mean of 105 and interpret it the way we wanted to about 10 pages back: it looks as if our pill increases IQ scores by about 5 points. But now, because we know that our sample mean is *significantly* different from a μ of 100, we are confident that we are not being misled by sampling error and that our sample does represent the population of higher IQ scores that would occur with the pill. Therefore, we are more confident that a relationship exists in the population and that we have discovered something about how nature operates. (But stay tuned, because we could be wrong.)

Interpreting z_{obt}: Retaining H_0

For the sake of illustration, let's say that our IQ pill had instead produced a sample mean of 99. Should we conclude that the pill decreases IQ scores, or should we conclude that our sample reflects sampling error from the population that occurs with no pill where μ is 100? Using the z-test, we compute the z-score for the sample as

$$z_{obt} = \frac{\overline{X} - \mu}{\sigma_{\overline{X}}} = \frac{99 - 100}{2.5} = \frac{-1.0}{2.5} = -.40$$

Now our sample mean has a z_{obt} of $-.40$. We again examine the sampling distribution, shown in Figure 10.6 on the next page.

The z_{obt} of $-.40$ is *not* beyond the z_{crit} of ± 1.96, so the sample does not lie in the region of rejection on our H_0 sampling distribution where μ is 100 (and $N = 36$ and $\sigma_{\overline{X}} = 2.5$). As the figure shows, when the bored statistician drew samples from this population, obtaining a mean of 99 through sampling error was a rather common, likely event. Thus, *our* sample mean of 99 was likely to have occurred because of sampling error when representing the population where μ is 100. Therefore, the null hypothesis—that our sample is merely a poor representation of the population where μ is 100—is a reasonable hypothesis. Given this situation, we certainly have no reason to think that H_0 is *not* true. In the language of statistics, we say that we have "failed to reject H_0" or we "retain H_0." Sampling error from the population where μ is 100 can explain these results just fine, thank you, so we will not reject this explanation.

> ***STAT ALERT*** We retain H_0 whenever z_{obt} does not lie beyond z_{crit} and thus does not lie in the region of rejection.

The shorthand way to communicate all of this is to say that we have *nonsignificant* results (note that we don't say *insignificant*). **Nonsignificant** means that our

FIGURE 10.6 Sampling Distribution of IQ Means

Our sample mean of 99 has a z_{obt} of $-.40$.

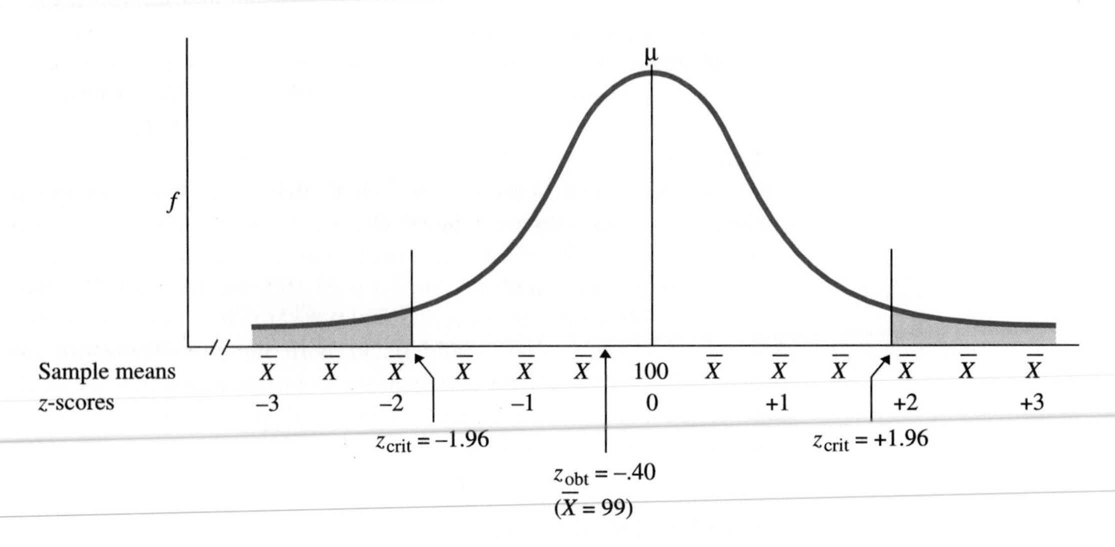

results are *not* too unlikely to accept as resulting from sampling error. *Nonsignificant* implies that the differences reflected by our results were likely to have occurred by chance if our data represent no difference in the population. Because we do not have *convincing* evidence of a real relationship, we play it safe and do not conclude that we have demonstrated a relationship.

When we decide that a result is not significant, we again report the α level used in making the decision. Thus, to report our nonsignificant z_{obt} when $\alpha = .05$, we would write

$$z_{obt} = -0.40, \quad p > .05$$

Notice that with nonsignificant results, we indicate that p is greater than .05.

Interpreting nonsignificant results When we retain H_0, we also retain the corresponding experimental hypothesis that our independent variable did not work as expected (we cannot yet become rich and famous). However, we have not proven that our pill does *not* work. The only thing we are sure of is that we have failed to find convincing evidence that the pill *does* work. Because we are likely to obtain a sample with a mean of 99 without the IQ pill, why should we think the pill works?

Recognize that failing to find evidence for something is not the same as proving that it is not there. For example, if you look outside on a cloudy night and do not see the moon, you have not proven that the moon is not there. You have simply failed to see it. You don't know if it is there (behind the clouds) or not there (below the horizon). In the same way, we have failed to obtain a sample mean that convincingly shows us that the pill works. Therefore, by failing to reject H_0, we *still* have two contradictory hypotheses that are both viable interpretations of our results: (1) H_0, that our data only reflect sampling error and show no real relationship, and (2) H_a, that our data do not reflect sampling error and do show a relationship. Thus, maybe in fact the pill does not work. Or maybe the pill does work, but it changes scores by so little that we do not see that it works. Or maybe the pill changes IQ scores greatly, but we don't see the change because we have so much sampling error in representing the different population μ that would be created. We simply don't know if the pill works or not.

Thus, when we do not reject H_0, we cannot say anything about whether the independent variable actually influences behavior or not. All that we can say is that we have failed to find a significant difference, and thus we have failed to demonstrate the predicted relationship in the population.

> ***STAT ALERT*** When we retain H_0, we do not know if the predicted relationship exists in the population or not.

Because of the various reasons that we might obtain nonsignificant results, we cannot design a study that is intended to show that no relationship exists. For example, we could not do a study to demonstrate that the pill does not work.

Summary of Statistical Hypothesis Testing

The steps and logic we have discussed above are used in all inferential procedures, so it is worthwhile to review them. In any research project, we do the following:

1. Create the experimental hypotheses predicting the relationship the study will or will not demonstrate.
2. Design the study to demonstrate the predicted relationship with the sample data.
3. Create the null hypothesis (H_0), which describes the value of the population parameter that the sample statistic represents if the predicted relationship does *not* exist. This value is also the mean of the sampling distribution.
4. Create the alternative hypothesis (H_a), which describes the value of the population parameter that the sample statistic represents if the predicted relationship *does* exist.
5. Select the value of α, which determines the size of the region of rejection.
6. Select the appropriate parametric or nonparametric procedure by matching the assumptions of the procedure to the study.
7. Collect the data, and compute the obtained value of the inferential statistic. This is analogous to finding a *z*-score for the sample data on the sampling distribution.
8. Set up the sampling distribution and, based on α and the way the test is set up, determine the critical value.
9. Compare the obtained value to the critical value.
10. If the obtained value lies beyond the critical value in the region of rejection, reject H_0, accept H_a, and describe the results as significant. Significant results provide greater confidence that your sample data represent the predicted relationship in the population and do not reflect mere sampling error in

representing no such relationship. Then describe the relationship in the population based on the sample data.

11. If the obtained value does not lie beyond the critical value in the region of rejection, do not reject H_0, and describe the results as nonsignificant. Nonsignificant results imply that you cannot eliminate the possibility that the data reflect sampling error, so you have failed to find convincing evidence that the sample represents the predicted relationship. Do *not* draw any conclusions about the possible relationship in the population.

THE ONE-TAILED TEST OF STATISTICAL HYPOTHESES

In some experiments we predict that scores will only increase or only decrease. In such situations we perform a one-tailed test. A **one-tailed test** is used when we predict the *direction* in which scores will change. For a one-tailed test, we set up our statistical hypotheses and the sampling distribution differently than for a two-tailed test.

The One-Tailed Test for Increasing Scores

Say that we have developed a "smart" pill, and we want to perform a single-sample experiment. Our experimental hypotheses are (1) we will demonstrate that the pill makes subjects smarter by increasing IQ scores or (2) we will not demonstrate that the pill makes subjects smarter.

Our alternative hypothesis again follows the experimental hypothesis that the independent variable works as predicted. Because the population without the pill has a μ of 100, if the smart pill worked and we gave it to everyone, it would *increase* IQ scores, so the population μ would be greater than 100. In symbols, our alternative hypothesis is

$$H_a\text{: } \mu > 100$$

H_a implies that our sample mean represents the *larger* population μ that would occur if the pill worked as predicted.

Our null hypothesis again implies that the independent variable does not work as predicted. Since we supposedly have a smart pill, the pill does not work if it either leaves IQ scores unchanged or *decreases* IQ scores (making subjects dumber). Therefore, if the smart pill did not work as predicted, the population would have a μ either equal to 100 or less than 100. Our null hypothesis is

$$H_0\text{: } \mu \leq 100$$

H_0 implies that our sample mean represents one of these populations.

We again test H_0 by examining the sampling distribution that describes the sample means we obtain when H_0 is true. The specific population of raw scores we use to create the sampling distribution is the one in which $\mu = 100$. This is because, if we determine that our subjects' IQ scores are above 100, then we automatically determine that they are above any value less than 100. Therefore, we test H_0: $\mu \leq 100$ by testing whether the sample represents the raw score population where μ is 100.

> ***STAT ALERT*** In a one-tailed test, the null hypothesis will always include a population parameter equal to some value. In testing H_0, we test whether the sample data represent that population.

We again set $\alpha = .05$. However, because we have a one-tailed test, we place our region of rejection in only *one tail* of the sampling distribution. You can identify which tail to put it in by identifying what data you must see in order to claim that your independent variable works as predicted. Here, for you to say that your smart pill works, you must conclude that your sample mean represents a μ larger than 100. To do that, first your sample mean itself must be larger than 100 (if not, then your sample is not smarter than people not given the pill). Second, your sample mean must be *significantly* larger than 100. The means that are significantly larger than 100 are in the region of rejection that is in the upper tail of the sampling distribution. Therefore, you place the entire region in the upper tail of the distribution, to form the sampling distribution shown in Figure 10.7. We don't place anything in the lower tail, because for our study, sample means down there are not different from means close to 100: they are just another result showing that the smart pill doesn't work, and we are not interested in distinguishing between whether it doesn't work because it has no influence on IQ or because it makes subjects dumber.

As in the previous chapter, the region of rejection in the upper tail of the distribution that constitutes 5% of the curve is marked by a z_{crit} of $+1.645$. The z_{crit} is only *plus* 1.645 so that we will reject H_0 only if the sample mean is greater

FIGURE 10.7 Sampling Distribution of IQ Means for a One-Tailed Test of Whether Scores Increase

The region of rejection is entirely in the upper tail.

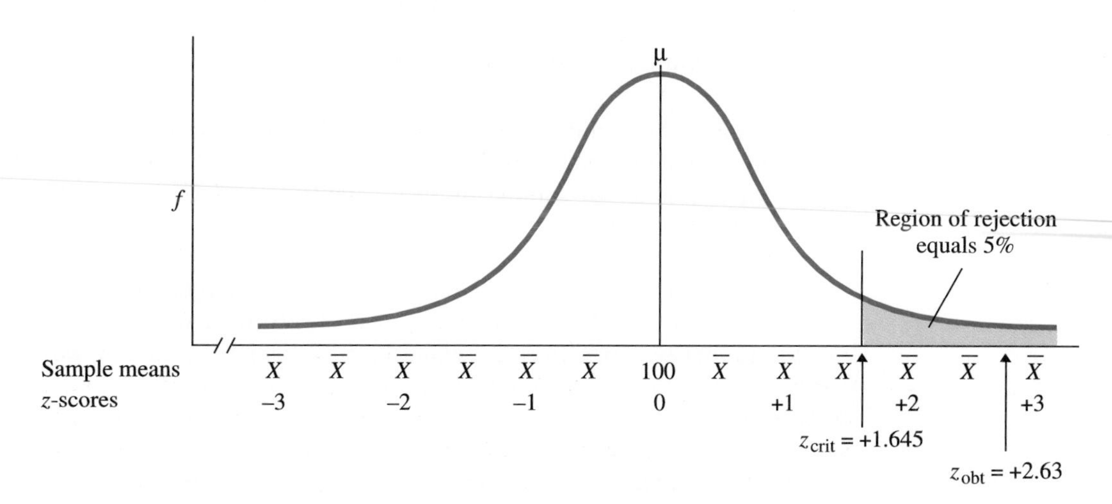

than 100. (However, notice that we have a *smaller* critical value than with the two-tailed test, and the edge of the region of rejection is closer to μ.)

Say that we test our smart pill on a sample of 36 subjects and find $\overline{X} = 106.58$. We are still basing our sampling distribution on the population of IQ scores where $\mu = 100$ and $\sigma_X = 15$, so using the previous formula for the z-test, we find that the sample mean has a z-score of $(106.58 - 100)/2.5$, which is $+2.63$. As shown in Figure 10.7, this z_{obt} is beyond our z_{crit}, so our sample mean is among those sample means that occur only 5% of the time when samples do represent the population where μ is 100. Thus, we conclude that our sample mean is too unlikely to accept as representing the population where $\mu = 100$. Further, if our sample is too unlikely to represent the population where μ is 100, then our sample is far too unlikely to represent a population where μ is < 100. Therefore, we reject the null hypothesis that $\mu \leq 100$, and accept the alternative hypothesis, that $\mu > 100$. We are now confident that our results do not reflect sampling error, but rather represent the predicted relationship where the smart pill works. We conclude that the pill produces a significant increase in IQ scores, and we estimate that with the pill, μ would equal about 106.58 (keeping in mind all of the cautions and qualifications we discussed previously for interpreting significant results).

If our z_{obt} did not lie in the region of rejection, we would retain H_0, and we would have no evidence as to whether the smart pill works or not.

The One-Tailed Test for Decreasing Scores

We can also arrange a one-tailed test using the lower tail. Say that we created a pill to lower IQ scores. If we gave this pill to the entire population, we would expect to find a population μ *less than* 100. On the other hand, if the pill did not work, it would produce the same population as no pill (with $\mu = 100$) or it would make subjects smarter (with $\mu > 100$). Thus the hypotheses we would be testing are

Pill does not work: H_0: $\mu \geq 100$

Pill works: H_a: $\mu < 100$

We again test our null hypothesis, using the sampling distribution from the raw score population where μ is 100. The only way for us to conclude that the pill lowers IQ is if our sample mean is significantly *less* than 100. Therefore, we place the entire region in the lower tail of the distribution, to form the sampling distribution in Figure 10.8.

With $\alpha = .05$, our z_{crit} is now *minus* 1.645. If our sample produces a *negative* z_{obt} beyond -1.645 (for example, $z_{obt} = -1.69$), then we reject the idea that our sample mean represents a μ equal to or greater than 100. If we reject H_0, we accept H_a (that the sample represents a μ less than 100). This significant result would then increase our confidence that we have demonstrated a relationship in nature where the pill works to lower IQ scores. If our z_{obt} did not fall in the region of rejection (for example, $z_{obt} = -1.25$), we would not reject H_0 and would have no evidence as to whether the pill works or not.

FIGURE 10.8 Sampling Distribution of IQ Means for a One-Tailed Test of Whether Scores Decrease

The region of rejection is entirely in the lower tail.

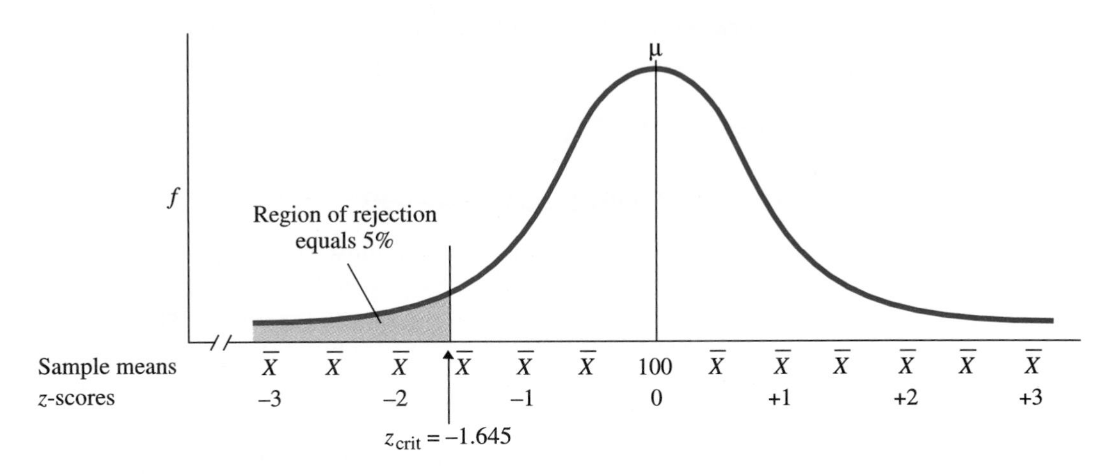

Choosing One-Tailed Versus Two-Tailed Tests

Look again at the previous sampling distributions for the one-tailed tests. Because there is no region of rejection in one tail of each distribution, this means that if our pills work in the opposite way from the way we predict, regardless of how well they work, we *cannot* reject H_0. If our smart pill produces a mean that is below 100, then the smart pill does not *increase* scores as predicted. Likewise, if our pill to lower IQ produces a mean above 100, then that pill does not *decrease* scores as predicted. In either case, regardless of the size of z_{obt} we would retain H_0 and the corresponding experimental hypothesis that we have not demonstrated that the pill works *as predicted.*

> ***STAT ALERT*** In a one-tailed test, z_{obt} is significant only if it lies beyond z_{crit} *and* has the same sign as z_{crit}.

You cannot switch the region of rejection and the sign of z_{crit} after the results are in. If you predict that the pill will increase IQ but it actually appears to decrease IQ, you can't switch and say, "Whoops, I meant to say it decreased scores." Remember, the tail you use is determined by the hypothesis of your experiment. After years of developing the theoretical and biochemical basis for a "smart pill," it would make no sense to suddenly say that the same basis leads you to predict it is a "dumb pill." Likewise, it makes no sense to switch between a one-tailed and a two-tailed test after the fact. Further, statistical hypothesis testing is based on testing random chance events, and switching after the results are in does not produce a fair test. Because of this, a one-tailed test should be used *only* if, before the experiment, you have a *convincing* reason for predicting the direction in which the independent variable will change scores. Otherwise, use a two-tailed test. This is safer because it allows you to determine whether an independent variable works, even if you cannot accurately predict whether it will work by increasing or decreasing scores.

STAT ALERT You should use a two-tailed test unless you have a specific prediction regarding the direction in which scores will change.

ERRORS IN STATISTICAL DECISION MAKING

There is one other major issue we must consider when performing statistical hypothesis testing, and it involves our potential errors. These are not errors in our calculations, but rather errors in our decisions: regardless of whether we conclude that our sample data do or do not represent the predicted relationship, we may be wrong.

Type I Errors: Rejecting H_0 When H_0 Is True

In previous examples where we rejected H_0 and claimed that our various pills worked, it is still possible that our sample was an infrequent, unrepresentative sample from the population where μ is 100. That is, it is possible that our sample so poorly represented the situation where the pill did not work, that we mistakenly thought the pill did work. If so, then we made a Type I error. A **Type I error** is defined as rejecting H_0 when H_0 is true. In a Type I error, we have so much sampling error that we—and our statistical procedures—are fooled into concluding that the predicted relationship exists when it really does not.

A Type I error can possibly occur *only* when we reject H_0. Then, the probability of making a Type I error is the probability that we have rejected a true H_0. This probability is determined by the size of our region of rejection, so, to use the lingo, it is our α that determines the probability of a Type I error. For example, Figure 10.9 shows the region of rejection in a one-tailed test when α is .05. When we claim that our results are significant, we are betting that sampling error didn't produce our sample mean. Making a Type I error is losing this bet. Therefore, we determine the probability of making a Type I error by using the same logic we used in the previous chapter to determine the probability of losing any bet. As the sampling distribution shows, sample means that are in the region of rejection occur 5% of the time when H_0 is true. But we reject H_0 100% of the time when our sample mean falls in the region of rejection. The result is that sample means that cause us to reject H_0 occur 5% of the time when H_0 is true. Therefore, 5% of the time when H_0 is true and sampling error produced our sample mean, we will reject H_0. In other words, 5% of the time when H_0 is true, we will make a Type I error. Thus, over the long run, the relative frequency of Type I errors is .05. Since relative frequency equals probability, the theoretical probability of making a Type I error is .05 anytime we reject H_0. (For a two-tailed test, when α is .05, the total region of rejection is still .05, so again the theoretical probability of making a Type I error is .05.)

Recall that our study must meet the assumptions of our statistical procedure. This is because if we violate the assumptions, then the true theoretical probability

FIGURE 10.9 Sampling Distribution of Sample Means Showing That 5% of All Sample Means Fall in the Region of Rejection When H_0 Is True

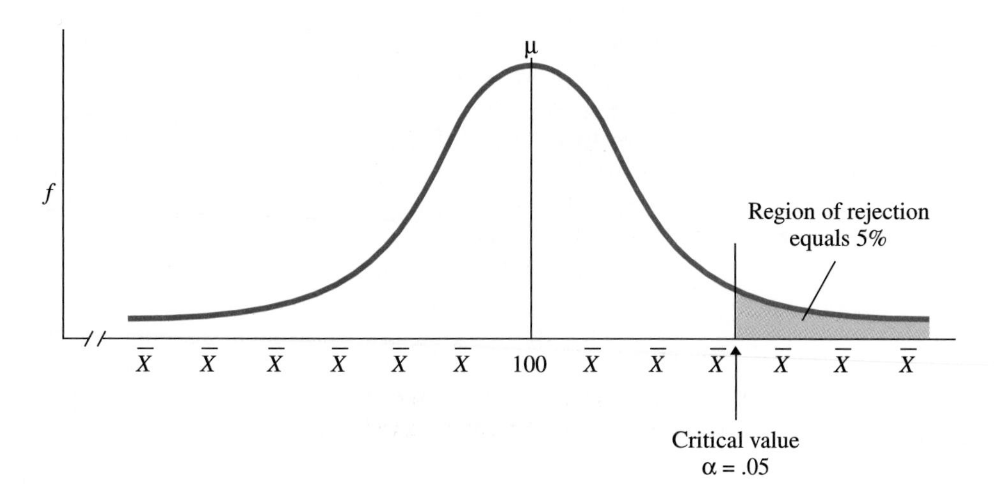

of a Type I error will be *greater* than we think it is. Luckily, as mentioned previously, parametric tests are robust. This means that if we violate their assumptions somewhat, the probability of a Type I error will still be close to our α. For example, if we set α at .05 and violate the assumptions of a parametric test somewhat, then the theoretical probability of a Type I error will be close to .05, such as .051. However, if we greatly violate the assumptions, we may think that α is .05 when in fact it is much larger, such as .20!

Here is an important distinction: Although the theoretical probability of a Type I error equals α, the actual probability of a Type I error is slightly less than α. This is because in figuring the size of the region of rejection, we include the critical value: in Figure 10.9, 5% of the curve is *at* or above $+1.645$. Yet for us to reject H_0, the z_{obt} must be *larger* than the critical value. We cannot determine the precise area under the curve for the infinitely small point located at 1.64500, so we can't remove it from our 5%. All that we can say is that when α is .05, any z_{obt} that allows us to reject H_0 falls in a region of rejection that is slightly less than 5% of the curve. Since the actual area of the region of rejection is less than α, the actual probability of a Type I error is also slightly less than α.

Thus, in any of our previous examples when we rejected H_0, the probability that we made a Type I error was slightly less than .05. That is why we reported our significant result as $z_{obt} = +2.0$, $p < .05$. Think of $p < .05$ as a shortened form of $p(\text{Type I error}) < .05$. This indicates that the probability is slightly less than .05 that we made a Type I error by calling this result significant when we shouldn't have. It is very important to always report the probability that you have made a Type I error.

On the other hand, we reported our nonsignificant result as $z_{obt} = -.40$, $p >$ .05. Here, $p > .05$ communicates that the reason we did not call this result significant is because if we did, then the probability that we made a Type I error would be greater than our α of .05 (and that's unacceptable).

We typically set α at .05 because .05 is an acceptably low probability of making a Type I error. If α is larger than .05, then we are too likely to conclude that a relationship exists when in fact it does not. This may not sound like a big deal, but the next time you fly in an airplane, consider that the designer's belief that the wings will stay on may actually be a Type I error: he's been misled by sampling error into erroneously thinking the wings will stay on. A 5% chance of this being the case is scary enough—we certainly don't want more than a 5% chance that the wings will fall off. In science, we are skeptical and careful, so we make our decisions like a jury: we want to be convinced "beyond a reasonable doubt" that sampling error did not produce our results, and having only a 5% chance that it did is reasonably convincing.

Sometimes, however, we want to reduce the probability of making a Type I error even further, and then we usually set alpha at .01. For example, we might have set alpha at .01 if our smart pill had some dangerous side effects. We would be especially concerned about needlessly subjecting the public to these side effects, so we would want to make it even less likely that we will conclude that the pill works when it really does not. When α is .01, the region of rejection is in the extreme 1% of the sampling distribution, so we have a larger absolute critical value than when α is .05. This means that our z_{obt} must be larger in order for us to reject H_0. Intuitively, it takes even more to convince us that the pill works, and thus there is a lower probability that we will make an error. Statistically, since we will reject H_0 only 1% of the time when H_0 is true, the probability of making a Type I error is now $p < .01$.

However, remember that we use the term *significant* in an all-or-nothing fashion: a result is *not* "more" significant when $\alpha = .01$ than when $\alpha = .05$. If z_{obt} falls in the region of rejection that we have used to define significant, then the result is significant, period! The *only* difference is that when $\alpha = .01$, there is a smaller probability that we have decided incorrectly and made a Type I error.

Finally, recognize that instead of comparing our z_{obt} to the z_{crit} for our alpha level, many computer programs directly compute the probability of a Type I error for a particular result. For example, we might see "$p = .02$." This indicates that the sample mean falls in the extreme 2% of the sampling distribution, and thus there is a .02 probability that we'll make a Type I error here. If our alpha is .05, then this result is even more convincing than we need, so it is significant. However, if the result produced "$p = .07$," then to call the result significant we'd need a region of rejection that is the extreme 7% of the sampling distribution, and there would be a .07 probability that we'll make a Type I error here. Because this implies an alpha level of .07, which is greater than our alpha of .05 and thus too big, this result is not significant.

Avoiding Type I errors If we *retain* H_0 when H_0 is true, then we avoid making a Type I error by making the correct decision: we conclude that we have no evidence that the pill works, and it doesn't. Since α is the theoretical probability of making a Type I error, $1 - \alpha$ is the probability of avoiding a Type I error. In other words, if 5% of the time we obtain samples in the region of rejection when H_0 is true, then 95% of the time we obtain samples that are not in the region of rejection when H_0 is true. Thus, 95% of the time when H_0 is true, we will not reject H_0 and so will avoid making a Type I error.

STAT ALERT We make a Type I error if we reject H_0 when H_0 is true. The theoretical probability of a Type I error is equal to α. We do not make a Type I error if we do not reject H_0 when H_0 is true. The theoretical probability of correctly retaining H_0 is equal to $1 - \alpha$.

Type II Errors: Retaining H_0 When H_0 Is False

In addition to Type I errors, it is possible to make a totally different kind of error. This error can only occur when we *retain* H_0, and it is called a Type II error. A **Type II error** is retaining H_0 when H_0 is false (and H_a is true). With a Type II error, we conclude that we have no evidence for the predicted relationship in the population, when, in fact, the relationship exists.

Anytime we discuss Type II errors, it is a given that H_0 is false and H_a is true. We make a Type II error because our sample mean is so close to the mean of the population described by H_0 that we conclude that the sample reflects sampling error from that population. We might have made a Type II error in our previous example when we obtained a sample mean of 99 and did not reject H_0. Perhaps the μ of the population with the pill would be 99, and our sample represents this μ. Or perhaps our pill would increase IQ scores greatly, say to a μ of 115, but we obtained a very unrepresentative sample with a mean of 99. In either case, because our sample mean of 99 was so close to 100 (the population μ without the pill), we retained H_0 and maybe made a Type II error.

The computation of the probability of Type II errors is beyond the scope of this discussion, but you should know that the symbol for the theoretical probability of a Type II error is β, the Greek letter **beta.** Whenever we retain H_0, β is the probability that we have made a Type II error.

Avoiding Type II errors We can potentially make a Type II error only when H_0 is false. If we reject H_0 when H_0 is false, we have avoided a Type II error and made a correct decision: for example, we conclude that the IQ pill works, and the pill does work. If β is the probability of making a Type II error, then $1 - \beta$ is the probability of avoiding a Type II error. Thus, anytime we reject H_0, the probability is $1 - \beta$ that we have made a correct decision and rejected a false H_0.

STAT ALERT We make a Type II error when we retain H_0 and H_0 is false. The theoretical probability of a Type II error is equal to β. We do not make a Type II error when we reject H_0 and H_0 is false. The theoretical probability of correctly rejecting H_0 is equal to $1 - \beta$.

Comparing Type I and Type II Errors

There is no doubt that Type I and Type II errors are two of the most confusing inventions ever devised. So, first recognize that Type I and Type II errors are mutually exclusive: if you have possibly made one type of error, then you cannot have made the other. Sometimes (although we never know when) the null hypothesis really is true. In that case, rejecting null is a Type I error and retaining it is the correct decision. But sometimes (we also never know when) the null

TABLE 10.1 Possible Results of Rejecting or Retaining H_0

		Our decision	
		We reject H_0 (claim H_a is true)	*We retain H_0 (claim H_0 may be true)*
The truth about H_0	*H_0 is true (H_a is false: no relationship exists)*	We make a Type I error ($p = \alpha$)	We are correct, avoiding a Type I error ($p = 1 - \alpha$)
	H_0 is false (H_a is true: a relationship exists)	We are correct, avoiding a Type II error ($p = 1 - \beta$)	We make a Type II error ($p = \beta$)

hypothesis is really false. In this case, retaining null is a Type II error and rejecting it is the correct decision.

Second, recognize that when we consider the possible true state of affairs regarding H_0, there are actually four possible outcomes in any study. To better understand them and their probabilities, look at Table 10.1. The first row describes the situation when H_0 is really true. If we reject H_0 when H_0 is true, we make a Type I error, and the probability of this is α. If we retain H_0 when H_0 is true, we avoid a Type I error by making a correct decision, and the probability of this is $1 - \alpha$. The second row describes the situation when H_0 is really false. If we retain H_0 when H_0 is false, we make a Type II error, and the probability of this is β. If we reject H_0 when H_0 is false, we avoid a Type II error by making a correct decision, and the probability of this is $1 - \beta$.

In any experiment, the results of our hypothesis testing procedures will place us in one of the two columns. Anytime we reject H_0, either we've made a Type I error or we've made the correct decision and avoided a Type II error. Anytime we retain H_0, either we've made a Type II error or we've made the correct decision and avoided a Type I error.

Statistical procedures are designed to minimize the probability of Type I errors because they are the more serious for science: a Type I error means that we conclude that an independent variable works when really it does not. Basing scientific "facts" on what are actually Type I errors can cause untold damage. On the other hand, Type II errors are also important. In order for us to accurately learn about nature, we must avoid making Type II errors and conclude that an independent variable works when it really does.

Power

Of the various outcomes in Table 10.1, the ideal situation is to reject H_0 when H_0 is false: our results lead us to conclude that our pill works, and the truth is that the pill does work. Not only have we avoided making an error, but we have also learned something about nature and our study has paid off (it's nice to be right). Thus, the goal of any scientific research is to reject H_0 when H_0 is false. This ability is so important that it has a special name: power. The **power** of a statistical test is the probability that we will reject H_0 when it is actually false, correctly concluding that the sample data reflect a real relationship. When null is false, the only other thing we might do is retain it, making a Type II error. Therefore, in other words, power is the probability that we will not make a Type II error, so power equals $1 - \beta$. Think of power as it is used with microscopes: the more powerful a microscope, the better we can see differences when they exist. The more powerful a statistical test, the better we can see differences that reflect a relationship, so the more likely we are to conclude that one is present when it is.

As researchers, we want to maximize our power. After all, why bother to conduct a study if we are unlikely to reject the null hypothesis even when the predicted relationship really *does* exist? If we don't have sufficient power, we will be uncertain about our decision anytime we retain H_0. For example, previously, when we failed to conclude that our pill produced significant differences when the sample mean was 99, maybe the problem was that we did not have much power. Maybe the probability was not high that we would reject H_0, *even if the pill really worked.* Therefore, in order to avoid this dilemma and to have confidence in our decision if we end up retaining H_0, we want to maximize our power. Then the probability of avoiding a Type II error will be high (and the probability of making a Type II error will be low).

As researchers, we cannot do anything to alter the probability that H_0 is false and that a relationship really exists in the population. However, we can do things to increase the probability that we will reject H_0 when it is false. The logic behind maximizing power is this: when H_0 is false, we "should" reject it. In other words, when H_0 is false, our results "should" be significant. Thus, we maximize our power by maximizing the probability that our results will be significant. (If it sounds as if we are cheating by trying to rig the decision to reject H_0, remember: by setting α at .05 or less, we control the probability of making the *wrong* decision when H_0 is true. At the same time, by maximizing power, we maximize the probability of making the *correct* decision when H_0 is false.)

As we have seen, our results are significant if our z_{obt} is larger than our z_{crit}. Therefore, we maximize power by maximizing the size of our obtained value relative to the critical value. This increases the probability that our results will be significant and thus increases the probability of rejecting H_0 when H_0 is false.

STAT ALERT The larger our obtained value, the more likely it is to be significant and thus the greater the power.

There are several things we can do to maximize power. At the beginning of this chapter, I said that we use nonparametric procedures only if we cannot use parametric procedures. This is because parametric tests are more powerful than nonparametric tests: if we analyze data using a parametric test, we are more likely to reject H_0 when it is false than if we analyze the same data using a nonparametric test.

Likewise, a one-tailed test is more powerful than a two-tailed test. This is because the z_{crit} for a one-tailed test is smaller (closer to the mean of the sampling distribution) than the z_{crit} for a two-tailed test. Previously, we used a z_{crit} of 1.645 for a one-tailed test and 1.96 for a two-tailed test. All other things being equal, our z_{obt} is more likely to be beyond 1.645 than beyond 1.96. Thus, we are more likely to conclude that our results are significant with a one-tailed test than with a two-tailed test. (With $\alpha = .05$ for both, the probability that we will incorrectly call the result significant is the same.) Of course, remember that we can reject H_0 only if our z_{obt} has the same sign as z_{crit}.

As we shall see in later chapters, there are also ways to design and conduct research so as to maximize the power of statistical procedures.

FINALLY

Understand that essentially, the purpose of inferential statistics is to minimize the probability that researchers will make Type I and Type II errors. If we had not performed the z-test for the studies we discussed in this chapter, we might have concluded that our IQ pills worked, even though we were actually being misled by sampling error. Saying that a variable works when it really does not is, to use our terminology, rejecting H_0 when it is true and making a Type I error. We would have no idea if this occurred with our pill, nor would we even know the chances that it occured. However, after finding that a pill produced significant results, we are now confident that we have not made a Type I error, because the probability of doing so is less than .05. Likewise, if our results are not significant, through the concept of power we can minimize the probability that we have made a Type II error, so we are confident that we are not concluding that a pill does not work when in fact it does.

Conceptually, all statistical hypothesis testing procedures follow the logic described here: H_0 is the hypothesis that says our data represent the populations we would find if the predicted relationship does not exist; H_a says that our data represent the predicted relationship. We then compute something like a z-score for our experiment's results on the sampling distribution when H_0 is true. If our z-score is larger than the critical value, it is unlikely that the results represent the populations described by H_0, so we reject H_0 and accept H_a. That's it! That's inferential statistics (well, not quite).

Each of the following chapters will describe procedures for testing statistical hypotheses from different kinds of experiments using different kinds of sampling distributions. In each, we will discuss the specific calculations we perform. However, the calculations should not be your primary concern. (After all, that's why they invented computers.) Remember, your overriding goal in this course is to learn when a particular procedure is used and what the answer means. You now know what the ultimate answer from any inferential procedure means: the results either are or are not significant, meaning that we either do or do not believe that the relationship portrayed by the sample data reflects a real relationship in the population (in nature). Your remaining goal is to learn when to use each specific procedure so that you can accurately determine this answer.

CHAPTER SUMMARY

1. *Sampling error* occurs when random chance produces a sample statistic that is not equal to the population parameter it represents.

2. *Inferential statistics* are procedures that allow us to decide whether sample data represent a particular relationship in the population.

3. *Parametric statistics* are inferential procedures in which we make assumptions about the parameters of the raw score populations our data represent. They are usually performed when it is appropriate to compute the mean.

4. *Nonparametric statistics* are inferential procedures that do not require such stringent assumptions about the population parameters represented by our sample. They are usually performed when it is appropriate to compute the median or mode.

5. The *alternative hypothesis,* H_a, is the statistical hypothesis that describes the population μ's we would find if an experiment were performed on the entire population and the predicted relationship existed. H_a implies that the sample mean represents one of these μ's.

6. The *null hypothesis,* H_0, is the statistical hypothesis that describes the population μ's we would find if the experiment were performed on the entire population and the predicted relationship did not exist. H_0 implies that the sample mean represents one of these μ's.

7. When we predict that changing the independent variable will change scores, but do not predict the direction in which the scores will change, we perform a *two-tailed test.* When we do predict the direction of the relationship, we perform a *one-tailed test.*

8. *Alpha,* α, is the theoretical size of the region of rejection. Typically we set α at .05 and then determine the appropriate critical value for the one- or two-tailed test.

9. The *z-test* is the parametric procedure for testing the statistical hypotheses from a single-sample experiment if (a) the population of raw scores is normally distributed and contains interval or ratio scores, and (b) the standard deviation of the raw score population, σ_X, is *known.* The value of z_{obt} indicates the location of the sample mean on the sampling distribution of means when H_0 is true.

10. If z_{obt} lies beyond z_{crit}, then the corresponding sample mean lies in the region of rejection. This indicates that such a sample mean is unlikely to occur when samples are randomly selected from the population described by H_0. Therefore, we reject the idea that our sample represents such a population, so we *reject* H_0 and *accept* H_a. This is called a *significant* result and is taken as evidence that the predicted relationship exists in the population.

11. If z_{obt} does not lie beyond z_{crit}, then the corresponding sample mean is *not* located in the region of rejection. This indicates that such a sample mean is likely to occur when randomly sampling the population described by H_0. Therefore, we *retain* H_0. This is called a *nonsignificant* result and is taken as a failure to obtain evidence that the predicted relationship exists in the population.

12. A *Type I error* occurs when we reject a true H_0. The theoretical probability of a Type I error is equal to α. If a result is significant, the probability that we have made a Type I error is $p < \alpha$. The theoretical probability of avoiding a Type I error by retaining a true H_0 is $1 - \alpha$.

13. A *Type II error* occurs when we retain a false H_0. The theoretical probability of making a Type II error is β. The theoretical probability of avoiding a Type II error by rejecting a false H_0 is $1 - \beta$.

14. When we reject H_0, either we have committed a Type I error or we have avoided a Type II error. When we retain H_0, either we have committed a Type II error or we have avoided a Type I error.

15. The *power* of a statistical test is the probability of rejecting a false H_0, and it equals $1 - \beta$. A powerful statistic has a high probability of detecting a relationship when one exists. When used appropriately, parametric procedures are more powerful than nonparametric procedures, and one-tailed tests are more powerful than two-tailed tests. The manner in which a study is designed and conducted also influences power.

PRACTICE PROBLEMS

(Answers for odd-numbered problems are provided in Appendix E.)

1. *a.* What is sampling error?

b. Why does the possibility of sampling error present a problem to researchers when inferring a population μ from a sample $\bar{X}$?

2\. What are inferential statistics used for?

3. *a.* What is the difference between a real relationship and one seen in a sample that results from sampling error?

b. By a real relationship, do we mean that our independent variable is necessarily related to our dependent variable?

4. For each of the following, decide whether the researcher should perform parametric or nonparametric procedures.

a. When ranking the intelligence of a group of subjects given a smart pill.

b. When comparing the median income for a group of college professors to that of the national population, containing all incomes.

c. When comparing the mean reading speed for a sample of deaf children to the average reading speed of the population of hearing children.

d. When measuring interval scores from a personality test given to a group of emotionally troubled people, and comparing them to the normal distribution of such scores found in the population of emotionally healthy people.

5. *a.* Why do researchers prefer parametric procedures?

b. Why can we use parametric procedures even if we cannot completely meet their assumptions?

6. What four things must a researcher do prior to collecting data for a study?

7. What are experimental hypotheses?

8. *a.* What does H_0 communicate?

b. What does H_a communicate?

9. When do we use a one-tailed test? When do we use a two-tailed test?

10. *a.* What is the advantage and the disadvantage of two-tailed tests?

b. What is the advantage and the disadvantage of one-tailed tests?

11. For each of the following experiments, describe the experimental hypotheses (identifying the independent and dependent variables):

a. A study to determine whether the amount of pizza consumed by college students during finals week increases relative to the rest of the semester.

b. A study to determine whether performing breathing exercises alters blood pressure.

c. A study to determine whether our sensitivity to pain is affected by increased levels of hormones.

d. A study to determine whether frequency of dreaming decreases as a function of more light in the room while sleeping.

12. For each study in problem 11, indicate whether a one- or a two-tailed test should be used, and state the H_0 and $H_{a.}$ Assume that $\mu = 50$ when the amount of the independent variable is zero.

13. *a.* What does α stand for, and what two things does it determine?

b. How does the size of α affect whether a result is significant or nonsignificant?

14. What does it mean to have significant results in terms of

a. The obtained and critical value?

b. The region of rejection of the sampling distribution?

c. Our likelihood of obtaining our sample mean when H_0 is true?

d. α?

15. A researcher predicts that listening to music while taking a test is beneficial. He obtains a sample mean of 54.63 when 49 subjects take a test while listening to music. The mean of the population of students who have taken this test without music is 50 ($\sigma_X = 12$).

a. Should he use a one-tailed or two-tailed test? Why?

b. What are H_0 and H_a for this study?

c. Compute z_{obt}.

d. With $\alpha = .05$, what is z_{crit}?

e. Does the researcher have evidence of a relationship in the population? If so, describe the relationship?

16. A researcher wonders whether attending a private high school leads to higher or lower performance on a test of social skills. A random sample of 100 students from a private school produces a mean score of 71.30 on the test, and the national mean score for students from public schools is 75.62 ($\sigma_X = 28.0$).

a. Should she use a one-tailed or a two-tailed test? Why?

b. What are H_0 and H_a for this study?

c. Compute z_{obt}.

d. With $\alpha = .05$, what is z_{crit}?

e. What should the researcher conclude about this relationship in the population?

17. *a.* What is the probability that the researcher in problem 15 made a Type I error, and what would the error be in terms of the independent and dependent variables?

b. What is the probability that the researcher in problem 15 made a Type II error, and what would the error be in terms of the independent and dependent variables?

18. *a.* What is the probability that the researcher in problem 16 made a Type I error? What would the error be in terms of the independent and dependent variables?

b. What is the probability that the researcher in problem 16 made a Type II error? What would the error be in terms of the independent and dependent variables?

19. *a.* What is power?

b. Why do researchers want to maximize power?

c. Why is a one-tailed test more powerful than a two-tailed test?

20. A report indicates that Brand X toothpaste significantly reduced tooth decay relative to other brands, with $p < .44$.

a. What does this indicate about the researcher's decision about Brand X?

b. What makes you suspicious of the claim that Brand X works better than other brands?

21. Foofy claims that using a one-tailed test is cheating. She reasons that a one-tailed test produces a smaller absolute value of z_{crit}, and therefore it is easier for us to reject H_0 than it is with a two-tailed test. If the independent variable doesn't work, she claims, we are more likely to make a Type I error. Why is she correct or incorrect?

22. Poindexter claims that the real cheating occurs when we increase power by increasing the likelihood that our results will be significant. He reasons that if we are generally more likely to reject H_0, then we are even more likely to do so when H_0 is true, and therefore we are more likely to make a Type I error. Why is he correct or incorrect?

23. Bubbles reads a report of Study A, in which using a two-tailed test, the results are significant: $z_{obt} = +1.97, p < .05$. She also reads about Study B, in which $z_{obt} = +14.21, p < .0001$.
    a. She concludes that the results of Study B are way beyond the critical value used in Study A, falling in a region of rejection containing only .0001 of the sampling distribution. Why is she correct or incorrect?
    b. She concludes that the results of Study B are more significant than those of Study A, both because the z_{obt} is so much larger and because α is so much smaller. Why is she correct or incorrect?
    c. In terms of their conclusions, what is the difference between the two studies?

24. A researcher measures the self-esteem scores of a sample of statistics students, reasoning that their frustration with this course may lower their self-esteem relative to that of the typical college student (where $\mu = 55$ and $\sigma = 11.35$). He obtains the following scores.
    44, 55, 39, 17, 27, 38, 36, 24, 36
    a. Should he use a one-tailed or two-tailed test? Why?
    b. What are H_0 and H_a for this study?
    c. Compute z_{obt}.
    d. With $\alpha = .05$, what is z_{crit}?
    e. What should the researcher conclude about the relationship between the self-esteem of statistics students and that of other students?

SUMMARY OF FORMULAS

1. *The computational formula for the z-test is*

$$z_{obt} = \frac{\overline{X} - \mu}{\sigma_{\overline{X}}}$$

where $\overline{X}$ is the value of our sample mean and μ is the mean of the sampling distribution when H_0 is true (the μ of the raw score population described by H_0).

2. *The computational formula for the standard error of the mean, $\sigma_{\bar{X}}$, is*

$$\sigma_{\bar{X}} = \frac{\sigma_X}{\sqrt{N}}$$

where N is the N of our sample and σ_X is the *true* population standard deviation.

11

Significance Testing of a Single Sample Mean or a Correlation Coefficient: The *t*-Test

To understand the upcoming chapter:

- From Chapter 5, recall that s_X is the *estimated* population standard deviation, that s_X^2 is the *estimated* population variance, and that both involve dividing by the degrees of freedom, or *df,* which equals $N - 1$.
- From Chapter 7, recall the uses and interpretation of r, r_s, and r_{pb}.
- From Chapter 10, recall the elements of significance testing, including one- and two-tailed tests, H_0 and H_a, Type I and Type II errors, and power.

Then your goals in this chapter are to learn:

- The difference between the z-test and the t-test.
- How to conceptualize the sampling distributions for t, r, r_s, and r_{pb}.
- How to perform hypothesis testing using the t-test.
- What is meant by the confidence interval for μ, and how it is computed.
- How to perform significance testing of r, r_s, and r_{pb}.
- How to maximize the power of t and r.

Statistical hypothesis testing is second nature to behavioral researchers. Different statistical procedures are used in virtually every type of research design. Although they have different names, in each one we always compute a statistic (like a *z*-score) that summarizes the location of our sample data on a sampling distribution when H_0 is true. The larger the value of the statistic, the less likely that H_0 is true for our study. A significant statistic indicates that it is too unlikely that H_0 is true, so we conclude that our data represent the predicted relationship. If you understand this simple summary, then all you need to do now is to learn to perform the mechanics of each specific procedure.

This chapter first introduces a procedure known as the *t*-test. Like the *z*-test, the *t*-test is used for significance testing of a single sample experiment involving one sample mean. As we will then see, the *t*-test also forms the basis for significance testing of a sample correlation coefficient. Finally, this chapter introduces the confidence interval, a new procedure for describing a population μ.

MORE STATISTICAL NOTATION

Officially the *t*-test is known as Student's *t*-test (although it was developed by a statistician named W. S. Gosset). The answer we obtain when we perform the *t*-test is symbolized by t_{obt}. The critical value of *t* is symbolized by t_{crit}. (Do not confuse *t*-tests—with a lowercase *t*—with *T*-scores—with a capital *T*—which we met in Chapter 6: *t* has nothing to do with *T*-scores.)

UNDERSTANDING THE *t*-TEST FOR A SINGLE SAMPLE MEAN

In Chapter 10 an assumption of the *z*-test was that we know the true standard deviation of the raw score population, σ_X, so that we can compute the true standard error of the mean, $\sigma_{\bar{X}}$, which is the standard deviation of the sampling distribution of means. Then z_{obt} is a *z*-score indicating the location of our sample mean on the sampling distribution of means. However, I have a confession: in most research we do *not* know the standard deviation of the raw score population, so we seldom perform the *z*-test. Instead, we estimate σ_X by computing s_X. Then we use this estimated population standard deviation to compute an *estimate* of the standard error of the mean. With this estimated standard error of the mean, we again compute something *like* a *z*-score to locate our sample mean on the sampling distribution of means. However, because this location is based on an estimate of the standard error, we have not computed a *z*-score and performed the *z*-test. Instead we have computed t_{obt} and performed the single-sample *t*-test. The **single-sample *t*-test** is the parametric procedure used to test the null hypothesis from a single-sample experiment when the standard deviation of the raw score population must be estimated.

We are still asking the same question here that we asked previously in Chapter 10, but the characteristics of our design simply dictate that we use a slightly different procedure to answer it. To see how this works, here is an example of an experiment that calls for the *t*-test. Some people cram just prior to taking a test. Presumably, this is so that they take the test before anything leaks out. To study this suicidal strategy, say that we randomly select a sample of subjects who have not read Chapter 12 in this book. Instead of having them study using their usual strategies, we make these students cram for one hour, and then we give them a test on the chapter. We assume that the resulting sample of test scores represents the population of test scores for crammers in statistics. To demonstrate a relationship, we must compare the population represented by our sample to some other population. From research by the book's author (me), we determine that for the population of students who do not cram for Chapter 12, the μ of scores on the same test is 75. Therefore, in our study we will compare the μ of the population of crammers represented by our sample to the μ of 75 for the population of noncrammers. Thus, the conditions of our independent variable are the presence or absence of cramming, and our dependent variable is test scores.

Some students maintain that cramming is a good way to study, actually improving their grades. We are open-minded, so our experimental hypotheses will not predict the direction in which cramming will change test grades. Therefore, we have the two-tailed experimental hypotheses that (1) the experiment will demonstrate that cramming either increases or decreases test grades or (2) the experiment will not demonstrate that cramming affects grades.

The null hypothesis always indicates that the experiment does not demonstrate the predicted relationship, so here it will indicate that cramming has no effect on grades. If cramming has no effect on grades, μ should be the same for crammers as for noncrammers: 75. Thus, our null hypothesis is

$$H_0: \mu = 75$$

H_0 implies that our sample of crammers represents the population where $\mu = 75$.

The alternative hypothesis always indicates that the experiment demonstrates the predicted relationship, so here it will indicate that cramming does affect grades. If cramming affects grades, the population μ of crammers should be different from that of noncrammers, so it should not be equal to 75. Therefore, the alternative hypothesis is

$$H_a: \mu \neq 75$$

H_a implies that our sample represents a population where $\mu \neq 75$.

Our next step is to select and set up the appropriate statistical test. First, we choose alpha: we think $\alpha = .05$ is acceptable (of course). Second, we check that our study meets the assumptions of the single-sample *t*-test.

The Assumptions of the Single-Sample *t*-Test

To perform the single sample *t*-test, we should be able to assume the following:

1. We have one random sample of interval or ratio scores.
2. The raw score population forms a normal distribution for which the mean is the appropriate measure of central tendency.
3. The standard deviation of the raw score population is estimated by the s_X computed from our sample. (We assume that the spread in the test scores of crammers and noncrammers is the same, so regardless of which population our sample represents, s_X is an estimate of that population's σ_X.)

If our data reasonably meet these three assumptions, we can perform the *t*-test. Because the *t*-test is a parametric test, it is robust, producing minimal error if we violate these assumptions somewhat. This is especially true if N is at least 30.

In our cramming study, we have a single-sample experiment measuring ratio scores, for which we can assume (1) the distribution of test scores is reasonably normal, (2) it is appropriate to summarize the scores using a mean, and (3) we can estimate the population standard deviation by computing s_X. Everything looks good, so we collect the data.

The Logic of the Single-Sample *t*-Test

Say that we test nine cramming subjects. (As we shall see, for maximum power, we never collect so few scores in an actual experiment.) The sample mean turns out to be $\overline{X} = 65.67$. If cramming makes no difference in scores, then the sample mean should be 75, representing the μ of the population of scores without cramming. However, our sample mean is 65.67, so it looks as if your parents were right, and cramming for an exam results in lower grades than not cramming. Thus, we might conclude that we have demonstrated a relationship in which increasing the amount of cramming results in lower grades.

But hold on! We must always consider sampling error. Maybe cramming has no effect on grades, and we are being misled by sampling error: maybe our sample represents the population of noncrammers, but by chance we selected subjects who just happened to have low scores on the exam. If so, then our sample poorly represents the population where $\mu = 75$. Maybe our null hypothesis is true, and the study did not demonstrate a relationship.

To be confident in our conclusions about the effect of cramming, we will first test this null hypothesis by performing the *t*-test. We use exactly the same logic as in the *z*-test: H_0 says that our mean of 65.67 is different from the μ of 75 because of sampling error. By computing t_{obt}, we determine the location of our sample mean on the sampling distribution of means that occurs when a sample *does* represent the population described by H_0. This allows us to determine the likelihood of obtaining our sample mean through sampling error if H_0 is true and the sample represents a population where μ is 75. If our t_{obt} is beyond our t_{crit}, our sample mean lies in the region of rejection, so we will reject the idea that our sample represents this population.

As we shall see, the only differences between the *z*-test and the *t*-test are that t_{obt} is not calculated in the same way as z_{obt} and that the value of t_{crit} is obtained from the *t*-distribution instead of from the *z*-distribution.

CALCULATING THE SINGLE-SAMPLE *t*-TEST

Let's first look at the definitional formulas for computing t_{obt}.

The computation of t_{obt} consists of three steps that parallel the three steps we previously performed in the *z*-test. There, we first determined the true standard deviation (σ_X) of the raw score population. For the *t*-test, we must compute an estimated standard deviation. Therefore, using the scores from our sample, we find the estimated standard deviation of the raw score population, s_X, using the formula

$$s_X = \sqrt{\frac{\Sigma X^2 - \frac{(\Sigma X)^2}{N}}{N - 1}}$$

The second step of the z-test was to compute the true standard error of the mean by dividing the true standard deviation (σ_X) by the square root of N:

$$\sigma_{\overline{X}} = \frac{\sigma_X}{\sqrt{N}}$$

For the t-test, we compute the estimated standard error of the mean. The symbol for the **estimated standard error of the mean** is $s_{\overline{X}}$. (The s stands for an estimate of the standard deviation, and the subscript $\overline{X}$ indicates that it is for a distribution of means.) We compute the estimated standard error of the mean by dividing the estimated standard deviation by the square root of N.

THE DEFINITIONAL FORMULA FOR THE ESTIMATED STANDARD ERROR OF THE MEAN IS

$$s_{\overline{X}} = \frac{s_X}{\sqrt{N}}$$

Notice the similarity between the preceding formulas for $\sigma_{\overline{X}}$ and $s_{\overline{X}}$. The only difference is whether we divide into σ_X or s_X.

The third step in the z-test was to compute z_{obt} using the formula

$$z_{obt} = \frac{\overline{X} - \mu}{\sigma_{\overline{X}}}$$

Similarly, the final step in the t-test is to compute t_{obt}.

THE DEFINITIONAL FORMULA FOR THE t-TEST FOR A SINGLE SAMPLE MEAN IS

$$t_{obt} = \frac{\overline{X} - \mu}{s_{\overline{X}}}$$

$\overline{X}$ is the sample mean, μ is the mean of the H_0 sampling distribution (which also equals the value of μ described in the null hypothesis), and $s_{\overline{X}}$ is the estimated standard error of the mean.

Notice the similarity between the formulas for z_{obt} and t_{obt}. The z_{obt} indicates the distance that the sample mean is from the μ of the sampling distribution, measured in units called the standard error of the mean. The t_{obt} measures this distance in estimated standard error units.

Computational Formulas for the Single-Sample *t*-Test

We can compute t_{obt} using the three steps given above, or we can use one of the following computational formulas. First, we can replace the symbol $s_{\overline{X}}$ with a formula for computing $s_{\overline{X}}$ and rewrite our definitional formula for t_{obt} as

$$t_{obt} = \frac{\overline{X} - \mu}{\frac{s_X}{\sqrt{N}}}$$

This simply shows the computation of $s_{\overline{X}}$ in the denominator of the *t*-test by dividing s_X by the square root of *N*.

To shorten the computations, we will not take the square root when computing the standard deviation, s_X. Instead, we will replace the standard deviation with the estimated variance, s_X^2, and then take the square root of the entire denominator. Thus, we have the following two formulas.

THE COMPUTATIONAL FORMULAS FOR THE SINGLE SAMPLE t-TEST ARE

$$t_{obt} = \frac{\overline{X} - \mu}{\sqrt{\frac{s_X^2}{N}}} \quad \text{and} \quad t_{obt} = \frac{\overline{X} - \mu}{\sqrt{(s_X^2)\left(\frac{1}{N}\right)}}$$

We can use either of these formulas. The formula on the left computes $s_{\overline{X}}$ by first dividing s_X^2 by *N*, and the formula on the right computes $s_{\overline{X}}$ by first multiplying s_X^2 times the quantity 1/*N*. With either formula, you perform all operations inside of the square root sign and then find the square root. Remember, the final number obtained in the denominator of any of these formulas is still the estimated standard error, $s_{\overline{X}}$. Then, dividing $s_{\overline{X}}$ into the difference found in the numerator gives t_{obt}.

TABLE 11.1 Grades of Nine Subjects Who Crammed for the Exam

Subject	*Grades (X)*	X^2
1	50	2500
2	75	5625
3	65	4225
4	72	5184
5	68	4624
6	65	4225
7	73	5329
8	59	3481
9	64	4096
$N = 9$	$\Sigma X =$ 591	$\Sigma X^2 =$ 39289
	$(\Sigma X)^2 =$ 349281	
	$\overline{X} =$ 65.67	

For our cramming study, say that we obtained the data in Table 11.1. First we compute s_X^2. Substituting our values from Table 11.1 into the formula for s_X^2, we have

$$s_X^2 = \frac{\Sigma X^2 - \frac{(\Sigma X)^2}{N}}{N - 1} = \frac{39289 - \frac{349281}{9}}{9 - 1} = 60.00$$

Thus, we estimate that the variance of the population of test grades represented by our sample is 60.

Now we compute t_{obt} for our mean of 65.67 when the population μ is 75, s_X^2 is 60, and N is 9. Filling in a computational formula, we have

$$t_{obt} = \frac{\overline{X} - \mu}{\sqrt{(s_X^2)\left(\frac{1}{N}\right)}} = \frac{65.67 - 75}{\sqrt{(60)\left(\frac{1}{9}\right)}}$$

In the denominator, 1/9 is .11, which multiplied times 60 is 6.667. So we have

$$t_{obt} = \frac{65.67 - 75}{\sqrt{6.667}} = \frac{-9.33}{2.582} = -3.61$$

The square root of 6.667 is 2.582. Because the denominator in the formula is the estimated standard error of the mean, $s_{\overline{X}}$ is 2.582. Dividing -9.33 by 2.582, we arrive at a t_{obt} of -3.61.

Thus, our sample mean produced a t_{obt} of -3.61 on the sampling distribution of means where $\mu = 75$. This is very similar to having a z-score of -3.61. According to H_0, if our sample is perfectly representative of the population where μ is 75, then the sample mean "should" be 75, producing a t_{obt} equal to 0.0. The question now is, "Is our t_{obt} of -3.61 (and the underlying difference between our mean and μ) significant?" To answer this question, we compare our t_{obt} to the appropriate t_{crit}, and for that we must examine the t-distribution.

THE *t*-DISTRIBUTION

In previous chapters we described the sampling distribution of means using z-scores, because the z-distribution provides us with a model for finding the probability of obtaining a particular sample mean from a particular raw score population when σ_X is known. Now, the t-distribution is our model for finding the probability of obtaining a sample mean from a particular raw score population when σ_X is *estimated.* Think of the t-distribution in the following way. One last time, we hire our by now *very* bored statistician. She infinitely draws samples of the same size N from the raw score population described by H_0, where μ is 75. For each sample, she computes $\overline{X}$, s_X, $s_{\overline{X}}$, and t_{obt}. She then plots the frequency distribution of the different means, labeling the X axis with the corresponding values of t_{obt} as well. Thus, the ***t*-distribution** is the distribution of all possible values of t

FIGURE 11.1 Example of a t-Distribution of Random Sample Means

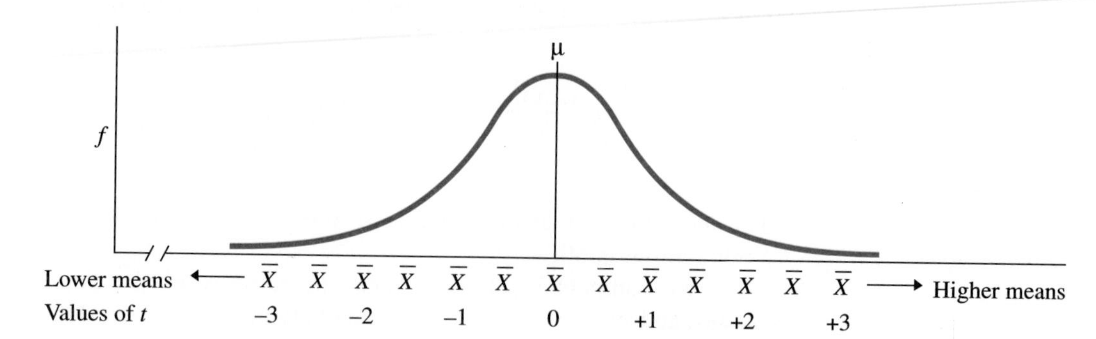

computed for random sample means having the same N that are selected from the raw score population described by H_0.

You can envision a t-distribution as shown in Figure 11.1. The values of t are analogous to z-scores. A sample mean equal to μ has a t equal to zero. Means greater than μ have positive values of t. Means less than μ have negative values of t. The larger the absolute value of t_{obt}, the farther it and the corresponding sample mean are from the μ of the distribution. Therefore, the larger the t, the lower the relative frequency and thus the lower the probability of obtaining the sample mean by chance when the sample represents the underlying raw score population.

Our t_{obt} locates our sample mean on this model, telling us the probability of obtaining such a mean when H_0 is true. To complete the t-test, we find t_{crit} and create the region of rejection. If our t_{obt} is beyond t_{crit}, then our sample mean is too unlikely a result to accept as representing the population described by H_0.

There is one important novelty here: there are actually *many* versions of the t-distribution, each having a slightly different shape. The shape of a particular t-distribution depends on the size of the sample N that the statistician uses when creating the t-distribution. If she selects samples with a small N, the t-distribution will be only a rough approximation to the standard normal curve. This is because s_X often contains large sampling error, so it is often a very rough estimate of σ_X. Further, she will often obtain a different value of s_X for each sample, and inconsistency in s_X will produce a t-distribution that is only a rough approximation to a normal curve. However, if she selects larger samples, the t-distribution will conform more closely to the true normal curve. A larger sample tends to more accurately represent the population, so s_X is closer to the one true value of σ_X. As we saw with the z-test, consistently using the true value of σ_X produces a sampling distribution that closely conforms to the true normal z-distribution. Therefore, as the sample size increases, each t-distribution is a successively closer approximation to the true normal curve.

The fact that there are differently shaped t-distributions is important for one reason: when we set up the region of rejection, we want it to contain precisely that portion of the curve defined by our α. If $\alpha = .05$, then we want to mark off precisely the extreme 5% of the curve. On distributions that are shaped differently, we mark off that 5% at different locations. Because the size of the

FIGURE 11.2 Comparison of Two t-Distributions Based on Different Sample N's

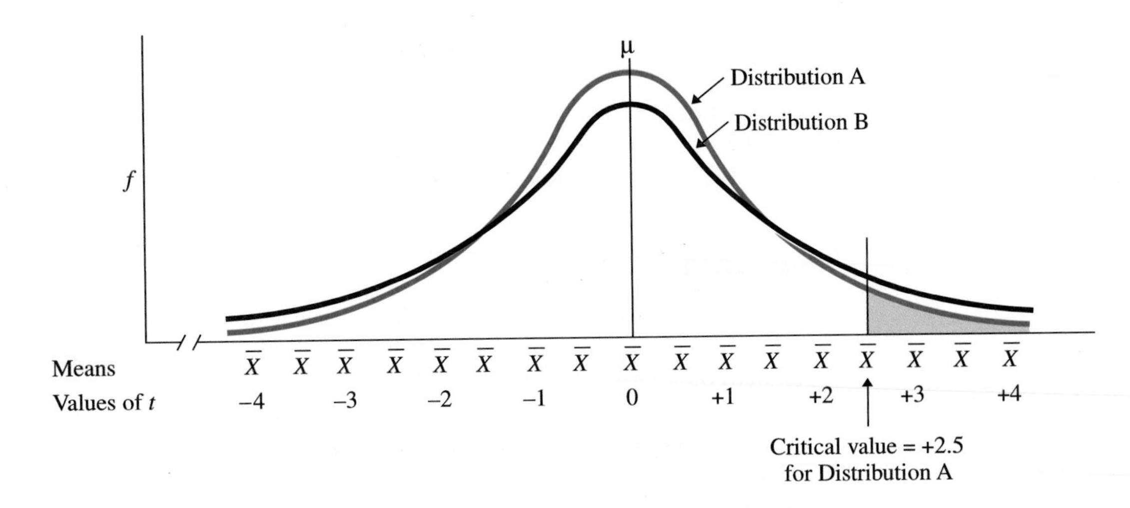

region of rejection is marked off by the critical value, *with differently shaped t-distributions we will have different critical values.* For example, Figure 11.2 shows a one-tailed region of rejection in two t-distributions. Say that the extreme 5% of Distribution A is beyond a t_{crit} of $+2.5$. If we use $+2.5$ as t_{crit} on Distribution B, the region of rejection will contain *more* than 5% of the distribution. Conversely, the t_{crit} marking off 5% of Distribution B will mark off *less* than 5% of Distribution A. (The same problem also exists for a two-tailed test.)

This issue is important because not only is α the size of the region of rejection, but it is also the theoretical probability of making a Type I error. Unless we use the correct t_{crit} from the t-distribution that is appropriate for our sample's N, the probability of a Type I error in our experiment will not equal our α (and that's not supposed to happen!). Thus, there is only one version of the t-distribution that we should use when testing a particular t_{obt}: the one that our bored statistician would have created by using the same N as in our sample.

The Degrees of Freedom

I have another confession: to identify the appropriate t-distribution for our study, we do not actually use N. Instead, the shape of a particular t-distribution is determined by the size of $N - 1$, what we call the degrees of freedom, or *df*, of the sample. Because we compute s_X using $N - 1$, it is the *df* that determines how consistently s_X estimates the true σ_X. Therefore, the larger the *df*, the closer the value of s_X in each sample is to σ_X, so the closer the t-distribution is to forming a normal curve.

It does not take a tremendously large *df*, however, to produce a truly normal t-distribution. When *df* is greater than 120, the t-distribution is virtually identical to the standard normal curve. But for each value of *df* between 1 and, 120 we have a differently shaped t-distribution. Thus, for samples having a *df* between 1 and 120 we determine our critical value by first identifying the appropriate sampling distribution using the particular *df* that we have in our sample. Only then will t_{crit} accurately mark off the region of rejection so that the true theoretical probability of a Type I error is equal to our α.

STAT ALERT The appropriate t_{crit} for the single-sample *t*-test comes from the *t*-distribution that has *df* equal to $N - 1$, where *N* is the number of scores in our sample.

The *t*-Tables

The good news is that we can obtain critical values from Table 2 in Appendix D, entitled "Critical Values of *t*." Take a look at these "*t*-tables". You'll find separate tables for two-tailed and one-tailed tests. Then identify the appropriate column for your value of α, and find the value of t_{crit} in the row opposite the *df* of your sample. For example, in our cramming study, *N* is 9, so *df* is $N - 1 = 8$. For our two-tailed test with $\alpha = .05$ and $df = 8$, t_{crit} is 2.306.

As usual, the table contains no positive or negative signs. In a one-tailed test, you must decide whether t_{crit} is positive or negative. Also, notice that the table uses the symbol for infinity (∞) for *df* greater than 120. This means that when a sample has a *df* greater than 120, using the sample to estimate the population's standard deviation is virtually the same as using the infinite population to calculate the true standard deviation. Then the *t*-distribution matches the standard normal curve, and the critical values are those we saw with the *z*-test.

Using the *t*-Tables

If you peruse the *t*-tables (a little light reading), you will *not* find a critical value for every *df* between 1 and 120. When the *df* of your sample does not appear in the table, there are two approaches you can take.

First, remember that all we need to know is whether or not t_{obt} lies beyond t_{crit} (and thus lies in the region of rejection). Often we can determine this by examining the critical values for the *df* above and below the *df* of our sample. For example, say that we perform a one-tailed *t*-test at $\alpha = .05$ with 49 *df*. The *t*-tables only give t_{crit} for 40 *df* (+1.684) and for 60 *df* (+1.671). Because 49 *df* lies between 40 *df* and 60 *df*, the critical value we seek lies between +1.671 and +1.684. It's a good idea to draw a picture of this, as shown in Figure 11.3.

Notice that as the *df increases,* the absolute size of t_{crit} *decreases,* and that our actual region of rejection starts at a point in between these two critical values. Therefore, if our t_{obt} lies beyond the t_{crit} of +1.684, then it is already in the region of rejection for 49 *df,* and so it is significant. On the other hand, if our t_{obt} is *not* beyond the t_{crit} of +1.671, then it is way short of the region of rejection for 49 *df,* and so it is not significant. In the same way, we can evaluate any obtained value that falls *outside* of the bracketing critical values given in the tables.

If t_{obt} falls *between* the bracketing values of t_{crit} in the table, then you should use the second approach, which is to use the interpolation procedure described in Appendix A.

FIGURE 11.3 One t-Distribution Showing the Location of Three Values of t_{crit}

The t_{crit} of +1.684 is for 40 df (dashed line), the t_{crit} of +1.671 is for 60 df (dotted line), and the t_{crit} for 49 df (solid line) is between them.

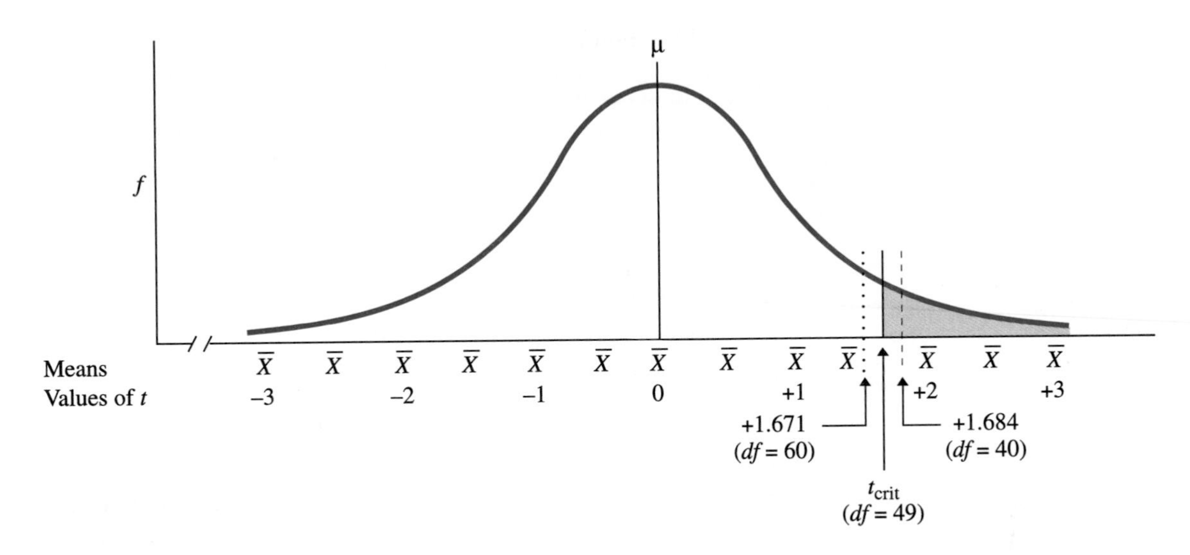

INTERPRETING THE SINGLE-SAMPLE t-TEST

Remember our cramming experiment? Our purpose is to decide whether cramming results in a different population of test scores than does not cramming. Once we have our values of t_{obt} and t_{crit}, the single-sample t-test is identical to the z-test.

In the cramming study, t_{obt} is -3.61 and the two-tailed t_{crit} is ± 2.306. With this information, we envision the sampling distribution shown in Figure 11.4.

FIGURE 11.4 Two-Tailed t-Distribution for $df = 8$ When H_0 Is True and $\mu = 75$

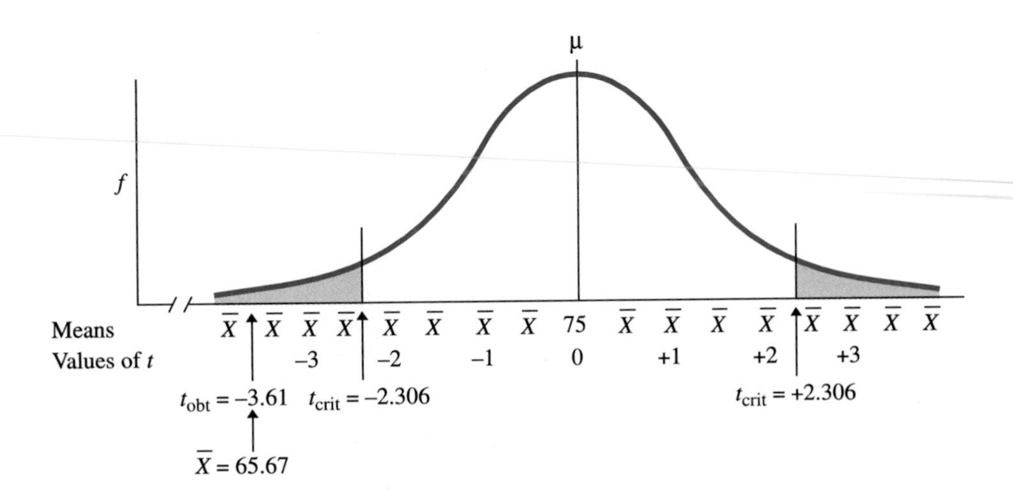

Remember, this is the sampling distribution when H_0 is true and the samples *do* represent the population where μ is 75 (and $s^2_{\bar{X}} = 60$ and $N = 9$). H_0 implies that because of sampling error, our sample mean of 65.67 poorly represents this μ. But, as the t-distribution shows, sampling error seldom produces a mean *that* poor at representing this μ. In fact, our t_{obt} lies beyond the t_{crit}, so our t_{obt} is in the region of rejection. Thus, when α is .05, a t_{obt} of -3.61 (and a $\bar{X}$ of 65.67) are too unlikely to occur if H_0 is true. Therefore, we reject H_0, rejecting the idea that our sample represents the population where μ is 75. We report the results of our t-test as

$$t(8) = -3.61, \quad p < .05$$

This statement communicates four facts:

1. We performed the t-test.
2. We had 8 degrees of freedom.
3. Our t_{obt} is -3.61.
4. We judged this t_{obt} and the underlying $\bar{X}$ to be significant, with the probability less than .05 that we made a Type I error (rejected a true H_0).

[In fact, from the t-tables, when α is .01, t_{crit} is ± 3.355. Because our t_{obt} is -3.61, it would be significant if we had used the .01 level. Therefore, instead of saying $p < .05$, we gain more information by reporting that $p < .01$, because now we know that the probability of a Type I error is not in the neighborhood of .04, .03, or .02.]

Thus, we conclude that our sample mean of 65.67 is significantly different from the population mean of 75. In other words, we accept our alternative hypothesis that our sample mean represents a μ that is not equal to 75. Because a sample mean of 65.67 is most likely to occur when the sample represents the population where μ is 65.67, our best estimate is that our sample represents a population of scores for crammers located at around 65.67. Because we expect one population of scores for crammers located at around 65.67 and a different population of scores for noncrammers located at a μ of 75, we conclude that our results demonstrate a relationship in the population between the independent variable (the presence or absence of cramming) and the dependent variable (test scores).

Of course, if our t_{obt} were a number that did not fall beyond our t_{crit} (for example, $t_{obt} = +1.32$), then it would not lie in the region of rejection and we would not reject H_0. We would conclude that our sample was likely to represent the population where μ is 75, so we would not have convincing evidence for a relationship between cramming and test scores. We would report such a nonsignificant result as $t(8) = +1.32$, $p > .05$.

Testing One-Tailed Hypotheses in the Single-Sample *t*-Test

We would perform a one-tailed test if, for example, we had predicted that cramming only *increases* test scores. Then H_a would be that the sample represents a

FIGURE 11.5 H_0 Sampling Distributions of t for a One-Tailed Test

On the left, we predict an increase in scores. On the right, we predict a decrease in scores.

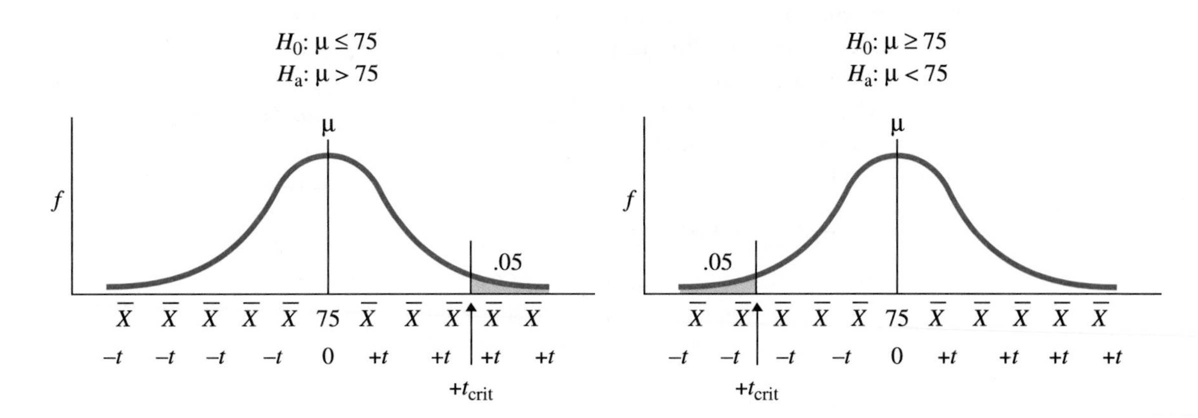

population μ greater than 75, or H_a: $\mu > 75$. H_0 would be that the sample represents a population μ less than or equal to 75, or H_0: $\mu \le 75$. We would obtain the one-tailed t_{crit} from the t-tables for our *df* and α. For us to conclude that the sample represents the predicted population where cramming increases scores, the sample mean must be significantly *larger* than 75. Therefore, we place all of the region of rejection in the upper tail of the sampling distribution, as shown in the left-hand graph in Figure 11.5. For t_{obt} to be significant, it must be positive and beyond t_{crit}. If it is, then our mean is too unlikely to represent a μ equal to 75, and even less likely to represent a μ less than 75. Therefore, we reject H_0 and accept H_a—that the sample mean represents a μ greater than 75.

We would also perform a one-tailed test if we had predicted that cramming *decreases* scores, using the sampling distribution on the right in Figure 11.5. Now H_a is that the sample mean represents a cramming μ that is less than 75. H_0 is that the sample mean represents a μ greater than or equal to 75. Because we predict a sample mean less than 75, to be significant t_{obt} must be negative and beyond $-t_{crit}$. If it is, our mean is too unlikely to represent a μ equal to 75, and even less likely to represent a μ greater than 75. Therefore, we reject H_0 and accept H_a—that the sample mean represents a μ less than 75.

In any of the above examples, when we reject H_0, we conclude that our sample mean represents a μ that is different from the one described by H_0. The next important step is to get a better estimate of the value of that μ by computing a confidence interval.

ESTIMATING THE POPULATION μ BY COMPUTING A CONFIDENCE INTERVAL

There are two ways to estimate a population μ. When we say that the population μ is equal to the sample mean, we are performing **point estimation.** Earlier we estimated that the μ of the population of crammers is located on the dependent variable of test grades at the *point* identified as 65.67. However, no one really believes that if we actually tested the entire population, μ would be *exactly* 65.67. The problem with point estimation is that it is extremely vulnerable to sampling error. Cramming may lower test grades as our study suggests, but the sample probably does not *perfectly* represent the population of crammers. Realistically, therefore, we can only say that the population μ of crammers will be *around* 65.67.

The other, better way to estimate a population μ is to include the possibility of sampling error and perform **interval estimation.** With interval estimation, we specify an interval, or range, of values within which we expect the population parameter to fall. You often encounter such intervals in real life, and they are usually phrased in terms of "plus or minus" some amount. For example, on the evening news when Dan Blather reports that a sample survey showed that 45% of the voters support the President, he may also report that there is plus or minus 3% error. What he means is that the surveyors have created an interval around 45%. They expect that if they asked the entire population, the μ would be within ±3% of 45%. In other words, they believe that between 42% and 48% of all voters in the population actually support the President.

We perform interval estimation by creating a confidence interval. Confidence intervals can be used to describe various population parameters, but the most common is the confidence interval for a single population μ. The **confidence interval for a single population μ** describes an interval containing values of μ, any one of which our sample mean is likely to represent. Previously, we said that our sample mean of crammers probably represents a μ *around* 65.67. A confidence interval is our way of statistically defining "around." As shown in the following diagram, by creating a confidence interval, we identify those values of μ around 65.67 that our sample mean is likely to represent.

$$\underbrace{\mu_{low} \; \ldots \; \mu \; \mu \; \mu \; \mu \; 65.67 \; \mu \; \mu \; \mu \; \mu \; \ldots \; \mu_{high}}_{\text{values of } \mu \text{, one of which is likely to be represented by our sample mean}}$$

The symbol μ_{low} stands for the lowest value of μ that our sample mean is likely to represent, and μ_{high} stands for the highest value of μ that the mean is likely to represent. When we compute these two values of μ, we have the confidence interval.

How do we know if a sample mean is likely to represent a particular value of μ? It depends on sampling error. For example, we intuitively know that sampling error is too unlikely to produce a sample mean of 65.67 if the sample represents a population μ of 500. In other words, 65.67 is significantly different from 500. On the other hand, sampling error is likely to produce a sample mean of 65.67 if the sample represents a population μ of 65 or 66. In other words, a mean of 65.67 is not significantly different from these μ's. Thus, a sample mean is likely to represent a particular value of μ if the sample mean is *not* significantly different from that μ. The logic behind computing a confidence interval is simply to compute the highest and lowest values of μ that are not significantly different from our sample mean. All values of μ between these two values are also not significantly different from the sample mean, so it is likely that the sample mean represents one of them.

> ***STAT ALERT*** A confidence interval contains the values between the highest and lowest values of μ that are not significantly different from our sample mean.

Computing the Confidence Interval for a Single μ

Because the *t*-test was appropriate for testing the significance of our sample mean in our hypothesis testing, the *t*-test also forms the basis for computing the confidence interval. However, previously we set up our *t*-test to compare the one value of μ described by H_0 to any possible sample mean we might obtain. Now we set it up to compare the one value of our sample mean to any possible μ. Then we find the highest and lowest possible values of μ that are not significantly different from our sample mean.

Recall that for a sample mean to differ significantly from μ, its t_{obt} must be *beyond* t_{crit}. Therefore, the most a sample mean can differ from μ and still not differ significantly is when its t_{obt} *equals* t_{crit}. We can state this using our formula for the *t*-test:

$$t_{obt} = \frac{\overline{X} - \mu}{s_{\overline{X}}} = t_{crit}$$

To find the largest and smallest values of μ that do not differ significantly from our sample mean, we simply determine the values of μ that we can put into this formula, with our sample mean and our $s_{\overline{X}}$, so that t_{obt} equals t_{crit}. Because we want to describe values above and below our sample mean, we use the two-tailed value of t_{crit}. Thus, we first want to find the value of μ that produces a $-t_{obt}$ equal to $-t_{crit}$. Rearranging the above formula, we have the formula for finding this value of μ:

$$\mu = (s_{\overline{X}})(+t_{crit}) + \overline{X}$$

We also want to find the value of μ that produces a $+t_{obt}$ equal to $+t_{crit}$. The formula for finding this value of μ is

$$\mu = (s_{\overline{X}})(-t_{crit}) + \overline{X}$$

Our sample mean represents a μ *between* these two values of μ, so we put the above formulas together into one formula.

> *THE COMPUTATIONAL FORMULA FOR THE CONFIDENCE INTERVAL FOR A SINGLE SAMPLE MEAN IS*
>
> $$(s_{\overline{X}})(-t_{crit}) + \overline{X} \leq \mu \leq (s_{\overline{X}})(+t_{crit}) + \overline{X}$$

The symbol μ stands for the unknown value represented by our sample mean. We replace the other symbols with the values of $\bar{X}$ and $s_{\bar{X}}$ that we computed from the sample data. We find the two-tailed value of t_{crit} in the *t*-tables at our α for $df = N - 1$, where N is the sample N.

> ***STAT ALERT*** In computing a confidence interval, use the two-tailed critical value, even if you have performed one-tailed hypothesis testing.

We can use the above formula to compute the confidence interval for our cramming study. There, $\bar{X} = 65.67$ and $s_{\bar{X}} = 2.582$. The two-tailed t_{crit} for $df = 8$ and $\alpha = .05$ is ± 2.306. Filling in the above formula for the confidence interval, we have

$$(2.582)(-2.306) + 65.67 \leq \mu \leq (2.582)(+2.306) + 65.67$$

After multiplying 2.582 times -2.306 and $+2.306$ we have

$$-5.954 + 65.67 \leq \mu \leq +5.954 + 65.67$$

Adding -5.954 is the same as subtracting 5.954, so the formula at this point tells us that our sample mean represents a μ that is greater than or equal to the quantity $65.67 - 5.954$, but less than or equal to the quantity $65.67 + 5.954$. In other words, our mean represents a μ of 65.67, plus or minus 5.954.

After adding ± 5.954 to 65.67, we have

$$59.72 \leq \mu \leq 71.62$$

This is the finished confidence interval for the μ represented in our cramming study. We can now return to our previous diagram of the confidence interval and replace the symbols μ_{low} and μ_{high} with the numbers 59.72 and 71.62, respectively.

$$\underbrace{59.72 \quad \ldots \quad \mu \quad \mu \quad \mu \quad \mu \quad 65.67 \quad \mu \quad \mu \quad \mu \quad \mu \quad \ldots \quad 71.62}$$

values of μ, one of which is likely to be represented by our sample mean

As shown, the confidence interval says that our sample mean probably represents a μ around 65.67, such that it is greater than or equal to 59.72, but less than or equal to 71.62.

Confidence Intervals and the Size of Alpha

Why do we call this a "confidence" interval? We defined this interval using $\alpha = .05$, so .05 is the theoretical probability of making a Type I error. Thus, 5% of the time the interval will be in error and will not contain the μ represented by our $\bar{X}$. However, recall that the quantity $1 - \alpha$ is the probability of avoiding a Type I error. Thus, $1 - .05$, or 95%, of the time the interval will contain the μ represented by our $\bar{X}$. Therefore, the probability is .95 that the interval contains the μ. Recall that probability is our way of expressing our confidence in an event. In our cramming study, we are 95% confident that the interval between 59.72 and 71.62 contains the μ represented by our sample mean.

STAT ALERT The amount of confidence we have that a confidence interval contains the μ represented by our sample mean is always equal to the quantity $1 - \alpha$ multiplied times 100.

The smaller our α, the smaller the probability of an error, so the greater our confidence. Had we set α equal to .01 in our cramming study, we would have $1 - .01(100)$, or a 99% confidence interval. With $\alpha = .01$ and $df = 8$, our t_{crit} would be ± 3.355, and the 99% confidence interval based on the sample mean of 65.67 would be

$$57.01 \leq \mu \leq 74.33$$

When we compare our 95% confidence interval to our 99% confidence interval, we see that the 99% confidence interval spans a wider range of values of μ:

95% confidence interval   59.72 . . . μ μ μ μ 65.67 μ μ μ μ . . . 71.62

99% confidence interval   57.01 . . . μ μ μ μ μ 65.67 μ μ μ μ μ . . . 74.33

values of μ, one of which is likely to be represented by our sample mean

Logically, the larger the range of values within the interval, the greater our confidence that the interval contains the μ represented by the sample mean. (Think of a confidence interval as a fishing net. The larger the net, the more confident we are that we will catch the μ represented by the sample mean.) There is, however, an inevitable tradeoff: when we increase confidence by including a wider range of possible values of μ, we less precisely identify the specific value of μ represented by our sample. Usually we compromise between sufficient confidence and sufficient precision by using $\alpha = .05$ and creating the 95% confidence interval.

Thus, we conclude our single-sample t-test by saying, with 95% confidence, that our sample of crammers represents a μ between 59.72 and 71.62. Because the center of the interval is at 65.67, we have defined how our mean represents a μ of *around* 65.67, providing much more information than when we merely say that μ is somewhere around 65.67. Therefore, anytime you wish to describe a population μ represented by a sample mean, you should compute a confidence interval.[1]

SIGNIFICANCE TESTS FOR CORRELATION COEFFICIENTS

It's time to shift mental gears. Another type of single-sample study arises when we compute a correlation coefficient. (Remember that the correlation coefficient

[1]We can also compute a confidence interval when performing the z-test. We use the formula given above, except we use the critical values from the z-tables in Appendix D. If $\alpha = .05$, then $z_{crit} = \pm 1.96$. If $\alpha = .01$, then $z_{crit} = \pm 2.575$.

describes the strength and direction of a linear relationship formed by pairs of X and Y scores. A coefficient of 0 indicates no relationship, while the larger the coefficient—the closer it is to ± 1.0—the stronger the relationship.) For example, say we studied the relationship between cramming and test scores using a correlational research design. For a sample of 25 subjects, we measured their test scores and how crammed their studying was. Using the formula given in Chapter 7, say we then computed a Pearson correlation coefficient of $r = -.45$. This indicates that the more a subject crammed, the lower his or her test score.

Remember, though, that this correlation coefficient is a sample statistic that only describes the relationship found in the sample. Ultimately we want to describe the relationship in the population. Therefore, we use the sample coefficient to estimate the population parameter we expect we'd find if we computed the correlation for the entire population. Recall that the population correlation coefficient is called rho and its symbol is ρ. Thus, for our cramming study, we want to conclude that if we measured the entire population and computed the correlation, ρ would equal $-.45$.

But hold on, there is a problem here: that's right, it's sampling error. The problem of sampling error applies to *all* sample statistics. Thus, even though our r suggests a relationship, it is possible that, because of sampling error, the sample actually represents either no relationship or a different relationship in the population. Therefore, regardless of the value of any sample correlation coefficient we compute, before we can be confident that the relationship exists in the population, we must answer the question "Is the correlation coefficient significant?"

STAT ALERT We never accept that a sample correlation coefficient demonstrates a real relationship until we decide that it is significant.

Statistical Hypotheses for the Correlation Coefficient

As usual, we should create our experimental and statistical hypotheses before we collect the data. We then perform either a one-tailed or a two-tailed test, depending on our hypotheses. We use a two-tailed test if we do not predict the direction of the relationship. For example, in our cramming study, we might have the experimental hypothesis that the study demonstrates either (1) that the more students cram, the higher their grades (a positive correlation), or (2) that the more students cram, the lower their grades (a negative correlation). Our other experimental hypothesis is that the study does not demonstrate any relationship.

This latter hypothesis translates into the null hypothesis, because H_0 always implies that the predicted relationship does not occur in the population. If there is neither the predicted positive nor negative correlation, then there is zero correlation. Most of the time, behavioral researchers test a null hypothesis involving zero correlation in the population. Therefore, in this book,

THE TWO-TAILED NULL HYPOTHESIS FOR SIGNIFICANCE TESTING OF A CORRELATION COEFFICIENT IS ALWAYS

$$H_0: \rho = 0$$

FIGURE 11.6 Scatterplot of a Population for Which $\rho = 0$, as Described by H_0.

Any r *is a result of selecting a sample scatterplot from within this scatterplot.*

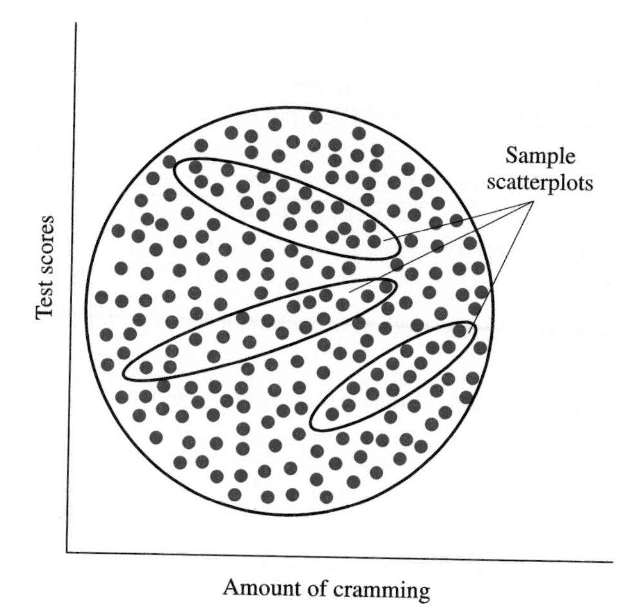

H_0 implies that the sample r represents a ρ equal to zero. If r does not equal zero, the difference is written off as sampling error. We can understand how such sampling error can occur by examining the hypothetical scatterplot in Figure 11.6, for a population for which H_0 is true and ρ is 0. Recall that a circular scatterplot reflects zero correlation, while a slanting elliptical scatterplot reflects an r not equal to zero. In any study, the null hypothesis implies that, by chance, we have selected an elliptical sample scatterplot from the circular population plot. Thus, our sample r may not equal 0, but this is because it is a poor representation of the population where ρ equals 0.

On the other hand, the alternative hypothesis always implies that the predicted relationship does occur in the population. If we predict that there is either a positive or a negative relationship, we predict that ρ does *not* equal zero.

THE TWO-TAILED ALTERNATIVE HYPOTHESIS FOR SIGNIFICANCE TESTING OF A CORRELATION COEFFICIENT IS ALWAYS

$$H_a : \rho \neq 0$$

H_a implies that our sample represents a population where ρ is not zero and thus represents a relationship in the population. We assume that the scatterplot for the population would then be similar to the scatterplot we've found for our sample.

As usual, we test our data by testing H_0. Here we test whether our sample correlation is likely to be a poor representation of a population ρ of zero. If our sample r is too unlikely to be accepted as representing $\rho = 0$, then we reject H_0 and accept H_a (that the sample represents a population where ρ is not equal to zero).

Recall from Chapter 7 that we may compute one of three types of correlation coefficients: the Pearson r, the Spearman r_s, and the point-biserial r_{pb}. The logic and format of the above statistical hypotheses are used regardless of which correlation coefficient we compute (we merely change the subscripts). The following sections discuss the particulars of statistical hypothesis testing for each type of correlation coefficient.

The Significance Test for the Pearson *r*

As usual, before we collect any data, we should make sure our study meets the assumptions of our statistical procedure. *The assumptions for hypothesis testing of the Pearson correlation coefficient are:*

1. We have a random sample of X-Y pairs, and each variable is an interval or ratio variable.

2. The Y scores and the X scores in the sample each represent a normal distribution. Further, they represent a *bivariate* normal distribution. This means that the Y scores at each value of X form a normal distribution and that the X scores at each value of Y form a normal distribution. (If N is larger than 25, however, violating this assumption is of little consequence.)

3. For the procedures discussed here, we are testing the null hypothesis that the population correlation is zero. (When H_0 states that ρ is some value other than 0.0, a different statistical procedure from the one presented here is used.)

Our cramming correlation meets the above assumptions, so we set α at .05 and test our r of $-.45$. To do that, we set up the sampling distribution.

The sampling distribution of *r* As with previous procedures, we test H_0 by examining the H_0 sampling distribution. Here the H_0 sampling distribution of r shows the values of r that occur when we randomly sample the population where ρ is 0. Our bored statistician has quit, but by now you could create the sampling distribution yourself. Using the same N as in our study, you would select an infinite number of samples of X-Y pairs from the population where $\rho = 0$. For each sample, you would compute r. If you then plotted the frequency of the various values of r, you would have the sampling distribution of r. The **sampling distribution of a correlation coefficient** is a frequency distribution showing all possible values of the coefficient that can occur when samples of size N are drawn from a population where ρ is zero. Such a sampling distribution is shown in Figure 11.7.

The only novelty here is that instead of plotting different sample means along the horizontal axis, we have plotted the different values of r. As shown, when H_0 is true and ρ is 0, the most frequent sample r will equal zero, so the mean

FIGURE 11.7 Distribution of Random Sample r's When $\rho = 0$

It is an approximately normal distribution, with values of r plotted along the X axis.

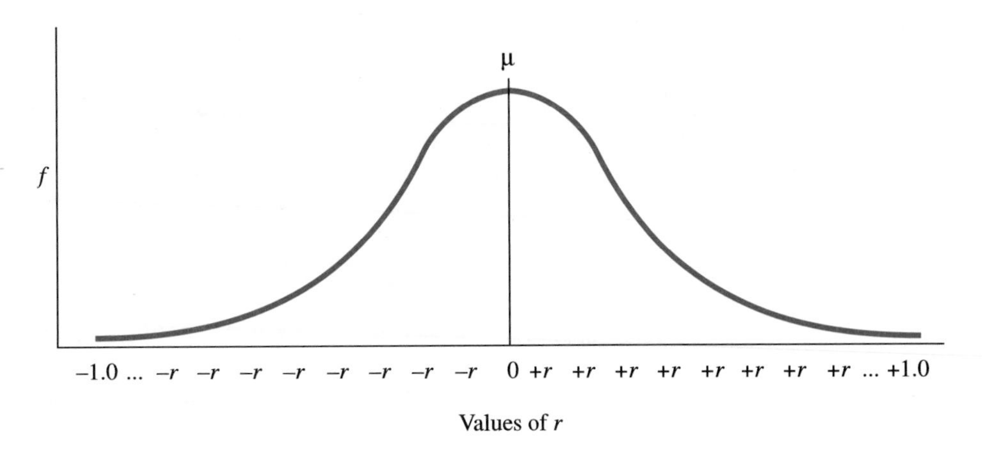

of the sampling distribution is 0. However, because of sampling error, sometimes the sample r will not equal zero. The larger the r (whether positive or negative), the less frequently it occurs, and thus the less likely it is to occur when the sample represents a population where ρ is zero.

To test H_0, we simply determine where on this distribution our sample r lies. To do so, we could perform a variation of the t-test, but luckily that is not necessary. Instead, the value of our sample r is just like a z-score, directly communicating its location on the sampling distribution relative to the mean of the distribution. The mean of the sampling distribution is always an r of zero. So, for example, our sample r of $-.45$ is a distance of .45 below the mean. Therefore, we test H_0 simply by examining the value of our obtained sample r. The symbol for an obtained r is r_{obt}. To determine whether r_{obt} lies in the region of rejection, we compare it to the critical value of r, which we identify as r_{crit}.

Determining the significance of the Pearson *r* As with the t-distribution, the shape of the sampling distribution of r is slightly different for each df, so there is a different value of r_{crit} for each df. Table 3 in Appendix D gives the critical values of the Pearson correlation coefficient. To use these "r-tables," we first need the appropriate degrees of freedom. *But,* here's a new one: with the Pearson correlation coefficient, the degrees of freedom equals $N - 2$, where N is the number of pairs of scores in the sample.

> ***STAT ALERT*** In significance testing of r, the degrees of freedom equals $N - 2$, where N is the number of pairs of scores.

To find r_{crit}, enter Table 3 for either a one- or a two-tailed test at the appropriate α and df. For our cramming correlation, N was 25, so $df = 23$. We are performing a two-tailed test with $\alpha = .05$ and $df = 23$, so r_{crit} is $\pm.396$. Armed with this

FIGURE 11.8 H_0 Sampling Distribution of r When H_0: $\rho = 0$

For the two-tailed test, there is a region of rejection for positive values of r_{obt} *and for negative values of* r_{obt}.

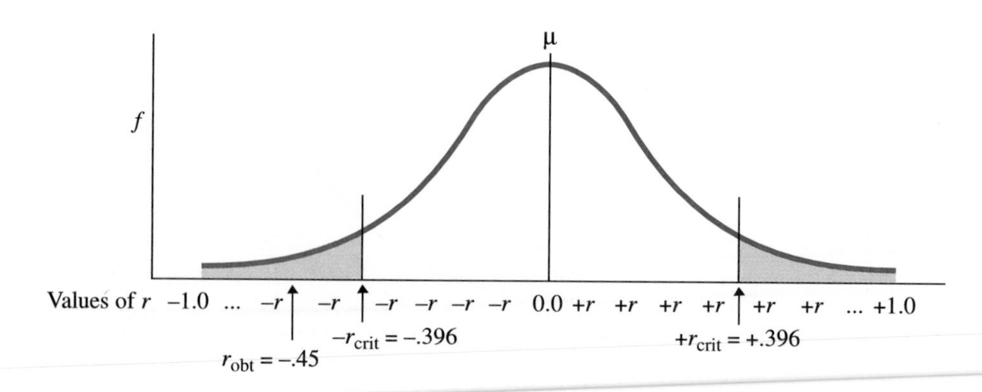

information, we set up the sampling distribution as in Figure 11.8. Our r_{obt} of $-.45$ is beyond the r_{crit} of $\pm.396$, so r_{obt} is in the region of rejection. Looking at the sampling distribution, we see that when samples *do* represent the population where ρ is 0, we seldom have an r_{obt} that falls in the region of rejection. Thus, it is too unlikely that our sample poorly represents the population where $\rho = 0$, because samples are seldom *that* poor. Therefore, we reject the hypothesis that the r_{obt} from our cramming study represents the population ρ of 0, and we conclude that our r_{obt} is significantly different from 0.

As usual, α is the theoretical probability that we'll make a Type I error. As the sampling distribution shows, the values of r in the region of rejection do occur when H_0 is true and ρ is 0. With $\alpha = .05$, over the long run we will obtain values of r_{obt} that cause us to erroneously reject H_0 a total of 5% of the time. Thus, the probability that we have made a Type I error this time is slightly less than .05. We report our significant r_{obt} as

$$r(23) = -.45, \quad p < .05$$

(Note the *df* in parentheses.)

Remember that by rejecting H_0, we have not proven anything. In particular, we conducted a correlational study, so we have not proven that changes in cramming *cause* test scores to change. In fact, we have not even proven that there is a relationship in nature (we may have made a Type I error). Instead, we are simply more confident that our r_{obt} does not merely reflect some quirk of sampling error but rather represents a "real" relationship in the population.

Because the sample r_{obt} is $-.45$, our best estimate is that if we performed this study on the population, we would find a ρ equal to $-.45$. However, recognizing that our sample may contain sampling error, we expect that ρ is probably *around* $-.45$. (We could more precisely identify the value of ρ by computing a confidence interval, which describes the values of ρ that r_{obt} is likely to represent. However, confidence intervals for ρ are computed using a very different procedure from the one we discussed previously.)

In Chapter 8 we saw that we further describe a relationship by computing the linear regression equation and r^2. However, we only do this when r_{obt} is significant, because only then are we confident that we are describing a "real" relationship. Thus, for our cramming study we would now compute the linear regression equation, which allows us to predict test scores if we know subjects' cramming scores. We also compute r^2, which equals $-.45^2$ or .20. Recall that this is the proportion of variance in our sample's Y scores that is accounted for by the relationship with the X scores. It tells us that we are 20% more accurate when we use the relationship between cramming and test scores to explain and predict test scores than we are when we do not use the relationship.

Remember that the term *significant* does not indicate that we have an *important* relationship. Although a relationship *must* be significant in order to be important, it can be significant and still be unimportant. We determine the importance of a relationship by computing r^2. The greater the proportion of variance accounted for, the more that knowing subjects' X scores improves our accuracy in predicting their Y scores—their behavior—and so the greater the scientific importance of the relationship. For example, we might find an r_{obt} of $+.10$ that is significant. In saying that r_{obt} is significant, we are saying *only* that it is unlikely to occur by chance. However, a relationship where r_{obt} equals $+.10$ is *not* statistically important, because $+.10^2$ is only .01: the relationship accounts for only 1% of the variance, and thus it is virtually useless in explaining differences in Y scores. Therefore, although we have demonstrated a relationship that is not likely to occur through sampling error, at the same time we have demonstrated an unimportant and not-very-useful relationship.

Of course, if r_{obt} does not lie beyond r_{crit}, then we retain H_0 and conclude that our sample may represent a population where $\rho = 0$. As usual, we have not proven that there is *not* a relationship in the population—we have simply failed to convincingly demonstrate that there *is* one. Therefore, we make no claims about the relationship that may or may not exist in the population, nor do we attempt to describe it using the regression equation or r^2. We report nonsignificant results as above, except that $p > .05$.

One-tailed tests of *r* If our experimental hypothesis predicted only a positive correlation or only a negative correlation, we would perform a one-tailed test and have either of the following:

THE ONE-TAILED HYPOTHESES FOR SIGNIFICANCE TESTING OF A CORRELATION COEFFICIENT ARE

Predicting positive correlation	***Predicting negative correlation***
H_0: $\rho \leq 0$	H_0: $\rho \geq 0$
H_a: $\rho > 0$	H_a: $\rho < 0$

We test each H_0 by again testing whether our sample represents a population where there is zero relationship—so we again examine the sampling distribution

FIGURE 11.9 H_0 Sampling Distribution of r Where $\rho = 0$ for One-Tailed Test

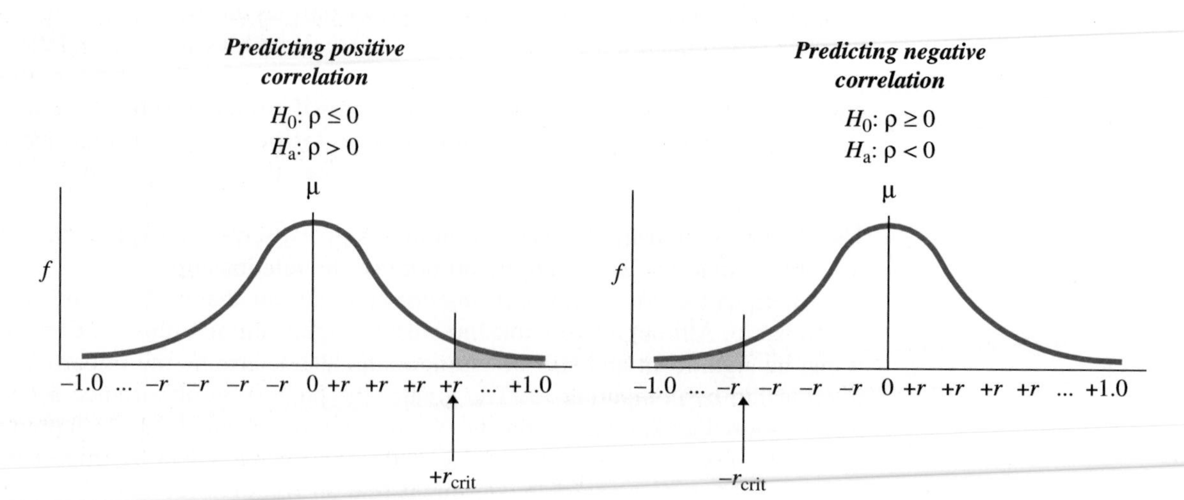

for $\rho = 0$. From the r-tables, we find the one-tailed critical value for our *df* and α, and we set up one of the sampling distributions shown in Figure 11.9.

If we predict a positive correlation, we use the left-hand sampling distribution. Our r_{obt} is significant if it is positive and falls beyond the positive r_{crit}. If it is, then we accept H_a (that the sample represents a ρ greater than 0).

Conversely, if we predict a negative correlation, we use the right-hand sampling distribution in Figure 11.9. Here, r_{obt} is significant if it is negative and falls beyond the negative r_{crit}. Again, if it is significant, we accept H_a (here, that the sample represents a ρ less than 0).

If r_{obt} is not beyond the appropriate r_{crit}, then r_{obt} is not significant, and we conclude that the study failed to demonstrate the predicted relationship.

Significance Testing of the Spearman r_s and the Point-Biserial r_{pb}

We also perform a similar form of significance testing when we compute the Spearman r_s or the point-biserial correlation coefficient, r_{pb}. These correlations describe relationships in our *sample,* but—that's right—perhaps they merely reflect sampling error. Perhaps if we computed the correlation in the population, we would find that our r_s actually represents a population correlation, symbolized by ρ_s, that is 0. Likewise, perhaps our r_{pb} actually represents a population correlation, symbolized by ρ_{pb}, that is 0. Therefore, before we can conclude that either of these sample correlations represents a relationship in nature, we must perform the appropriate hypothesis testing.

To test each sample correlation coefficient, we perform the following steps:

1. Set alpha: how about .05?

2. Consider the assumptions of the test. The Spearman r_s assumes that we have a random sample of pairs of *ranked* (ordinal) scores. The point-biserial r_{pb}

assumes that we have random scores from one dichotomous variable and one interval or ratio variable. (Note: Because of the type of data involved and the lack of parametric assumptions, technically the procedures for testing r_s and r_{pb} are nonparametric procedures. But because they are otherwise the same as for r, we discuss them here.)

3. Create the statistical hypotheses. We can test either the one- or the two-tailed hypothesis we had with ρ, except now we substitute ρ_s or ρ_{pb}.

The only new aspect in testing r_s or r_{pb} is how we conceptualize their respective sampling distributions.

Significance testing of r_s For significance testing of r_s, we have a new family of differently shaped sampling distributions and a different table of critical values. Table 4 in Appendix D, entitled "Critical Values of the Spearman Rank-Order Correlation Coefficient," contains the critical values of r_s for one- and two-tailed tests for α levels of .05 and .01. We obtain the critical value from this table in the same manner as we did from previous tables, except that we use N, *not* degrees of freedom.

> ***STAT ALERT*** The critical value of r_s is obtained using N, the number of pairs of scores in the sample.[2]

As an example, in Chapter 7 we correlated the nine rankings of each of two observers and found that $r_s = +.85$. Is this value significant? We performed this correlation assuming the observers' rankings would agree, so we were predicting a positive correlation. Therefore, we have a one-tailed test with the hypotheses H_0: $\rho_s \leq 0$ and H_a: $\rho_s > 0$. From Table 4, with $\alpha = .05$ and $N = 9$, the critical value for the one-tailed test is $+.600$. We can envision the H_0 sampling distribution of r_s when $N = 9$ as shown in Figure 11.10.

FIGURE 11.10 One-Tailed H_0 Sampling Distribution of Values of r_s When H_0 Is $\rho_s = 0$

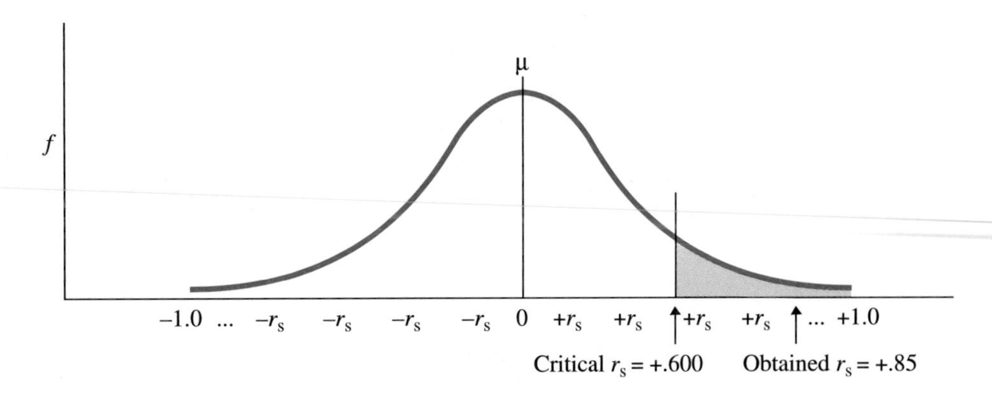

[2]Table 4 contains critical values for N up to 30. When N is greater than 30, transform r_s to a z-score using the formula $z_{obt} = (r_s)(\sqrt{N - 1})$. For $\alpha = .05$, the two-tailed $z_{crit} = \pm 1.96$ and the one-tailed $z_{crit} = 1.645$.

We determine whether r_s is significant in the same way we have determined significance for previous statistics. In our example, the obtained r_s of $+.85$ is beyond the critical value of $+.600$, so we reject H_0: an r_s of $+.85$ is too unlikely to represent the population where ρ_s is zero or less than zero, so we accept H_a (that $\rho_s > 0$). We have a significant r_s, and we estimate that the value of the correlation in the population of such rankings, ρ_s, is around $+.85$. We report our results as

$$r_s(9) = +.85, \quad p < .05$$

(Note that the N of the sample is given in parentheses.)

In a different study, with different predictions, we might have performed a one-tailed test using the other tail, or a two-tailed test (with the appropriate two-tailed critical value).

Significance testing of r_{pb} We test r_{pb} using the same logic as above. The H_0 sampling distributions of r_{pb} are identical to the distributions for the Pearson r, so critical values of r_{pb} are obtained from Table 3 in Appendix D, "Critical Values of the Pearson Correlation Coefficient." Again we use degrees of freedom, which equals $N - 2$.

> ***STAT ALERT*** The critical value of r_{pb} is found in Appendix D, Table 3, for df equal to $N - 2$, where N is the number of pairs of scores.

As an example, in Chapter 7 we computed the r_{pb} between subjects' scores on the dichotomous variable of gender (male or female) and their scores on a personality test. We obtained $r_{pb} = +.46$, with $N = 10$. Say we perform a two-tailed test with the hypotheses H_0: $\rho_{pb} = 0$ and H_a: $\rho_{pb} \neq 0$. From Table 3, with $\alpha = .05$ and $df = 8$, the critical value for the two-tailed test is $\pm.632$. We can envision the sampling distribution of r_{pb} as shown in Figure 11.11. Our obtained r_{pb} of $+.46$ is not beyond the critical value, so we do not reject H_0 and conclude

FIGURE 11.11 H_0 Sampling Distribution of r_{pb} When H_0 Is $\rho_{pb} = 0$

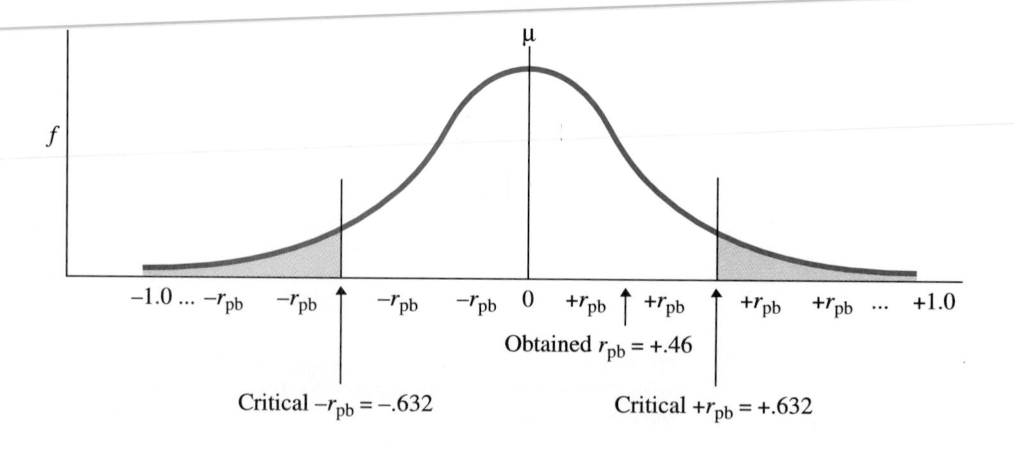

that our sample may represent the population where ρ_{pb} equals zero. We report our results as

$$r_{pb}(8) = +.46, \quad p > .05$$

In a different study, if we had predicted only a positive or only a negative correlation, then we would perform a one-tailed test. Remember, though, as discussed in Chapter 7, *we* determine whether r_{pb} is positive or negative by how we arrange the categories of the dichotomous variable. Therefore, one-tailed tests of r_{pb} must be set up so that they are consistent with how we arrange the data.

MAXIMIZING THE POWER OF A STATISTICAL TEST

In Chapter 10 we discussed the *power* of a statistical procedure, which is the probability of *not* committing a Type II error. We avoid this error by rejecting H_0 when H_0 is false (concluding that the data represent a relationship when it actually does exist). Any discussion of power assumes we have the situation where H_0 *is* false, so our job is to reject it. Therefore, we maximize the power of a statistical procedure by maximizing the probability that we will obtain significant results. To do that, we make various decisions about how to conduct a study, each of which translates into maximizing the absolute size of the obtained statistic relative to the critical value. The larger the obtained value, the more likely it is to fall beyond the critical value and thus the more likely it is to be significant, causing us to reject H_0 when it is false.[3]

The following sections discuss how researchers maximize power in the *t*-test and in correlations.

Maximizing Power in the *t*-Test

To maximize power, we want to maximize the size of t_{obt} relative to t_{crit}. Looking at the formula

$$t_{obt} = \frac{\overline{X} - \mu}{\frac{s_X}{\sqrt{N}}}$$

we see that there are three aspects of a study that increase power.

1. *The greater the differences produced by changing the independent variable, the greater the power.* Remember our cramming study? The greater the difference between the sample mean for crammers and the μ for noncrammers, the greater our power. Conceptually, the larger the difference between test scores with cramming and scores without cramming, the less likely we are to erroneously accept that the difference is due to sampling error. Mathematically, in the above formula, a larger difference between $\overline{X}$ and μ produces

[3]More advanced textbooks give computational procedures for determining the amount of power that is present in a given study.

a larger numerator. All other things (N and s_X) being equal, dividing into a larger numerator results in a larger t_{obt}. Therefore, when we design our research, we try to select conditions of the independent variable that are substantially different from one another. Thus, we would want to have our cramming subjects cram "hard" and for a sufficient amount of time, and have the noncrammers definitely not cram at all. Then we are likely to see a big difference in test scores when comparing the two.

2. *The smaller the variability of the raw scores, the greater the power.* The t-test measures the difference between $\overline{X}$ and μ relative to the standard error of the mean, $s_{\overline{X}}$. The size of the standard error is determined by the variability (the standard deviation or variance) of the raw scores. Conceptually, the smaller the raw score variability, the more easily we can detect a relationship. For example, if every student who crammed obtained a score of exactly 65 on the test and those who did not cram consistently scored exactly 75, we could more easily detect the relationship, because it would be a clearer, more consistent one. In fact, this would produce a perfectly consistent relationship, which would be very hard to mistake as resulting from sampling error. Mathematically, in the above formula, the smaller the variability of the raw scores, s_X, the smaller the denominator. Dividing by a smaller denominator produces a larger t_{obt}. Therefore, when conducting research, we want to test our subjects and measure their scores in a very consistent way so that we minimize the variability of the scores within a sample. (In our cramming study, we would want to control anything that might produce differences in test scores. Therefore, all subjects would have the same amount of time to take the test, would take it in the same room with the same observer, and so on.)

3. *For small samples, the greater the N, the greater the power.* Conceptually, the larger the N, the more accurately we will represent the true situation in the population and the less likely we are to make any type of error. Mathematically, the size of N influences the results in two ways. First, in the above formula we divide s_X by $\sqrt{N}$. Dividing s_X by a larger number produces a smaller final denominator in the t-test, which results in a larger t_{obt}. Second, a larger N produces a larger df, and the larger the df, the smaller the absolute value of t_{crit}. The smaller the t_{crit}, the more likely that the t_{obt} lies beyond it, so the more likely that t_{obt} is significant. Therefore, in research, we select as large an N as we can.

   Notice, however, that we are discussing *small* samples. Generally, an N of at least 30 is needed for minimal power, and increasing N up to 121 adds substantially to the power. However, an N such as 500 is generally not substantially more powerful than an N of, say, 450.

Maximizing Power in the Correlation Coefficient

To maximize power when testing a correlation coefficient, we essentially focus on the same three factors as we do to maximize power in the t-test, except that we want to maximize the size of the correlation coefficient.

1. *Avoiding the restriction of range problem maximizes power.* One way to obtain a large correlation coefficient is to avoid the restriction of range problem. Recall from Chapter 7 that restriction of range occurs when we have a small range of scores on either the X or the Y variable. A restricted range produces a sample correlation that is smaller than it would be without a restricted range. Thus, for greatest power, we want to avoid a restriction of range. This boils down to again saying that we want large differences on the X variable that are consistently paired with Y scores. Therefore, as with the t-test, we'd want large differences in cramming in our study, starting with subjects who did not cram at all and ranging all the way up to those who crammed very much. This will maximize the size of the correlation coefficient, so we are less likely to erroneously retain the null hypothesis. Mathematically, a larger obtained correlation coefficient is more likely to fall beyond the critical value and thus be significant.

2. *Minimizing the variability of the* Y *scores at each* X *increases power.* Recall that a correlation coefficient describes the extent to which one value of Y is consistently paired with one and only one value of X, and that variability in Y works against the strength of the relationship. Therefore, we want to minimize the variability so that subjects scoring a particular X all score at or close to the same Y score. Again the goal is to strive for perfectly consistent responses from subjects. Thus, as with the t-test, subjects are tested in a very consistent fashion so that no extraneous variables occur that might produce variability in their test scores. This maximizes the size of the correlation coefficient so that we are less likely to erroneously retain the null hypothesis. Mathematically, again a larger coefficient is more likely to be significant.

3. *Increasing the N of small samples maximizes power.* As with the t-test, the larger the sample N, the greater the probability that we will accurately represent the population and avoid any error. In addition, with a larger N and thus a larger df, the critical value is smaller, and thus a given coefficient is more likely to be significant.

FINALLY

I hope you found this chapter rather boring—not because it *is* boring, but because, for each statistic, we performed virtually the same operations. In significance testing of *any* statistic, we ultimately do and say the same things. In all cases, if the obtained statistic is out there far enough in the H_0 sampling distribution, it is too unlikely for us to accept as occurring if H_0 is true, so we reject that H_0 is true. Any H_0 implies that the sample does not represent the predicted relationship, so by rejecting H_0 we become more confident that the data do represent the predicted relationship, with the probability of a Type I error equal to $p < \alpha$.

CHAPTER SUMMARY

1. The t-test is used to test the significance of a single sample mean when (a) there is one random sample of interval or ratio data, (b) the raw score population is a normal distribution for which the mean is the appropriate measure of central tendency, and (c) the standard deviation of the raw score population is estimated by computing s_X using the sample data.

2. A t-distribution is a theoretical sampling distribution of all possible values of t when a raw score population is infinitely randomly sampled using a particular N. The shape of a t-distribution is determined by the *degrees of freedom,* or *df.* The appropriate t-distribution for a single-sample t-test is the distribution identified by $N - 1$ degrees of freedom.

3. In *point estimation,* the value of the population parameter is assumed to equal the value of the corresponding sample statistic. Because the sample statistic probably contains sampling error, any point estimate is likely to be incorrect.

4. In *interval estimation,* the value of a population parameter is assumed to lie within a specified interval. Interval estimation is performed by computing a confidence interval. The *confidence interval for a single population μ* describes an interval containing values of μ, any one of which our sample mean is likely to represent. The interval is computed by determining the highest and lowest values of μ that are not significantly different from the sample mean. Our confidence that the interval actually contains the value of μ represented by the sample $\overline{X}$ is equal to $(1 - \alpha)100$.

5. The *sampling distribution of a correlation coefficient* is a frequency distribution showing all possible values of the coefficient that occur when samples of size N are drawn from a population where the correlation coefficient is zero.

6. Significance testing of the *Pearson r* assumes that (a) we have a random sample of pairs of scores from two interval or ratio variables and (b) the Y scores are normally distributed at each value of X and the X scores are normally distributed at each value of Y.

7. Significance testing of the *Spearman* r_s assumes that we have a random sample of pairs of ranked-order (ordinal) scores.

8. Significance testing of the *point-biserial* r_{pb} assumes that we have a random sample of pairs of scores where one score is from a dichotomous variable and one score is from an interval or ratio variable.

9. Anything that increases the probability of rejecting H_0 increases the *power* of a statistical test. Therefore, anything that increases the size of the obtained value relative to the critical value increases power.

10. We maximize the power of the t-test by (a) creating large differences when changing the conditions of the independent variable, (b) minimizing the variability of the scores in a sample, and (c) increasing the N of small samples.

11. We maximize the power of a correlation coefficient by (a) avoiding the restriction of range problem, (b) minimizing the variability in Y at each X, and (c) increasing the N of small samples.

PRACTICE PROBLEMS

(Answers for odd-numbered problems are provided in Appendix E.)

1. A scientist has conducted a single-sample experiment.

   a. What two parametric procedures are available to her?

   b. What is the deciding factor for selecting between them?

2. What are the other assumptions of the t-test?

3. *a.* What is the difference between $s_{\bar{X}}$ and $\sigma_{\bar{X}}$?

   b. How is their use the same?

4. *a.* Why must we obtain different values of t_{crit} when samples have different N's?

   b. What must we compute prior to finding t_{crit}?

5. You wish to compute the 95% confidence interval for a sample with a *df* of 80. Using interpolation, determine the value of t_{crit} you should use.

6. Say you have a sample mean of 44 in a study.

   a. How do you estimate the corresponding μ using point estimation? ✓

   b. What does a confidence interval computed for this μ tell you?

   c. Why is computing a confidence interval a better approach than using a point estimate?

7. *a.* What is power?

   b. Why is power especially important when we fail to reject H_0?

   c. What do researchers do to avoid this dilemma?

8. Poindexter performed a two-tailed experiment in which $N = 20$. He couldn't find his t-tables, but somehow he remembered that the t_{crit} at $df = 10$. He decided to compare his t_{obt} to this t_{crit}. Why is this a correct or incorrect approach? (*Hint:* Consider whether t_{obt} turns out to be significant or nonsignificant at this t_{crit}.)

9. You wish to determine whether this textbook is beneficial or detrimental to students learning statistics. On a national statistics exam, $\mu = 68.5$ for students who have used other textbooks. A random sample of students who have used this book has the following scores:

   64 69 92 77 71 99 82 74 69 88

   a. What are H_0 and H_a for this study?

   b. Compute t_{obt}.

   c. With $\alpha = .05$, what is t_{crit}?

   d. What do you conclude about the use of this book?

   e. Compute the confidence interval for μ.

10. A researcher predicts that smoking cigarettes decreases a person's sense of smell. On a standard test of olfactory sensitivity, the μ for nonsmokers is 18.4. By giving this test to a random sample of subjects who smoke a pack a day, the researcher obtains the following scores:

    16 14 19 17 16 18 17 15 18 19 12 14

    a. What are H_0 and H_a for this study?

    b. Compute t_{obt}.

    c. With $\alpha = .05$, what is t_{crit}?

    d. What should the researcher conclude about this relationship?

    e. Compute the confidence interval for μ.

11. Foofy conducts a study to determine if hearing an argument in favor of an issue alters subjects' attitudes toward the issue one way or the other. She presents a thirty-second speech in favor of an issue to 8 subjects. In a national survey, the mean attitude score in favor of the issue was $\mu = 50$. Testing her subjects with this survey, she obtains the following scores:

    10 33 86 55 67 60 44 71

    a. What are H_0 and H_a?

    b. What is the value of t_{obt}?

    c. With $\alpha = .05$, what is the value of t_{crit}?

    d. What are the statistical results?

    e. If appropriate, compute the confidence interval for μ.

    f. Using the preceding statistics, what conclusions should Foofy draw about the relationship between strong arguments and their impact on attitudes?

12. For the study in problem 11,

    a. What statistical principle should Foofy be concerned with?

    b. Identify three problems with her study from a statistical perspective.

    c. Why would correcting the problems identified in part *b* improve her study?

13. Poindexter examined the relationship between the quality of sneakers worn by volleyball players and their average number of points scored per game. Studying 20 subjects who owned sneakers of good to excellent quality, he computed $r = +.41$. Without further ado, he immediately claimed to have support for the notion that better-quality sneakers are related to better performance on a somewhat consistent basis. He then computed r^2 and the regression equation. Do you agree or disagree with his approach? Why?

14. Eventually, for the study in problem 13, Poindexter reported that $r(18) = +.41$, $p > .05$.

a. What should he conclude about this relationship?

b. What other computations should he perform to describe the relationship in these data?

c. What statistical principle should he be concerned with?

d. What aspects of the study can he improve to better deal with this principle?

e. What will correcting these things from part *d* do in regard to his finding significant results?

15. A scientist suspects that as his subjects' stress level changes, so does the amount of their impulse buying. He collects data from 72 subjects and obtains an r of $+.38$.

a. What are H_0 and H_a?

b. With $\alpha = .05$, what is r_{crit}?

c. What are the statistical results of this study?

d. What conclusions should be drawn about the relationship in the population?

e. What other calculations should be performed to describe the relationship in these data?

16. Foofy conducts a correlational study examining the relationship between an individual's physical strength and his or her college grade point average. Using a computer, she computes the correlation for a sample of 2,000 subjects and obtains $r(1998) = +.08$, $p < .0001$. She claims she has uncovered a useful tool for predicting which college applicants are likely to succeed academically. Do you agree or disagree? Why?

17. A researcher investigates the relationship between handedness and strength of personality. She tests 42 subjects, assigning left-handers a score of 1 and right-handers a score of 2. She obtains a correlation coefficient of $+.33$ between subjects' handedness and their scores on a personality test.

a. Which type of correlation coefficient did she compute?

b. What are H_0 and H_a?

c. With $\alpha = .05$, what is the critical value?

d. What should the researcher conclude about the strength of the relationship in the population? What should she conclude about the direction of the relationship in the population?

e. Are the results of this study relatively useful?

18. A newspaper article claims that for all U.S. colleges, the academic rank of the college is negatively related to the rank of its football team. You examine the accuracy of this claim. From a sample of 28 colleges, you obtain a correlation coefficient of $-.32$.

a. Which type of correlation coefficient did you compute?

b. What are H_0 and H_a?

c. With $\alpha = .05$, what is the critical value?

d. What are the statistical results?

e. What should you conclude about the accuracy of the newspaper claim for all colleges in the United States?

f. In trying to determine a particular school's academic ranking in your sample, how important is it that you look at the school's football ranking?

SUMMARY OF FORMULAS

1. *The definitional formula for the single sample* t*-test is*

$$t_{obt} = \frac{\overline{X} - \mu}{s_{\overline{X}}}$$

The value of $s_{\overline{X}}$ is computed as

$$s_{\overline{X}} = \frac{s_X}{\sqrt{N}}$$

and s_X is computed as

$$s_X = \sqrt{\frac{\Sigma X^2 - \frac{(\Sigma X)^2}{N}}{N - 1}}$$

2. *The computational formulas for the single sample* t*-test are*

$$t_{obt} = \frac{\overline{X} - \mu}{\sqrt{\frac{s_X^2}{N}}} \quad \text{and} \quad t_{obt} = \frac{\overline{X} - \mu}{\sqrt{(s_X^2)\left(\frac{1}{N}\right)}}$$

where s_X^2 is the estimated variance computed for the sample.

Values of t_{crit} *are found in Table 2 of Appendix D, "Critical Values of* t*," for* df $= N - 1$.

3. *The computational formula for a confidence interval for a single population μ is*

$$(s_{\bar{X}})(-t_{\text{crit}}) + \bar{X} \leq \mu \leq (s_{\bar{X}})(+t_{\text{crit}}) + \bar{X}$$

where t_{crit} is the two-tailed value for $df = N - 1$ and $\bar{X}$ and $s_{\bar{X}}$ are computed using the sample data.

4. *To test the significance of a correlation coefficient,* compare the obtained correlation coefficient to the critical value.

a. *Critical values of r* are found in Appendix D, Table 3, for $df = N - 2$, where N is the number of pairs of scores in the sample.

b. *Critical values of r_s* are found in Appendix D, Table 4, for N, the number of pairs in the sample.

c. *Critical values of r_{pb}* are found in Appendix D, Table 3, for $df = N - 2$, where N is the number of pairs in the sample.

12

Significance Testing of Two Sample Means: The *t*-Test

To understand the upcoming chapter:

- From Chapter 4, recall how we graph experimental results.
- From Chapters 5 and 8, recall how we conceptualize the proportion of variance accounted for.
- From Chapter 7, understand how we use the point-biserial correlation coefficient, r_{pb}.
- And, of course, don't forget what you've learned about inferential statistics.

Then your goals in this chapter are to learn:

- The logic of a two-sample experiment.
- The difference between independent samples and related samples.
- How to perform the independent-samples and related-samples *t*-tests.
- How to compute a confidence interval for the difference between two μ's and for the μ of difference scores.
- How r_{pb} is used to describe effect size in the two-sample experiment.

So far we've limited our discussions to statistical procedures for a single-sample experiment. In this chapter we discuss the major parametric statistical procedures used when an experiment involves *two* samples. These procedures center around the two-sample *t*-test. As the name implies, this test is similar to the single-sample *t*-test we saw in Chapter 11. We will again determine whether we have a significant relationship, but the characteristics of a two-sample design dictate that we use slightly different formulas to do so. In addition, we'll discuss procedures for more completely describing any significant relationship we may find.

MORE STATISTICAL NOTATION

It is time to pay very close attention to subscripts. We will compute the mean of each of our two samples, identifying one as $\overline{X}_1$ and the other as $\overline{X}_2$. Likewise, we will compute an estimate of the variance of the raw score population represented by each sample, identifying the variance from one sample as s_1^2 and the variance from the other sample as s_2^2. Recall that we compute the estimated population variance using the general formula

$$s_X^2 = \frac{\Sigma X^2 - \frac{(\Sigma X)^2}{N}}{N - 1}$$

Finally, instead of using N to indicate the number of scores in a sample, we will use the lowercase n with a subscript to indicate the number of scores in each sample. Thus, n_1 is the number of scores in Sample 1, and n_2 is the number of scores in Sample 2.

UNDERSTANDING THE TWO-SAMPLE EXPERIMENT

To perform the single-sample experiments discussed in previous chapters, we must already know the value of μ of the population of raw scores under one condition of our independent variable. However, I have another confession: Usually behavioral researchers study variables for which they do not know any values of μ ahead of time. In such cases, they may design a two-sample experiment. In a two-sample experiment, we measure subjects' scores under two conditions of the independent variable. Condition 1 produces sample mean $\overline{X}_1$ that represents μ_1, the population μ we would find if we tested everyone in the population under Condition 1. Condition 2 produces sample mean $\overline{X}_2$ that represents μ_2, the population μ we would find if we tested everyone in the population under Condition 2. A possible outcome from such an experiment is shown in Figure 12.1. For statistical purposes, the specific values of μ_1 and μ_2 are not our first concern. What *is* important is that μ_1 and μ_2 are *different* from each other. If the sample means represent a different population for each condition, then our experiment has demonstrated a relationship in nature: as we change the conditions of the independent variable, the scores in the population change in a consistent fashion.

Of course, the problem is that even though we may have different sample means, the independent variable may not really work, and so the relationship may not really exist in the population. Therefore, if we could test the entire population, we might actually find the same population of scores under both conditions of our independent variable. In Figure 12.1, for example, you might find only the lower or upper distribution, or you might find one in the middle.

FIGURE 12.1 Relationship in the Population in a Two-Sample Experiment

As the conditions change, the population tends to change in a consistent fashion.

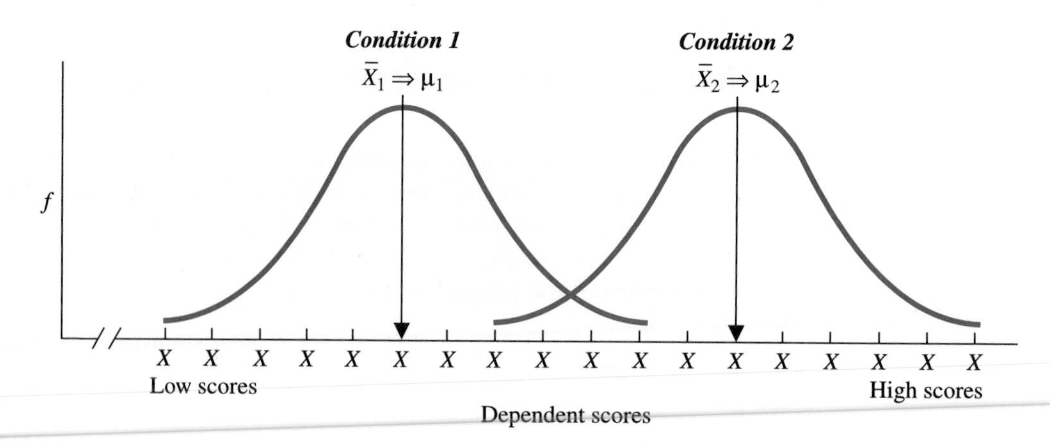

Then the conditions of the independent variable literally would not make a difference in the population and there would be only one value of μ: you could call it μ_1 or μ_2, it wouldn't matter because it would be the *same* μ. Thus, it is possible that our different sample means in Figure 12.1 poorly represent the same population μ and that we are being misled by sampling error. After all, if we draw two random samples from one population, we hardly expect both sample means to equal each other *and* the population μ. Therefore, before we make any conclusions about our experiment, we must perform hypothesis testing to determine whether the difference between our sample means is likely to merely reflect sampling error.

Hypothesis testing for two sample means is performed with the two-sample *t*-test. However, as we shall see, there are two distinctly different ways that we can create our samples, so there are two very different versions of the *t*-test, one for independent samples and one for related samples.

STAT ALERT There are two versions of the two-sample *t*-test, depending on how the researcher creates the samples.

First we will discuss the *t*-test for independent samples.

THE *t*-TEST FOR TWO INDEPENDENT SAMPLES

The **t-test for independent samples** is the parametric procedure used for significance testing when we have sample means from two *independent* samples. Two samples are independent when we randomly select and assign a subject to a sample, without regard to who else has been selected for either sample. Then the samples are composed of independent events, which, as we saw in Chapter 9, means that the probability of a particular score's occurring in one sample is not influenced by a particular score's occurring in the other sample.

Here is a study that calls for the independent samples t-test. Say that we wish to study the effects of hypnosis on memory. We know that hypnosis tends to help people recall past events, but we want to determine how well people recall an event when they are mildly hypnotized compared to when they are deeply hypnotized. We will randomly select two samples of subjects, who will each read a story containing 25 details. Later, one sample will be mildly hypnotized and then recall the story, while the other sample will be deeply hypnotized and then recall the story. Each subject's score will be the number of details that he or she recalls from the story. Thus, the conditions of our independent variable are mild and deep hypnosis, and our dependent variable is the number of details subjects recall.

By now you know the routine: (1) we check the assumptions of the statistical procedure and create our statistical hypotheses, (2) we set up and perform the statistical test, and (3) if our results are significant, we describe the relationship we have demonstrated.

Assumptions of the *t*-Test for Two Independent Samples

In addition to requiring independent samples, the independent samples t-test assumes the following.

1. The two random samples of dependent scores measure an interval or ratio variable.
2. The population of raw scores represented by each sample forms a normal distribution, and the mean is the appropriate measure of central tendency. (If each sample n is greater than 30, the populations need only form roughly normal distributions.)
3. We do not know the variance of any raw score population and must estimate it from the sample data.
4. And here's a new one: the populations represented by our samples have homogeneous variance. **Homogeneity of variance** means that if we could compute the true variance of the populations represented by our samples, the value of σ_X^2 in one population would equal the value of σ_X^2 in the other population. (Although it is not required that each sample have the same n, the more the n's differ from each other, the more important it is to have homogeneity of variance.[1])

Our hypnosis study meets the above assumptions, because we will randomly assign subjects to either condition, and our recall scores are ratio scores that we think form approximately normal distributions with homogeneous variance.

Now we derive our statistical hypotheses.

Statistical Hypotheses for the *t*-Test for Two Independent Samples

Depending on our experimental hypotheses, we may have a one- or a two-tailed test. For our hypnosis study, say that we have no reason to predict whether deep

1. The next chapter introduces a test for determining whether we can assume homogeneity of variance.

hypnosis will produce higher or lower recall scores than mild hypnosis. This is a two-tailed test because we do not predict the direction of the relationship: we merely predict that the sample from each condition represents a different population of recall scores, having a different value of μ. Therefore, our experimental hypotheses are that the experiment either does or does not demonstrate a relationship between degree of hypnosis and recall.

First, we write our alternative hypothesis. Because we do not predict which of the two μ's (and corresponding $\overline{X}$'s) will be larger, the predicted relationship exists if one population mean (μ_1) is larger or smaller than the other (μ_2)—in other words, if μ_1 does not equal μ_2. We could state the alternative hypothesis as H_a: $\mu_1 \neq \mu_2$, but there is a better way to do it. If the two μ's are not equal, then their *difference* is not equal to zero. Thus, the two-tailed alternative hypothesis for our study is

$$H_a: \mu_1 - \mu_2 \neq 0$$

H_a implies that the means from our two conditions each represent a different population of recall scores, having a different μ.

Of course, we must also deal with our old nemesis, the null hypothesis. Perhaps there is no relationship, so if we tested the entire population under the two conditions, we would find the same population μ. Any difference in our sample means may be due to sampling error, and we can call μ either μ_1 or μ_2—it doesn't matter, because μ_1 *equals* μ_2. We could state our null hypothesis as H_0: $\mu_1 = \mu_2$, but again there is a better way. If the two μ's are equal, then their difference is zero. Thus, our two-tailed null hypothesis is

$$H_0: \mu_1 - \mu_2 = 0$$

H_0 implies that both sample means represent the same population of recall scores, having the same μ.

Notice that we derived these hypotheses without specifying the value of either μ, so we have the same hypotheses regardless of the dependent variable we are measuring. Therefore, the above hypotheses are the two-tailed hypotheses for *any* independent samples t-test when we are testing whether there is no relationship in the population.

As usual, we test our null hypothesis. To see how this is done, let's jump ahead and say that our hypnosis study produced the sample means in Table 12.1. We can summarize our results by looking at the *difference* between these means. Apparently, changing the condition of the independent variable from mild hypnosis to deep hypnosis results in a difference in mean recall scores of 3 points. However, H_0 says that these sample means represent the same population μ. We always test H_0 by

TABLE 12.1 Sample Means from Hypnosis Study

Degree of hypnosis	*Mean number of details recalled*
Mild hypnosis	20
Deep hypnosis	23

finding the probability of obtaining our results when H_0 is true. Here we will determine the probability of obtaining a *difference* of 3 between our $\overline{X}$'s when both sample means actually represent the same population μ. H_0 says that the difference between μ_1 and μ_2 is zero, so if it weren't for sampling error, the difference between the sample means would be zero. Therefore, to test H_0 we determine whether a difference of 3 is significantly different from zero. If it is, then there is a significant difference *between* the two sample means: sampling error is too unlikely to produce such different sample means if they represent the same population μ. Then we can be confident that the sample means represent two different μ's.

> ***STAT ALERT*** In the independent-samples t-test, we always test whether the difference between $\overline{X}_1$ and $\overline{X}_2$ is significantly different from the difference between μ_1 and μ_2 described by H_0.

As usual, to test H_0 we need a sampling distribution.

The Sampling Distribution When H_0 Is $\mu_1 - \mu_2 = 0$

Any sampling distribution is the distribution of sample statistics when H_0 is true. Here we can think of creating our sampling distribution as follows. Using the same n's as in our study, we select *two* random samples at a time from one raw score population. We compute the two sample means and arbitrarily subtract one from the other. The result is the *difference between the means,* which we symbolize by $\overline{X}_1 - \overline{X}_2$. We do this an infinite number of times and plot a frequency distribution of all the differences. We then have the **sampling distribution of differences between the means,** which is the distribution of all possible differences between two means when they are drawn from the raw score populations described by H_0. We can envision the sampling distribution of differences between the means as shown in Figure 12.2.

FIGURE 12.2 Sampling Distribution of Differences Between Means When H_0: $\mu_1 - \mu_2 = 0$

The mean of this distribution is zero. Larger positive differences are to the right, and larger negative differences are to the left.

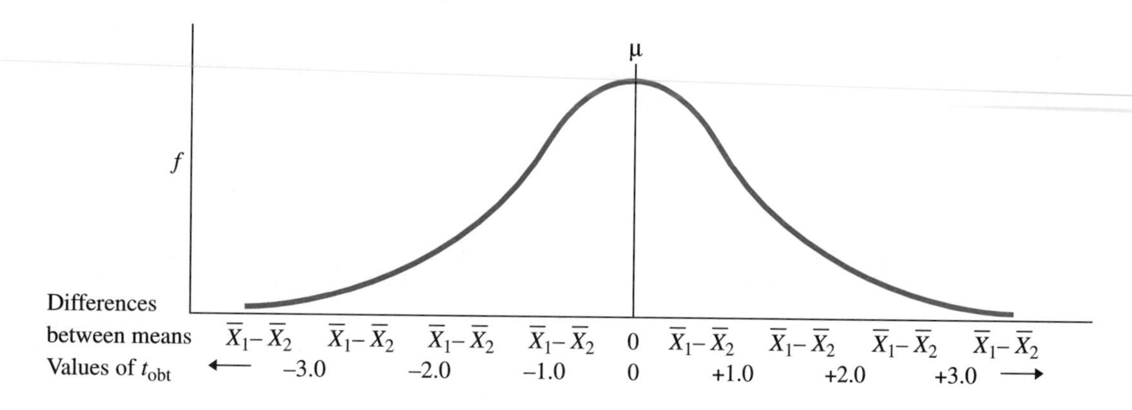

This distribution is just like any other sampling distribution except that we have plotted the *differences* between two sample means along the X axis, each labeled $\overline{X}_1 - \overline{X}_2$. As usual, the mean of the sampling distribution is the value stated in H_0, and here H_0 is $\mu_1 - \mu_2 = 0$. The mean of the distribution is zero because most often each sample mean will equal the raw score population μ, in which case the difference between the means will be zero. However, sometimes by chance both sample means will not equal μ or each other. Depending on whether $\overline{X}_1$ or $\overline{X}_2$ is larger, the difference will be greater than zero (positive) or less than zero (negative). The larger the absolute difference between the means, the farther into either tail of the distribution the difference falls, so the less frequent and thus the less likely such a difference is when H_0 is true.

To test H_0, we simply determine where the difference between our sample means lies on this sampling distribution. Our model of the sampling distribution of differences between the means is the t-distribution. We will locate the difference between our means on the sampling distribution by computing t_{obt}. As shown in Figure 12.2 the larger the absolute value of $\pm t_{obt}$, the less likely it is that the corresponding difference between the means occurs when H_0 is true.

Of course, we must compute t_{obt}.

Computing the *t*-Test for Two Independent Samples

We will arbitrarily label the deep hypnosis condition as Sample 1, so its mean, variance, and n are $\overline{X}_1$, s_1^2, and n_1, respectively. For the mild hypnosis condition, we have $\overline{X}_2$, s_2^2, and n_2.

In the previous chapter we computed t_{obt} by performing three steps: we computed the estimated variance of the raw score population, then we computed the estimated standard error of the sampling distribution, and then we computed t_{obt}. In performing the two-sample t-test, we complete the same three steps.

Estimating the population variance First we compute s_X^2 for each sample, using the familiar formula given at the beginning of this chapter. Each s_X^2 estimates the population variance, but each estimate may contain sampling error. (Because of this, if s_1^2 does not equal s_2^2, it does not necessarily violate our assumption of homogeneity of variance in the population.) To obtain the best estimate of the population variance, we compute a weighted average of the two values of s_X^2. Each variance is weighted based on the size of *df* in the sample. The weighted average of the sample variances is called the *pooled variance* (we pool our resources). The symbol for the pooled variance is s_{pool}^2.

THE COMPUTATIONAL FORMULA FOR THE POOLED VARIANCE IS

$$s_{pool}^2 = \frac{(n_1 - 1)s_1^2 + (n_2 - 1)s_2^2}{(n_1 - 1) + (n_2 - 1)}$$

TABLE 12.2 Data from the Hypnosis Study

	Sample 1: subjects under deep hypnosis	***Sample 2: subjects under mild hypnosis***
Mean details recalled	$\overline{X}_1 = 23$	$\overline{X}_2 = 20$
Number of subjects	$n_1 = 17$	$n_2 = 15$
Sample variance	$s_1^2 = 9.0$	$s_2^2 = 7.5$

This formula says that we multiply the s_X^2 from each sample times $n - 1$ for that sample, then add the results together and divide by the sum of $(n_1 - 1) + (n_2 - 1)$.

As shown in Table 12.2, for our hypnosis study s_1^2 is 9.0 and n_1 is 17; s_2^2 is 7.5 and n_2 is 15. Filling in the above formula, we have

$$s_{\text{pool}}^2 = \frac{(17 - 1)9.0 + (15 - 1)7.5}{(17 - 1) + (15 - 1)}$$

In the numerator, 16 times 9 is 144, and 14 times 7.5 is 105. In the denominator, 16 plus 14 is 30. Now we have

$$s_{\text{pool}}^2 = \frac{144 + 105}{30} = \frac{249}{30} = 8.30$$

We have $s_{\text{pool}}^2 = 8.30$. Thus, we estimate that the variance of any of the populations of recall scores represented by our samples is 8.30.

With the value of s_{pool}^2, we can compute the standard error of the sampling distribution.

Computing the standard error of the difference The standard error of the sampling distribution of differences between the means is called the standard error of the difference. The **standard error of the difference** is the estimated standard deviation of the sampling distribution of differences between the means, indicating how spread out the values of $(\overline{X}_1 - \overline{X}_2)$ are when the distribution is created using samples having our n and our value of s_{pool}^2. The symbol for the standard error of the difference is $s_{\overline{X}_1 - \overline{X}_2}$. (The subscript $\overline{X}_1 - \overline{X}_2$ indicates that we are dealing with differences between pairs of means.)

In the previous chapter, we saw that for the single-sample *t*-test, a formula for the standard error of the mean was

$$s_{\overline{X}} = \sqrt{(s_X^2)\left(\frac{1}{N}\right)}$$

The formula for the standard error of the difference is very similar.

THE DEFINITIONAL FORMULA FOR THE STANDARD ERROR OF THE DIFFERENCE IS

$$s_{\bar{X}_1-\bar{X}_2} = \sqrt{(s^2_{\text{pool}})\left(\frac{1}{n_1} + \frac{1}{n_2}\right)}$$

To compute $s_{\bar{X}_1-\bar{X}_2}$, it is easiest if you first reduce the fractions $1/n_1$ and $1/n_2$ to decimals; then add them together and multiply the sum times s^2_{pool}. Then find the square root.

For our hypnosis study, we computed s^2_{pool} as 8.30, and n_1 is 17 and n_2 is 15. Filling in the above formula, we have

$$s_{\bar{X}_1-\bar{X}_2} = \sqrt{8.3\left(\frac{1}{17} + \frac{1}{15}\right)}$$

Since 1/17 is .059 and 1/15 is .067, their sum is .126. Then, we have

$$s_{\bar{X}_1-\bar{X}_2} = \sqrt{8.3(.126)} = \sqrt{1.046} = 1.02$$

Thus, for our data the standard error of the difference, $s_{\bar{X}_1-\bar{X}_2}$, equals 1.02.

Note that we can take the above definitional formula for $s_{\bar{X}_1-\bar{X}_2}$ and replace the symbol for s^2_{pool} with the formula for s^2_{pool}. Then,

THE COMPUTATIONAL FORMULA FOR THE STANDARD ERROR OF THE DIFFERENCE IS

$$s_{\bar{X}_1-\bar{X}_2} = \sqrt{\left(\frac{(n_1 - 1)s_1^2 + (n_2 - 1)s_2^2}{(n_1 - 1) + (n_2 - 1)}\right)\left(\frac{1}{n_1} + \frac{1}{n_2}\right)}$$

In the left-hand parentheses we compute s^2_{pool}, and by multiplying it times the right-hand parentheses and then taking the square root, we compute $s_{\bar{X}_1-\bar{X}_2}$.

Now that we have determined the standard error of the difference, we can finally compute t_{obt}.

Computing t_{obt} for two independent samples In previous chapters we found how far the result of our study ($\bar{X}$) was from the mean of the H_0 sampling distribution (μ), measured in standard error units. In general, this formula is

$$t_{\text{obt}} = \frac{(\text{result of the study}) - (\text{mean of } H_0 \text{ sampling distribution})}{\text{standard error}}$$

Now we will perform the same computation, but here the "result of the study" is the *difference* between the two sample means. So in place of "result of the study" we put the quantity $\bar{X}_1 - \bar{X}_2$. Also, now the mean of the H_0 sampling distribution is the *difference* between μ_1 and μ_2 described by H_0. Thus, we replace "mean of H_0 sampling distribution" with the quantity $\mu_1 - \mu_2$. Finally, we replace "standard error" with $s_{\overline{X}_1-\overline{X}_2}$. Putting this all together, we have our formula.

THE DEFINITIONAL FORMULA FOR THE INDEPENDENT-SAMPLES t-TEST IS

$$t_{\text{obt}} = \frac{(\overline{X}_1 - \overline{X}_2) - (\mu_1 - \mu_2)}{s_{\overline{X}_1-\overline{X}_2}}$$

$\overline{X}_1$ and $\overline{X}_2$ are our sample means, $s_{\overline{X}_1-\overline{X}_2}$ is computed as shown above, and the value of $\mu_1 - \mu_2$ is the difference specified by the null hypothesis. The reason we write H_0 as $\mu_1 - \mu_2 = 0$ is so that it directly tells us the mean of the sampling distribution and thus the value of $\mu_1 - \mu_2$ to put in the above formula. (Later this becomes important when $\mu_1 - \mu_2$ is not zero.)

Now we can compute t_{obt} for our hypnosis study. Our sample means were 23 and 20, the difference between μ_1 and μ_2 specified by our H_0 is 0, and we computed $s_{\overline{X}_1-\overline{X}_2}$ to be 1.02. Putting these values into the above formula, we have

$$t_{\text{obt}} = \frac{(23 - 20) - 0}{1.02}$$

Then we have

$$t_{\text{obt}} = \frac{(+3) - 0}{1.02} = \frac{+3}{1.02} = +2.94$$

Our t_{obt} is +2.94. Like a *z*-score, it tells us how far our sample mean difference $(\overline{X}_1 - \overline{X}_2)$ lies from the mean of the H_0 sampling distribution $(\mu_1 - \mu_2)$, measured in units of the standard error of the difference. Thus, the difference of +3 between our sample means is located at something like a *z*-score of +2.94 on the sampling distribution of differences when H_0 is true and both samples represent the same population μ.

Computational formula for t_{obt} for two independent samples We can save a little paper by combining the steps of computing s^2_{pool}, $s_{\overline{X}_1-\overline{X}_2}$, and t_{obt} into one formula.

THE COMPUTATIONAL FORMULA FOR THE t-TEST FOR INDEPENDENT SAMPLES IS

$$t_{\text{obt}} = \frac{(\overline{X}_1 - \overline{X}_2) - (\mu_1 - \mu_2)}{\sqrt{\left(\frac{(n_1 - 1)s_1^2 + (n_2 - 1)s_2^2}{(n_1 - 1) + (n_2 - 1)}\right)\left(\frac{1}{n_1} + \frac{1}{n_2}\right)}}$$

The numerator is the same as in the previous formula for t_{obt}. In the denominator, however, we have replaced the symbol for the standard error, $s_{\overline{X}_1-\overline{X}_2}$, with its previous computational formula.

Thus, for our hypnosis study, substituting the data into the above formula, we have

$$t_{obt} = \frac{(23 - 20) - 0}{\sqrt{\left(\frac{(17 - 1)9.0 + (15 - 1)7.5}{(17 - 1) + (15 - 1)}\right)\left(\frac{1}{17} + \frac{1}{15}\right)}}$$

which becomes

$$t_{obt} = \frac{+3}{\sqrt{(8.3)(.126)}} = \frac{+3}{1.02} = +2.94$$

Thus, again we find that $t_{obt} = +2.94$. Note that the denominator in the t-test is the standard error of the difference, so $s_{\overline{X}_1-\overline{X}_2}$ is again 1.02.

Interpreting t_{obt} in the Independent Samples *t*-Test

To determine if t_{obt} is significant, we compare it to t_{crit}, which is found in Table 2 in Appendix D. We again obtain t_{crit} using degrees of freedom, but we have two samples, so we compute df differently: now degrees of freedom equals $(n_1 - 1) + (n_2 - 1)$.

> ***STAT ALERT*** Critical values of t for the independent samples t-test have $df = (n_1 - 1) + (n_2 - 1)$.

Another way of expressing df is $(n_1 + n_2) - 2$.

For our hypnosis study, $n_1 = 17$ and $n_2 = 15$, so df equals $(17 - 1) + (15 - 1)$, which is 30. As usual, we set alpha at .05. The t-tables show that for a two-tailed test, t_{crit} is ± 2.042. Figure 12.3 locates these values on our sampling distribution of differences.

FIGURE 12.3 H_0 Sampling Distribution of Differences Between Means When $\mu_1 - \mu_2 = 0$

The t_{obt} shows the location of $\overline{X}_1 - \overline{X}_2 = +3.0$.

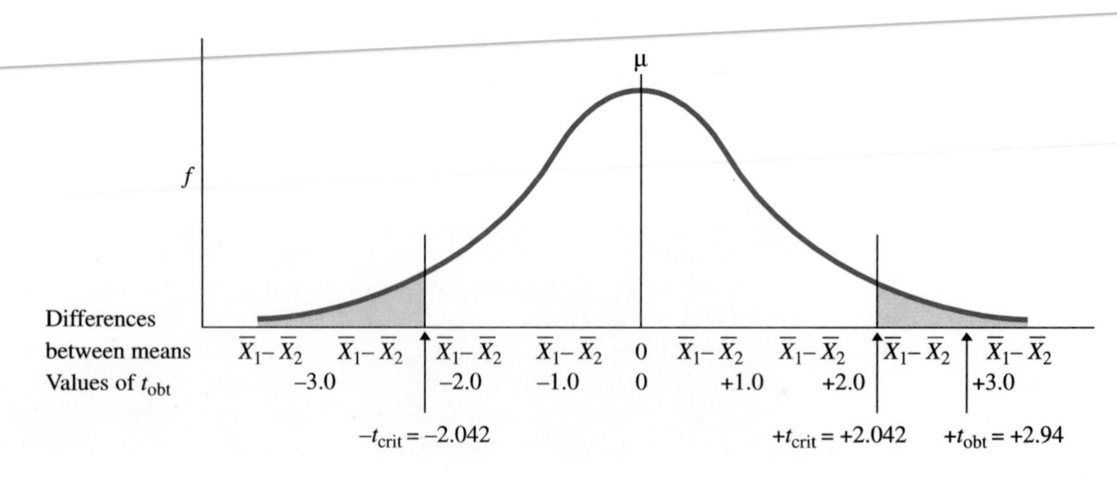

Although H_0 says that the difference between our sample means is merely a poor representation of no difference between μ_1 and μ_2, the location of our t_{obt} on the sampling distribution shows that samples such as ours are seldom *that* poor at representing no difference. In fact, our t_{obt} lies beyond t_{crit}, so it—and the corresponding difference of +3—is in the region of rejection. Therefore, we reject H_0 and conclude that the difference between our two sample means is too unlikely for us to accept as occurring by chance if the sample means represent the same population μ. In other words, the difference between our means is significantly different from zero. We communicate our results as

$$t(30) = +2.94, \quad p < .05$$

As usual, *df* is reported in parentheses, and because $\alpha = .05$, the probability is less than .05 that we have made a Type I error (rejected a true H_0).

We can now accept the alternative hypothesis, which says that $\mu_1 - \mu_2 \neq 0$: the difference between our sample means represents a difference between two population μ's that is not zero. Now we work backwards from this difference to the individual populations of recall scores. Because the sample means produced a significant difference, we can also say that our sample means differ significantly from each other. Further, the mean for deep hypnosis (23) is larger than the mean for mild hypnosis (20). So, to be precise, we conclude that deep hypnosis leads to significantly higher recall scores than mild hypnosis. Thus, we have evidence of a relationship in the population where increasing the degree of hypnosis is associated with higher recall scores.

If our t_{obt} were not beyond t_{crit}, we would not reject H_0. Then we would not have convincing evidence that the difference between our sample means was anything other than sampling error, so we could not say that there was a relationship between degree of hypnosis and recall scores.

Because we did find a significant difference, we now describe the relationship. From the previous chapter you already know that we could compute a confidence interval for the μ that is likely to be represented by each of our sample means. However, there is another way to describe the populations represented by our samples. We found a significant difference of +3 between our sample means, so we expect that if we performed this experiment on the entire population, we would find that the μ for deep hypnosis differed from the μ for mild hypnosis by +3. Of course, our samples may contain a little sampling error, so the actual difference between the μ's is probably *around* +3. We create a confidence interval to more specifically define this difference.

Confidence Interval for the Difference Between Two μ's

As you know, a confidence interval describes a range of population values, any one of which is likely to be represented by our sample data. The **confidence interval for the difference between two μ's** describes a range of *differences* between two population μ's, any one of which is likely to be represented by the *difference* between our two sample means. To compute the interval, we find the largest and smallest values of the quantity $\mu_1 - \mu_2$ that are not significantly different from the difference between our sample means $(\overline{X}_1 - \overline{X}_2)$.

THE COMPUTATIONAL FORMULA FOR THE CONFIDENCE INTERVAL FOR THE DIFFERENCE BETWEEN TWO POPULATION μ'S IS

$$(s_{\bar{X}_1-\bar{X}_2})(-t_{crit}) + (\bar{X}_1 - \bar{X}_2) \leq \mu_1 - \mu_2 \leq (s_{\bar{X}_1-\bar{X}_2})(+t_{crit}) + (\bar{X}_1 - \bar{X}_2)$$

Here $\mu_1 - \mu_2$ stands for the unknown difference we are estimating, t_{crit} is the two-tailed value found for the appropriate α at $df = (n_1 - 1) + (n_2 - 1)$, and the values of $s_{\bar{X}_1-\bar{X}_2}$ and $(\bar{X}_1 - \bar{X}_2)$ are computed in the *t*-test from the sample data.

In our hypnosis study, the two-tailed t_{crit} for $df = 30$ and $\alpha = .05$ is ± 2.042. We computed that $s_{\bar{X}_1-\bar{X}_2}$ is 1.02 and $\bar{X}_1 - \bar{X}_2$ is $+3$. Filling in the above formula, we have

$$(1.02)(-2.042) + (+3) \leq \mu_1 - \mu_2 \leq (1.02)(+2.042) + (+3)$$

Multiplying 1.02 times ± 2.042 gives

$$-2.083 + (+3) \leq \mu_1 - \mu_2 \leq +2.083 + (+3)$$

By adding and subtracting 2.083 from $+3$, we obtain the final confidence interval:

$$.0917 \leq \mu_1 - \mu_2 \leq 5.083$$

Because we used the .05 alpha level, we have created the 95% confidence interval. It indicates that, if we were to perform this experiment on the entire population of subjects, we are 95% confident that the interval betweeen .0917 and 5.083 contains the difference we would find between the μ's under mild and deep hypnosis. In essence, if someone asked us how big a difference mild versus deep hypnosis makes for everyone in recalling the story, our answer would be that we are 95% confident that the difference in number of details recalled is, on average, between about .90 and 5.08.

Performing One-Tailed Tests on Independent Samples

We could have performed our hypnosis study using a one-tailed test if we had predicted the direction of the difference between the two conditions. Say that we predicted a positive relationship where the greater the degree of hypnosis, the higher the recall scores. Everything we have discussed above applies to the one-tailed test, but beware: one-tailed tests can produce serious confusion! This is because we *arbitrarily* call one mean $\bar{X}_1$ and one mean $\bar{X}_2$ and then we subtract $\bar{X}_1 - \bar{X}_2$. How we assign the subscripts determines whether we have a positive or a negative difference, and thus whether we have a positive or a negative t_{obt}. As you know, the sign of t_{obt} is very important in a one-tailed test.

To prevent confusion, it is helpful to use more meaningful subscripts than 1 and 2. For example, we could use a subscript *d* for the deep hypnosis condition and an *m* for the mild hypnosis condition. Then there are two important steps. First, we must decide which sample $\bar{X}$ and corresponding population μ we expect to be larger and then carefully formulate our hypotheses. If we think that the μ

FIGURE 12.4 Sampling Distribution for One-Tailed Independent Samples *t*-Test

Above each figure is the appropriate H_a, along with the appropriate way to subtract the sample means to obtain the predicted t_{obt}.

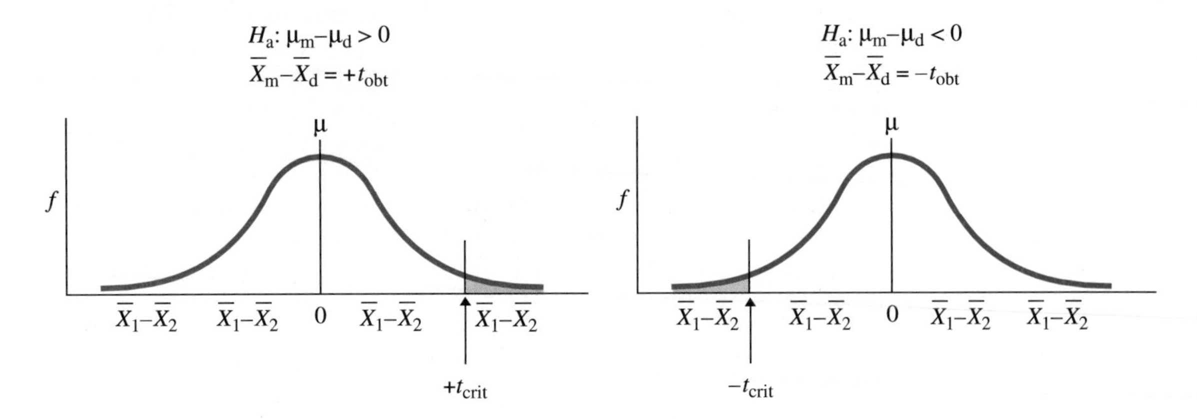

for mild hypnosis (μ_m) is *smaller* than the μ for deep hypnosis (μ_d) and then subtract the smaller μ_m from the larger μ_d, we predict a difference that is greater than zero. Our alternative hypothesis would then be written as H_a: $\mu_d - \mu_m > 0$ (and H_0 as $\mu_d - \mu_m \leq 0$). The second important step is to subtract the sample means in the *same* way that we subtract the μ's in our hypotheses. If we subtract $\mu_d - \mu_m$, then we must subtract $\overline{X}_d - \overline{X}_m$. The third step is to locate the region of rejection. If our means represent the predicted relationship, then subtracting the smaller $\overline{X}_m$ from the larger $\overline{X}_d$ should yield a difference that is positive, producing a positive t_{obt}. For us to reject H_0, our t_{obt} must lie beyond our t_{crit}, so t_{crit} must be a *positive* number. Therefore, as shown on the left-hand graph in Figure 12.4, we place the region of rejection in the positive tail of the sampling distribution.

One-tailed tests are confusing because we could have reversed how we stated H_a, expressing it as H_a: $\mu_m - \mu_d < 0$. This again predicts a larger μ_d, but now subtracting the larger μ_d from the smaller μ_m produces a difference that is less than zero. Subtracting the sample means this way, we predict that the difference will be negative, producing a negative t_{obt}. Therefore, we place the region of rejection in the negative tail of the distribution, and we have a negative t_{crit}, as shown in the right-hand graph in Figure 12.4.

You must pay attention and think through each step. If you subtract the sample means in the opposite way from the way you subtract the μ's, the sign of your t_{obt} will be the opposite of what it should be for testing your H_0. Then you may erroneously conclude that there is not a significant difference when there is, or vice versa.

Testing Hypotheses About Nonzero Differences

Regardless of whether we have used a one- or two-tailed test, so far we have tested H_0: $\mu_1 - \mu_2 = 0$, which says that there is no difference between the populations. Looking at H_0 from another perspective, we can interpret it as saying that there is a difference between μ_1 and μ_2, and that difference equals 0. Using the same procedure, we can also test an H_0 that says that the difference between the two populations equals any other amount.

For example, here's a new experiment. Say that on a test of flying ability, the population of Navy pilots differs from the population of Marine pilots by 10 points. However, recently the Navy developed a new training method that has been used with all its pilots. To test the effectiveness of this method, we randomly select a sample of Navy pilots and compare their flying ability to the flying ability of a sample of Marine pilots. Now we want to test whether our samples represent populations that still differ by 10 points. In a two-tailed test, our null hypothesis is that the training program has no effect, so with or without it, the difference between Navy and Marine pilots should still be 10. Thus, we have

$$H_0: \mu_1 - \mu_2 = 10$$

On the other hand, our alternative hypothesis is that the training program has an effect, so the difference will no longer be 10. Thus, we have

$$H_a: \mu_1 - \mu_2 \neq 10$$

In the t-test for this study, we do everything we did previously, except wherever we said that the quantity $\mu_1 - \mu_2$ was zero, we now say that $\mu_1 - \mu_2$ is 10. Now we have a sampling distribution of differences created when the μ's of the two populations differ by 10, so the mean of the sampling distribution is 10. Comparing t_{obt} to t_{crit} indicates whether the difference between our sample means is significantly different from 10. If it is, then we conclude that with the training, the populations of Navy and Marine pilots no longer differ by 10, but rather by the difference represented by our samples.

THE *t*-TEST FOR RELATED SAMPLES

Now we will discuss the other version of the t-test for two samples. This version is used when, instead of performing experiments containing two independent samples, we perform experiments containing two related samples. The ***t*-test for related samples** is the parametric procedure used for significance testing when we have two sample means from two related samples. Two samples are *related* when we pair each score in one sample with a particular score in the other sample. There are two general types of research designs that produce related samples: matched samples designs and repeated measures designs.

In a **matched samples design,** the researcher matches each subject in one sample with a subject in the other sample. For example, say that we want to measure how well subjects in two samples shoot baskets, where one sample uses a standard basketball and the other uses a new type of basketball (with handles). Ideally, we want the subjects in one sample to be the same height as those in the other sample. If they are not, then differences in shooting baskets may be due to the differences in height instead of the different balls. We can create two samples containing subjects who are the same height by forming matching pairs of subjects who are the same height and then assigning each member of the pair to one condition. Thus, if we randomly select two subjects who are six feet tall, we will randomly assign one subject to one condition and the other subject to the other condition. Likewise, a four-foot subject in one condition is matched with a four-footer in the other condition, and so on. Now any differences in basket shooting between the two samples cannot be due to differences in height, because the same heights are present in both samples. In the same way, we may match subjects using any relevant variable, such as weight, age, physical ability, or the school they attend. (We may also rely on natural pairs to match our subjects. For example, we might study pairs of identical twins, placing one member of a pair in each condition.)

The ultimate form of matching is to match each subject with himself or herself. This is the other, much more common, way of producing related samples, called repeated measures. In a **repeated measures design,** we repeatedly measure the same subjects under all conditions of an independent variable. For example, we might first measure a sample of subjects when they were using the standard basketball, and then measure the same subjects again when they were using the new basketball. Here, although we have one sample of subjects, we again have two samples of scores. And, of course, any differences in basket shooting between the samples cannot be due to differences in height or to any other attribute of the subjects.

What makes matched and repeated measures samples *related* is the fact that each score in one sample is related to the paired score in the other sample. Related samples are also called *dependent samples.* As we saw in Chapter 9, two events are dependent when the probability of one event is influenced by the occurrence of the other event. Related samples are dependent because the probability that one score in a pair is a particular value is influenced by the value of the paired score. For example, if a four-foot-tall male scored close to 0 in one sample, the probability is high that the other matching four-footer also scored close to 0. This is not the case when we have independent samples, as in our hypnosis study. There, the fact that a subject scored 0 in the mild hypnosis condition did not influence the probability of any subject's scoring a 0 in the deep hypnosis condition.

In the previous *t*-test for independent samples, our sampling distribution described the probability of obtaining a particular difference between two means from independent samples of scores. With related samples, we must compute this probability differently, so we create our sampling distribution differently and we compute t_{obt} differently.

Assumptions of the *t*-Test for Related Samples

Except for requiring related samples, the assumptions of the *t*-test for related samples are the same as those for the *t*-test for independent samples: (1) the dependent variable involves an interval or ratio scale, (2) the population represented by either sample forms a normal distribution, (3) the variance of the raw score populations is estimated by s_X^2, and (4) the populations represented by our samples have homogeneous variance. Because related samples form pairs of scores, the *n* in each sample must be equal.

If the data meet the assumptions, then it's onward and upwards: (1) we create our statistical hypotheses, (2) we set up and perform the statistical test, and (3) if our results are significant, we describe the relationship we have demonstrated.

The Logic of Hypotheses Testing in the Related Samples *t*-Test

Let's say that we are interested in phobias (irrational fears of objects or events). We have a new therapy we want to test on spider-phobics—people who are frightened by big, black, hairy, spiders (which doesn't sound all that irrational). From the membership of the local phobia club, we randomly select a decidedly unpowerful N of five spider-phobics, and we test our therapy using repeated measures of two conditions: before therapy and after therapy. Before therapy we will measure each subject's fear response to a picture of a spider. We will do so by measuring heart rate, breathing rate, perspiration, etc., and then compute a "fear" score between 0 (no fear) and 20 (holy terror!). After providing the therapy, we will again measure the subject's fear response to the picture. (Anytime we have a before-and-after, or pre-test/post-test, design such as this, we use the *t*-test for related samples.) Our study meets the assumptions of the related samples *t*-test, and we set alpha at .05.

Obviously, we expect to demonstrate that the therapy will decrease subjects' fear of spiders, so our experimental hypotheses are one-tailed: (1) the experiment will demonstrate a relationship in which the population μ for scores after therapy, is lower than the μ for scores before therapy, or (2) the experiment will not work, either showing no difference between the μ's for before and after scores or showing that the μ for scores after therapy is larger than the μ for scores before therapy. So far, it sounds as if we are testing the same type of hypotheses we had in the independent samples *t*-test. However, instead of directly comparing the means from our samples to determine the population μ's they represent, we must first transform our data. Then we test our hypotheses using these transformed scores.

We transform the raw scores in related samples by finding the *difference scores*. A difference score is the difference between the two raw scores in a pair. The symbol for a difference score is D. Say that for our phobia study we collected the data shown in Table 12.3. We find each difference score by arbitrarily subtracting each subject's after-therapy score from the corresponding before-therapy score. (We could subtract the before scores from the after scores, but we must subtract all scores in the same way.)

TABLE 12.3 Scores for the Before-Therapy and After-Therapy Conditions

Each D equals (before − after).

Subject	*Before therapy*	−	*After therapy*	=	*Difference, D*
1 (Dorcas)	11	−	8	=	+3
2 (Biff)	16	−	11	=	+5
3 (Millie)	20	−	15	=	+5
4 (Attila)	17	−	11	=	+6
5 (Slug)	10	−	11	=	−1
					$\Sigma D = 18$

Now we summarize the sample of difference scores by computing the mean difference score. The symbol for the mean difference is $\bar{D}$. We add the positive and negative differences to find the sum of the differences, symbolized by ΣD. Then we divide this amount by N, the number of difference scores. For our phobia data, the mean difference, $\bar{D}$, equals 18/5, which is +3.6. This indicates that the before scores were, on average, 3.6 points higher than the after scores.

Now here's the strange part: forget about the before and after scores for the moment, and consider only the difference scores. From a statistical standpoint, we have *one* sample mean ($\bar{D}$) from *one* random sample of (difference) scores. As we saw in the previous chapter, when we have one sample mean, we perform the single-sample *t*-test! The fact that we have difference scores in no way violates this *t*-test, so we will create our statistical hypotheses and then test them in virtually the same way we did with the single-sample *t*-test.

> ***STAT ALERT*** The *t*-test for two related samples is performed by conducting the single-sample *t*-test on the sample of difference scores.

Statistical Hypotheses for the Related Samples *t*-Test

Now we can create our statistical hypotheses. Our sample of difference scores represents the population of difference scores that would result if we could measure the population of raw scores under each of our conditions and then subtract the scores in one population from the corresponding scores in the other population. The population of difference scores has some μ that we identify as μ_D. To create our statistical hypotheses, we simply determine the predicted values of μ_D in H_0 and H_a.

In our one-tailed phobia study, we predict that the population of scores after therapy will contain lower fear scores than the population of scores before therapy. If we subtract the after scores from the before scores, as we did in our sample, then we should have a population of difference scores containing positive numbers. The resulting μ_D should also be a positive number. Our alternative hypothesis always implies that the predicted relationship exists, so here we state it as

$$H_a: \mu_D > 0$$

H_a implies that our sample $\bar{D}$ represents a population of differences having a μ_D greater than zero, and thus demonstrates that after-therapy fear scores are lower than before-therapy scores in the population.

On the other hand, according to the null hypothesis, there are two ways we may fail to demonstrate the predicted relationship. First, the therapy may do nothing to fear scores, so that the population of before scores contains the same scores as the population of after scores. If we subtract the after scores from the before scores, the population of difference scores will have a μ_D of zero. Note that every difference need not equal zero. Because of random physiological or psychological factors, all subjects may not produce the same fear score on two observations, and thus their difference scores may not be zero. On average, however, the positive and negative differences will cancel out to produce a μ_D of zero. Second, the therapy may increase fear scores, so that subtracting larger after scores from smaller before scores produces a population of difference scores consisting of negative numbers, with a μ_D that is less than zero. Thus, given the way we are subtracting to find our difference scores, our null hypothesis is

$$H_0\text{: } \mu_D \leq 0$$

H_0 implies that our sample $\bar{D}$ represents such a population μ_D, thus demonstrating that the predicted relationship between therapy and fear scores does not exist.

As usual, we test H_0 by testing whether our sample mean is likely to represent the μ that is described by H_0. Here, H_0 says that our sample mean represents the population of difference scores where μ_D equals zero. If the sample perfectly represents this population, then $\bar{D}$ "should" equal zero. However, because of those chance fluctuations in fear scores, all of the difference scores in our sample may not equal zero, so neither will $\bar{D}$. Thus, H_0 says that $\bar{D}$ represents a population where μ_D is zero, and if $\bar{D}$ is not zero, it is because of sampling error.

To test H_0, we will examine the sampling distribution of means, which here is the sampling distribution of $\bar{D}$. This is the frequency distribution of the different values of $\bar{D}$ that occur when H_0 is true. We simply locate our $\bar{D}$ on this sampling distribution by computing t_{obt}.

Computing the *t*-Test for Related Samples

To see how to compute t_{obt}, we will use the data from our phobia study, presented in Table 12.4. Note that we need N, ΣD, $\bar{D}$, and ΣD^2.

Computing t_{obt} here is identical to computing the single-sample *t*-test discussed in Chapter 11. In the following formulas I have simply changed the symbols from X to D. To compute the t_{obt} for related samples, we perform the following three steps.

The first step is to find s_D^2, which is the estimated variance of the population of difference scores.

THE FORMULA FOR s_D^2 IS

$$s_D^2 = \frac{\Sigma D^2 - \frac{(\Sigma D)^2}{N}}{N - 1}$$

TABLE 12.4 Summary of Data from Phobia Study

Subject	*Before therapy*	−	*After therapy*	=	*Difference, D*	*D^2*
1	11		8		+3	9
2	16		11		+5	25
3	20		15		+5	25
4	17		11		+6	36
5	10		11		−1	1
	$\bar{X} = 14.80$		$\bar{X} = 11.20$		$\Sigma D = +18$	$\Sigma D^2 = 96$
$N = 5$					$\bar{D} = +3.6$	

Using our phobia data, we fill in the formula:

$$s_D^2 = \frac{\Sigma D^2 - \frac{(\Sigma D)^2}{N}}{N - 1} = \frac{96 - \frac{(18)^2}{5}}{4} = 7.80$$

The second step is to find $s_{\bar{D}}$. This is the **standard error of the mean difference,** or the "standard deviation" of the sampling distribution of $\bar{D}$.

THE FORMULA FOR THE STANDARD ERROR OF THE MEAN DIFFERENCE, $s_{\bar{D}}$, IS

$$s_{\bar{D}} = \sqrt{(s_D^2)\left(\frac{1}{N}\right)}$$

For our phobia study, with $s_D^2 = 7.80$ and $N = 5$, we have

$$s_{\bar{D}} = \sqrt{(s_D^2)\left(\frac{1}{N}\right)} = \sqrt{(7.80)\left(\frac{1}{5}\right)} = \sqrt{1.56} = 1.25$$

The third step is to find t_{obt}.

THE DEFINITIONAL FORMULA FOR THE t-TEST FOR RELATED SAMPLES IS

$$t_{obt} = \frac{\bar{D} - \mu_D}{s_{\bar{D}}}$$

For our phobia study, $\bar{D}$ is $+3.6$, $s_{\bar{D}}$ is 1.25, and H_0 says that μ_D equals 0. Putting these values into the above formula, we have

$$t_{obt} = \frac{\bar{D} - \mu_D}{s_{\bar{D}}} = \frac{+3.6 - 0}{1.25} = +2.88$$

This tells us that our sample $\bar{D}$ is located at a t_{obt} of $+2.88$ on the sampling distribution of $\bar{D}$ when $\mu_D = 0$.

Computational formula for the related samples *t*-test We can combine the above steps of computing $s_{\bar{D}}$ and t_{obt} into one formula.

THE COMPUTATIONAL FORMULA FOR THE t-TEST FOR RELATED SAMPLES IS

$$t_{obt} = \frac{\bar{D} - \mu_D}{\sqrt{(s_D^2)\left(\frac{1}{N}\right)}}$$

The numerator is the same as we saw in the definitional formula. The denominator simply contains the formula for the standard error of the mean difference, $s_{\bar{D}}$.

Combined formula for t_{obt} in the related samples t-test When we are testing an H_0 that contains $\mu_D = 0$, then μ_D is always zero in the above formulas for t_{obt}, so the numerator always equals the value of $\bar{D}$. Using this fact, we have combined all computations into one combined formula for t_{obt}.

STAT ALERT We can use the combined formula only when our one- or two-tailed null hypothesis includes $\mu_D = 0$.

THE COMBINED FORMULA FOR t_{obt} WHEN H_0 INCLUDES $\mu_D = 0$ IS

$$t_{obt} = \frac{\Sigma D}{\sqrt{[N(\Sigma D^2) - (\Sigma D)^2]\left(\frac{1}{N-1}\right)}}$$

Putting the data for our phobia study into the above formula, we have

$$t_{obt} = \frac{+18}{\sqrt{[5(96) - (18)^2]\left(\frac{1}{5-1}\right)}}$$

Thus, we have

$$t_{obt} = \frac{+18}{\sqrt{(156)(.25)}} = \frac{+18}{\sqrt{39.00}} = \frac{+18}{6.245} = +2.88$$

So t_{obt} is again $+2.88$.

STAT ALERT Usually the final answer in the denominator of the t-test is the standard error. However, when we use the combined formula above, the final answer is *not* $s_{\bar{D}}$. You must compute $s_{\bar{D}}$ using the formula given previously.

Interpreting t_{obt} in the Related Samples t-Test

To interpret t_{obt}, we compare it to t_{crit}. We find t_{crit} in the t-tables (Table 2 in Appendix D) for $df = N - 1$, where N is the number of difference scores. For our phobia study, with $\alpha = .05$ and $df = 4$, the one-tailed value of t_{crit} is $+2.132$. Our t_{crit} is positive, because we predicted that therapy works to decrease fear. When we subtract the (hopefully) smaller after scores from the larger before scores, we should obtain a positive value of $\bar{D}$, producing a positive t_{obt}.

FIGURE 12.5 One-Tailed Sampling Distribution of Random $\bar{D}$'s When $\mu_D = 0$

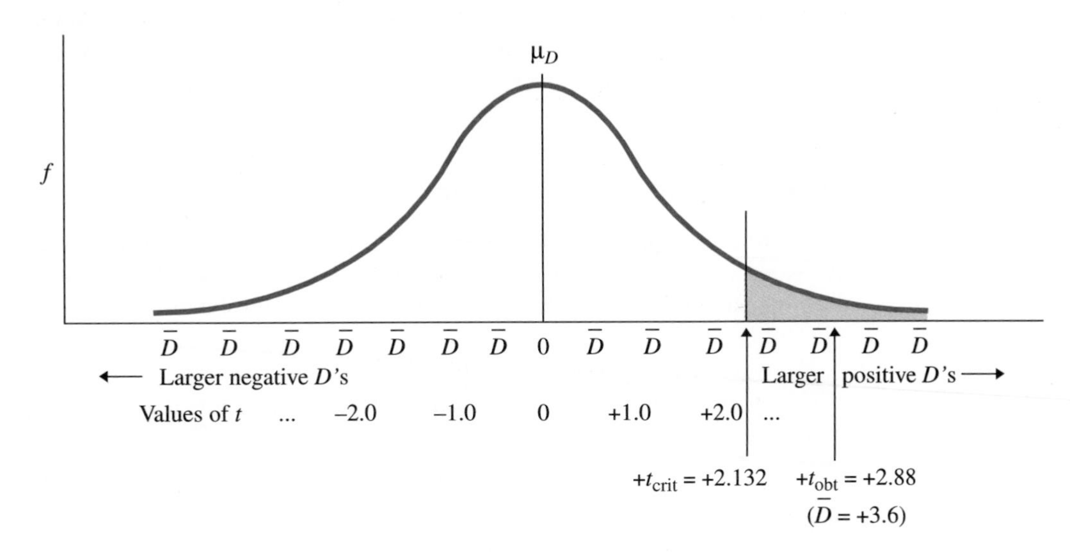

Figure 12.5 shows our completed sampling distribution. This is another sampling distribution of means, except that each mean is the mean of a sample of difference scores drawn from the population of difference scores where μ_D equals zero. The average sample mean will be 0, so the μ of the distribution is zero. The values of $\bar{D}$ that are farther from zero (whether positive or negative) are less likely to occur when H_0 is true and the sample represents a μ_D of zero. As shown, our t_{obt} of $+2.88$ is in the region of rejection, so sampling error is too unlikely to have produced our $\bar{D}$ of $+3.6$ if the sample actually represents a μ_D of zero (and even less likely if the sample represents a μ_D less than zero). Therefore, we reject the null hypothesis that our sample represents $\mu_D \leq 0$. Our results are significant, and we report them as

$$t(4) = +2.88, \quad p < .05$$

By rejecting H_0, we accept the H_a that our sample represents a μ_D that is greater than zero. We would see such a population of difference scores only if the population of before-therapy scores contained scores that were larger than those in the population of after-therapy scores. Therefore, we conclude that we have demonstrated a significant difference between the before and after scores, a difference that represents a relationship in the population. Thus, we are confident that our therapy works. (If t_{obt} had not been beyond t_{crit}, then we would retain H_0, and we would not have convincing evidence that our therapy reduces fear scores.)

Way back in Table 12.4, we saw that the mean of the before scores was 14.80 and the mean of the after scores was 11.20. Because we have determined that the before and after conditions produce significant differences in fear scores, we can also say that these means differ significantly. Our best estimate is that if we measured the fear scores of all subjects in the population, before therapy the μ would be around 14.80, and after therapy the μ would be around 11.20. Although it would be nice to more precisely define "around" here, we *cannot* use our previous formulas for computing a confidence interval because our sample means are from related samples. Instead, we can only deal with μ_D. Since our sample mean, $\bar{D}$, is $+3.6$, it probably represents a population of difference scores where μ_D is "around" $+3.6$. We compute a confidence interval to describe this μ_D.

Computing the Confidence Interval for μ_D

The **confidence interval for μ_D** describes a range of values of μ_D, one of which our sample mean is likely to represent. We compute the interval by computing the highest and lowest values of μ_D that are not significantly different from $\bar{D}$.

THE COMPUTATIONAL FORMULA FOR THE CONFIDENCE INTERVAL FOR μ_D OF THE POPULATION OF DIFFERENCE SCORES IS

$$(s_{\bar{D}})(-t_{crit}) + \bar{D} \leq \mu_D \leq (s_{\bar{D}})(+t_{crit}) + \bar{D}$$

The value of t_{crit} is the two-tailed value for $df = N - 1$, where N is the number of difference scores, $s_{\bar{D}}$ is the standard error of the mean difference as computed above, and $\bar{D}$ is the mean of our difference scores.

To compute the confidence interval for our phobia study, we have $s_{\bar{D}} = 1.25$ and $\bar{D} = +3.6$. Because we seek values above and below $+3.6$, we switch to the two-tailed critical value here, and with $\alpha = .05$ and $df = 4$, t_{crit} is ± 2.776. Filling in the above formula, we have

$$(1.25)(-2.776) + 3.6 \leq \mu_D \leq (1.25)(+2.776) + 3.6$$

which becomes

$$0.13 \leq \mu_D \leq 7.07$$

Thus, we are 95% confident that our sample mean of $+3.6$ represents a population μ_D within this interval. In other words, if we performed this study on the entire population, we would expect the average difference in before and after scores to be between 0.13 and 7.07.

Testing Other Hypotheses with the Related Samples *t*-Test

In the preceding study we could have reversed how we computed the difference scores, subtracting the predicted larger, before scores from the predicted smaller, after scores. Then, if the therapy did not work as predicted, we would expect a $\bar{D}$ of zero or greater, and we'd have H_0: $\mu_D \geq 0$. If the therapy worked, we would expect a negative $\bar{D}$, representing a μ_D less than zero, and we'd have H_a: $\mu_D < 0$

In an experiment where we do not predict which sample of raw scores will produce the higher scores, we have two-tailed hypotheses: our sample of difference scores either does or does not reflect differences between the populations for our two conditions. If the populations of raw scores do not differ, then μ_D is zero, so H_0: $\mu_D = 0$. If the populations of raw scores differ, then μ_D is not zero, so H_a: $\mu_D \neq 0$.

Finally, we can also test an H_0 that the populations differ by some amount other than zero. For example, previously we discussed a study where Navy and Marine pilots differed by 10 points in flying ability. Suppose we want to perform a related samples study in which we match each Navy pilot with a Marine pilot who has had the same amount of flying experience. Our null hypothesis is that the training does not work, so the two samples still represent populations that differ by 10. When we subtract the two populations, the population of differences has a μ_D equal to 10, so our H_0 is $\mu_D = 10$. Our alternative hypothesis is that the training alters the difference between the two raw score populations so that, when we subtract the two populations, the difference is not 10. Thus we have H_a: $\mu_D \neq 10$.

To test any of the above null hypotheses, we select the appropriate t_{crit} and then calculate and test t_{obt} using the same procedure we used in the phobia study.

POWER AND THE TWO-SAMPLE t-TEST

Remember *power*—the probability of *not* making a Type II error, or the probability of rejecting H_0 when it is false? We maximize power by maximizing the size of t_{obt} relative to t_{crit}. Looking at the formulas

$$t_{obt} = \frac{(\bar{X}_1 - \bar{X}_2) - (\mu_1 - \mu_2)}{\sqrt{\left(\frac{(n_1 - 1)s_1^2 + (n_2 - 1)s_2^2}{(n_1 - 1) + (n_2 - 1)}\right)\left(\frac{1}{n_1} + \frac{1}{n_2}\right)}} \quad \text{and}$$

$$t_{obt} = \frac{\bar{D} - \mu_D}{\sqrt{(s_D^2)\left(\frac{1}{N}\right)}}$$

we see that anything that increases the size of the numerator or decreases the size of the denominator produces a larger t_{obt}. Therefore, as we saw in Chapter 11, you maximize the size of t_{obt} by designing a study in which you:

1. *Maximize the difference produced by the two conditions.* To do this, select two very different amounts as the conditions of your independent variable, so that they are likely to produce a large difference in dependent scores. With the independent samples test, this produces a larger value of $\bar{X}_1 - \bar{X}_2$. With the related samples test, this produces a larger value of $\bar{D}$ and a larger difference between $\bar{D}$ and μ_D. In either case, the numerator in the formula will be larger, so t_{obt} will be larger.
2. *Minimize the variability of the raw scores.* To do this, conduct the study as consistently as possible, eliminating any extraneous variables that might produce differences in scores among subjects *within* a particular condition. In the independent samples test, this minimizes each s_X^2, producing a

TABLE 12.5 Scores of Subjects 2 and 3 from Phobia Study

Subject	*Before therapy*	−	*After therapy*	=	*Difference, D*
2	16		11		+5
3	20		15		+5

smaller s^2_{pool} and thus a smaller $s_{\bar{X}_1 - \bar{X}_2}$. In the related samples test, this minimizes the differences between the subjects' difference scores, producing a smaller s^2_D. In either case, dividing by a smaller denominator produces a larger t_{obt}.

3. *Maximize the sample* n*'s.* The larger the number of subjects in an independent samples or related samples design, the smaller the denominator when calculating t_{obt}. In addition, larger n's give a larger *df*, resulting in a smaller value of t_{crit}. Therefore, whenever you can easily test more subjects per group (within reason), you should do so.

In all of the above, you increase your power because you increase the probability of rejecting H_0 (and so you are more likely to make the correct decision at those times when H_0 is false). However, recognize that, in addition, a related samples design is intrinsically more powerful than an independent samples design. For example, say we reanalyzed our phobia study, violating the assumptions and treating the two samples as independent samples of fear scores. Comparing the two procedures, we would find that the variability of the original fear scores was larger than the variability of the difference scores. We eliminate some of the variability in the data by transforming them to D's. For example, Table 12.5 presents the scores of Subjects 2 and 3 (from Table 12.4). Although there is variability (differences) in their before scores and in their after scores, there is no variability in their difference scores. Reducing variability in scores produces a larger t_{obt}, so the t_{obt} for related samples will be larger than the t_{obt} for independent samples. Thus, by choosing a related samples design when it is appropriate, you obtain a larger t_{obt} and increase your power.

DESCRIBING THE RELATIONSHIP IN A TWO-SAMPLE EXPERIMENT

By now you understand how to determine whether results are significant. However, the fact that t_{obt} is significant is not the end of the story. If you stop after hypothesis testing, then you have found a "real" relationship but you have not described it. Essentially, it's the same as saying, "I've computed a correlation coefficient, but I'm not going to tell you what it is." (Frustrating, isn't it?) Remember, the ultimate purpose of research is to understand the laws of behavior that your results reflect. Understanding the characteristics of the relationship will help you when performing this final task of a study.

Your focus should always be on your sample means (or other measures of central tendency, when they are used). Then you can begin to understand the behavior by summarizing the typical score—and typical behavior—found in each condition. Confidence intervals then allow you to generalize your findings, describing the range of typical scores—and typical behaviors—you expect in the population (in nature).

In addition, as we see in the following sections, we also describe the relationship by graphing the results and computing the effect size.

Graphing the Results of a Two-Sample Experiment

As you know, we graph experimental results by plotting the mean of each condition on the Y axis and the conditions of the independent variable on the X axis. Thus, we would plot the results of our hypnosis study as shown in the left-hand graph in Figure 12.6 and the results of our phobia study as shown in the right-hand graph. (Note that for the phobia study, we plot the mean of the original fear scores from the before and after conditions, and not the D's.) Such slanting lines indicate a relationship in the samples. Because the sample means are significantly different, we expect that if we measured and then plotted the population μ's, we would obtain similar graphs.

Our descriptive procedures then allow us to summarize and envision the relationship. First, we use the line graph to envision the scatterplot that the data would form if we plotted it (as if we plotted each subject's data point using the condition under which the subject was tested as the X score and the subject's score on the dependent variable as the Y score. Because the mean is the center of the distribution, we envision data points above and below the mean Y score at each X. For example, in the hypnosis graph in Figure 12.6, we envision data points above and below the mean data point for mild hypnosis, and likewise we envision data points around the mean data point for deep hypnosis. Thus, we literally expect subjects to score around the mean of 20 in the mild hypnosis

FIGURE 12.6 Line Graphs of the Results of the Hypnosis Study and the Phobia Study

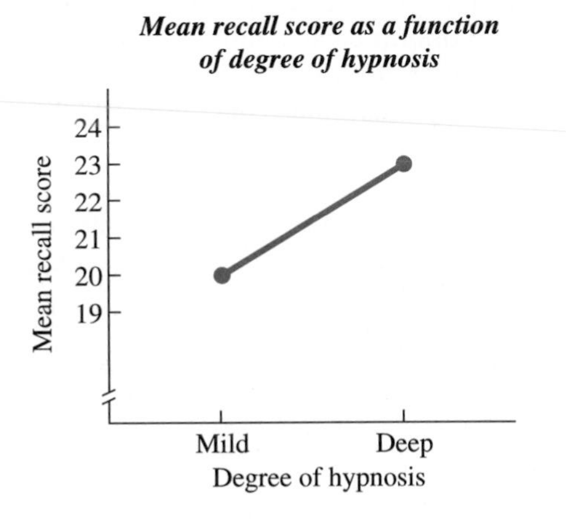

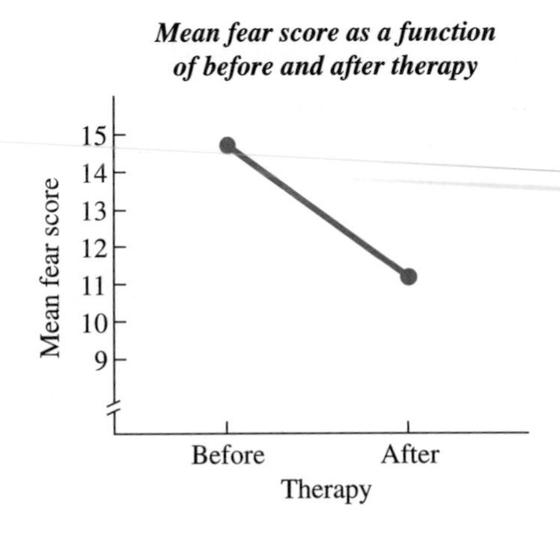

condition and around the mean of 23 in the deep hypnosis condition. Further, notice that we can use this approach both to describe our subjects' actual scores and to predict scores for other subjects we might test under these conditions. And we can predict scores for other conditions we might create. For example, say we want to predict the recall scores for subjects who would undergo medium hypnosis. At the point midway between mild and deep hypnosis on the X axis, travel up to the line on the graph and then left to the Y axis, arriving at a predicted score of around 21.5.

If this all sounds familiar, it's because it's the same procedure we used in Chapter 8 when we performed the linear regression procedure. Recall that a regression line is the straight line that summarizes a scatterplot by passing as close as possible to the mean of Y scores at each X. A subject's predicted Y score is the value of Y lying on the regression line at that subject's X. Note, however, that when there are only two values of X, the regression line will connect the means of Y at each X. Therefore, if we were to use the procedures in Chapter 8 to produce the linear regression line for each experiment, we would again produce the lines in Figure 12.6, and our best prediction of a subject's Y score in a particular condition would again be the mean of that condition.

Further, recognize that the line graphs clearly show that the relationship involving hypnosis is a positive relationship and that the relationship involving therapy is a negative relationship. By knowing the *type* of relationship produced, in addition to the scores we predict for subjects in each condition, we have more information to use when we try to understand the behaviors reflected by our variables. What we're missing, of course, is how much error we have in our predictions or, conversely, what the *strength* of the relationship is.

As you may suspect, it is at this point that correlational statistics come together with our previous analysis of means. A significant relationship is a significant relationship, regardless of whether we summarize it using means or using a correlation coefficient. We performed the t-test because it is the more powerful way to test H_0 in a two-sample experiment. However, recall that the advantage of a correlation coefficient is that it describes the strength of a relationship, indicating how consistently one value of Y is associated with one and only one value of X. Therefore, whenever the results of an experiment are significant, our next step is to correlate the scores on the dependent variable with the conditions of the independent variable. This describes how consistently subjects scored at or close to the mean score for each condition. Further, when we view a line graph as summarizing the scatterplot, recall that the correlation coefficient indicates the extent to which the data points "hug" the regression line and produce a "skinny" scatterplot. Thus, computing the correlation coefficient in the hypnosis study, for example, will allow us to better envision this relationship as well as tell us how consistently close subjects' recall scores were to the mean score of 20 under mild hypnosis and the mean score of 23 under deep hypnosis.

STAT ALERT Whenever we obtain significant results by testing means, the first step in describing the strength of the relationship is to compute the appropriate correlation coefficient.

Describing the Strength of the Relationship in a Two-Sample Experiment Using r_{pb}

The point-biserial correlation coefficient (r_{pb}) is the appropriate coefficient for describing a two-sample experiment. This is because r_{pb} is used when we have one dichotomous X variable (that is, consisting of two categories) and one continuous interval or ratio Y variable. That's precisely what we have in a two-sample experiment: the conditions of the independent variable are a dichotomous X variable, and scores on the dependent variable are a continuous interval or ratio Y variable. We could compute r_{pb} using the formula in Chapter 7, but instead we can compute r_{pb} directly from our t_{obt}.

THE FORMULA FOR COMPUTING r_{pb} FROM t_{obt} IS

$$r_{pb} = \sqrt{\frac{(t_{obt})^2}{(t_{obt})^2 + df}}$$

Although this formula can be used for either independent or related samples, there is one important distinction:

> **For independent samples, $df = (n_1 - 1) + (n_2 - 1)$, where each n is the number of scores in a sample.**
>
> **For related samples, $df = N - 1$, where N is the number of difference scores.**

In our hypnosis study, we found $t_{obt} = +2.94$ with $df = 30$, so

$$r_{pb} = \sqrt{\frac{(2.94)^2}{(2.94)^2 + 30}} = \sqrt{\frac{8.64}{8.64 + 30}} = \sqrt{\frac{8.64}{38.64}} = \sqrt{.224} = .47$$

Thus, the relationship between recall scores and degree of hypnosis can be summarized as $r_{pb} = .47$. Likewise, for our phobia study, with $t_{obt} = +2.88$ and $df = 4$, the relationship between fear scores and before or after therapy can be summarized as $r_{pb} = .82$.

Notice that the final step in the formula for computing r_{pb} is to find the square root, so our answer will always be positive. Depending on the data, we decide whether r_{pb} is positive or negative. We saw that our hypnosis study produced a positive relationship, so $r_{pb} = +.47$. Our phobia study produced a negative relationship, so $r_{pb} = -.82$. Although both describe significant relationships, the larger r_{pb} for the phobia study indicates that there we demonstrated a stronger, more consistent relationship than in the hypnosis study.

Recall, however, that a problem with a correlation coefficient is that it can be difficult to interpret: it does not measure in units of consistency, so we can only subjectively evaluate it relative to 0 and ± 1.0. Instead, as we all remember, the most direct way to evaluate a relationship is to square the correlation coefficient, computing the proportion of variance accounted for.

Describing the Effect Size in a Two-Sample Experiment

The ultimate way to describe the relationship in a two-sample experiment is to compute the proportion of variance accounted for by squaring the value of r_{pb} (to do this, simply don't find the square root in the formula). The resulting value also implies how consistent or strong the relationship is, but mathematically it is more directly interpretable. Essentially, r^2_{pb} indicates how much more accurately we can predict the different dependent scores when we know the condition under which subjects were tested, compared to when we are unaware of the conditions and instead predict each subject's score using the overall mean of all scores. For our hypnosis study, squaring r_{pb} gives $(.47)^2$, or .22. Thus, on average we are 22% closer to predicting subjects' actual recall scores when we use the mean score of each hypnosis condition to predict their scores rather than when we use the overall mean recall score of the experiment. Likewise, in the phobia study, $r^2_{pb} = (.82)^2$, or .67, so we can account for 67% of the variance in fear scores by knowing whether or not subjects have undergone the therapy.

In an experiment, the proportion of variance accounted for goes by a different name: it is called the "effect size." This is because when an experiment demonstrates a significant relationship, our explanation is that the independent variable has an *effect* on the dependent variable. That is, we argue that by changing the conditions of the independent variable, we "cause" scores on the dependent variable to change. (The word "cause" is in quotes because remember, we never *prove* that changing the independent variable causes the scores to change: a change in some other unknown variable may actually have caused subjects' scores to change.) Thus, we explain the variability in our recall scores, for example, as being "caused" by changing the degree of hypnosis, and so how consistently hypnosis does this is called its effect size. The **effect size** indicates how consistently differences in the dependent scores are "caused" by changes in the independent variable. The larger the effect size, the more consistently the raw scores for each condition are located at or close to the mean score for that condition, so the more consistent is the effect of the independent variable.

The greater the effect size, the more that knowing the condition of the independent variable improves our accuracy in describing and predicting subjects' dependent scores, and so the greater the scientific importance of the relationship involving our independent variable. For example, say that, in a different study, a given degree of hypnosis caused everyone to have the same recall score: Increase the degree of hypnosis, and every recall score increased by the same amount. There would be no differences in recall scores except when the degree of hypnosis changed, so there would be no variability in Y scores not associated with changes in X. In this hypothetical case, r_{pb} would equal $+1.0$, so 100% of the variance in recall scores would be accounted for by hypnosis. Here, degree of hypnosis seems to completely determine recall scores, so it is a very important variable.

On the other hand, we actually found that r_{pb} was only $+.47$. Now, at one degree of hypnosis we get more or less the same recall scores, but the Y scores differ when X has not changed, so that only 22% of the differences in recall scores are accounted for by changing degree of hypnosis. We assume that everything has a cause, so there must be *other* variables that are causing the unaccounted-for differences in recall scores (perhaps differences in subjects' IQ, memory ability, or motivation play a role). Therefore, because of these other influences, the variable of degree of hypnosis accounts for only 22% of the variance, and so it is only somewhat important in determining variability in recall scores in this study.

> ***STAT ALERT*** The effect size indicates how big a role the conditions of the independent variable play in determining subjects' scores on the dependent variable.

Recognize that although a large effect size indicates an important relationship in a statistical sense, it does not indicate importance in a practical sense: statistics will never tell you if you have performed a silly study. In the real world, our conclusion that memory was improved by deep hypnosis has little practical importance. (To improve my memory, I should walk around under deep hypnosis all the time?) Statistical importance addresses a different issue: if you want to understand how nature works when it comes to hypnosis and memory, *then* this relationship is relevant and important. It is relevant because our results were significant, so we are confident that it is a "real" relationship found in nature. It is important to the extent that degree of hypnosis has an influence on memory that accounts for 22% of the differences in recall scores.

FINALLY

Two-sample experiments are fairly common in psychological research, so you are likely to use two-sample *t*-tests in your own research, and you'll need to understand them when reading the research of others. Effect size is another procedure you will frequently encounter, because it is the only way to determine whether a relationship is important or is much ado about nothing. In fact, the American Psychological Association now requires all published research reports to include a measure of effect size.[2] Because effect size has not always been reported in psychological research, great elaborate experiments have often been performed to study what are actually very minor variables. (Researchers sometimes forget the old adage "Things that are not worth doing are not worth doing well!") Therefore, compute effect size whenever you have significant results in an experiment. As we've seen, to describe a significant relationship you should think "correlation coefficient," but because this can be difficult to interpret, think "squared correlation coefficient." Then, by using r^2_{pb}—or other measures you will learn about—you'll know whether your variable is worth studying, and you'll be on the cutting edge of statistical sophistication (and that's not bad).

CHAPTER SUMMARY

1. Two samples are *independent* when we randomly select subjects and assign them to a sample without regard to who else has been selected for either sample.

2. See the fourth edition of the *Publication Manual of the American Psychological Association*, 1994, published by the American Psychological Association, Washington, D.C.

2. The *independent samples* t-*test* assumes that (a) the two random, independent samples of scores measure an interval or ratio variable, (b) the population of raw scores represented by each sample forms a normal distribution for which it is appropriate to compute the mean, and (c) the populations represented by the samples have homogeneous variance.

3. *Homogeneity of variance* means that if we could compute the variance of the populations represented by our samples, the value of σ_X^2 in one population would equal the value of σ_X^2 in the other population.

4. A significant t_{obt} from the independent samples *t*-test indicates that the difference between our sample means, $\bar{X}_1 - \bar{X}_2$, is significantly different from, and thus unlikely to represent, the difference between μ_1 and μ_2 described by H_0. Therefore, the results are assumed to represent the predicted relationship in the population.

5. The *confidence interval for* $\mu_1 - \mu_2$ contains a range of differences between two population μ's, any one of which is likely to be represented by the difference between our independent sample means.

6. Two samples are *related samples* when each score in one sample is paired with a particular score in the other sample. We pair the scores either by *matching* each subject in one condition with a subject in the other condition or by *repeated measures* of the same subjects under both conditions.

7. In performing the *related samples* t-*test,* we first find the difference between the two scores in each pair. Then we perform the *t*-test for a single sample, using the difference scores. A significant t_{obt} indicates that the mean of the difference scores, $\bar{D}$, is significantly different from the μ_D described by H_0. This implies that the means of the raw scores in each condition differ significantly and thus represent a relationship in the population.

8. The *confidence interval for* μ_D contains a range of values of μ_D, any one of which is likely to be represented by the sample mean, $\bar{D}$, from our related samples.

9. The power of the two-sample *t*-test increases with (a) larger differences in scores between the conditions, (b) smaller variability of scores within each condition with independent samples and smaller variability in the difference scores with related samples, and (c) larger *N*. All other things being equal, the *t*-test for related samples is more powerful than the *t*-test for independent samples.

10. We describe the strength of a significant relationship between the independent and dependent variables in a two-sample experiment by computing the *point-biserial correlation coefficient,* r_{pb}.

11. The squared point-biserial coefficient, r_{pb}^2, measures the proportion of variance in the dependent scores that is accounted for, or explained, by changing the conditions of the independent variable. The proportion of variance accounted for in an experiment is called the *effect size.* The larger the effect size, the more consistently the dependent scores change as we change the conditions of the independent variable.

PRACTICE PROBLEMS

(Answers for odd-numbered problems are provided in Appendix E.)

1. A scientist has conducted a two-sample experiment.

a. What two versions of a parametric procedure are available to him?

b. What is the deciding factor for selecting between them?

2. *a.* How does a researcher create independent samples?

b. What are the two ways that a researcher can create related samples?

c. What other assumptions must be met before using the two-sample *t*-test?

3. All other things being equal, should we create a study as a related samples or independent samples design? Why?

4. For each of the following, which type of *t*-test is required?

a. An investigation of the effects of a new memory-enhancing drug on the memory of Alzheimer's patients, testing a group of patients before and after administration of the drug.

b. An investigation of the effects of alcohol on motor coordination, comparing one group of subjects given a moderate dose of alcohol to the population μ for subjects given no alcohol.

c. An investigation of whether males and females rate the persuasiveness of an argument delivered by a female speaker differently.

d. The study described in part *c*, but with the added requirement that for each male of a particular age, there is a female of the same age.

5. *a.* What is $s_{\bar{X}_1 - \bar{X}_2}$?

b. What is $s_{\bar{D}}$?

6. What is homogeneity of variance?

7. Foofy has obtained a statistically significant two-sample t_{obt}. Now, what three things should she do to complete her analysis?

8. *a.* What does a measure of effect size tell us?

b. How is it computed in a two-sample experiment?

9. What does the confidence interval for μ_D indicate?

10. What does a confidence interval for the difference between two μ's indicate?

11. In an experiment, a researcher seeks to demonstrate a relationship between hot or cold baths (the independent variable) and the amount of relaxation they produce (the dependent variable). He obtains the following relaxation scores:

Sample 1 (hot): $\bar{X} = 43$, $s_X^2 = 22.79$, $N = 15$
Sample 2 (cold): $\bar{X} = 39$, $s_X^2 = 24.6$, $N = 15$

a. What are H_0 and H_a for this study?

b. Compute t_{obt}.

c. With $\alpha = .05$, what is t_{crit}?

d. What should the researcher conclude about this relationship?

e. Compute the confidence interval for the difference between the μ's.

f. How big of an effect does bath temperature have on relaxation?

g. Describe how you would graph these results.

12. A researcher investigates whether a period of time feels longer when people are bored than when they are not bored. The researcher obtains the following estimates of the time period (in minutes):

Sample 1 (bored): $\overline{X} = 14.5$, $s_X^2 = 10.22$, $N = 28$
Sample 2 (not bored): $\overline{X} = 9.0$, $s_X^2 = 14.6$, $N = 34$

a. What are H_0 and H_a for this study?

b. Compute t_{obt}.

c. With $\alpha = .05$, what is t_{crit}?

d. What should the researcher conclude about this relationship?

e. Compute the confidence interval for the difference between the μ's.

f. How important is boredom in determining how quickly time seems to pass?

13. Foofy predicts that students who use a computer program that corrects spelling errors will receive higher grades on a term paper. She uses an independent samples design in which Group A uses a spelling checker and Group B does not. She tests H_0: $\mu_A - \mu_B \leq 0$ and H_a: $\mu_A - \mu_B > 0$. She obtains a negative value of t_{obt}.

a. What should she conclude about this outcome?

b. Assuming that her sample means actually support her predictions, what miscalculation is she likely to have made?

14. To increase the power of the hypnosis study discussed in this chapter, how could you design the study to

a. Maximize the difference in the recall scores between the conditions?

b. Minimize the variability in the scores?

15. A researcher predicts that subjects will score higher on a questionnaire measuring well-being when they are exposed to lots of sunshine than they will when not exposed to much sunshine. A sample of 8 subjects is first measured after low levels of sunshine exposure and then again after high levels of exposure. The researcher collects the following well-being scores:

Low:	14	13	17	15	18	17	14	16
High:	18	12	20	19	22	19	19	16

a. Subtracting low from high, what are H_0 and H_a?

b. Compute the appropriate *t*-test.

c. With $\alpha = .05$, report your results.

d. Compute the appropriate confidence interval.

e. What is the predicted well-being score for a subject tested under low sunshine? Under high sunshine?

f. On average, how much more accurate are these predictions than if we did not know how much sunshine subjects experience?

g. What should the researcher conclude about these results?

16. A researcher investigates whether classical background music is more or less soothing to air-traffic controllers than Top-40 background music. He plays classical background music to one group and Top-40 music to another. At the end of the day, he gives each subject an irritability questionnaire and obtains the following data:

Sample A (classical): $n = 6$, $\overline{X} = 14.69$, $s_X^2 = 8.4$
Sample B (Top-40): $n = 6$, $\overline{X} = 17.21$, $s_X^2 = 11.6$

After computing the independent samples *t*-test, he finds $t_{obt} = +1.38$.

a. With $\alpha = .05$, report the statistical results.

b. What should the researcher conclude about these results?

c. What other statistics should be computed?

d. What statistical flaw is likely in the experiment?

e. What could the researcher do to improve the experiment?

f. What effect might this have?

17. A researcher investigates whether children exhibit a higher number of aggressive acts after watching a violent television show. The number of aggressive acts for the same ten subjects before and after watching the show are as follows:

Sample 1 (After)	*Sample 2 (Before)*
5	4
6	6
4	3
4	2
7	4
3	1
2	0
1	0
4	5
3	2

a. Subtracting before scores from after scores, what are H_0 and H_a for this study?

b. Compute t_{obt}.

c. With $\alpha = .05$, what is t_{crit}?

d. What should the researcher conclude about this relationship?

e. Compute the confidence interval for μ_D.

f. If you want to understand children's aggression, how important is it to consider whether they watch violent television shows?

18. You investigate whether the older or younger male in pairs of brothers tends to be more extroverted. You obtain the following extroversion scores:

Sample 1 (Younger)	***Sample 2 (Older)***
10	18
11	17
18	19
12	16
15	15
13	19
19	13
15	20

a. What are H_0 and H_a for this study?

b. Compute t_{obt}.

c. With $\alpha = .05$, what is t_{crit}?

d. What should you conclude about this relationship?

e. Is this a scientifically informative relationship?

19. To increase the power of the phobia study discussed in this chapter, how would you design the study to

a. Maximize the difference between the before and after anxiety scores?

b. Minimize variability in the difference scores?

SUMMARY OF FORMULAS

1. *Formulas for independent samples*

A. *The computational formula for the t-test for independent samples is*

$$t_{obt} = \frac{(\overline{X}_1 - \overline{X}_2) - (\mu_1 - \mu_2)}{\sqrt{\left(\frac{(n_1 - 1)s_1^2 + (n_2 - 1)s_2^2}{(n_1 - 1) + (n_2 - 1)}\right)\left(\frac{1}{n_1} + \frac{1}{n_2}\right)}}$$

Values of t_{crit} are found in Table 2 in Appendix D for $df = (n_1 - 1) + (n_2 - 1)$.

In the formula, $(\overline{X}_1 - \overline{X}_2)$ is the difference between the sample means, $(\mu_1 - \mu_2)$ is the difference described in H_0, s_1^2 and n_1 are from one sample, and s_2^2 and n_2 are from the other sample. Values of s_1^2 and s_2^2 are found using the formula

$$s_X^2 = \frac{\Sigma X^2 - \frac{(\Sigma X)^2}{N}}{N - 1}$$

B. *The computational formula for the confidence interval for the difference between two population μ's is*

$$(s_{\bar{X}_1-\bar{X}_2})(-t_{\text{crit}}) + (\bar{X}_1 - \bar{X}_2) \leq \mu_1 - \mu_2 \leq (s_{\bar{X}_1-\bar{X}_2})(+t_{\text{crit}}) + (\bar{X}_1 - \bar{X}_2)$$

where t_{crit} is the two-tailed value for $df = (n_1 + n_2) - 2$, the quantity $(\bar{X}_1 - \bar{X}_2)$ is the difference between the sample means, and $s_{\bar{X}_1-\bar{X}_2}$ is the standard error of the difference found using the formula

$$s_{\bar{X}_1-\bar{X}_2} = \sqrt{\left[\frac{(n_1 - 1)s_1^2 + (n_2 - 1)s_2^2}{(n_1 - 1) + (n_2 - 1)}\right]\left[\frac{1}{n_1} + \frac{1}{n_2}\right]}$$

2. *Formulas for related samples*

A. *The computational formula for the* t-test *for related samples is*

$$t_{\text{obt}} = \frac{\bar{D} - \mu_D}{\sqrt{(s_D^2)\left(\frac{1}{N}\right)}}$$

Values of t_{crit} are found in Table 2 in Appendix D for $df = N - 1$, where N is the number of difference scores.

In the formula, $\bar{D}$ is the mean of the difference scores, μ_D is the value described by H_0, and s_D^2 is the variance of the difference scores, found using the formula

$$s_D^2 = \frac{\Sigma D^2 - \frac{(\Sigma D)^2}{N}}{N - 1}$$

where D is each difference score and N is the number of difference scores.

B. *The combined formula for the* t-test *for related samples when $\mu_D = 0$ is*

$$t_{\text{obt}} = \frac{\Sigma D}{\sqrt{[N(\Sigma D^2) - (\Sigma D)^2]\left(\frac{1}{N - 1}\right)}}$$

Values of t_{crit} are found in Table 2 in Appendix B for $df = N - 1$, where N is the number of difference scores.

C. The computational formula for the confidence interval for μ_D is

$$(s_{\bar{D}})(-t_{\text{crit}}) + \bar{D} \leq \mu_D \leq (s_{\bar{D}})(+t_{\text{crit}}) + \bar{D}$$

t_{crit} is the two-tailed value for $df = N - 1$, where N is the number of difference scores, and $s_{\bar{D}}$ is the standard error of the mean difference, found using the formula

$$s_{\bar{D}} = \sqrt{(s_D^2)\left(\frac{1}{N}\right)}$$

where s_D^2 is the variance of difference scores and N is the number of difference scores.

3. *The formula for computing r_{pb} from t_{obt} is*

$$r_{\text{pb}} = \sqrt{\frac{(t_{\text{obt}})^2}{(t_{\text{obt}})^2 + df}}$$

With independent samples, $df = (n_1 - 1) + (n_2 - 1)$, where each n is the number of scores in a sample. With related samples, $df = N - 1$, where N is the number of difference scores.

4. *The proportion of variance accounted for* in a two-sample experiment equals the squared value of r_{pb}, or r_{pb}^2.

13

One-Way Analysis of Variance: Testing the Significance of Two or More Sample Means

To understand the upcoming chapter:

- From Chapter 5, understand that variance is a way of measuring the differences between scores by measuring their differences from some central score.
- From Chapter 10, understand why researchers limit the probability of a Type I error to .05.
- From Chapter 12 recall why we "pool" the sample variances to estimate the variance in the population and why we compute effect size for significant results.
- And from our various discussions, understand that all inferential procedures simply involve testing H_0 by calculating an obtained statistic that locates our data on the appropriate sampling distribution.

Then your goals in this chapter are to learn:

- The terminology of analysis of variance.
- Why we compute F_{obt}, and why *post hoc* tests are needed.
- What is meant by treatment variance and error variance.
- Why F_{obt} should equal 1.0 if H_0 is true, and why F_{obt} is greater than 1.0 if H_0 is false.
- When to compute Fisher's protected t-test or Tukey's *HSD*.
- How "eta squared" and "omega squared" describe the effect size of the independent variable.

You may have noticed that each new procedure we discuss involves a more complex experiment than the previous one (it's a plot!). In Chapter 12 we saw how to analyze a two-sample experiment involving two conditions of an independent variable. However, researchers often conduct experiments involving more than two conditions. The parametric procedure used in such experiments is called *analysis of variance.* Analysis of variance is perhaps *the* most common inferential statistical procedure because it can be used with many different experimental designs: it can be applied to independent samples or to related samples, to any number of conditions of an independent variable, and to any number of independent variables.

In this chapter we will discuss analysis of variance for one independent variable. At first glance, this procedure may appear to be very different from previous procedures. This is because analysis of variance is calculated differently and has its own language. However, ultimately the logic of hypothesis testing for this procedure is identical to that for previous procedures we have discussed.

MORE STATISTICAL NOTATION

There are several new terms used in analysis of variance. First, analysis of variance is abbreviated as **ANOVA.** Second, a **one-way ANOVA** is performed when only one independent variable is involved in the experiment. Third, an independent variable is called a **factor.** Thus, an experiment with one independent variable has one factor and we perform a one-way ANOVA.

As usual, the specific procedure and formulas we employ in a particular study depend on the type of data we have and on the study's design. ANOVA is a parametric procedure, so we must have data like those we had in previous chapters. Then the type of design we have depends upon whether we have independent or related samples. As in the previous chapter, we have independent samples when we randomly select subjects and assign them to one condition without regard to who has been selected for any other condition. In ANOVA, an independent variable that is studied using independent samples in all conditions is called a **between-subjects factor.** Throughout this chapter we will discuss the between-subjects, one-way ANOVA. However, when an independent variable is studied using related samples in all conditions, we have a **within-subjects factor,** and so we perform a within-subjects ANOVA (discussed in Appendix C).

Finally, each condition of the independent variable, or factor, is called a **level.** It is important to know the number of levels in a factor, and the symbol for the number of levels in a factor is k. These conditions, or levels, of an independent variable are also called **treatments.** When the different conditions of the independent variable produce significant differences in scores, we call this the *treatment effect* of the factor.

To illustrate all of these terms, let's say we are interested in how subjects' performance on a task is influenced by how difficult they perceive the task to be (the perceived difficulty of the task). We select three random samples containing five subjects each and provide them with the same easy math problems. However, to influence their perceptions, we tell subjects in Sample 1 that the problems are easy, subjects in Sample 2 that the problems are of medium difficulty, and subjects

TABLE 13.1 Diagram of a Study Having Three Levels of One Factor

Factor A: independent variable of perceived difficulty			
Level A_1: easy	*Level A_2: medium*	*Level A_3: difficult*	← *Conditions $k = 3$*
X	X	X	
X	X	X	
X	X	X	
X	X	X	
X	X	X	
$\bar{X}_1$	$\bar{X}_2$	$\bar{X}_3$	Overall $\bar{X}$
$n_1 = 5$	$n_2 = 5$	$n_3 = 5$	$N = 15$

in Sample 3 that the problems are difficult. Our dependent measure is the number of problems that subjects correctly solve within an allotted time. Thus, we seek to demonstrate a treatment effect on these scores by telling subjects the problems are either easy, of medium difficulty, or difficult. We have three conditions—easy, medium, and difficult—for our independent variable of perceived difficulty. Therefore, we will perform a one-way ANOVA with the three *levels* of easy, medium, and difficult for our *factor* of perceived difficulty. If subjects are tested under only one condition, and we do not match subjects, then this is a one-way, between-subjects design.

A good way to see the layout of a one-way ANOVA is to diagram it as shown in Table 13.1. Each column is a level of the factor, containing the scores of the subjects tested under that condition. (Here they are symbolized by X, but they will be the number of problems solved by a subject.) The symbol n stands for the number of scores in each level, and here there are five scores per level. The mean of each level is the mean of the scores from that condition. We again identify the n and $\bar{X}$ for each level using a subscript. Thus, $\bar{X}_1$ and n_1 are the mean and n from level 1, the easy condition. Because there are three levels in this factor, $k = 3$. (Notice that the general format is to label the factor as factor A, with levels A_1, A_2, and A_3.)

The total number of scores in the experiment is symbolized by N, and here $N = 15$. Further, the overall mean of all scores in the experiment will be the mean of all 15 scores.

AN OVERVIEW OF ANOVA

As with all experiments, the purpose of the above study is to demonstrate a relationship between the independent variable and the dependent variable. The only difference from previous studies is that now we have three samples of subjects. Ideally we will find a different sample mean for each condition. Then we want to conclude that if we tested the entire population under each level of the factor (each level of difficulty), we would find three different populations of scores located at three different μ's. But there is the usual problem: differences between the sample means may reflect sampling error, in which case we would actually find the same population of scores, having the same μ, for each level of difficulty. As usual, before we can conclude that a relationship exists, we must eliminate the idea that the differences between our sample means reflect sampling error. Thus, **analysis of variance** is the parametric statistical procedure for determining whether significant differences exist in an experiment containing two or more sample means. (When there are only two levels of the independent variable, you can perform either the two-sample t-test or ANOVA.)

Why Perform ANOVA? Experiment-wise Error Rate

You might think that we could use the independent samples t-test to determine whether there are significant differences between our three means above. That is, we might perform "multiple t-tests," testing whether $\overline{X}_1$ differs from $\overline{X}_2$, then whether $\overline{X}_2$ differs from $\overline{X}_3$, and finally whether $\overline{X}_1$ differs from $\overline{X}_3$. However, we cannot use this approach because of the probability of making a Type I error (rejecting a true H_0). With $\alpha = .05$, the theoretical probability of a Type I error in a *single* t-test is .05. However, we can make a Type I error when comparing $\overline{X}_1$ to $\overline{X}_2$, or when comparing $\overline{X}_2$ to $\overline{X}_3$, or when comparing $\overline{X}_1$ to $\overline{X}_3$. Therefore, the *overall* probability of making a Type I error somewhere in the experiment is considerably greater than .05.

This overall probability of making a Type I error is called the experiment-wise error rate. The **experiment-wise error rate** is the probability of making a Type I error in an experiment. When we compare only two means in an experiment, we can use the t-test, because with only one comparison, our experiment-wise error rate is equal to our α. However, when there are more than two levels in a factor, so that we make more than one comparison between the means, performing multiple t-tests results in an overall alpha that is greater than the one we selected: if we think $\alpha = .05$, it will actually be greater than .05. Because of the importance of avoiding Type I errors, we cannot afford to have our actual alpha greater than .05. Therefore, we *must* perform ANOVA. Only then will the experiment-wise error rate after we have compared all sample means equal the alpha we have chosen.

As usual, before we proceed with a study, we must check that it meets the assumptions of our statistical procedure.

Assumptions of the One-Way, Between-Subjects ANOVA

In a one-way, between-subjects ANOVA, the experiment has only one independent variable, and all of the conditions contain independent samples. We assume that

1. Each condition contains a random sample of interval or ratio scores.
2. The population represented by the scores in each condition forms a normal distribution, and the mean is the appropriate measure of central tendency.
3. The variances of all populations represented in the study are homogeneous.

Like other parametric procedures, ANOVA is robust, so we can use it even if we do not have perfectly normal or homogeneous populations. Although the number of subjects in each condition (n) need not be equal, violations of the assumptions are less serious when all n's are equal. Also, certain procedures are *much* easier to perform with equal n's.

If the study meets the assumptions of ANOVA, we set alpha (usually at .05) and create our statistical hypotheses.

Statistical Hypotheses of ANOVA

ANOVA tests only two-tailed hypotheses. Our null hypothesis is that there are no differences between the populations represented by the conditions. Thus, for our perceived difficulty study with the three conditions of easy, medium, and difficult, we have

$$H_0: \mu_1 = \mu_2 = \mu_3$$

In general, when we perform ANOVA on a factor with k levels, the null hypothesis is

$$H_0: \mu_1 = \mu_2 = \cdot \cdot \cdot = \mu_k$$

The "$. . . = \mu_k$" indicates that there are as many μ's as there are levels.

H_0 implies that our sample means for all conditions represent the same population mean, and therefore the sample means should be equal. If the means are not equal, it is because of sampling error in representing the one value of μ.

You might think that the alternative hypothesis would be that the various μ's are not equal, or $\mu_1 \neq \mu_2 \neq \mu_3$. However, a study may demonstrate differences between *some* but not *all* conditions. Perhaps our data represent a difference between μ_1 and μ_2, but not between μ_1 and μ_3, or perhaps only μ_2 and μ_3 differ. To communicate this idea, we write our alternative hypothesis as

$$H_a: \text{not all } \mu\text{'s are equal}$$

H_a implies that there is a relationship in the population involving at least two of our conditions: the population mean represented by one of the sample means is different from the population mean represented by at least one other sample mean.

As usual, we test H_0, so in ANOVA, we always test whether all sample means represent the same population mean.

The Order of Operations in ANOVA: The *F* Statistic and *Post Hoc* Comparisons

The statistic that forms the basis for ANOVA is called F. The ***F* statistic** simultaneously compares all sample means in a factor to determine whether two or more sample means represent different μ's. We will call the F we calculate our F_{obt}. We compare F_{obt} to the critical value of F, our F_{crit}.

When F_{obt} is not significant, it indicates that there are no significant differences between any of our sample means and that all means are likely to represent the same μ. When this occurs, the experiment has failed to demonstrate a relationship, we are finished with our statistical analyses, and it's back to the drawing board.

When F_{obt} is significant, it indicates that *somewhere* among our means there is *at least* one significant difference, so that *at least two* sample means are likely to represent different μ's. Therefore, we are not finished. The problem is that F_{obt} does not indicate *which* specific means differ significantly, and maybe more than two means are significantly different, or maybe all of them are. Thus, for example, if F_{obt} for our perceived difficulty study is significant, then it will indicate only that there is at least one significant difference somewhere between the means of the easy, medium, and difficult levels.

Obviously, to understand the relationship between our variables, we must determine which levels differ significantly from which. Therefore, we must perform a second statistical procedure, called *post hoc* comparisons. ***Post hoc* comparisons** are like *t*-tests, in which we compare one *pair* of sample means from our factor at a time. By comparing all possible pairs, we can determine which means differ significantly from each other. Thus, if F_{obt} for our perceived difficulty study is significant, we will perform *post hoc* comparisons to compare the means from easy and medium, easy and difficult, and medium and difficult. The results of these comparisons will indicate which means differ significantly from each other.

Note that we perform *post hoc* comparisons *only* when F_{obt} is significant. (*Post hoc* means "after the fact," which here is after F_{obt} is significant.) By finding a significant F_{obt} and then performing *post hoc* comparisons, we ensure that the experiment-wise probability of a Type I error in all of the significant differences together will actually be less than .05 (or whatever alpha we have selected).

STAT ALERT If F_{obt} is significant, we perform *post hoc* comparisons to determine which specific means differ significantly.

There is one exception to this rule. When there are only two levels in the factor, the significant difference indicated by F_{obt} must be between the only two means in the study, so it is unnecessary to perform *post hoc* comparisons.

In later sections we shall see how to perform *post hoc* comparisons. Our first procedure, though, is to compute F_{obt}. Therefore, in the following sections we discuss the statistical basis for ANOVA and the logic of its computation. The remainder of this chapter will then deal with getting through these computations, so hold on . . . here we go.

COMPONENTS OF THE *F* STATISTIC

Analysis of variance does just that: it analyzes variance. Recall that computing variance is simply a way to measure the *differences* between scores. (As you read the following, keep saying to yourself: "Variance is differences.") ANOVA involves partitioning the variance. That is, we take the total variability of the scores in an experiment and break it up, or partition it, in terms of its source. As we shall see, there are two potential sources of variance. First, scores may differ from each other even when subjects are in the same condition. We call this variability the variance *within groups.* Second, scores may differ from each other because they are from different conditions. We call this variability the variance *between groups.* Thus, in a one-way ANOVA, we partition the variance as shown in this diagram:

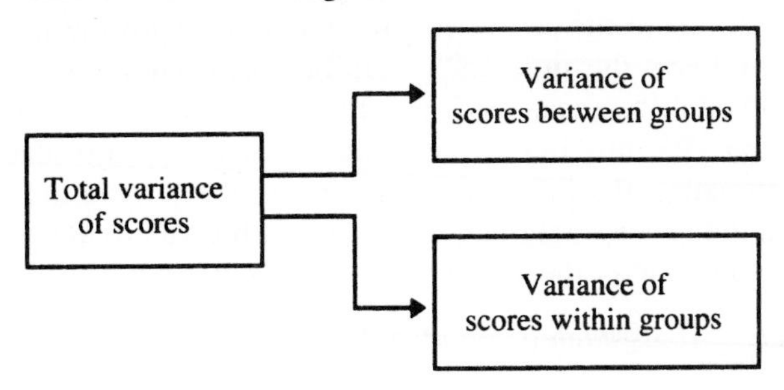

Using our sample data, we compute estimates of the value that each of these variances would have in the population. But we do not *call* each an estimated variance. Instead, we call each a **mean square.** This is a shortened name for the *mean squared deviation,* which is variance. The symbol for a mean square is *MS.* Because we estimate the variance within groups and the variance between groups, we compute two mean squares: the mean square within groups and the mean square between groups.

The Mean Square Within Groups

The **mean square within groups** is an estimate of the variability of scores as measured by differences within the conditions of an experiment. The symbol for the mean square within groups is MS_{wn}. The word *within* says it all: think of MS_{wn} as the "average variability" of the scores within, or inside, each condition around the mean of that condition. Table 13.2 illustrates how to conceptualize the computation of MS_{wn}. You can think of MS_{wn} in this way: it is as if we first find the variance in level 1 (finding the squared differences between the scores in level 1 and $\bar{X}_1$), then we find the variance of scores in level 2 around $\bar{X}_2$, and then we find the variance of the scores in level 3 around X_3. Then we "pool"

TABLE 13.2 How to Conceptualize the Computation of MS_{wn}

	Factor A	
Level A_1*:*	*Level* A_2*:*	*Level* A_3*:*
X	X	X
X	X	X
X	X	X
X	X	X
X	X	X
$\bar{X}_1$	$\bar{X}_2$	$\bar{X}_3$

the variances together, just like we did in the *t*-test. Thus, the MS_{wn} is the "average" variability of the scores in each condition around the mean of that condition. Because we assume homogeneity of variance, pooling the variances gives us the best, most representative estimate of the variance among the scores in any population that may be represented by our samples.

We compare the scores in each condition with the mean for that condition, so MS_{wn} reflects the inherent variability between subjects' scores that arises from individual differences (or from other random factors) when subjects are all treated the same. We saw in Chapter 5 that such variance is also called error. Therefore, by estimating the inherent variability in the population, MS_{wn} estimates what is called the **error variance,** which is symbolized by σ^2_{error}. (For this reason, MS_{wn} is also known as the *error term.*) In symbols,

Sample	***Estimates***	***Population***
MS_{wn}	$\rightarrow$	σ^2_{error}

We assume that MS_{wn} is an estimate of the error variance found in any of the populations represented by our samples. Thus, for example, an MS_{wn} equal to 4 indicates that we estimate the variance of the scores in any of the populations represented by our samples to be 4.

> ***STAT ALERT*** The MS_{wn} is an estimate of the error variance, the inherent variability within any population represented by our samples.

The Mean Square Between Groups

The other variance we compute in ANOVA is the mean square between groups. Here, the word *between* says it all. The **mean square between groups** is an estimate of the variability in scores that occurs *between* the levels in a factor. The mean square between groups is symbolized by MS_{bn}. We can conceptualize the computation of MS_{bn} as shown in Table 13.3. We summarize the scores in each level by computing the mean of that level, and then we determine how much it deviates from the overall mean of all scores in the experiment. In the same way that the squared deviations of raw scores around their mean describe

TABLE 13.3 How to Conceptualize the Computation of MS_{bn}

Factor A			
Level A_1	*Level A_2*	*Level A_3*	
X	X	X	
X	X	X	
X	X	X	
X	X	X	
X	X	X	
$\bar{X}_1$	$\bar{X}_2$	$\bar{X}_3$	Overall $\bar{X}$

how different the scores are from each other, the squared deviations of the sample means from the overall mean indicate how different the sample means are from each other. *Thus, MS_{bn} is our way of measuring how much the means in a factor differ from each other.*

The key to understanding ANOVA is to understand what MS_{bn} represents when H_0 is true and when it is false. First, consider the situation where the H_0 is true: the scores in our various conditions all come from the same population. When we sample the population, not every score will equal μ or each other. You can see this represented in the distribution in Figure 13.1. First look at the two scores (X's). Because of inherent variability, any two scores may differ. Here, for example, by chance one score is above μ and the other is below μ. If the population was available, we would measure such differences between all scores using the differences between the scores and μ to calculate σ^2_{error}. With our sample data, we instead estimate it using MS_{wn}.

Here's the crucial part: if H_0 is correct and our treatment does not work, then the reason we get a difference between any two *sample means* for our conditions is also due to this inherent variability. For example, as shown in Figure 13.1, by chance we may get a *batch* of scores above μ in one condition that produce a relatively high $\overline{X}$, but in another condition we may get a batch of scores below μ that produce a relatively low $\overline{X}$. If the population was available, we would measure the differences between all means. Instead, from our experiment we can estimate this using MS_{bn}.

Now, here's the *really* crucial part: the differences between sample means—their variability—is determined by how variable the raw scores are. If most scores are rather close to each other, then we will tend to get very similar scores in each

FIGURE 13.1 The One Population of Scores When H_0 is True, Showing Logic of MS_{wn} and MS_{bn}

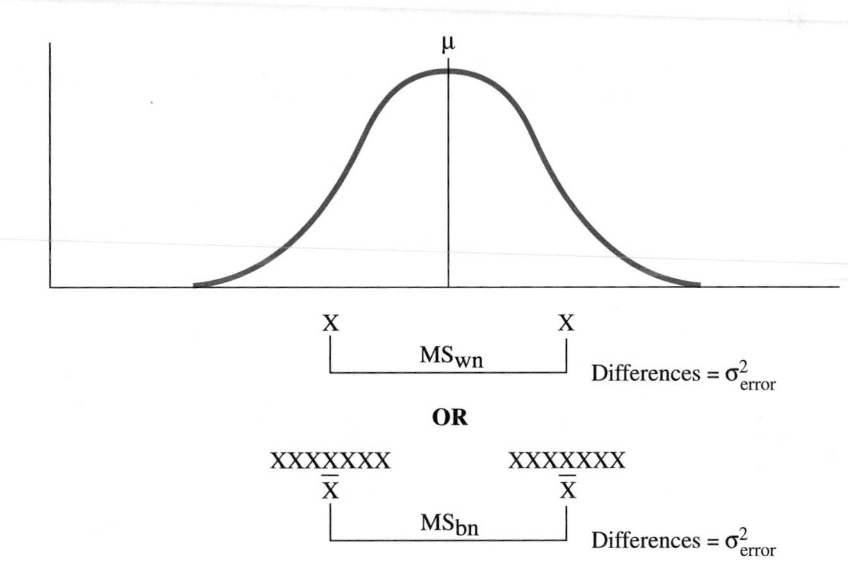

sample, so our means will be close to each other. However, if the scores are highly variable, then we are likely to get a very different batch of scores from one sample to the next, producing very different $\overline{X}$'s. Because the differences between the means are related to the differences between the scores, we could calculate σ^2_{error} in the population either by finding how much randomly selected raw scores differ or by finding how much random sample means differ. It doesn't matter, because both reflect the inherent differences between scores. Likewise, when we use our sample data to calculate MS_{bn} and MS_{wn}, they are (1) both estimating the one value of σ^2_{error}, so (2) they should be *equal.*

STAT ALERT If H_0 is true, then ideally MS_{bn} and MS_{wn} should be equal.

However, now consider when H_0 is false and the scores in our various conditions do not come from the same population. Now there are two reasons that combine to determine the value of MS_{bn}. First, any two sample means differ because we have a *treatment effect* (what we have called a "real" relationship, in which different conditions produce different populations of scores and each sample mean represents a different population μ). Differences due to a treatment effect are called treatment variance, which we symbolize by σ^2_{treat}. **Treatment variance** reflects differences between scores that occur because the scores are from different populations. In Figure 13.2, the distance separating the μ's of the distributions can be seen as reflecting treatment variance. We summarize these differences as the differences between the means of our levels, which we determine when calculating MS_{bn}. Therefore, MS_{bn} contains an estimate of the treatment variance.

Second, the difference between any two means also reflects the inherent variability among the scores in each population. For example, as shown in Figure 13.2, we might obtain by chance a $\overline{X}$ below μ in one population, and a $\overline{X}$ above μ in the other. Thus, the total distance between the sample means reflects two things: differences due to treatment, σ^2_{treat} (the solid line) *plus* differences due to the inherent variability of the scores in each population, σ^2_{error}, (the dotted lines). So

FIGURE 13.2 Two Populations of Scores When H_0 Is False, Showing Logic of MS_{bn}

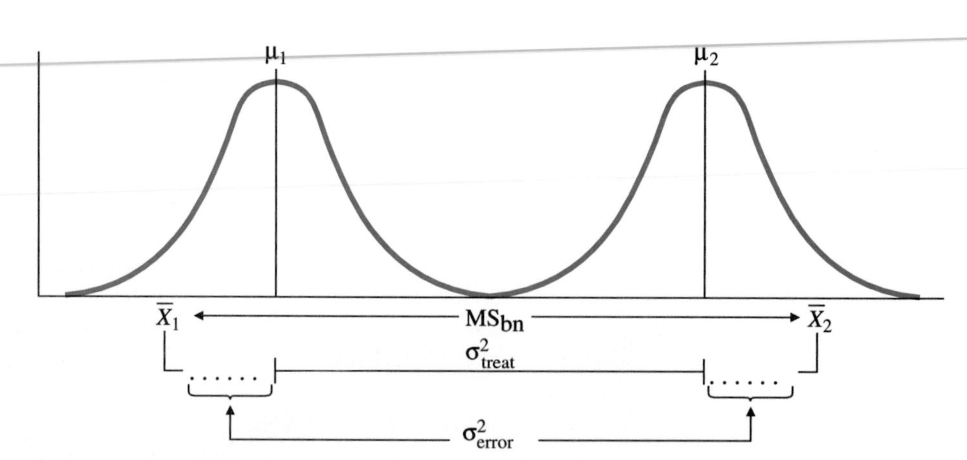

here's the *most* crucial part: because we calculate MS_{bn} as the differences between our sample means, the MS_{bn} contains an estimate of *both* error variance plus treatment variance. In symbols, this is

Sample	***Estimates***	***Population***
MS_{bn}	$\rightarrow$	$\sigma^2_{error} + \sigma^2_{treat}$

> **STAT ALERT** If H_0 is false, then MS_{bn} contains estimates of both error variance (σ^2_{error}), which measures differences *within* each population, and treatment variance (σ^2_{treat}), which measures differences *between* the populations.

The σ^2_{error} component is again the same value as estimated by MS_{wn}. But with the added σ^2_{treat} component, MS_{bn} will now be *larger* than MS_{wn}. Thus, for example, say that when MS_{wn} equals 4, we compute that MS_{bn} equals 10. We can think of MS_{bn} as indicating that, in addition to an error variance of 4, there is an "average difference" of 6 between the populations.

It's nice to think of MS_{bn} and MS_{wn} as breaking down like this, but of course the problem of sampling error makes the issue a little more complicated.

Comparing the Mean Squares: The Logic of the *F*-Ratio

For our purposes, we are not interested in directly comparing the actual values of MS_{bn} and MS_{wn}. Instead, in computing F_{obt}, we are interested in their ratio, called the F-ratio. The ***F*-ratio** is the mean square between groups divided by the mean square within groups.

> *THE COMPUTATIONAL FORMULA FOR THE* F*-RATIO IS*
>
> $$F_{obt} = \frac{MS_{bn}}{MS_{wn}}$$

> **STAT ALERT** In the F-ratio, MS_{bn} is always on top!

We can conceptualize what the F-ratio represents as follows:

	Sample	***Estimates***	***Population***
$F_{obt} =$	MS_{bn}	$\rightarrow$	$\sigma^2_{error} + \sigma^2_{treat}$
	MS_{wn}	$\rightarrow$	σ^2_{error}

The MS_{bn} represents the inherent differences between scores in any population (σ^2_{error}) *plus* whatever differences there are between the populations represented by our samples (σ^2_{treat}). This value is divided by the MS_{wn}, which is only an estimate of the error variance in the populations (σ^2_{error}).

Now we can understand what the F-ratio tells us about our null hypothesis. If H_0 is true and all conditions represent the same population, then there are zero differences due to treatment. Instead, any differences between the sample means totally reflect the inherent variability of scores that occurs naturally in the one population. Thus, when H_0 is true, MS_{bn} contains solely σ^2_{error}, and the σ^2_{treat} component of MS_{bn} equals zero. In symbols, when H_0 is true, we have

	Sample	***Estimates***	***Population***

$$F_{obt} = \frac{MS_{bn}}{MS_{wn}} \quad \begin{matrix} \rightarrow \\ \rightarrow \end{matrix} \quad \frac{\sigma^2_{error} + 0}{\sigma^2_{error}} = \frac{\sigma^2_{error}}{\sigma^2_{error}} = 1$$

Both mean squares are merely estimates of the one value of σ^2_{error}. If our data represent the situation where H_0 is true, then the mean square between groups should *equal* the mean square within groups. When two numbers are equal, their ratio equals 1. *Therefore, when H_0 is true and all conditions represent one population, the F-ratio should equal 1.*

On the other hand, H_a says that not all μ's are equal. If H_a is true, then at least two conditions represent different populations, and there are differences between at least two of our means that are due to treatment. Therefore, when H_a is true, the σ^2_{treat} component of MS_{bn} does not equal zero, so we have

	Sample	***Estimates***	***Population***

$$F_{obt} = \frac{MS_{bn}}{MS_{wn}} \quad \begin{matrix} \rightarrow \\ \rightarrow \end{matrix} \quad \frac{\sigma^2_{error} + \text{some amount of } \sigma^2_{treat}}{\sigma^2_{error}} = F > 1$$

Here MS_{bn} contains error variance *plus* some amount of treatment variance, so MS_{bn} will be *larger* than MS_{wn}, which contains only error variance. Placing a larger number in the numerator of the F-ratio produces an F_{obt} greater than 1. *Thus, when H_a is true, MS_{bn} is larger than MS_{wn}* and F_{obt} *is greater than 1.* The larger the differences between the populations represented by our sample means, the larger the σ^2_{treat} component, and thus the larger MS_{bn} will be. Regardless of the effect of the independent variable, however, the size of MS_{wn} remains constant. Therefore, the larger the differences between means, the larger F_{obt} will be.

Recognize that regardless of whether we have a positive, negative, or curvilinear relationship, MS_{bn} is a variance reflecting squared differences between the means. Therefore, any type of relationship produces an MS_{bn} larger than MS_{wn}, and F_{obt} is greater than 1.0. (This is why we have only two-tailed hypotheses in ANOVA.) An F_{obt} between 0 and 1 is possible, but it occurs when the denominator of the F-ratio, MS_{wn}, is larger than the numerator, MS_{bn}. Here we assume that MS_{bn} and/or MS_{wn} are merely poor estimates of σ^2_{error} and therefore are not equal. F_{obt} cannot be less than zero, because the mean squares are variances and therefore cannot be negative numbers.

Thus, on the one hand, F_{obt} should equal 1.0 if our sample means represent the same μ. On the other hand, F_{obt} should be greater than 1.0 if our means represent two or more different μ's. But hold on! There is one other reason F_{obt} might be greater than 1.0, and that is (here we go again) sampling error! When H_0 is true, F_{obt} "should" equal 1.0 *if* the mean squares are perfectly representative. *But,* through sampling error with *one* population, we might obtain differences between our sample means that are larger than the differences between the scores in the population. This will produce an MS_{bn} that is larger than MS_{wn}, so F_{obt} will be larger than 1.0 simply because of sampling error.

This all boils down to the same old problem of significance testing. An F_{obt} greater than 1.0 may accurately reflect the situation where two or more conditions of the independent variable represent different populations of raw scores (and there is a "real" treatment effect). Or, because of sampling error, an F_{obt} greater than 1.0 may inaccurately reflect the situation where all conditions represent the same population (and there only *appears* to be a treatment effect). Therefore, whenever we obtain an F_{obt} greater than 1.0, we must test H_0: we determine the probability of obtaining such an F_{obt} if H_0 is true and all sample means do represent one μ. To do this, we examine the F-distribution.

The *F*-Distribution

The ***F*-distribution** is the sampling distribution showing the various values of F that occur when H_0 is true and all conditions represent one population μ. We could create a sampling distribution in the following way: Using the same number of levels (k) and the same n's as in our study, we randomly sample one raw score population repeatedly. Each time, we compute MS_{bn} and MS_{wn}, form the F-ratio, and compute F_{obt}. After doing this an infinite number of times, we plot the various values of F_{obt}. The resulting distribution can be envisioned as shown in Figure 13.3. This is more or less the same old H_0 sampling distribution, except that now the X axis shows the values of F that occur when H_0 is true and all sample means *do* represent the same μ. The F-distribution is skewed, because there is no limit to how large F_{obt} can be, but it cannot be less than zero. The mean of the distribution is 1.0 because, most often, MS_{bn} will equal MS_{wn} and F will equal 1.0. We are concerned with the tail, which shows that sometimes the means are unrepresentative, producing an MS_{bn} that is larger than MS_{wn}, which results in an F greater than 1.0. However, as shown, the larger the F, the less frequent and thus the less likely it is when H_0 is true.

The F_{obt} we compute in a study can reflect a relationship in the population only when it is greater than 1.0, so we place the entire region of rejection in the upper tail of the F-distribution. (With ANOVA, we always have two-tailed

FIGURE 13.3 Sampling Distribution of F When H_0 Is True

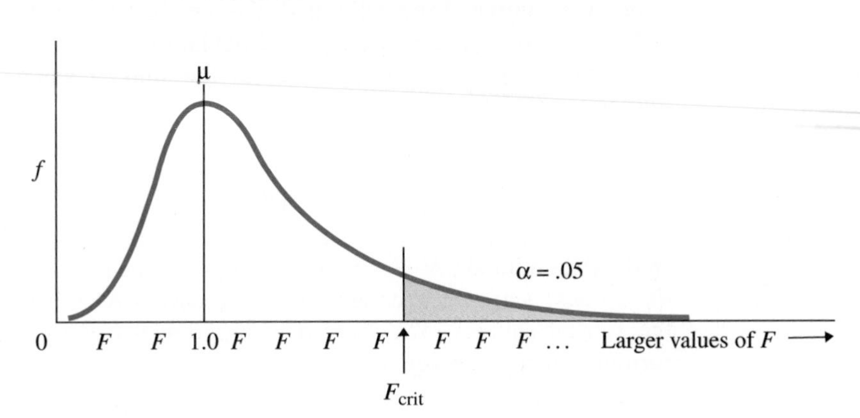

hypotheses, yet we test them using a one-tailed test!) If our F_{obt} is larger than F_{crit}, then F_{obt}—and the differences between the sample means that produced it—is too unlikely for us to accept as having occurred by chance when H_0 is true and all conditions of the factor represent the same population. Therefore, we reject H_0 and have a significant F_{obt}.

Like the *t*-distribution, the *F*-distribution consists of a family of curves. Each *F*-distribution has a slightly different shape depending on the degrees of freedom in the data, and thus there is a different value of F_{crit} for each *df*. However, there are *two* values of *df* that determine the shape of each *F*-distribution: the *df* for the mean square between groups and the *df* for the mean square within groups. The symbol for the *df* between groups is df_{bn}, and the symbol for the *df* within groups is df_{wn}. We use both df_{bn} and df_{wn} when finding F_{crit}.

To obtain F_{crit}, turn to Table 5 in Appendix D, entitled "Critical Values of *F*." Across the top of these *F*-tables, the columns are labeled "*df* between groups," and along the left-hand side, the rows are labeled "*df* within groups." We locate the appropriate column and row using the *df*'s from our study. The critical values of *F* in dark type are for $\alpha = .05$, and the values in light type are for $\alpha = .01$. For example, say we eventually determine that $df_{bn} = 2$ and $df_{wn} = 12$. Then for $\alpha = .05$, the F_{crit} is 3.88.

STAT ALERT Be careful to keep your "withins" and "betweens" straight: $df_{bn} = 2$ and $df_{wn} = 12$ is very different from $df_{bn} = 12$ and $df_{wn} = 2$.

Do not be overwhelmed by all the details of ANOVA. Buried in here is the simple idea that the larger the differences between our means for the conditions of the independent variable, the larger the MS_{bn} and thus the larger the F_{obt}. If the F_{obt} is larger than F_{crit}, then the F_{obt} (and the corresponding differences between our means) is too unlikely to occur if our means represent the same population μ. Therefore, we can reject H_0 and confidently conclude that our data represent a "real" relationship.

COMPUTING THE *F*-RATIO

When we computed the estimated variance in Chapter 5, we called the quantity $\Sigma(X - \bar{X})^2$ the sum of the squared deviations. In ANOVA, we shorten this to the **sum of squares.** The symbol for the sum of squares is *SS*. Then, in the numerator of the following formula for variance, we can replace the sum of the squared deviations with *SS*:

$$s^2_X = \frac{\Sigma(X - \bar{X})^2}{N - 1} = \frac{SS}{df} = MS$$

In the denominator, $N - 1$ is the degrees of freedom, so we replace $N - 1$ with *df*. Because variance is called a mean square in ANOVA, the fraction formed by the sum of squares (*SS*) divided by the degrees of freedom (*df*) is the general formula for a mean square.

Adding our subscripts, we compute the mean square between groups (MS_{bn}) by computing the sum of squares between groups (SS_{bn}) and then dividing by the degrees of freedom between groups (df_{bn}). Likewise, we compute the mean square within groups (MS_{wn}) by computing the sum of squares within groups (SS_{wn}) and then dividing by the degrees of freedom within groups (df_{wn}). Once we have MS_{bn} and MS_{wn}, we compute F_{obt}.

If all this strikes you as the most confusing thing ever devised by humans, you will find creating an ANOVA summary table very helpful. The general format of the summary table for a one-way ANOVA is:

Summary Table of One-Way ANOVA

Source	*Sum of squares*	*df*	*Mean square*	*F*
Between	SS_{bn}	df_{bn}	MS_{bn}	F_{obt}
Within	SS_{wn}	df_{wn}	MS_{wn}	
Total	SS_{tot}	df_{tot}		

The source column identifies each component. Eventually we will fill in the values in the other columns. Notice that as we do, the computations become built in. Dividing SS_{bn} by df_{bn} produces the MS_{bn}. Dividing SS_{wn} by df_{wn} produces the MS_{wn}. Finally, the fraction formed by putting MS_{bn} "over" MS_{wn} produces F_{obt}. [Along the way, we will also compute the total sum of squares (SS_{tot}) and the total *df* (df_{tot}).] The F_{obt} is always placed in the row labeled "Between." (In place of the word "Between," we can use the name of the factor or independent variable. Also, in place of the word "Within," you will sometimes see "Error.")

Computational Formulas for the One-Way, Between-Subjects ANOVA

To provide some example data to work with, say that we actually performed our study of the effects of perceived difficulty that we discussed earlier: we told three samples of five subjects each that some math problems were easy, of medium difficulty, or difficult, and we measured the number of problems they correctly solved. The data are presented in Table 13.4 on the next page.

As shown in the table, the first step in performing ANOVA is to compute ΣX, ΣX^2, and $\bar{X}$ for each level. By adding the ΣX from each level, we compute the total ΣX, and by adding the ΣX^2 from each level, we compute the total ΣX^2. Then, as shown in the following sections, we compute the sum of squares, the degrees of freedom, the mean squares, and then F_{obt}. So that you do not get lost, as you complete each step, fill in the results in the ANOVA summary table created above. (There *will* be a test later.)

Computing the sums of squares The first task is to compute the sum of squares. We do this in three steps, as shown on the next page.

TABLE 13.4 Data from Perceived Difficulty Experiment

Factor A: perceived difficulty			
Level A_1: easy	*Level A_2: medium*	*Level A_3: difficult*	
9	4	1	
12	6	3	
4	8	4	
8	2	5	
7	10	2	*Totals*
$\Sigma X = 40$	$\Sigma X = 30$	$\Sigma X = 15$	$\Sigma X = 85$
$\Sigma X^2 = 354$	$\Sigma X^2 = 220$	$\Sigma X^2 = 55$	$\Sigma X^2 = 629$
$n_1 = 5$	$n_2 = 5$	$n_3 = 5$	$N = 15$
$\bar{X}_1 = 8$	$\bar{X}_2 = 6$	$\bar{X}_3 = 3$	$k = 3$

Step 1 is to compute the total sum of squares (SS_{tot}).

THE COMPUTATIONAL FORMULA FOR SS_{tot} IS

$$SS_{tot} = \Sigma X^2_{tot} - \left(\frac{(\Sigma X_{tot})^2}{N}\right)$$

Here we treat the entire experiment as if it were one big sample. Thus, ΣX_{tot} is the sum of all X's, and ΣX^2_{tot} is the sum of all squared X's. N is the total N in the study.

Using the data from Table 13.4, we have $\Sigma X^2_{tot} = 629$, $\Sigma X_{tot} = 85$, and $N = 15$, so

$$SS_{tot} = 629 - \frac{(85)^2}{15}$$

$$SS_{tot} = 629 - \frac{7225}{15}$$

$$SS_{tot} = 629 - 481.67$$

Thus, $SS_{tot} = 147.33$.

Step 2 is to compute the sum of squares between groups (SS_{bn}).

THE COMPUTATIONAL FORMULA FOR SS_{bn} IS

$$SS_{bn} = \Sigma\left(\frac{(\text{sum of scores in the column})^2}{n \text{ of scores in the column}}\right) - \left(\frac{(\Sigma X_{tot})^2}{N}\right)$$

When we diagram the study, each column represents a level of the factor, a separate sample of subjects tested under a particular condition. Thus, the formula says to find the ΣX for each level, to square ΣX, and then to divide by the n in that level. After doing this for all levels, we add the results together and subtract the quantity $(\Sigma X_{tot})^2/N$. From Table 13.4, we have

$$SS_{bn} = \left(\frac{(40)^2}{5} + \frac{(30)^2}{5} + \frac{(15)^2}{5}\right) - \left(\frac{(85)^2}{15}\right)$$

so

$$SS_{bn} = (320 + 180 + 45) - 481.67$$

and

$$SS_{bn} = 545 - 481.67$$

Thus, we have $SS_{bn} = 63.33$.

Step 3 is to compute the sum of squares within groups (SS_{wn}). It was relatively painless to directly compute SS_{tot} and SS_{bn} above, but it is not so painless to directly compute SS_{wn}. Instead, we use a shortcut. Mathematically, SS_{tot} equals SS_{bn} plus SS_{wn}. Therefore, the total minus the between leaves the within.

THE COMPUTATIONAL FORMULA FOR SS_{wn} IS

$$SS_{wn} = SS_{tot} - SS_{bn}$$

In our example, SS_{tot} is 147.33 and SS_{bn} is 63.33, so

$$SS_{wn} = 147.33 - 63.33 = 84.00$$

so $SS_{wn} = 84.00$. (Now that's painless.)

Filling in the first column of our ANOVA summary table, we have

Summary Table of One-Way ANOVA

Source	*Sum of squares*	*df*	*Mean square*	*F*
Between	63.33	df_{bn}	MS_{bn}	F_{obt}
Within	84.00	df_{wn}	MS_{wn}	
Total	147.33	df_{tot}		

As a double check, make sure that the total equals the sum of the between plus the within. Here, $63.33 + 84.00 = 147.33$.

Now compute the degrees of freedom.

Computing the degrees of freedom We compute df_{bn}, df_{wn}, and df_{tot}. Again, we use three steps.

1. *The degrees of freedom between groups equals* $k - 1$, where k is the number of levels in the factor. In our example, there are three levels in the factor

of perceived difficulty (easy, medium, and difficult), so $k = 3$. Thus, $df_{\text{bn}} = 2$.

2. *The degrees of freedom within groups equals $N - k$,* where N is the total N of the study and k is the number of levels in the factor. In our example, the total N is 15 and k equals 3, so df_{wn} equals $15 - 3$, or 12.
3. *The degrees of freedom total equals $N - 1$,* where N is the total N in the experiment. In our example, we have a total of 15 scores, so $df_{\text{tot}} = 15 - 1 = 14$.

The df_{tot} is useful because it equals the sum of the df_{bn} plus the df_{wn}. Thus, to check our work in the example, we find that $df_{\text{bn}} + df_{\text{wn}} = 2 + 12$, which equals 14, our df_{tot}.

After *SS* and *df* are recorded, our summary table looks like this:

Summary Table of One-Way ANOVA

Source	*Sum of squares*	*df*	*Mean square*	*F*
Between	63.33	2	MS_{bn}	F_{obt}
Within	84.00	12	MS_{wn}	
Total	147.33	14		

Now find each mean square.

Computing the mean squares To compute the mean squares, we work directly from our summary table. Any mean square equals the appropriate sum of squares divided by the corresponding *df*.

THE COMPUTATIONAL FORMULA FOR MS_{bn} IS

$$MS_{\text{bn}} = \frac{SS_{\text{bn}}}{df_{\text{bn}}}$$

From the summary table for our example, we see that

$$MS_{\text{bn}} = \frac{63.33}{2} = 31.67$$

so MS_{bn} is 31.67.

THE COMPUTATIONAL FORMULA FOR MS_{wn} IS

$$MS_{\text{wn}} = \frac{SS_{\text{wn}}}{df_{\text{wn}}}$$

For our example,

$$MS_{wn} = \frac{84}{12} = 7.00$$

so MS_{wn} is 7.00.

Notice that we do *not* compute the mean square for SS_{tot}.

Placing these values in the summary table, we have

Summary Table of One-Way ANOVA

Source	*Sum of squares*	*df*	*Mean square*	*F*
Between	63.33	2	31.67	F_{obt}
Within	84.00	12	7.00	
Total	147.33	14		

Computing the F_{obt} Last but not least, we compute F_{obt}.

THE COMPUTATIONAL FORMULA FOR F IS

$$F_{obt} = \frac{MS_{bn}}{MS_{wn}}$$

In our example, MS_{bn} is 31.67 and MS_{wn} is 7.0, so

$$F_{obt} = \frac{MS_{bn}}{MS_{wn}} = \frac{31.67}{7.00} = 4.52$$

Thus F_{obt} is 4.52.

Now we have the completed ANOVA summary table:

Summary Table of One-Way ANOVA

Source	*Sum of squares*	*df*	*Mean square*	*F*
Between	63.33	2	31.67	4.52
Within	84.00	12	7.00	
Total	147.33	14		

Interpreting F_{obt} in a One-Way ANOVA

To interpret F_{obt}, we must have F_{crit}, so we turn to the *F*-tables in Appendix D. In our example, df_{bn} is 2 and df_{wn} is 12. With $\alpha = .05$, F_{crit} is 3.88.

FIGURE 13.4 Sampling Distribution of F When H_0 Is True for $df_{bn} = 2$ and $df_{wn} = 12$

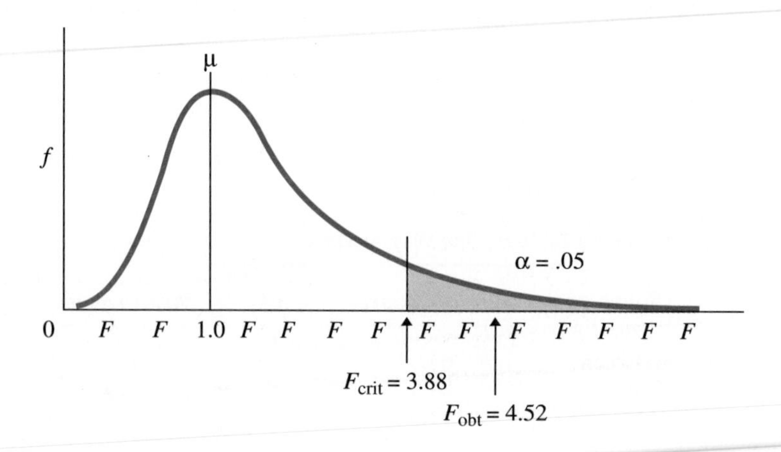

Thus, our F_{obt} is 4.52 and our F_{crit} is 3.88. Lo and behold, as shown in Figure 13.4, we have a significant F_{obt}. The null hypothesis (H_0: $\mu_1 = \mu_2 = \mu_3$) says that the differences between our sample means are due to sampling error and that all means poorly represent one population mean. However, our F_{obt} is out there in the region of rejection, telling us that such differences between $\bar{X}$'s hardly ever happen when H_0 is true. Because F_{obt} is larger than F_{crit}, we reject H_0: we conclude that the differences between our sample means are too unlikely for us to accept as having occurred by chance if all means represent one population μ. That is, we conclude that our F_{obt} is significant and that the factor of perceived difficulty does produce a significant difference in mean performance scores. We report our results as

$$F(2, 12) = 4.52, \quad p < .05$$

Notice that in the parentheses we report df_{bn} and then df_{wn}. (Get in the habit of saying the df_{bn} first and the df_{wn} second.) As usual, because $\alpha = .05$, the probability that we have incorrectly rejected H_0 (made a Type I error) is $p < .05$.

Of course, had F_{obt} been less than F_{crit}, then the corresponding differences between our means would *not* be too unlikely to accept as occurring by chance if H_0 is true, so we would not reject H_0.

Because we rejected H_0 and accepted H_a, we return to the means from the levels of our factor:

Perceived difficulty

Easy	*Medium*	*Difficult*
$\bar{X}_1 = 8$	$\bar{X}_2 = 6$	$\bar{X}_3 = 3$

We are confident that these means represent a relationship in the population: changing the level of perceived difficulty would produce different populations of performance scores, having different μ's. However, we do not know whether *every* sample mean represents a different μ. Remember: a significant F_{obt} merely indicates that there is *at least* one significant difference between two of these means. Now we must determine which specific means differ significantly, and to do that we perform *post hoc* comparisons.

PERFORMING *POST HOC* COMPARISONS

Statisticians have developed a variety of *post hoc* procedures which differ in how likely they are to produce Type I or Type II errors. We shall discuss two procedures that have acceptably low error rates.[1] Depending on whether or not our n's are equal, we compute either Fisher's protected t-test or Tukey's *HSD* test.

Fisher's Protected *t*-Test

You should perform Fisher's protected t-test when the n's in all levels of the factor are not equal.

THE COMPUTATIONAL FORMULA FOR THE PROTECTED t*-TEST IS*

$$t_{obt} = \frac{\bar{X}_1 - \bar{X}_2}{\sqrt{MS_{wn}\left(\frac{1}{n_1} + \frac{1}{n_2}\right)}}$$

This is basically the formula for the independent samples t-test, except that we have replaced the pooled variance (s^2_{pool}) computed in the t-test with the MS_{wn} computed in the ANOVA. We are testing H_0: $\mu_1 - \mu_2 = 0$, where $\bar{X}_1$ and $\bar{X}_2$ are the means for any two levels of the factor and n_1 and n_2 are the corresponding n's in those levels. The t_{crit} is the two-tailed value found in Appendix D, Table 2, for $df = df_{wn} = N - k$.

We can perform the protected t-test even when the n's in all the conditions are equal. Thus, for example, we can compare the mean from our easy level ($\bar{X} = 8.0$) to the mean from our difficult level ($\bar{X} = 3.0$). Each n is 5, and from our ANOVA summary table we know that MS_{wn} is 7.0. Filling in the formula, we have

$$t_{obt} = \frac{8.0 - 3.0}{\sqrt{7.0\left(\frac{1}{5} + \frac{1}{5}\right)}}$$

[1]Carmer, S. G., and Swanson, M. R. (1973). An evaluation of ten multiple comparison procedures by Monte Carlo methods. *Journal of the American Statistical Association, 68,* pp. 66–74.

Then we have

$$t_{obt} = \frac{+5.0}{\sqrt{7.0(.4)}}$$

which becomes

$$t_{obt} = \frac{+5.0}{\sqrt{2.8}} = \frac{+5.0}{1.67} = +2.99$$

We then compare our t_{obt} to the two-tailed value of t_{crit} found in the t-tables. For our example, with $\alpha = .05$ and $df_{wn} = 12$, t_{crit} is ± 2.179. Because our t_{obt} of $+2.99$ is beyond the t_{crit} of ± 2.179, we conclude that the means from the easy and difficult levels differ significantly (that they do not represent the same μ).

To complete our *post hoc* comparisons, we perform the protected t-test on all possible pairs of means in the factor. Thus, after comparing the means from easy and difficult, we would perform the protected t-test comparing the means from easy and medium and the means from medium and difficult. When we are finished, we will have *protected* the experiment-wise error rate, so that the probability of a Type I error for all of these comparisons together is $p < .05$.

If a factor contains many levels, then the protected t-test becomes very tedious. If you think there has *got* to be a better way, you're right.

Tukey's *HSD* Multiple Comparisons Test

You should perform the Tukey *HSD* multiple comparisons procedure *only* when the n's in all levels of the factor are equal. The *HSD* is a convoluted variation of the t-test in which we compute the minimum difference between any two means that is required for the means to differ significantly (*HSD* stands for the Honestly Significant Difference). There are four steps in performing the *HSD* test.

Step 1 is to find q_k. The value of q_k is a number found in Table 6 in Appendix D, entitled "Values of the Studentized Range Statistic, q_k." In the table, locate the column labeled with the k corresponding to the number of means in the factor. Next, find the row labeled with the df_{wn} used to compute your previous significant F_{obt}. Then find the value of q_k for the appropriate α. For our study above, $k = 3$, $df_{wn} = 12$, and $\alpha = .05$, so $q_k = 3.77$. By using the appropriate q_k in our computations, we protect our experiment-wise error rate for the number of means we are comparing.

Step 2 is to compute the *HSD*.

THE COMPUTATIONAL FORMULA FOR THE HSD IS

$$HSD = (q_k)\left(\sqrt{\frac{MS_{wn}}{n}}\right)$$

MS_{wn} is the denominator from your significant F-ratio, and n is the number of scores in each level of the factor.

In our example, MS_{wn} was 7.0 and n was 5, so

$$HSD = (q_k)\left(\sqrt{\frac{MS_{\text{wn}}}{n}}\right) = (3.77)\left(\sqrt{\frac{7.0}{5}}\right) = 4.46$$

Thus, our HSD is 4.46.

Step 3 is to determine the differences between all means. Simply subtract each mean from every other mean. Ignore whether differences are positive or negative (this is a two-tailed test of the H_0 that $\mu_1 - \mu_2 = 0$). For our perceived difficulty study, we can diagram the differences as shown below:

Perceived difficulty

Easy	*Medium*	*Difficult*
$\bar{X}_1 = 8$	$\bar{X}_2 = 6$	$\bar{X}_3 = 3$

2.0 3.0

5.0

$HSD = 4.46$

On the line connecting any two levels is the absolute difference between their means.

Step 4 is to compare each difference between two means to the HSD. If the absolute difference between the two means is *greater than* the HSD, then these means differ significantly. (It is as if we performed the protected t-test on these two means and t_{obt} was significant.) If the absolute difference between the two means is less than or equal to the HSD, then it is *not* a significant difference (and would not produce a significant t_{obt}).

Above, our HSD was 4.46. Of our three means, the means from the easy level ($\bar{X}_1 = 8$) and the difficult level ($\bar{X}_3 = 3$) differ by more than 4.46, so they differ significantly. The mean from the medium level ($\bar{X}_2 = 6$) differs from the other means by less than 4.46, so it does not differ significantly from them.

Thus, our final conclusion about our perceived difficulty study is that we demonstrated a relationship between subjects' scores and perceived difficulty, but only for the easy and difficult conditions. If these two conditions were given to the population, we would expect to find two different populations of scores, having two different μ's. We cannot say anything about whether the medium level would produce a different population and μ, because we failed to find that it produced a significant difference.

SUMMARY OF STEPS IN PERFORMING A ONE-WAY ANOVA

It has been a long haul, but here is everything we do when performing a one-way ANOVA:

1. Our null hypothesis is H_0: $\mu_1 = \mu_2 = \ldots \mu_k$, and our alternative hypothesis is H_a: not all μ's are equal. We choose our alpha level, check the assumptions, and then collect the data.

2. We first compute the sum of squares between groups (SS_{bn}) and the sum of squares within groups (SS_{wn}). Then we compute the degrees of freedom between groups (df_{bn}) and the degrees of freedom within groups (df_{wn}). Dividing the SS_{bn} by the df_{bn} gives the mean square between groups (MS_{bn}), and dividing the SS_{wn} by the df_{wn} gives the mean square within groups (MS_{wn}). Finally, dividing the MS_{bn} by the MS_{wn} gives the F_{obt}.

3. We find F_{crit} in Appendix D, Table 5, using the df_{bn} and the df_{wn}. If the null hypothesis is true, F_{obt} "should" equal 1. The larger the value of F_{obt}, the less likely it is that H_0 is true. If F_{obt} is larger than F_{crit}, we have a significant F_{obt}. We conclude that the means in at least two conditions differ significantly, representing at least two different population μ's.

4. If F_{obt} is significant and there are more than two levels of the factor, we determine which specific levels differ significantly by performing *post hoc* comparisons. We perform the protected *t*-test if the *n*'s in all levels of the factor are not equal or the *HSD* procedure if all *n*'s are equal.

If you followed all of that, then congratulations, you are getting *good* at this stuff. Of course, all of this merely determines whether we have a relationship. Now we must describe that relationship.

DESCRIBING THE RELATIONSHIP IN A ONE-WAY ANOVA

As we saw in previous chapters, we are not finished once we have demonstrated a significant relationship. Ultimately we want to understand how nature operates on behaviors in this situation (here, how subjects' perceived difficulty influences their math performance). To help us, we want to first understand the relationship we have demonstrated. Therefore, we describe the relationship by computing a confidence interval for each μ, by graphing the relationship, and by computing the proportion of variance accounted for.

The Confidence Interval for Each Population μ

In our example, the mean from the easy condition ($\overline{X} = 8.0$) differs significantly from the mean from the difficult condition ($\overline{X} = 3.0$). Therefore, we expect that the population means represented by these conditions would be "around" 8 and 3, respectively. As usual, to more clearly define "around," we compute a confidence interval for the μ represented by each sample mean. This confidence interval for a single μ is the same as the one we discussed in Chapter 11, except that here it is computed directly from the components of ANOVA.

THE COMPUTATIONAL FORMULA FOR COMPUTING THE CONFIDENCE INTERVAL FOR A SINGLE μ, USING THE RESULTS OF A BETWEEN-SUBJECTS ANOVA, IS

$$\left(\sqrt{\frac{MS_{\text{wn}}}{n}}\right)(-t_{\text{crit}}) + \bar{X} \leq \mu \leq \left(\sqrt{\frac{MS_{\text{wn}}}{n}}\right)(+t_{\text{crit}}) + \bar{X}$$

The value of t_{crit} is the two-tailed value found in the t-tables, using the appropriate α and using our df_{wn} as the df. We find MS_{wn} in our ANOVA, and $\bar{X}$ and n are from the level we are describing.

For example, in our easy condition, $\bar{X} = 8.0$, we have $MS_{\text{wn}} = 7.0$, $df_{\text{wn}} = 12$, and $n = 5$. The two-tailed t_{crit} (at $df = 12$ and $\alpha = .05$) is ± 2.179. Placing these values in the above formula, we have

$$\left(\sqrt{\frac{7.0}{5}}\right)(-2.179) + 8.0 \leq \mu \leq \left(\sqrt{\frac{7.0}{5}}\right)(+2.179) + 8.0$$

This becomes

$$(-2.578) + 8.0 \leq \mu \leq (+2.578) + 8.0$$

Finally we have

$$5.42 \leq \mu \leq 10.58$$

Because $\alpha = .05$, this is the 95% confidence interval: if we were to test the entire population of subjects under our easy condition, we are 95% confident that the population mean would fall between 5.42 and 10.58.

We follow the same procedure to describe the μ from any other significant level of the factor.

Graphing the Results in ANOVA

As usual, we graph our results with the dependent variable on the Y axis and the independent variable (levels of the factor) on the X axis. Then we plot the mean for each condition. Figure 13.5 shows the line graph for our perceived difficulty study. Note that even though the medium level of difficulty did not produce significant differences, we still include it in the graph.

Eta Squared: The Proportion of Variance Accounted For in the Sample Data

Remember that so far, by only saying that the relationship is significant, it's as if you're saying "I've found a correlation, but I'm not telling what it is." Therefore, first think "correlation coefficient" to describe the strength of the relationship between your independent and dependent variables. We use the computations from an ANOVA to compute a new correlation coefficient called eta (pronounced

FIGURE 13.5 Mean Number of Problems Correctly Solved as a Function of Perceived Difficulty

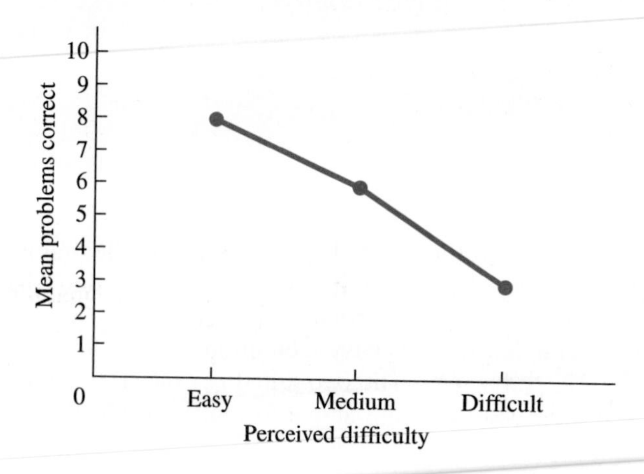

"ay-tah"). **Eta** is analogous to r_{pb}, except that eta can be used to describe any linear or nonlinear relationship containing two or more levels of a factor.

But then remember that we get to the heart of describing a relationship by thinking "squared correlation coefficient." This tells us the *effect size* of the independent variable—the proportion of variance in dependent scores that is associated with changing our conditions. With ANOVA, we compute effect size by squaring the value of eta: **eta squared** indicates the proportion of variance in the dependent variable that is accounted for by changing the levels of a factor. The symbol for eta squared is η^2, and it is calculated as follows:

THE COMPUTATIONAL FORMULA FOR η^2 *IS*

$$\eta^2 = \frac{SS_{bn}}{SS_{tot}}$$

The SS_{bn} is a way of measuring the differences between the means from our various levels. The SS_{tot} is a way of measuring the total differences between all scores in the experiment. Thus, η^2 reflects the proportion of the total differences in the scores that is associated with differences between the sample means. This is interpreted in the same way that we previously interpreted $r_{pb}{}^2$. Therefore, η^2 indicates how consistently the differences in scores can be predicted using the means of the levels of our factor. The larger the value of η^2, the more that knowing the subjects' condition of the independent variable improves our accuracy in describing and predicting their dependent scores. In turn this means that the factor is more consistent in "causing" subjects to have a particular dependent score in a particular condition, and thus the factor is more scientifically important in explaining differences in subjects' underlying behavior.

For example, from the ANOVA summary table for our study of perceived difficulty, we see that SS_{bn} was 63.33 and SS_{tot} was 147.33. So

$$\eta^2 = \frac{SS_{bn}}{SS_{tot}} = \frac{63.33}{147.33} = .43$$

This tells us that we are .43, or 43%, more accurate at predicting subjects' scores when we predict for them the mean score for the particular difficulty level subjects were tested under, rather than using the overall mean of the study. Thus, 43% of all the variance in our scores is accounted for, or explained, as being the result of changing the levels of perceived difficulty. Because 43% is a substantial amount, the variable of perceived difficulty is relatively important in determining subjects' subsequent performance.

Eta squared can be used with either equal or unequal n's. However, eta squared is a *descriptive* statistic that only describes the effect size in our sample data. Usually this is adequate. However, we can estimate the effect size in the *population* by computing omega squared.

Omega Squared: The Proportion of Variance Accounted For in the Population

Omega squared is an estimate of the proportion of variance in the population that would be accounted for by our relationship. The symbol for omega squared is ω^2. There are different formulas for omega squared, depending on the particular design of the experiment, but when we have a between-subjects factor with equal n's in all levels of the factor,

THE COMPUTATIONAL FORMULA FOR ω^2 IS

$$\omega^2 = \frac{SS_{bn} - (df_{bn})(MS_{wn})}{SS_{tot} + MS_{wn}}$$

Filling in this formula from the summary table of the one-way ANOVA for our perceived difficulty study, we have

$$\omega^2 = \frac{63.33 - (2)(7.0)}{147.33 + 7.0} = \frac{49.33}{154.33} = .32$$

Thus, if we could conduct this study on the entire population, we would expect perceived difficulty to account for .32, or 32%, of the variance in performance scores.

OTHER CONSIDERATIONS IN USING ANOVA

Before we leave the one-way ANOVA, there are three topics to consider: the within-subjects ANOVA, power, and homogeneity of variance.

The Within-Subjects ANOVA

Recall that sometimes we conduct research that requires related samples, and then we perform the *within-subjects* ANOVA. This occurs when we match each subject in one condition with a subject in every other condition or, more commonly, when we employ repeated measures (where we measure the same group of subjects under all the levels of a factor). A within-subjects F_{obt} is conceptually identical to the F_{obt} we discussed in this chapter (as are *post hoc* tests, η^2, and so on,) except that the calculations are slightly different. The computational formulas for a one-way, within-subjects ANOVA are presented in Appendix C. Because these computations are similar to those you will see in the next chapter, you'll understand them better if you read that chapter first.

Power and ANOVA

Recall that we always want to maximize our *power,* the probability of rejecting H_0 when H_0 is false. Here we do this by maximizing the size of F_{obt}. Looking at the F-ratio,

$$F_{obt} = \frac{MS_{bn}}{MS_{wn}}$$

we see that if we increase the size of the numerator or decrease the size of the denominator, the resulting F_{obt} will be larger. As we have done in previous chapters, we attempt to increase the power of an ANOVA by designing an experiment that (1) maximizes the size of the differences between means, which increases the size of MS_{bn}, (2) minimizes the variability of scores within the conditions, which reduces the size of MS_{wn}, and (3) maximizes the n of each condition, which increases df_{wn} and thus also minimizes MS_{wn}. (A larger df_{wn} also results in a smaller value of F_{crit}.) Any of the above will increase the probability that F_{obt} is significant, thereby giving us greater power to reject H_0 when it is false. The same considerations influence the power of our *post hoc* comparisons to detect a difference between each pair of sample means.

The Test for Homogeneity of Variance: The F_{max} Test

Throughout our discussion of t-tests and ANOVA, we have assumed that our samples represent populations that have homogeneous, or equal, variance (the value of σ_X^2 is the same for each population). This is important, because when we violate this assumption, the actual probability of a Type I error will be greater than our α. Therefore, if we are unsure that the data meet this assumption, we first perform a homogeneity of variance test. A **homogeneity of variance test** determines whether two estimated variances (s_X^2) are significantly different from each other. A significant difference between the values of s_X^2 for two samples indicates that they are likely to represent populations having different values of σ_X^2. Therefore, if our values of s_X^2 differ significantly, we cannot assume that the variances of our populations are homogeneous, and we should not perform the parametric procedure. Instead, we perform the appropriate nonparametric procedure (discussed in Chapter 15).

Although there are several tests of homogeneity, a simple version is Hartley's F_{max} test. This test is used for independent samples when the n's in all conditions are equal. To perform the F_{max} test, first compute the estimated population variance (s_X^2) for each condition, using the formula you learned in Chapter 5. Then select the largest value of s_X^2 and the smallest value of s_X^2. As usual, the null hypothesis says there is no difference, so here H_0 says that the two sample variances represent the same value of σ_X^2. Then,

THE COMPUTATIONAL FORMULA FOR THE HOMOGENEITY OF VARIANCE TEST IS

$$F_{max} = \frac{\text{largest } s_X^2}{\text{smallest } s_X^2}$$

Critical values for F_{max} are found in Table 7 in Appendix D, "Critical Values of the F_{max} Test." In the table, columns across the top are labeled "k" for the number of levels in the factor. The rows are labeled "$n - 1$," where n is the number of scores in each level.

The logic of the F_{max} test is the same as the logic of the F-ratio. If both values of s_X^2 perfectly represent the same σ_X^2, then the two s_X^2's should be equal, so F_{max} should equal 1. If they are not equal, then F_{max} will be larger than 1. The larger the F_{max}, the greater the difference between the largest and smallest s_X^2. If F_{max} is significant, the difference between the two values of s_X^2 is significant.

For example, from our perceived difficulty study's data back in Table 13.4, we calculate that for the easy condition, $s_X^2 = 8.5$; for the medium condition, $s_X^2 = 10$; and for the difficult condition, $s_X^2 = 2.5$. Placing the largest and smallest values (10 and 2.5) in the formula for F_{max} gives

$$F_{max} = \frac{10}{2.5} = 4.0$$

With three levels in our factor, $k = 3$, and because n is 5 in each condition, $n - 1$ is 4. From Appendix D's Table 7, the critical value of F_{max} is 15.50. Because our obtained F_{max} of 4.0 is less than the critical value of 15.50, the two variances are *not* significantly different. Therefore, we do not have evidence that the sample variances represent different population variances, so we can assume that our data meet the homogeneity of variance assumption, and it is acceptable to perform ANOVA.

FINALLY

The ANOVA is a *very* common procedure in psychological research, so be sure that you are comfortable with the logic and terminology of this chapter. When all is said and done, the F-ratio is a convoluted way of measuring the differences between sample means and then fitting those differences to a sampling distribution. The larger the F_{obt}, the less likely that the differences between the sample means are the result of sampling error. A significant F_{obt} indicates that the sample means are unlikely to represent one population mean. Once we determine that our F_{obt} is significant, we then determine which of our sample means actually differ significantly and describe the relationship. That's all there is to it.

There is, however, one other type of procedure that you should be aware of. Everything in our discussions so far has involved *one* dependent variable, and the statistics we have performed are called *univariate statistics.* We can, however, measure subjects on two or more dependent variables in one experiment. Statistics for multiple dependent variables are called *multivariate statistics.* These include the multivariate t-test and the multivariate analysis of variance (MANOVA). Even though these are very complex procedures, the basic logic still holds: the larger the t_{obt} or F_{obt}, the less likely it is that the samples represent the same population. But to discuss multivariates further would require another book, so we'll leave it at that.

CHAPTER SUMMARY

1. Analysis of variance, or *ANOVA,* has its own vocabulary. Here are the general terms we have used previously, along with the corresponding ANOVA terms:

General term	=	*ANOVA term*
independent variable	=	factor
condition	=	level
sum of squared deviations	=	sum of squares (*SS*)
variance (s_X^2)	=	mean square (*MS*)
effect of independent variable	=	treatment effect

2. A *one-way* analysis of variance tests for significant differences between the means from two or more levels of a factor. In a *between-subjects factor,* each independent sample is tested under only one level. In a *within-subjects* factor, we create related samples (either by matching different subjects or by repeatedly measuring the same subjects) in all conditions.

3. The *experiment-wise error rate* is the probability that a Type I error will occur when all pairs of means in an experiment are compared. We perform ANOVA instead of multiple t-tests because with ANOVA the experiment-wise error rate will equal our α.

4. The *assumptions of the one-way, between-subjects ANOVA* are that (a) the scores in each condition are independent random samples, (b) each sample represents a normally distributed population of interval or ratio scores, and (c) all populations represented in the study have homogeneous (equal) variance.

5. ANOVA tests two-tailed hypotheses. H_0 indicates that the mean of each condition represents the same population mean, so H_0 is $\mu_1 = \mu_2 = \mu_3 = \ldots \mu_k$, where k is the number of levels of the factor. H_a indicates that the means from at least two of the conditions represent different population means, so H_a is not all μ's are equal.

6. The *mean square within groups, MS_{wn},* is an estimate of *error variance,* the inherent variability among scores *within* each population. The *mean square between groups, MS_{bn},* is an estimate of the error variance plus the treatment variance. The *treatment variance* reflects differences in scores between the populations represented by the conditions of the independent variable.

7. F_{obt} is computed using the *F-ratio,* which equals the mean square between groups divided by the mean square within groups.

8. F_{obt} may be greater than 1 because either (a) there is no treatment effect, but the sample means are not perfectly representative of this, or (b) two or more sample means represent different population means.

9. The *F-distribution* is the sampling distribution of all possible values of F_{obt} when H_0 is true.

10. The larger the value of F_{obt}, the less likely it is that the sample means from all levels represent one population mean. If the F_{obt} is greater than F_{crit}, then the F_{obt} is significant. Therefore, the corresponding differences between sample means are too unlikely to accept as having occurred by chance if all means actually represent the same population mean.

11. All of the levels in a factor may not produce significant differences. Therefore, if (a) F_{obt} is significant and (b) there are more than two levels of the factor, we perform *post hoc comparisons* to determine which specific means differ significantly. When the *n*'s of all conditions are *not* equal, we perform *Fisher's protected t-test* on each pair of means. If all *n*'s are equal, then we perform *Tukey's HSD test.* Any two means that differ from each other by more than the *HSD* are significantly different.

12. When a significant relationship exists between the factor and the dependent scores, *eta squared* (η^2) describes the *effect size*—the *proportion of variance accounted for* by the factor in the sample data. *Omega squared* (ω^2) estimates the proportion of variance accounted for in the population.

13. The *power* of ANOVA increases as we increase the differences between the conditions of the independent variable, decrease the variability of scores within each condition, or increase the *n* of small samples.

14. The F_{max} *test* is a test of homogeneity of variance. It determines whether the variances from the levels in an experiment differ significantly. If they do not, then the data meet the assumption of homogeneity of variance.

PRACTICE PROBLEMS

(Answers for odd-numbered problems are provided in Appendix E.)

1. What does each of the following terms mean?
   a. ANOVA
   b. one-way design

c. factor

d. level

e. between subjects

f. within subjects

2. A researcher conducts an experiment in which scores are measured under two conditions of an independent variable.

   a. How will the researcher know whether to perform a parametric or nonparametric statistical procedure?

   b. Which two parametric procedures are available to her?

   c. If the researcher conducts an experiment with three levels of the independent variable, which two versions of a parametric procedure are available to her?

   d. How can she choose between them?

3. *a.* What are error variance and treatment variance?

   b. What are the two types of mean squares, and what does each one estimate?

4. *a.* What is the experiment-wise error rate?

   b. Why does performing ANOVA solve the problem of experiment-wise error rate created by performing multiple *t*-tests?

5. *a.* In a study comparing the effects of four conditions of the independent variable, what is H_0 for ANOVA?

   b. What is H_a for ANOVA in the same study?

   c. Describe in words what H_0 and H_a say for the study?

6. *a.* Why should F_{obt} equal 1.0 if the data represent the H_0 situation?

   b. Why is F_{obt} greater than 1.0 when the data represent the H_a situation?

   c. What does a significant F_{obt} indicate about differences between the levels of a factor?

7. *a.* Poindexter computes an F_{obt} of .63. How should this be interpreted?

   b. He computes another F_{obt} of -1.7. How should this be interpreted?

8. Foofy obtained a significant F_{obt} from an experiment with five levels. She concludes that she has demonstrated a relationship in which changing each condition of the independent variable results in a significant change in the dependent variable.

   a. Is she correct? Why or why not?

   b. What must she now do?

9. *a.* When is it necessary to perform *post hoc* comparisons? Why?

   b. When is it unnecessary to perform *post hoc* comparisons? Why?

10. Identify the two types of *post hoc* tests, and indicate when each is used.

11. *a.* What do η^2 and ω^2 have in common?

    b. For what purpose is each used?

12. *a.* In the perceived difficulty study discussed in this chapter, how can you maximize the power of the ANOVA?

    b. How do these strategies affect the size of F_{obt} and thus increase power?

    c. What does this do to *post hoc* tests?

13. Here are data from an experiment studying the effect of age on creativity scores:

Age 4	*Age 6*	*Age 8*	*Age 10*
3	9	9	7
5	11	12	7
7	14	9	6
4	10	8	4
3	10	9	5

    a. Compute F_{obt} and create an ANOVA summary table.

    b. With $\alpha = .05$, what do you conclude about F_{obt}?

    c. Perform the appropriate *post hoc* comparisons.

    d. What should you conclude about this relationship?

    e. Statistically, how important is the relationship in this study?

    f. Describe how you would graph these results.

14. In a study where $k = 3$, $n = 16$, $\overline{X}_1 = 45.3$, $\overline{X}_2 = 16.9$, and $\overline{X}_3 = 8.2$, you compute the following sums of squares:

Source	*Sum of squares*	*df*	*Mean square*	*F*
Between	147.32	___	___	___
Within	862.99	___	___	
Total	1010.31	___		

    a. Complete the ANOVA summary table.

    b. With $\alpha = .05$, what do you conclude about F_{obt}?

    c. Perform the appropriate *post hoc* comparisons. What do you conclude about this relationship?

    d. What is the effect size in this study, and what does this tell you about the influence of the independent variable?

15. *a.* What is the F_{max} test used for?

    b. In the study in problem 14, $s_X^2 = 43.68$ in level 1, $s_X^2 = 23.72$ in level 2, and $s_X^2 = 9.50$ in level 3. With $\alpha = .05$, does this study meet the assumption of homogeneity of variance?

    c. What does this indicate about the statistical procedure we should employ in this study?

(The Summary of Formulas begins on the next page.)

SUMMARY OF FORMULAS

1. *The format for the summary table for a one-way ANOVA is as follows:*

Summary Table of One-Way ANOVA

Source	*Sum of squares*	*df*	*Mean square*	*F*
Between	SS_{bn}	df_{bn}	MS_{bn}	F_{obt}
Within	SS_{wn}	df_{wn}	MS_{wn}	
Total	SS_{tot}	df_{tot}		

2. Computing the sum of squares
   a. *The computational formula for* SS_{tot} *is*

$$SS_{tot} = \Sigma X^2_{tot} - \left(\frac{(\Sigma X_{tot})^2}{N}\right)$$

   All scores in the experiment are included, and N is the total number of scores.

   b. *The computational formula for* SS_{bn} *is*

$$SS_{bn} = \Sigma\left(\frac{(\text{sum of scores in the column})^2}{n \text{ of scores in the column}}\right) - \left(\frac{(\Sigma X_{tot})^2}{N}\right)$$

   where each column contains the scores from one level of the factor.

   c. *The computational formula for* SS_{wn} *is*

$$SS_{wn} = SS_{tot} - SS_{bn}$$

3. Computing the mean square
   a. *The computational formula for* MS_{bn} *is*

$$MS_{bn} = \frac{SS_{bn}}{df_{bn}}$$

   with $df_{bn} = k - 1$, where k is the number of levels in the factor.

   b. *The computational formula for* MS_{wn} *is*

$$MS_{wn} = \frac{SS_{wn}}{df_{wn}}$$

   with $df_{wn} = N - k$, where N is the total N of the study and k is the number of levels in the factor.

4. *The computational formula for the F-ratio is*

$$F_{\text{obt}} = \frac{MS_{\text{bn}}}{MS_{\text{wn}}}$$

Critical values of F are found in Table 5 in Appendix D for df_{bn} and df_{wn}.

5. When F_{obt} is significant and k is greater than 2, *post hoc* comparisons must be performed.

   a. *The computational formula for the protected* t-*test is*

$$t_{\text{obt}} = \frac{\overline{X}_1 - \overline{X}_2}{\sqrt{MS_{\text{wn}}\left(\frac{1}{n_1} + \frac{1}{n_2}\right)}}$$

   Values of t_{crit} are the two-tailed values found in the t-tables for $df = df_{\text{wn}} = N - k$.

   b. *When the n's in all levels are equal, the computational formula for the HSD is*

$$HSD = (q_k)\left(\sqrt{\frac{MS_{\text{wn}}}{n}}\right)$$

   Values of q_k are found in Table 6 for df_{wn} and k, where k equals the number of levels of the factor.

6. *The computational formula for computing the confidence interval for a single* μ, *using the results of a between-subjects ANOVA, is*

$$\left(\sqrt{\frac{MS_{\text{wn}}}{n}}\right)(-t_{\text{crit}}) + \overline{X} \leq \mu \leq \left(\sqrt{\frac{MS_{\text{wn}}}{n}}\right)(+t_{\text{crit}}) + \overline{X}$$

$\overline{X}$ and n are from the level being described, and t_{crit} is the two-tailed value of t_{crit} at the appropriate α for df_{wn}.

7. *The computational formula for eta squared is*

$$\eta^2 = \frac{SS_{\text{bn}}}{SS_{\text{tot}}}$$

8. *The computational formula for omega squared, when the factor is a between-subjects factor and all n's are equal, is*

$$\omega^2 = \frac{SS_{\text{bn}} - (df_{\text{bn}})(MS_{\text{wn}})}{SS_{\text{tot}} + MS_{\text{wn}}}$$

9. *The computational formula for the* F_{max} *test is*

$$F_{\text{max}} = \frac{\text{largest } s_X^2}{\text{smallest } s_X^2}$$

where each s_X^2 is computed using the scores in one level of the experiment. Critical values are found in Table 7, where k is the number of levels in the factor and $n - 1$ is found using n, the number of scores in each level.

14

Two-Way Analysis of Variance: Testing the Means from Two Independent Variables

To understand the upcoming chapter:

- From Chapter 13 you should understand the terms *factor* and *level,* how we calculate F, how a significant F indicates that somewhere among the means there is at least one significant difference, that we identify the specific differences by performing *post hoc* tests, and that η^2 indicates the effect size of the factor.

Then your goals in this chapter are to learn:

- What a two-way factorial ANOVA is and how to diagram it.
- How to collapse across a factor to find main effect means.
- How to calculate the cell means for the interaction.
- How the F's in a two-way ANOVA are computed.
- What a significant main effect indicates.
- What a significant interaction indicates.
- How to perform *post hoc* tests, compute η^2, and draw the graph for each effect.
- How to interpret the results of a two-way experiment.

In the previous chapter we saw that ANOVA simultaneously tests for significant differences between all means from one factor. The real beauty of ANOVA, however, is that it can be applied to even more complex experiments, containing more than one factor. We can change the conditions of as many independent variables as we wish and then examine the effect of each independent variable separately, as well as the effect of all possible combinations of the variables.

When an experiment contains more than one factor, we perform a multi-factor ANOVA. To introduce you to multi-factor ANOVA, this chapter will discuss experiments containing two factors, which we analyze using a two-way, between-subjects ANOVA. This is like the ANOVA of the previous chapter, except that we compute several values of F_{obt}. Therefore, be forewarned that the procedure is rather involved (although it is more tedious than it is difficult). As you read this chapter, do not be concerned about memorizing all of the specific formulas. Rather, understand the overall steps involved, so that you can see why we perform the computations.

MORE STATISTICAL NOTATION

A two-way ANOVA has two different independent variables, or factors. Each factor may contain any number of levels, so we use a code to describe the number of levels in each factor of a specific ANOVA. The generic format for describing a two-way ANOVA is to call one independent variable factor A and the other independent variable factor B. To describe a particular ANOVA, we use the number of levels in each factor. If, for example, factor A has two levels and factor B has two levels, we have a two-by-two ANOVA, which is written as 2×2. Or if one factor has four levels and the other factor has three levels, we have a 4×3 ANOVA, and so on.

To understand what goes on in a two-way design and how we envision it, say that we are again interested in the effects of a "smart pill" on subjects' IQ. To begin, we will call the number of smart pills we will give to subjects factor A, and will test two levels (1 or 2 pills). We will then measure subjects' IQ scores. The basic design of this factor is shown in Table 14.1. Each column in the diagram represents a level of factor A and, within a column, each X represents a subject's IQ score. Averaging the scores in each column (vertically) yields the mean IQ for each pill level, showing the effect of factor A: how the typical IQ score changes as a function of increasing the dosage.

TABLE 14.1 Diagram of Factor of Number of Smart Pills

Each X represents a subject's IQ score, and each $\bar{X}$ is the mean for a level of factor A.

Factor A: Number of Pills	
Level A_1: One Pill	*Level A_2: Two Pills*
X	X
X	X
X	X
X	X
X	X
X	X
$\bar{X}$	$\bar{X}$

TABLE 14.2 Diagram of Factor of Age

Each row represents a level of age. Each X represents a subject's IQ score, and each $\bar{X}$ represents the mean IQ score in a level.

Factor B: Age	*Level B_1: 10-year-olds*	X X X X X X	$\bar{X}$
	Level B_2: 20-year-olds	X X X X X X	$\bar{X}$

Say that we were also interested in studying the influence of the subjects' age. We will call this variable factor B, and will test two levels (10- and 20-year-olds). We can envision this portion of the design as illustrated in Table 14.2. For reasons that will become clear in a moment, the novelty in this diagram is that the two conditions are arranged horizontally, so that each *row* represents a different level of the factor. Within the rows, each *X* again represents a subject's IQ score. Here, averaging the scores in each row (horizontally) yields the mean IQ for each age level, showing the effect of factor B: how the typical IQ score changes as a function of increasing subjects' age.

To create a two-way design, we would simultaneously manipulate both the subjects' age and the number of pills we give them. A good way to visualize a two-way ANOVA is to draw a diagram that shows how we combine the levels of both factors. Table 14.3 shows the diagram of this 2 × 2 ANOVA. Each column of the diagram is still a level of factor A (number of pills). Each row is still a level of factor B (age). But now we have a new term: each small square produced by a particular combination of a level of factor A with a level of factor B is called a **cell.** In this design we have four cells, each containing a sample of subjects who are one age and are given one amount of our smart pill. For example,

TABLE 14.3 Two-way Design for Studying the Factors of Number of Smart Pills and Subjects' Age

		Factor A: number of pills		
		Level A_1: 1 pill	*Level A_2: 2 pills*	
Factor B: age	*Level B_1: 10-year-olds*	X X X $\bar{X}_{A_1B_1}$	X X X $\bar{X}_{A_2B_1}$	← scores
	Level B_2: 20-year-olds	X X X $\bar{X}_{A_1B_2}$	X X X $\bar{X}_{A_2B_2}$	← scores
		↑ one of the four cells		

the highlighted cell contains the scores of a sample of 20-year-olds who receive one pill.

We identify the levels of factor A as A_1 and A_2 and the levels of factor B as B_1 and B_2. Sometimes it is useful to identify each cell using the levels of the two factors. For example, the cell formed by combining level 1 of factor A and level 1 of factor B is cell A_1B_1. We can identify the mean and *n* from each cell in the same way, so, for example, in cell A_1B_1 we have $\overline{X}_{A1B1}$.

One final consideration: when we combine all levels of one factor with all levels of the other factor, we have a **complete factorial design.** The design in Table 14.3 is a complete factorial, because all of our levels of drug dose are combined with all of our age levels. On the other hand, in an **incomplete factorial design,** all levels of the two factors are not combined. For example, if for some reason we did not collect data for 20-year-olds given one smart pill, we would have an incomplete factorial design. Incomplete factorial designs require elaborate procedures not discussed here.

In this chapter we will focus on the two-way factorial ANOVA in which both factors are between-subjects factors (we have independent samples in all cells of the experiment).

OVERVIEW OF THE TWO-WAY ANOVA

As with all experiments, the purpose of a two-factor experiment is to determine whether there is a relationship between the independent variable and the dependent variable. The only difference from previous experiments is that now we have two independent variables.

Why would we want to combine two factors in one experiment? We could perform two separate studies, one testing the effect of factor A and one testing the effect of factor B. However, in a multi-factor design we can learn everything about the influence of each factor that we would learn about if it were the only independent variable. But we can also study something with a multi-factor design that we would otherwise miss—the *interaction effect* produced by the combined manipulation of factor A and factor B. The primary reason for conducting multi-factor studies is so that we can observe the interaction effect. Later we will discuss interaction effects in detail, but for now, think of an interaction as the effect produced by the particular combination of the levels from the two factors. Thus, in the previous example, the interaction would indicate the influence on scores as we combine a particular age of subjects with a particular number of smart pills.

As usual, regardless of whether we are talking about each factor separately or their interaction, we want to conclude that our sample means indicate that if we tested the entire population under the various conditions in our experiment, we would find different populations of scores located at different μ's. But there is the usual problem: differences between the sample means may simply reflect sampling error, so we might actually find the same population, having the same μ for all conditions. Therefore, once again we must eliminate the idea that the differences between our sample means merely reflect sampling error. To do this, we perform ANOVA. As usual, we first set our alpha, usually $\alpha = .05$, and then check the assumptions.

Assumptions of the Two-Way, Between-Subjects ANOVA

We can perform the two-way, between-subjects ANOVA when we have a complete factorial design and we can assume that

1. All cells contain independent random samples of subjects.
2. The dependent variable measures interval or ratio scores.
3. The populations represented by the data are approximately normally distributed, and the mean is the appropriate measure of central tendency.
4. The represented populations all have homogeneous variance (all have equal σ_X^2). (We can test whether the cells meet this assumption using the F_{max} test discussed in the previous chapter.)

If our experiment generally meets the assumptions, we perform the two-way ANOVA.

Logic of the Two-Way ANOVA

Here is a semi-fascinating idea for a study. Have you ever noticed that television commercials are much louder than the programs themselves? Many advertisers seem to believe that increased volume creates increased viewer attention and so makes the commercial more persuasive. To test whether louder messages are more persuasive, we conduct an experiment. We play a recording of an advertising message at each of three volumes. (Volume is measured in decibels, but to simplify things we'll refer to the three levels of volume as soft, medium, and loud.) We are also interested in the differences between how males and females are persuaded, so we have another factor: the gender of the listener. If, in one study, we examine both the volume of the message and the gender of the subjects hearing the message, we have a two-factor experiment involving three levels of volume and two levels of gender. We will test all conditions with independent samples, so we have a 3×2 between-subjects, factorial ANOVA. Our dependent variable is a persuasiveness score, indicating how persuasive a subject believes the message to be on a scale of 0 (not at all) to 25 (totally convincing).

Say we collect the scores and organize them as in Table 14.4. As usual, for simplicity we have a distinctly unpowerful *N:* nine men and nine women were randomly selected, and then three men and three women were randomly assigned to hear the message at each volume, so we have three persuasiveness scores per cell.

But now what? How do we make sense out of it all? To answer this question, stop and think about what the study is designed to investigate. We want to show the effects of (1) changing the levels of the factor of volume, (2) changing the levels of the factor of gender, and (3) changing the combination, or interaction, of the factors of volume and gender. As usual, we want to determine that each of these has a significant effect. Since we want to view each of these effects one at a time, *the way to understand a two-way ANOVA is to treat it as if it contained three one-way ANOVAs*. You already understand a one-way ANOVA, so the rest of this chapter simply provides a guide for computing the various *F*'s.

Any two-way ANOVA breaks down into finding the main effects and the interaction effect.

Main effects of factor A and factor B The **main effect** of a factor is the effect on the dependent scores of changing the levels of that factor while ignoring all other factors in the study. In our persuasiveness study, to find the main effect of factor A (volume), we simply ignore the levels of factor B (gender). Literally erase the horizontal line that separates the males and females in Table 14.4, and treat the experiment as if it were as follows:

Factor A: volume

Level A_1: *soft*	*Level* A_2: *medium*	*Level* A_3: *loud*
9	8	18
4	12	17
11	13	15
2	9	6
6	10	8
4	17	4
$\overline{X}_{A_1} = 6$	$\overline{X}_{A_2} = 11.5$	$\overline{X}_{A_3} = 11.33$
$n_{A_1} = 6$	$n_{A_2} = 6$	$n_{A_3} = 6$

We ignore the fact that we have some males and females in each condition: we simply have six subjects tested under each volume. In this diagram we have one factor, with three means from the three levels of volume, so $k = 3$, with $n = 6$ in each level.

In statistical terminology, we have collapsed across the factor of subject gender. **Collapsing** across a factor means averaging together all scores from all levels of the factor. When we collapse across one factor, we have the *main effect means*

TABLE 14.4 A 3 × 2 ANOVA for the Factors of Volume of Message and Gender of Subject

		Factor A: volume		
		Level A_1: *soft*	*Level* A_2: *medium*	*Level* A_3: *loud*
Factor B: gender	*Level* B_1: *male*	9 4 11	8 12 13	18 17 15
	Level B_2: *female*	2 6 4	9 10 17	6 8 4

$N = 18$

for the remaining factor. Thus, by collapsing across gender above we have the main effect means for the three levels of volume, $\bar{X}_{A_1} = 6$, $\bar{X}_{A_2} = 11.5$, and $\bar{X}_{A_3} = 11.33$.

Once we have collapsed, we then essentially perform a one-way ANOVA on the preceding diagram. We find the main effect of factor A by asking "Do these main effect means represent different μ's that would be found if we tested the entire population under each of these three volumes?" To answer this question, we first create our statistical hypotheses for factor A. The null hypothesis is

$$H_0\text{: } \mu_{A_1} = \mu_{A_2} = \mu_{A_3}$$

For our study, this says that changing volume has no effect, so the main effect means from the levels of volume represent the same population of persuasiveness scores. If we can reject H_0, then we will accept the alternative hypothesis, which is

$$H_{a}\text{: not all } \mu_A\text{'s are equal}$$

For our study, this says that at least two main effect means from our levels of volume represent different populations of persuasiveness scores, having different μ's.

We now follow the procedures of a one-way ANOVA. To test H_0, we compute an F_{obt} called F_A. If F_A is significant, it indicates that at least two of our main effect means differ significantly. Then we will describe this relationship by graphing the main effect means, performing *post hoc* comparisons to determine which of the specific means of factor A differ significantly, and determining the proportion of variance in dependent scores that is accounted for by changing the levels of factor A.

Once we have completed our analysis of the main effect of factor A, we move on to the main effect of factor B. To do this, we collapse across factor A (volume). Now we erase the vertical lines separating the levels of volume shown earlier in Table 14.4, obtaining this diagram of the main effect of factor B:

Factor B: gender	*Level B_1: male*	9	8	18	$\bar{X}_{B_1} = 11.89$
		4	12	17	
		11	13	15	$n_{B_1} = 9$
	Level B_2: female	2	9	6	$\bar{X}_{B_2} = 7.33$
		6	10	8	
		4	17	4	$n_{B_2} = 9$

Now we simply have the persuasiveness scores of males and females, ignoring the fact that some of each heard the message at different volumes. In this diagram we have a one-factor, two-sample design. We treat this as a one-way ANOVA to see if there are significant differences between the main effect means of persuasiveness scores for males ($\bar{X}_{B_1} = 11.89$) and females ($\bar{X}_{B_2} = 7.33$). Notice that here there are two levels of subject gender, so k is now 2 and the n of each level is 9, but for factor A, k was 3 and n was 6.

STAT ALERT In a two-way ANOVA, the values of n and k may be different for each factor.

Now we write our statistical hypotheses for the main effect of factor B. The null hypothesis is

$$H_0: \mu_{B_1} = \mu_{B_2}$$

For our study, this says that changing gender has no effect, so the sample mean for males represents the same population mean as the sample mean for females. If we can reject H_0, then we will accept the alternative hypothesis, which is

$$H_a: \text{not all } \mu_B\text{'s are equal}$$

For our study, this says that our sample means for males and females represent different populations of persuasiveness scores, having different μ's.

To test H_0 for factor B, we compute a separate F_{obt}, called F_B. If F_B is significant, it indicates that the main effect means for factor B differ significantly. Then we graph the main effect means for factor B, perform the *post hoc* comparisons, and compute the proportion of variance accounted for by factor B.

Interaction effects of factor A and factor B After we examine the main effects of factors A and B, we examine the interaction of the two factors. The interaction of two factors is called a two-way interaction. The **two-way interaction** is the influence on scores of combining the particular levels of factor A with the levels of factor B. In our example, the interaction effect is the effect of each particular volume when combined with each particular subject gender. An interaction is identified as A $\times$ B. Our factor A has 3 levels and our factor B has 2 levels, so we have a 3 $\times$ 2 interaction.

Because we examine the combinations of the levels of both factors in an interaction, we do not collapse across, or ignore, either factor. Instead, we treat each *cell* in the study as a level of the interaction and compare the cell mean. We can diagram the interaction effect this way:

A $\times$ B Interaction Effect

Male soft	*Male medium*	*Male loud*	*Female soft*	*Female medium*	*Female loud*
9	8	18	2	9	6
4	12	17	6	10	8
11	13	15	4	17	4
$\bar{X} = 8$	$\bar{X} = 11$	$\bar{X} = 16.67$	$\bar{X} = 4$	$\bar{X} = 12$	$\bar{X} = 6$

$k = 6$
$n = 3$

These are the original six cells, containing three scores per subject, that we saw back in Table 14.4. We can *think* of this as a one-way ANOVA for the six levels of the interaction, so k is now 6 and n is now 3. However, examining an interaction is not as simple as saying that the cell means are significantly different. Here we are testing the extent to which the cell means differ *after* we have removed those differences that are attributable to the separate main effects of factor A and factor B. Thus, consistent differences between scores that are not due to changing the levels of factor A alone or factor B alone are due to changing the combinations of the levels of A and B.

Understanding and interpreting an interaction is difficult, because both independent variables are changing, as well as the dependent scores. To simplify the process, we look at the influence of changing the levels of factor A under *one* level of factor B. Then we see if this effect of factor A is *different* when we look at the other level of factor B. For example, the means are grouped in the above diagram so that on the left we can see how the three means for males change as we change volume. On the right we can see how the means for females change as we change volume. For males, each time we increase volume, the mean score also appears to increase. However, this is not the case for females: increasing volume from soft to medium apparently increases the mean, but increasing volume from medium to loud apparently *decreases* the mean.

Thus, an interaction exists *when the influence that changing the levels of one factor has on scores depends on which level of the other factor you examine.* Above, whether increasing the volume of ads consistently increases their persuasiveness depends on whether the subjects are male or female. This dependency leads us to two other, equivalent ways of defining an interaction. First, an interaction occurs *when the effect of changing the levels of one factor is not consistent for each level of the other factor.* (Above, increasing volume does not have the same effect on the males' scores as on females'.) Second, we have an interaction *when the relationship between one factor and the dependent variable changes as we change the levels of the other factor.* (For males we have a positive linear relationship between volume and scores, but for females we have a curvilinear relationship.)

All of these statements describe the overall pattern produced by the influence of combining the levels of the two factors. Conversely, of course, this influence may not occur. Above, for example, if the scores for females had increased in the same way that the scores for males increased, then there would be no interaction. When we do not have an interaction, we see that (1) the influence of changing the levels of one factor does not depend on which level of the other variable we are talking about, (2) the effect of changing one factor is the same for all levels of the other factor, and (3) we have the same relationship between the scores and one factor at each level of the other factor.

> ***STAT ALERT*** A two-way interaction indicates that the influence of one factor on scores depends on the level of the other factor that is present.

As with other effects, our data may appear to represent an interaction, but this may be an illusion created by sampling error. Therefore, we determine whether there is a significant interaction by performing a procedure similar to a one-way ANOVA for the six cell means. First we create the statistical hypotheses for the $A \times B$ interaction. As usual, the null hypothesis implies there is no effect.

Therefore, in words the null hypothesis (H_0) is that any differences between our cell means do not represent an interaction between the population μ's.[1]

The alternative hypothesis, H_a, is that at least some differences between our cell means do represent an interaction between the population μ's.

To test H_0 for the interaction, we compute yet another separate F_{obt}, called $F_{A \times B}$. If $F_{A \times B}$ is significant, it indicates that at least two of the cell means differ significantly in a way that produces an interaction. Then we graph the interaction effect by graphing the cell means, we perform *post hoc* comparisons to determine which cell means differ significantly, and we compute the proportion of variance accounted for by the interaction.

Overview of the Computations of the Two-Way ANOVA

Thus, as we've seen, in a two-way ANOVA we compute three F's: one for the main effect of factor A, one for the main effect of factor B, and one for the interaction effect of A $\times$ B. The logic and calculations for each of these are basically the same as in the one-way ANOVA, because any F_{obt} is the ratio formed by dividing the mean square between groups (MS_{bn}) by the mean square within groups (MS_{wn}).

As usual, MS_{wn} is the variance within groups; and in a two-way ANOVA we compute MS_{wn} by computing the "average" variability of the scores in each *cell.* All subjects in a cell are treated identically, so any differences among the scores are due to the inherent variability of scores. Thus, the MS_{wn} is our estimate of the error variance in the population—the inherent variability within any of the raw score populations represented by our samples. This is our *one* estimate of the error variance, so we use it as the denominator in computing all three F ratios.

We measure the variance between groups by computing the MS_{bn}. This is our way of measuring the differences between our means, as an estimate of the treatment variance plus the error variance in the population. Thus, as usual, each F estimates the following:

	Sample	***Estimates***	***Population***
F_{obt} =	$\dfrac{MS_{bn}}{MS_{wn}}$	$\rightarrow$ $\rightarrow$	$\dfrac{\sigma^2_{error} + \sigma^2_{treat}}{\sigma^2_{error}}$

However, because we have two factors and the interaction, we have three sources of between-groups variance. We partition the between-groups variance into (1) variance between groups due to factor A, (2) variance between groups due to factor B, and (3) variance between groups due to the interaction. We will compute a separate mean square between groups for each of these as an estimate of the treatment variance each produces in the population, and then we will compute the appropriate F ratio.

Table 14.5 on the next page shows all the means from our persuasiveness study that we will compare. For each effect, we will compute a mean square between groups by dividing the appropriate sum of squares (SS) by the corresponding degrees of freedom (df). Thus, first we will collapse across factor B and examine

[1]Technically, H_0 says that differences between scores due to A at one level of B equal the differences between scores due to A at the other level of B. Thus, we have H_0: $\mu_{A1B1} - \mu_{A2B1} = \mu_{A1B2} - \mu_{A2B2} = \mu_{A2B1} - \mu_{A3B1} = \mu_{A2B2} - \mu_{A3B2}$. H_a is that not all differences are equal.

TABLE 14.5 Summary of Means in Persuasiveness Study

		Factor A: volume			
		A_1: *soft*	A_2: *medium*	A_3: *loud*	
Factor B: gender	B_1: *male*	$\bar{X} = 8$	$\bar{X} = 11$	$\bar{X} = 16.67$	$\bar{X}_{male} = 11.89$
	B_2: *female*	$\bar{X} = 4$	$\bar{X} = 12$	$\bar{X} = 6$	$\bar{X}_{fem} = 7.33$
		$\bar{X}_{soft} = 6$	$\bar{X}_{med} = 11.5$	$\bar{X}_{loud} = 11.33$	

the main effect means of factor A, volume ($\bar{X}_{A_1} = 6$, $\bar{X}_{A_2} = 11.5$, and $\bar{X}_{A_3} = 11.33$). We describe the differences between these means by computing the sum of squares between groups for factor A (called SS_A), and then, after dividing by the degrees of freedom for factor A (called df_A), we have the mean square between groups for factor A (called MS_A).

Likewise, we will collapse across factor A and examine the difference between the main effect means for factor B, gender ($\bar{X}_{B_1} = 11.89$ and $\bar{X}_{B_2} = 7.33$). We compute the sum of squares between groups for factor B, SS_B, and then, dividing by the degrees of freedom between groups for factor B, df_B, we have the mean square between groups for factor B, MS_B.

For the interaction, we do not collapse across either factor: we will compare the differences between the six cell means that are not attributable to factor A or factor B alone. We compute the sum of squares between groups for A × B, $SS_{A \times B}$, and after dividing by the degrees of freedom for A × B, $df_{A \times B}$, we have the mean square between groups for the interaction, $MS_{A \times B}$. We also compute MS_{wn} by computing SS_{wn} and then dividing by df_{wn}.

In the summary table in Table 14.6, we can see all of the components of the two-way ANOVA. To complete the summary table, for factor A we divide MS_A by MS_{wn} to produce F_A. For factor B, we divide MS_B by MS_{wn} to produce F_B. For the interaction we divide $MS_{A \times B}$ by MS_{wn} to produce $F_{A \times B}$.

Each F_{obt} in a two-way ANOVA is tested in the same way we tested F_{obt} in the previous chapter. F_{obt} may be larger than 1.0 because (1) H_0 is true but we

TABLE 14.6 Summary Table of Two-Way ANOVA

Source	*Sum of squares*	/	*df*	=	*Mean square*	*F*
Between						
Factor A (volume)	SS_A		df_A		MS_A	→ F_A
Factor B (gender)	SS_B		df_B		MS_B	→ F_B
Interaction (vol × gen)	$SS_{A \times B}$		$df_{A \times B}$		$MS_{A \times B}$	→ $F_{A \times B}$
Within	SS_{wn}		df_{wn}		MS_{wn}	
Total	SS_{tot}		df_{tot}			

have sampling error, or (2) H_0 is false and at least two sample means represent a "real" relationship in the population. The larger the value of F_{obt}, the less likely it is that H_0 is true. If any F_{obt} is larger than F_{crit}, then F_{obt} is significant and we reject the corresponding H_0.

COMPUTING THE TWO-WAY ANOVA

Having a computer perform the calculations is the best way to perform this ANOVA. However, whether or not you use a computer program, you should first organize the data in each cell. Table 14.7 shows our persuasiveness scores for the factors of volume and gender, as well as the various components we must compute.

To compute the components, first we compute ΣX and ΣX^2 for each cell and note the n of each cell. Thus, for the male-soft cell, $\Sigma X = 4 + 9 + 11 = 24$, $\Sigma X^2 = 4^2 + 9^2 + 11^2 = 218$, and $n = 3$. We also compute the mean for each cell (for the male-soft cell, $\overline{X} = 24/3 = 8$). These are the means we will test in the interaction.

Now we collapse across factor B, gender, and look only at the three volumes. We compute ΣX vertically for each column: the ΣX in a column is the sum of the ΣXs from the cells in that column (for soft, $\Sigma X = 24 + 12$). We note the n in each column (here $n = 6$) and compute the sample mean for each column ($\overline{X}_{soft} = 6$). These are the means we will test in the main effect of factor A.

TABLE 14.7 Summary of Data for 3 × 2 ANOVA

		Factor A: volume			
		A_1: soft	*A_2: medium*	*A_3: loud*	
Factor B: gender	*B_1: male*	4 9 11 $\overline{X} = 8$ $\Sigma X = 24$ $\Sigma X^2 = 218$ $n = 3$	8 12 13 $\overline{X} = 11$ $\Sigma X = 33$ $\Sigma X^2 = 377$ $n = 3$	18 17 15 $\overline{X} = 16.67$ $\Sigma X = 50$ $\Sigma X^2 = 838$ $n = 3$	$\overline{X}_{male} = 11.89$ $\Sigma X = 107$ $n = 9$
	B_2: female	2 6 4 $\overline{X} = 4$ $\Sigma X = 12$ $\Sigma X^2 = 56$ $n = 3$	9 10 17 $\overline{X} = 12$ $\Sigma X = 36$ $\Sigma X^2 = 470$ $n = 3$	6 8 4 $\overline{X} = 6$ $\Sigma X = 18$ $\Sigma X^2 = 116$ $n = 3$	$\overline{X}_{fem} = 7.33$ $\Sigma X = 66$ n = 9
		$\overline{X}_{soft} = 6$ $\Sigma X = 36$ $n = 6$	$\overline{X}_{med} = 11.5$ $\Sigma X = 69$ $n = 6$	$\overline{X}_{loud} = 11.33$ $\Sigma X = 68$ $n = 6$	$\Sigma X_{tot} = 173$ $\Sigma X^2_{tot} = 2075$ $N = 18$

Now we collapse across factor A, volume, and look only at males versus females. We compute ΣX horizontally for each row: the ΣX in a row equals the sum of the ΣX's from the cells in that row (for males, $\Sigma X = 24 + 33 + 50 = 107$). We note the n in each row (here $n = 9$) and compute the sample mean for each row ($\overline{X}_{male} = 11.89$). These are the means we will test in the main effect of factor B.

Finally, we compute the total ΣX (called ΣX_{tot}), by adding the ΣX from the three levels of factor A (the three column sums), so $\Sigma X_{tot} = 36 + 69 + 68 = 173$. Alternatively, we can add the ΣX from the two levels of factor B. We also find the total ΣX^2 (called ΣX^2_{tot}), by adding the ΣX^2 from each cell, so $\Sigma X^2_{tot} = 218 + 377 + 838 + 56 + 470 + 116 = 2075$. Note that the total N is 18.

As we'll see, we use these components to compute all of our sums of squares and degrees of freedom. Then we compute the mean squares and, finally, each F_{obt}. To keep track of your computations and prevent brain strain, fill in the ANOVA summary table as you go along.

Computing the Sums of Squares

First we compute the various sums of squares. We do this in five steps.

Step 1 is to compute the total sum of squares, SS_{tot}.

THE COMPUTATIONAL FORMULA FOR SS_{tot} *IS*

$$SS_{tot} = \Sigma X^2_{tot} - \left(\frac{(\Sigma X_{tot})^2}{N}\right)$$

This equation says to divide $(\Sigma X_{tot})^2$ by N and then subtract the answer from ΣX^2_{tot}.

For our persuasiveness study, in Table 14.7 we have $\Sigma X_{tot} = 173$, $\Sigma X^2_{tot} = 2075$, and $N = 18$. Filling in the formula, we have

$$SS_{tot} = 2075 - \left(\frac{(173)^2}{18}\right)$$

Which becomes

$$SS_{tot} = 2075 - 1662.72$$

so $SS_{tot} = 412.28$.

Note that the quantity $(\Sigma X_{tot})^2/N$ above is also used in the computation of most of the other sums of squares. It is called the *correction* (here the correction equals 1662.72).

Step 2 is to compute the sum of squares for factor A. In our diagram of the two-way ANOVA, the levels of factor A form the columns.

THE COMPUTATIONAL FORMULA FOR THE SUM OF SQUARES BETWEEN GROUPS FOR COLUMN FACTOR A IS

$$SS_{\text{A}} = \Sigma\left(\frac{(\text{sum of scores in the column})^2}{n \text{ of scores in the column}}\right) - \left(\frac{(\Sigma X_{\text{tot}})^2}{N}\right)$$

This equation says to square the ΣX for each column of factor A and divide by the n in the column. After doing this for all levels, add the answers together and then subtract the correction.

In our example, from Table 14.7 we found that the three columns produced sums of 36, 69, and 68 and n was 6. Filling in the above formula, we have

$$SS_{\text{A}} = \left(\frac{(36)^2}{6} + \frac{(69)^2}{6} + \frac{(68)^2}{6}\right) - \left(\frac{(173)^2}{18}\right)$$

$$SS_{\text{A}} = (216 + 793.5 + 770.67) - 1662.72$$

$$SS_{\text{A}} = 1780.17 - 1662.72$$

so $SS_{\text{A}} = 117.45$.

Step 3 is to compute the sum of squares between groups for factor B. In our diagram of the two-way ANOVA, the levels of factor B form the rows.

THE COMPUTATIONAL FORMULA FOR THE SUM OF SQUARES BETWEEN GROUPS FOR ROW FACTOR B IS

$$SS_{\text{B}} = \Sigma\left(\frac{(\text{sum of scores in the row})^2}{n \text{ of scores in the row}}\right) - \left(\frac{(\Sigma X_{\text{tot}})^2}{N}\right)$$

This equation says to square ΣX for each level of factor B and divide by the n in the level. After doing this for all levels, add the answers and then subtract the correction.

In our example, we found that the two rows produced sums of 107 and 66 and n was 9. Filling in the above formula gives

$$SS_{\text{B}} = \left(\frac{(107)^2}{9} + \frac{(66)^2}{9}\right) - 1662.72$$

$$SS_{\text{B}} = 1756.11 - 1662.72$$

so $SS_{\text{B}} = 93.39$.

Step 4 is to compute the sum of squares between groups for the interaction, $SS_{\text{A} \times \text{B}}$. To do this, we first compute something called the overall sum of squares between groups, which we identify as SS_{bn}.

THE COMPUTATIONAL FORMULA FOR SS_{bn} IS

$$SS_{bn} = \Sigma\left(\frac{(\text{sum of scores in the cell})^2}{n \text{ of scores in the cell}}\right) - \left(\frac{(\Sigma X_{tot})^2}{N}\right)$$

Here we find $(\Sigma X)^2$ for each cell, divide by the n of the cell, add the answers from all cells together, and subtract the correction.

In our example, filling in the formula gives

$$SS_{bn} = \left(\frac{(24)^2}{3} + \frac{(33)^2}{3} + \frac{(50)^2}{3} + \frac{(12)^2}{3} + \frac{(36)^2}{3} + \frac{(18)^2}{3}\right) - 1662.72$$

$$SS_{bn} = 1976.33 - 1662.72$$

so $SS_{bn} = 313.61$.

The reason we compute SS_{bn} is that it is equal to the sum of squares for factor A plus the sum of squares for factor B plus the sum of squares for the interaction. To find $SS_{A\times B}$, we subtract the sum of squares for both main effects (in steps 2 and 3) from the overall SS_{bn}. Thus,

THE COMPUTATIONAL FORMULA FOR THE SUM OF SQUARES BETWEEN GROUPS FOR THE INTERACTION, $SS_{A\times B}$, IS

$$SS_{A\times B} = SS_{bn} - SS_A - SS_B$$

In our example, $SS_{bn} = 313.61$, $SS_A = 117.45$, and $SS_B = 93.39$, so

$$SS_{A\times B} = 313.61 - 117.45 - 93.39$$

so $SS_{A\times B} = 102.77$.

Step 5 is to compute the sum of squares within groups, SS_{wn}. The sum of squares within groups plus the overall sum of squares between groups equals the total sum of squares. Therefore, when we subtract the overall SS_{bn} in step 4 from the SS_{tot} in step 1, we obtain the SS_{wn}.

THE COMPUTATIONAL FORMULA FOR THE SUM OF SQUARES WITHIN GROUPS, SS_{wn}, IS

$$SS_{wn} = SS_{tot} - SS_{bn}$$

In our example, $SS_{tot} = 412.28$ and $SS_{bn} = 313.61$, so

$$SS_{wn} = 412.28 - 313.61$$

Thus, $SS_{wn} = 98.67$.

Placing the various sums of squares in the ANOVA summary table, we get Table 14.8. Notice that we do not include the overall SS_{bn}.

TABLE 14.8 Summary Table of Two-Way ANOVA

Source	*Sum of squares*	*df*	*Mean square*	*F*
Between				
Factor A (volume)	117.45	df_A	MS_A	F_A
Factor B (gender)	93.39	df_B	MS_B	F_B
Interaction (vol × gen)	102.77	$df_{A \times B}$	$MS_{A \times B}$	$F_{A \times B}$
Within	98.67	df_{wn}	MS_{wn}	
Total	412.28	df_{tot}		

Computing the Degrees of Freedom

Now we must determine the various values of *df.*

1. *The degrees of freedom between groups for factor A is* $k_A - 1$, where k_A is the number of levels in factor A. (In our example, k_A is the three levels of volume, so $df_A = 2$.)
2. *The degrees of freedom between groups for factor B is* $k_B - 1$, where k_B is the number of levels in factor B. (In our example, k_B is the two levels of gender, so $df_B = 1$.)
3. *The degrees of freedom between groups for the interaction is the df for factor A multiplied times the df for factor B.* (In our example, $df_A = 2$ and $df_B = 1$, so $df_{A \times B} = 2$.)
4. *The degrees of freedom within groups equals* $N - k_{A \times B}$, where N is the total N of the study and $k_{A \times B}$ is the total number of cells in the study. (In our example, N is 18 and we have six cells, so $df_{wn} = 18 - 6 = 12$.)
5. *The degrees of freedom total equals* $N - 1$. Use this to check your previous calculations, because the sum of the above *df*'s should equal df_{tot}. (In our example, $df_{tot} = 17$.)

Computing the Mean Squares

It is easiest to perform the remainder of the computations by working directly from the summary table. So far, with the sums of squares and degrees of freedom we have Table 14.9, at the top of the next page.

Now compute the mean squares. Any mean square equals the appropriate sum of squares divided by the appropriate *df.*

THE COMPUTATIONAL FORMULA FOR THE MEAN SQUARE FOR FACTOR A IS

$$MS_A = \frac{SS_A}{df_A}$$

TABLE 14.9 Summary Table of Two-Way ANOVA

Source	*Sum of squares*	*df*	*Mean square*	*F*
Between				
Factor A (volume)	117.45	2	MS_A	F_A
Factor B (gender)	93.39	1	MS_B	F_B
Interaction (vol × gen)	102.77	2	$MS_{A \times B}$	$F_{A \times B}$
Within	98.67	12	MS_{wn}	
Total	412.28	17		

In our example,

$$MS_A = \frac{117.45}{2} = 58.73$$

THE COMPUTATIONAL FORMULA FOR THE MEAN SQUARE FOR FACTOR B IS

$$MS_B = \frac{SS_B}{df_B}$$

In our example,

$$MS_B = \frac{93.39}{1} = 93.39$$

THE COMPUTATIONAL FORMULA FOR THE MEAN SQUARE FOR THE INTERACTION IS

$$MS_{A \times B} = \frac{SS_{A \times B}}{df_{A \times B}}$$

Thus, we have

$$MS_{A \times B} = \frac{102.77}{2} = 51.39$$

THE COMPUTATIONAL FORMULA FOR THE MEAN SQUARE WITHIN GROUPS IS

$$MS_{wn} = \frac{SS_{wn}}{df_{wn}}$$

TABLE 14.10 Summary Table of Two-Way ANOVA

Source	*Sum of squares*	*df*	*Mean square*	*F*
Between				
Factor A (volume)	117.45	2	58.73	F_A
Factor B (gender)	93.39	1	93.39	F_B
Interaction (vol × gen)	102.77	2	51.39	$F_{A \times B}$
Within	98.67	12	8.22	
Total	412.28	17		

Thus, we have

$$MS_{wn} = \frac{98.67}{12} = 8.22$$

Putting these values in the summary table, we get Table 14.10.

Now, finally, we compute the F's.

Computing F_{obt}

Recall that to compute any F, we divide the MS_{bn} by the MS_{wn}. Therefore,

THE COMPUTATIONAL FORMULA FOR F_A FOR THE MAIN EFFECT OF FACTOR A IS

$$F_A = \frac{MS_A}{MS_{wn}}$$

In our example we have

$$F_A = \frac{58.73}{8.22} = 7.14$$

THE COMPUTATIONAL FORMULA FOR F_B FOR THE MAIN EFFECT OF FACTOR B IS

$$F_B = \frac{MS_B}{MS_{wn}}$$

TABLE 14.11 Summary Table of Two-Way ANOVA

Source	*Sum of squares*	*df*	*Mean square*	*F*
Between				
Factor A (volume)	117.45	2	58.73	7.14
Factor B (gender)	93.39	1	93.39	11.36
Interaction (vol × gen)	102.77	2	51.39	6.25
Within	98.67	12	8.22	
Total	412.28	17		

Thus, we have

$$F_B = \frac{93.39}{8.22} = 11.36$$

THE COMPUTATIONAL FORMULA FOR $F_{A \times B}$ FOR THE INTERACTION EFFECT IS

$$F_{A \times B} = \frac{MS_{A \times B}}{MS_{wn}}$$

Thus, we have

$$F_{A \times B} = \frac{51.39}{8.22} = 6.25$$

And now, the finished summary table is shown in Table 14.11.

Interpreting Each F_{obt}

Once we have completed the summary table, we determine whether each F_{obt} is significant in the same way we did in Chapter 13: we compare each F_{obt} to the appropriate value of F_{crit}. To find each F_{crit} in the F-tables (Table 5 in Appendix D), we need the df_{bn} and the df_{wn} used in computing the corresponding F_{obt}. (Notice that, in the F-tables, the columns labeled df_{bn} are also labeled "degrees of freedom in numerator of F ratio." To find each F_{crit}, use the df you used in calculating the numerator of the F as your df_{bn}.)

1. To find F_{crit} for testing F_A, we use df_A as the df between groups and df_{wn}. In our example, $df_A = 2$ and $df_{wn} = 12$. So, for $\alpha = .05$, the F_{crit} for 2 and 12 df is 3.88.
2. To find F_{crit} for testing F_B, we use df_B as the df between groups and df_{wn}. In our example, $df_B = 1$ and $df_{wn} = 12$. So, at $\alpha = .05$, the F_{crit} for 1 and 12 df is 4.75.

3. To find F_{crit} for the interaction, we use $df_{A \times B}$ as the *df* between groups and df_{wn}. In our example, $df_{A \times B} = 2$ and $df_{wn} = 12$. Thus, at $\alpha = .05$, the F_{crit} for 2 and 12 *df* is 3.88.

Note that since the df_{bn} for factor B is different from the df_{bn} for factor A, the F_{crit} for factor B is different from the F_{crit} for factor A.

STAT ALERT Each F_{crit} in the two-way ANOVA will be different if the degrees of freedom between groups are different.

Thus, we end up comparing our values of F_{obt} from our ANOVA summary table with our values of F_{crit}, as follows:

	F_{obt}	F_{crit}
Main effect of volume (A)	7.14	3.88
Main effect of gender (B)	11.36	4.75
Interaction (A × B)	6.25	3.88

By now you can do this with your eyes closed: imagine a sampling distribution with a region of rejection and F_{crit} in the positive tail. (If you can't imagine this, look back in Chapter 13 at Figure 13.3.) First, our obtained F_A of 7.14 is larger than the F_{crit}, so F_A falls in the region of rejection. Therefore, we conclude that differences between the means for the levels of factor A are significant: changing the volume of a message produced significant differences in persuasiveness scores. However, remember that a significant F_{obt} only indicates that, somewhere in the factor, at least two of the means differ significantly. Therefore, we are confident that changing volume results in at least two different population means. Since $\alpha = .05$, we report our results as

$$F(2, 12) = 7.14, \quad p < .05$$

Likewise, our F_B is significant, so we conclude that the males and females in our study represent different populations of scores. We report this result as

$$F(1, 12) = 11.36, \quad p < .05$$

Finally, our $F_{A \times B}$ of 6.25 is significant, so we conclude that the specific combinations of the levels of our factors produce means that represent an interaction in the population: The effect of changing the volume on the population *depends on* whether it is a population of males or a population of females. Or, we can say that the difference between the male and female populations of scores *depends on* whether a message is played at the soft, medium, or loud level. We report this result as

$$F(2, 12) = 6.25, \quad p < .05$$

Note: It is just a coincidence of the particular data in our example that all three values of F_{obt} were significant. Whether any one F_{obt} is significant does not influence whether any other F_{obt} is significant. With different data, any combination of the main effects and/or the interaction may or may not be significant.

At this point we have completed our ANOVA. However, we are a long way from being finished with our analysis. Because each significant F_{obt} indicates only that a difference exists somewhere among the means we tested, we now must examine those means.

INTERPRETING THE TWO-WAY EXPERIMENT

To understand and interpret the results of a two-way ANOVA, we must first examine each significant main effect and interaction by graphing their respective means and performing the *post hoc* comparisons on those means.

First we want to look at each effect by graphing it.

Graphing the Effects

To interpret our persuasiveness study, we again look at the various means, shown in Table 14.12.

Graphing main effects As usual, we plot the dependent variable along the Y axis and the levels of a factor along the X axis. We graph the main effect of factor A by plotting the main effect means from each level of factor A (the column means across the bottom of Table 14.12). We graph the main effects of factor B by separately plotting the main effect means from each level of factor B (the row means at the right of Table 14.12). Figure 14.1 shows the resulting graphs of our main effects. Note that since volume is measured in decibels, the X axis of the volume factor should be labeled in decibels. Also note that we graph the main effect of our gender variable as a bar graph because, as discussed in Chapter 4, this is how we graph means from a nominal independent variable.

The graphs reflect the same differences we see when we compare the means in Table 14.12. In the right-hand graph, we see that males scored higher than

TABLE 14.12 Summary of Means for Persuasiveness Study

		Factor A: volume			
		A_1: *soft*	A_2: *medium*	A_3: *loud*	
Factor B: gender	B_1: *male*	$\bar{X} = 8$	$\bar{X} = 11$	$\bar{X} = 16.67$	$\bar{X}_{male} = 11.89$
	B_2: *female*	$\bar{X} = 4$	$\bar{X} = 12$	$\bar{X} = 6$	$\bar{X}_{fem} = 7.33$
		$\bar{X}_{soft} = 6$	$\bar{X}_{med} = 11.5$	$\bar{X}_{loud} = 11.33$	

FIGURE 14.1 Graphs Showing Main Effects of Volume and Gender

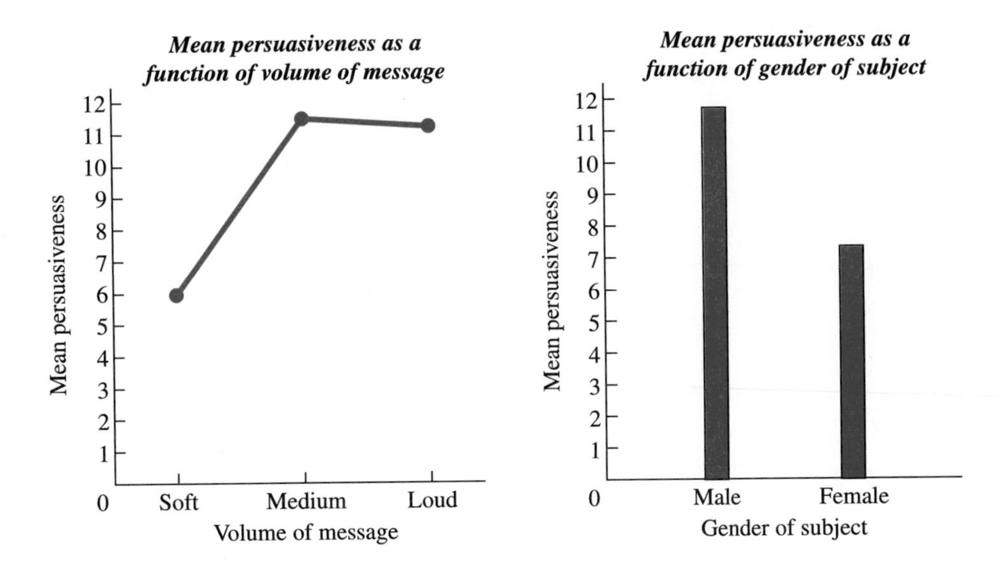

females. In the left-hand graph, the slanting line between soft and medium volume suggests a large (possibly significant) difference between those means. However, the line between medium and loud volume is close to horizontal and the means are close to equal, so there may not be a significant difference here. To specifically determine which means differ significantly, we perform *post hoc* comparisons. But first we will graph our significant interaction.

Graphing the interaction effect An interaction can be a beast to interpret, so always graph it! To graph the interaction, we plot all cell means on a *single* graph. As usual, we place the dependent variable along the *Y* axis. However, we have two independent variables (here volume and subject gender), but only one *X* axis. To solve this problem, we place the levels of one factor along the *X* axis. We show the second factor by drawing on the graph a separate line connecting the means for each level of the factor. Because we must draw a line for each level of the second factor, we typically place the factor with more levels on the *X* axis so that there are as few lines as possible. Thus, for our persuasiveness study, we label the *X* axis with the three volume levels. Then, plotting the cell means from Table 14.12 above, we connect the three means for males with one line and the three means for females with a different line. The graph of the interaction between volume and subject gender is shown in Figure 14.2 on the next page. Notice that we always provide a key to identify each line.

The way to read the graph is to look at one line at a time. Thus, for males (the dashed line), as we increased volume, mean persuasiveness scores increased. However, for females (the solid line), as we increased volume, persuasiveness scores first increased but then decreased. Thus, the effect of increasing volume was not the same for males as for females. Instead, the effect of increasing volume on persuasiveness scores *depends* on whether the subjects are male or female.

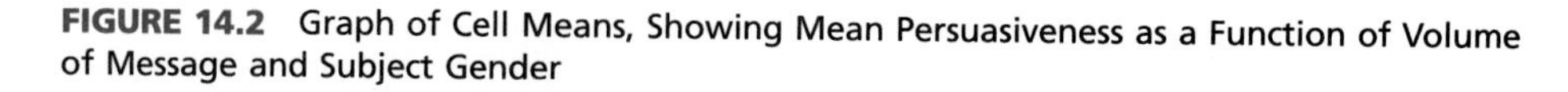

FIGURE 14.2 Graph of Cell Means, Showing Mean Persuasiveness as a Function of Volume of Message and Subject Gender

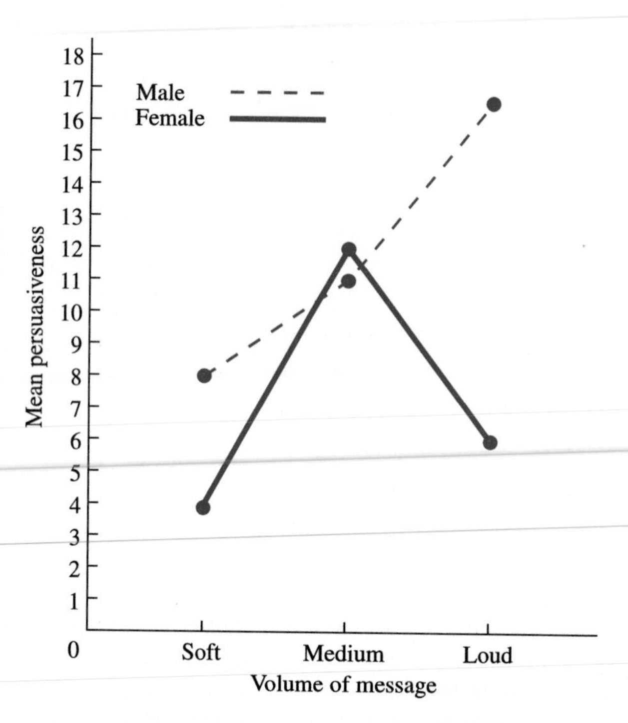

There is a positive linear relationship between scores and increasing volume for males and a nonlinear relationship between scores and increasing volume for females.

Of course, if the interaction were not significant, then we would have no reason to believe that changing the volume had different effects on males and females, regardless of what our graph suggested. Further, we do not know which of these cell means actually differ significantly, because we haven't performed the *post hoc* comparisons yet.

Before we get to the *post hoc* tests, however, note one final aspect of an interaction. An interaction can produce an infinite variety of different graphs, but the key is that *an interaction will produce a graph on which the lines are not parallel.* Remember that a line on a graph summarizes a relationship between the X and Y scores and that a line that is shaped or oriented differently from another line indicates a different relationship. Therefore, when the lines we produce by graphing the cell means are not parallel, each line depicts a *different* relationship. This indicates that the relationship between X and Y changes depending on the level of the second factor, so an interaction is present. Conversely, when an interaction is not present, the graph contains lines that are essentially parallel, with each line depicting essentially the same relationship. To see this distinction, say that our data had produced one of the two graphs in Figure 14.3. On the left-hand graph, we have the ultimate in nonparallel lines. Here, as we change the levels of A, the mean scores either increase or decrease, *depending*

FIGURE 14.3 Two Graphs Showing When an Interaction Is and Is Not Present

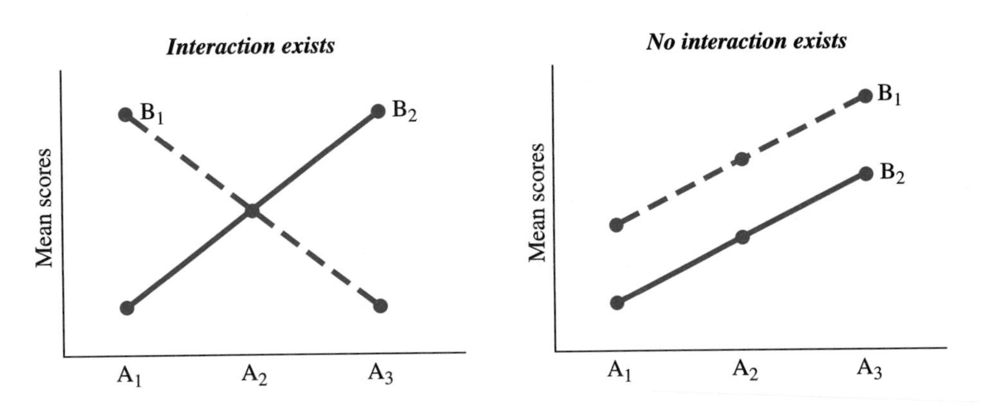

on the level of B we are talking about. Therefore, this graph depicts an interaction. However, in the right-hand graph, the lines are parallel. Here, as we change the levels of A, the scores increase, regardless of which level of factor B we examine. Therefore, this graph does not depict an interaction. (The fact that, *overall,* the scores are higher in B_1 than in B_2 is the main effect of factor B.)

Think of significance testing of the interaction $F_{A \times B}$ as testing whether the lines are significantly different from parallel. When an interaction is *not* significant, the lines on the graph are not significantly different from parallel, so they may represent parallel lines that would be found if we graphed the means of the populations. Conversely, when an interaction is significant, somewhere in the graph the lines *do* differ significantly from parallel. Therefore, if we could graph the means of the populations, the lines probably would not be parallel, and there would be an interaction in the population.

Performing the *Post Hoc* Comparisons

As usual, we perform *post hoc* comparisons on the means from any *significant* F_{obt}. If we have unequal n's between the levels of a factor, we perform Fisher's protected t-test, as described in Chapter 13. If the n's in all levels of a factor are equal, we perform Tukey's *HSD* procedure. However, recognize that the *post hoc* comparisons for an interaction effect are computed differently than those for a main effect.

Performing Tukey's *HSD* for main effects We perform the *post hoc* comparisons on *each* significant main effect, as if it were a one-way ANOVA. We compare all of the column means in factor A, then we compare all of the row means in factor B.

Recall that the computational formula for the *HSD* is

$$HSD = (q_k)\left(\sqrt{\frac{MS_{wn}}{n}}\right)$$

where MS_{wn} is the denominator of the F_{obt}, q_k is found in Table 6 of Appendix D for df_{wn} and k (where k is the number of levels in the factor), and n is the number of scores in a level. *But,* tread carefully here: for each factor, there may be a different value of n and k! In our persuasiveness study, six scores went into each mean for a level of volume, but nine scores went into each mean for a level of subject gender. *The n is always the number of scores used to compute each mean you are comparing.* Also, because q_k depends on k, different factors having a different number of levels will have different values of q_k.

> **STAT ALERT** We must compute a different *HSD* for each main effect when their k's or n's are different.

In our persuasiveness study, for the volume factor we have three means for the main effect, so $k = 3$, and the n of each mean is 6. In our ANOVA we found $MS_{wn} = 8.22$ and $df_{wn} = 12$. From Table 6, for $\alpha = .05$, we find $q_k = 3.77$. Placing these values in the above formula gives

$$HSD = (q_k)\left(\sqrt{\frac{MS_{wn}}{n}}\right) = (3.77)\left(\sqrt{\frac{8.22}{6}}\right) = 4.41$$

Thus, the *HSD* for factor A is 4.41.

Finding the differences between all pairs of means, we have

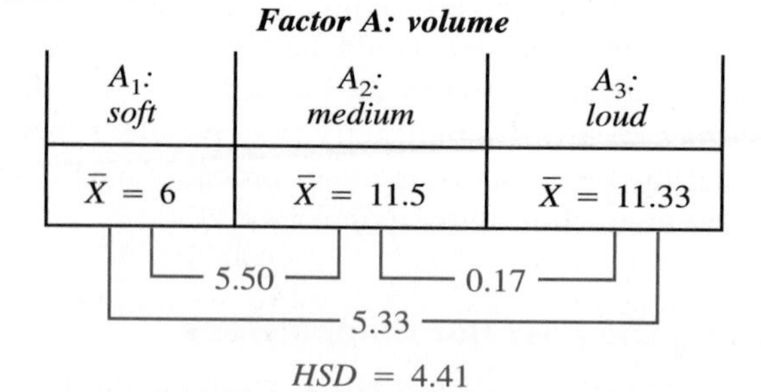

In the middle of each line connecting two means is the absolute difference between them. The mean for soft volume differs from the means for medium and loud by more than the *HSD* of 4.41. Thus, soft volume produces a significant difference from the other volumes. But, because the means for medium and loud volume differ by less than 4.41, these conditions do *not* differ significantly.

Because factor B (subject gender) contains only two levels, we do not perform *post hoc* comparisons (it must be that the mean for males differs significantly from the mean for females). If, however, there were more than two levels in a significant factor B, we would compute the appropriate *HSD* for the n and k in that factor and compare all of these main effect means as well.

Performing Tukey's *HSD* for the interaction In performing *post hoc* comparisons on a significant interaction, we compare the cell means. However, we do *not* compare every cell mean to every other cell mean. Look at the diagram of

TABLE 14.13 Summary of Interaction Means for Persuasiveness Study

Solid lines connecting two cells show examples of unconfounded comparisons; dashed lines connecting two cells show examples of confounded comparisons.

		Factor A: volume		
		A_1: soft	A_2: medium	A_3: loud
Factor B: gender	B_1: male	$\bar{X} = 8$	$\bar{X} = 11$	$\bar{X} = 16.67$
	B_2: female	$\bar{X} = 4$	$\bar{X} = 12$	$\bar{X} = 6$

the interaction means in Table 14.13. We would not, for example, compare the mean for males at the loud volume to the mean for females at the soft volume. This is because even if the means do differ significantly, we would not know what caused the difference. We would be comparing apples to oranges, because the two cells differ in terms of both subject gender *and* volume. Therefore, we would have a confused, or confounded, comparison. A **confounded comparison** occurs when two cells differ along more than one factor. When performing *post hoc* comparisons on an interaction, we perform only **unconfounded comparisons,** in which two cells differ along only one factor. Therefore, we compare only cell means within the same column (comparing means vertically), because we can explain the differences as resulting from factor B. We also compare means within the same row (comparing means horizontally), because we can explain these differences as resulting from factor A. We do not, however, make any diagonal comparisons, because these are confounded comparisons.

When we have equal n's in all cells, we can compare the means in the interaction using a slight variation of the Tukey *HSD.*[2] Previously when we computed the *HSD,* we found q_k in Table 6 using k, the number of means being compared. To compute the *HSD* for an interaction, we must first determine the *adjusted k.* This k adjusts for the actual number of unconfounded comparisons we will make out of all the cell means in our interaction. We obtain our adjusted k from Table 14.14 (on the next page and at the beginning of Table 6 of Appendix D). In the left-hand column, locate the design of the study you are examining. Do not be concerned about the order of the numbers. For example, we called our persuasiveness study a 3 × 2 design, so we look at the row labeled "2 × 3." Reading across that row, as a double check we confirm that the middle column contains the total number of cell means in the interaction (yup, we have 6). In the right-hand column is the adjusted value of k (for our example, it is 5).

[2]Adapted from Cicchetti, D.V. 1972, Extension of Multiple Range Tests to Interaction Tables in the Analysis of Variance, *Psychological Bulletin, 77,* 405–408.

TABLE 14.14 Values of Adjusted k

Design of study	*Number of cell means in study*	*Adjusted value of k*
2 × 2	4	3
2 × 3	6	5
2 × 4	8	6
3 × 3	9	7
3 × 4	12	8
4 × 4	16	10
4 × 5	20	12

The adjusted value of k is the value of k to use in obtaining q_k from Table 6. Thus, for our persuasiveness study, we find that, with $\alpha = .05$, $df_{\text{wn}} = 12$, and adjusted $k = 5$, our $q_k = 4.51$. Now we compute the *HSD* using the same old formula we used previously. Our MS_{wn} is 8.22, but now n is 3, the number of scores in each *cell.* We have

$$HSD = (q_k)\left(\sqrt{\frac{MS_{\text{wn}}}{n}}\right) = (4.51)\left(\sqrt{\frac{8.22}{3}}\right) = 7.47$$

Thus, the *HSD* for our interaction is 7.47.

Now we determine the differences between all cell means within each column and between all cell means within each row. To see these differences, we can arrange our interaction means as shown in Table 14.15. In the middle of each line connecting two cells is the absolute difference between the two means.

Any difference between two means that is larger than our HSD of 7.47 is a significant difference. We have only three significant differences here: (1) between the mean for females at the soft volume and the mean for females at the medium

TABLE 14.15 Table of Interaction Means

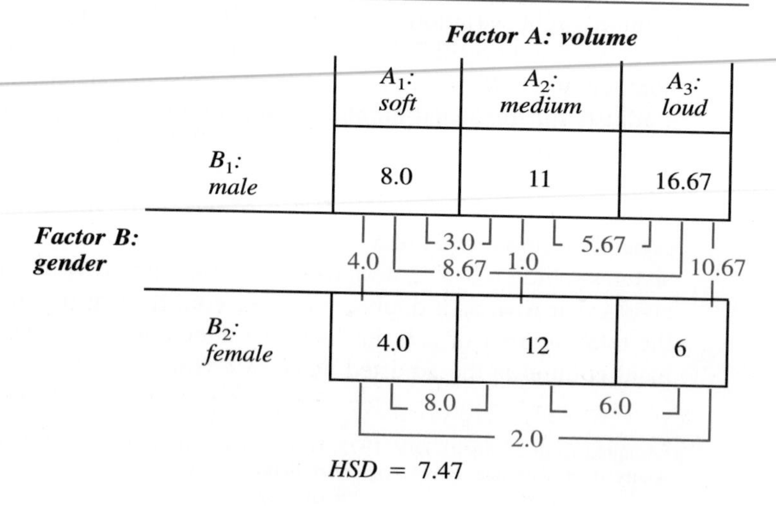

volume, (2) between the mean for males at the soft volume and the mean for males at the loud volume, and (3) between the mean for males at the loud volume and the mean for females at the loud volume.

Interpreting the Overall Results of the Experiment

There is no *one* way to interpret all experiments, because the data in each experiment indicate something different. Our goal is to come up with a complete, honest, and simplified description of the results of the study. We do that by looking at the significant *post hoc* comparisons within all significant main effects and interaction effects.

We can summarize the differences we found in the persuasiveness study using the diagram in Table 14.16. Outside of the diagram are the main effect means. Each line connecting two means indicates that they differ significantly. Inside the diagram, each line connecting two cell means indicates a significant difference within the interaction.

Both of the main effects and the interaction produced significant values of F_{obt}. However, recognize that often our conclusions about significant main effects must be qualified (or are downright untrue) given a significant interaction. For example, there is a significant difference between the main effect means of males and females. If there were not a significant interaction, we could conclude that, overall, males score higher than females. However, when we look at the cell means of the interaction, we see that gender differences *depend* on the volume: only in the loud condition is there a significant difference between males and females. (This difference is so large that it produces an overall mean for males that is larger than the overall mean for females.) Therefore, because our interaction contradicts the overall pattern suggested by the main effect, we *cannot* make an overall, general conclusion about differences between males and females.

Likewise, we cannot make an overall conclusion based on the main effect of volume, which showed that soft volume was significantly different from both the medium and loud volumes. The interaction indicates that increasing the volume

TABLE 14.16 Summary of Significant Differences in Our Persuasiveness Study

Each line connects two means that differ significantly.

		Factor A: volume			
		Level A_1: soft	*Level A_2: medium*	*Level A_3: loud*	
Factor B: gender	*Level B_1: male*	8.0	11	16.67	$\bar{X} = 11.89$
	Level B_2: female	4.0	12	6	$\bar{X} = 7.33$
		$\bar{X}_{soft} = 6$	$\bar{X}_{med} = 11.5$	$\bar{X}_{loud} = 11.33$	

from soft to medium produced a significant difference only in females, and increasing the volume from soft to loud produced a significant difference only in males.

Thus, as the above example illustrates, we usually do not draw any conclusions about significant main effects when the interaction is significant. After all, the interaction indicates that the influence of one factor depends on the levels of the other factor and vice versa, so we should not turn around and act like either factor has a consistent overall effect. Therefore, in such situations, we usually must limit our interpretation of a study to the interaction. When the interaction is not significant, then we focus on any significant main effects.

STAT ALERT The primary interpretation of a two-way ANOVA rests on the interpretation of the significant interaction.

Thus, we conclude that increasing volume beyond soft does tend to increase persuasiveness scores in the population, but this increase occurs for females with medium volume and for males with loud volume. Further, we conclude that differences in persuasiveness scores between males and females do occur in the population, but only if the volume of the message is loud.

Remember experiment-wise error, the probability of a Type I error somewhere in our conclusions? Well, after all of the above shenanigans, we have protected our experiment-wise error, so that for all of these conclusions together, the probability of a Type I error is still $p < .05$. Also, remember power—the probability of not making a Type II error? All that we said in previous chapters about power applies to the two-way ANOVA as well. Thus, for any differences that are not significant, we must be concerned about whether we have maximized our power by maximizing the difference between the means, minimizing the variability within each cell, and having a large enough n.

To round out our analysis of an experiment, we should consider two additional procedures: computing the effect size and computing confidence intervals.

Describing the Effect Size: Eta Squared

Remember that anytime you demonstrate a significant relationship, you should then think "squared correlation coefficient" to describe the relationship. Therefore, in the two-way ANOVA, we again compute eta squared (η^2) to describe effect size—the proportion of variance in the dependent scores that is accounted for by our manipulation. We compute a separate eta squared for each *significant* main and interaction effect. The formula for eta squared is again

$$\eta^2 = \frac{\text{sum of squares between for the effect}}{SS_{\text{tot}}}$$

To compute each eta squared, we divide the sum of squares for the factor, either SS_{A}, SS_{B}, or $SS_{\text{A}\times\text{B}}$, by SS_{tot}. For example, in our persuasiveness study, for factor A (volume), SS_{A} was 117.45 and SS_{tot} was 412.28. Therefore, we have

$$\eta^2_{\text{A}} = \frac{SS_{\text{A}}}{SS_{\text{tot}}} = \frac{117.45}{412.28} = .28$$

Thus, if we predict subjects' scores using the main effect mean of the volume condition they were tested under, we can account for .28, or 28%, of the total variance in persuasiveness scores in our data. Following the same procedure for the gender factor, we have an SS_B of 93.39, so η^2_B is .23: using the main effect mean for each subject's gender, we can account for an additional 23% of the variance in scores. Finally, for the interaction of volume and gender, $SS_{A \times B}$ is 102.77, so $\eta^2_{A \times B}$ is .25: using the mean of the cell that subjects were in to predict their scores, we can account for an additional 25% of the variance in scores.

Recall that the greater the proportion of variance accounted for, the more important the factor is in determining subjects' scores. Because each of the above effects has about the same size, they are all of equal importance in understanding differences in persuasiveness scores in our experiment. However, suppose that one effect accounted for only .01, or 1%, of the total variance. Such a small η^2 indicates that this relationship is very inconsistent. Therefore, it is not a very useful or informative relationship, and we are better served by emphasizing the other, larger significant effects. In essence, if eta squared tells us that an effect was not a big deal in our experiment, then we should not make a big deal out of it when interpreting the experiment.

Usually we are content to describe the effect size in our sample data by computing eta squared. However, we can also estimate the effect size in the population by computing omega squared (ω^2) for each factor or interaction (using the formula presented in Chapter 13).

Confidence Intervals for a Single μ

We can compute the confidence interval for the population μ that is represented by the mean of a level from a main effect or by a cell mean from the interaction. We use the formula presented in the previous chapter, which was

$$\left(\sqrt{\frac{MS_{\text{wn}}}{n}}\right)(-t_{\text{crit}}) + \overline{X} \leq \mu \leq \left(\sqrt{\frac{MS_{\text{wn}}}{n}}\right)(+t_{\text{crit}}) + \overline{X}$$

where t_{crit} is the two-tailed value at the appropriate α with $df = df_{\text{wn}}$, MS_{wn} is from our ANOVA summary table, $\overline{X}$ is the mean for the level or cell we are describing, and n is the number of scores that the mean is based on.

SUMMARY OF THE STEPS IN PERFORMING A TWO-WAY ANOVA

Everything we have discussed in this chapter is summarized in the following steps for performing a two-way ANOVA:

1. Design the experiment, check the assumptions, and collect the data.
2. Compute the sums of squares between groups for each main effect and for the interaction, and compute the sum of squares within groups. Dividing each sum of squares by the appropriate *df* produces the mean square between

groups for each main effect and the interaction, as well as the mean square within groups. Dividing each mean square between groups by the mean square within groups produces each F_{obt}.

3. Find F_{crit} in Table 5 of Appendix D, using the *df* between groups for each factor or interaction and the df_{wn}. If the F_{obt} is larger than F_{crit}, then there is a significant difference between two or more of the means for that factor or interaction.
4. Graph the main effects by plotting the mean of each level of a factor, with the dependent variable on the Y axis and the levels of the factor on the X axis. Graph the interaction by plotting the cell means. Label the X axis with the levels of one factor, and in the body of the graph use a separate line to connect the means from each level of the other factor.
5. Perform *post hoc* comparisons for each significant main effect or interaction.
6. Based on significant main and/or interaction effects and the specific means from cells and levels that differ significantly, develop an overall conclusion regarding the relationships demonstrated by the study.
7. Compute eta squared to describe the proportion of variance in dependent scores accounted for by each significant main effect or interaction in the experiment.
8. Compute the confidence interval for the value of μ represented by the mean in any relevant level or cell.

Congratulations, you are getting *very* good at this stuff.

A WORD ON WITHIN-SUBJECTS AND MIXED DESIGNS

Recall that sometimes an experiment involves related samples (usually because we have employed repeated measures, measuring the same group of subjects under all conditions of a factor). When both factors are repeated measures factors, we perform a two-way, within-subjects (or repeated measures) ANOVA. If a study has a mix of one repeated measures factor and one between-subjects factor, we perform a *mixed design* ANOVA. Although the computations for within-subjects and mixed designs are different from those for the between-subjects design discussed here, the logic is the same: the individual factors and the interaction are each treated as a one-way ANOVA, and a significant F_{obt} indicates that somewhere among the corresponding means we have a significant effect. Then, for each significant result we perform *post hoc* tests, and so on.

FINALLY

Technically there is no limit to the number of factors we can have in an ANOVA. There is, however, a practical limit to how many factors we can *interpret,* especially when we try to interpret the interaction. Say that we add a third factor—the sex of the speaker of the message—to our previous two-way persuasiveness study. We then have a three-way (3 × 2 × 2) ANOVA in which we compute an F_{obt} for the main effect of each factor: A (volume), B (subject gender), and C (speaker gender). We also have an F_{obt} for each of three two-way interactions (A × B, A × C, and B × C). In addition, we have an F_{obt} for a three-way interaction (A × B × C)! If it's significant, it indicates that the interaction between volume and subject gender changes, depending on the sex of the speaker.

If this sounds very complicated, it's because it *is* very complicated. Further, to graph a 3 × 2 × 2 interaction, you would have to draw at least four lines on *one* graph! Three-way interactions are very difficult to interpret, and interactions containing four or more factors are practically impossible to interpret. Therefore, unless you have a very good reason for including many factors in one study, it is best to limit a study to two or, at most, three factors. You may not learn about many variables at once, but what you do learn you will understand.

CHAPTER SUMMARY

1. In a two-way, between-subjects ANOVA, there are two independent variables, and all of the conditions of both factors contain independent samples. In a *complete factorial design,* each level of one factor is combined with all levels of the other factor. Each *cell* is formed by a particular combination of a level from each factor. In the ANOVA, we examine the *main effects* of manipulating each variable separately, as well as the *interaction effect* of manipulating both variables simultaneously.

2. The *assumptions* of the two-way, between-subjects ANOVA are that (a) each cell is a random independent sample of interval or ratio scores, (b) the populations represented in the study are normally distributed, and (c) the variances of all populations are homogeneous.

3. We perform a two-way ANOVA as three one-way ANOVAs, computing an F_{obt} for each main effect and for the interaction. To obtain the means in each main effect, we *collapse* across the levels of the other factor. For the interaction, we do not collapse: we examine all cell means in the experiment.

4. A significant F_{obt} for a main effect indicates that at least two main effect means from the factor represent significant differences in scores. We graph the effect of the factor by plotting the mean score for each level.

5. A significant F_{obt} for an interaction indicates that the effect on scores of changing the levels of one factor *depends on* which level of the other factor we examine. Therefore, the relationship between one factor and the dependent variable changes as we change the levels of the other factor. When graphed, an interaction produces *nonparallel lines.*

6. We perform *post hoc comparisons* on each significant effect having more than two levels to determine which specific means differ significantly. In performing *post hoc* comparisons on the interaction, we make only *unconfounded* comparisons. The means from two cells are unconfounded if the

cells differ along only one factor. Two means are *confounded* if the cells differ along more than one factor.

7. We draw conclusions from a two-way ANOVA by considering the significant main and interaction effects and which level or cell means differ significantly. Usually we cannot draw conclusions about the main effects when the interaction is significant.

8. *Eta squared* is computed to describe the effect size of each significant main effect and interaction. A confidence interval can be computed for the μ represented by any $\overline{X}$ in the study.

PRACTICE PROBLEMS

(Answers for odd-numbered problems are provided in Appendix E.)

1. *a.* A researcher conducts a study involving one independent variable. What are the two types of parametric procedures available to her?

   b. She next conducts a study involving two independent variables. What are the three versions of a parametric procedure available to her?

   c. In part *b*, what aspect of her design determines which version she should perform?

2. Identify the following terms

   a. Two-way design

   b. Complete factorial

   c. Cell

   d. Two-way, between-subjects design

3. *a.* What is meant by a main effect?

   b. What is meant by a two-way interaction effect?

   c. Why do we usually base the interpretation of a two-way design on the interaction when it is significant?

4. What does it mean to collapse across a factor?

5. For a 2 × 2 ANOVA, what are the following.

   a. The statistical hypotheses for factor A.

   b. The statistical hypotheses for factor B.

   c. The statistical hypotheses for A × B.

6. Describe how a two-way ANOVA is similar to three one-way ANOVAs.

7. Below are the cell means of three experiments. For each experiment, compute the main effect means and indicate whether there appears to be an effect of A, B, and/or A × B.

Study 1

	A_1	A_2
B_1	2	4
B_2	12	14

Study 2

	A_1	A_2
B_1	10	5
B_2	5	10

Study 3

	A_1	A_2
B_1	8	14
B_2	8	2

8. In problem 7, if we label the *X* axis with factor A and graph the cell means, what pattern will we see for each interaction?

9. After performing a 3 × 4 ANOVA with equal *n*'s, you find that all *F*'s are significant. What other procedures should you perform?

10. *a.* When is it appropriate to compute the effect size in a two-way ANOVA?
    b. For each effect, what does the effect size tell us?

11. *a.* How can we increase the power of a two-way ANOVA?
    b. Doing so will increase the power of F_{obt} and what other procedure?

12. In an experiment you measure the popularity of two brands of soft drinks (factor A), and for each brand you test males and females (factor B.) The following table shows the main effect and cell means from the study:

		Factor A		
		Level A_1 Brand X	*Level A_2 Brand Y*	
Factor B	*Level B_1 Males*	14	23	18.5
	Level B_2 Females	25	12	18.5
		19.5	17.5	

a. Describe the graph of the interaction means when factor A is on the *X* axis.
b. Does there appear to be an interaction? Why?
c. Why will a significant interaction prohibit you from making conclusions based on the main effects?

13. Why is it wise to limit a multi-factor experiment to two or three factors?

14. *a.* What is a confounded comparison, and when does it occur?
    b. What is an unconfounded comparison, and when does it occur?
    c. Why do we not perform confounded comparisons when performing *post hoc* comparisons on an interaction?

15. A study compared the performance scores of males and females tested either early or late in the day. Here are the data:

		Factor A	
		Level A_1: males	*Level A_2: females*
Factor B	*Level B_1: early*	6 11 9 10 9	8 14 17 16 19
	Level B_2: late	8 10 9 7 10	4 6 5 5 7

a. Using $\alpha = .05$, perform an ANOVA and complete the summary table.
b. Compute the main effect means and interaction means.
c. Perform the appropriate *post hoc* comparisons.
d. What can we conclude about the relationships this study demonstrates?
e. Compute the effect size where appropriate.

16. We conduct an experiment involving two levels of self-confidence (A_1 is low, and A_2 is high) and examine subjects' anxiety scores after they speak to one of four groups of differing sizes (B_1 through B_4 represent speaking to a small, medium, large, or extremely large group, respectively). We compute the following sums of squares ($n = 5$ and $N = 40$):

Source	*Sum of squares*	*df*	*Mean square*	*F*
Between				
Factor A	8.42	___	___	___
Factor B	76.79	___	___	___
Interaction	23.71	___	___	___
Within	110.72	___	___	
Total	219.64	___		

a. Complete the ANOVA summary table.
b. With $\alpha = .05$, what do you conclude about each F_{obt}?
c. Compute the appropriate values of *HSD*.
d. For the levels of factor B, the means are $\overline{X}_1 = 18.36$, $\overline{X}_2 = 20.02$, $\overline{X}_3 = 24.6$, and $\overline{X}_4 = 27.3$. What should you conclude about the main effect of B?
e. How important is the size of the audience in determining a subject's anxiety score? How important is the subjects' self-confidence?

SUMMARY OF FORMULAS

The general format for the summary table for a two-way between-subjects ANOVA is

Summary Table of Two-Way ANOVA

Source	*Sum of squares*	*df*	*Mean square*	*F*
Between				
Factor A	SS_A	df_A	MS_A	F_A
Factor B	SS_B	df_B	MS_B	F_B
Interaction	$SS_{A \times B}$	$df_{A \times B}$	$MS_{A \times B}$	$F_{A \times B}$
Within	SS_{wn}	df_{wn}	MS_{wn}	
Total	SS_{tot}	df_{tot}		

1. Computing the sums of squares

a. *The computational formula for SS_{tot} is*

$$SS_{tot} = \Sigma X^2_{tot} - \left(\frac{(\Sigma X_{tot})^2}{N}\right)$$

b. *The computational formula for the sum of squares between groups for the column factor A is*

$$SS_A = \Sigma\left(\frac{(\text{sum of scores in the column})^2}{n \text{ of scores in the column}}\right) - \left(\frac{(\Sigma X_{tot})^2}{N}\right)$$

c. *The computational formula for the sum of squares between groups for the row factor B is*

$$SS_B = \Sigma\left(\frac{(\text{sum of scores in the row})^2}{n \text{ of scores in the row}}\right) - \left(\frac{(\Sigma X_{tot})^2}{N}\right)$$

d. *The computational formula for the sum of squares between groups for the interaction, $SS_{A \times B}$, is*

$$SS_{A \times B} = SS_{bn} - SS_A - SS_B$$

where SS_{bn} is found using the formula

$$SS_{bn} = \Sigma\left(\frac{(\text{sum of scores in the cell})^2}{n \text{ of scores in the cell}}\right) - \left(\frac{(\Sigma X_{tot})^2}{N}\right)$$

e. *The computational formula for the sum of squares within groups, SS_{wn}, is*

$$SS_{wn} = SS_{tot} - SS_{bn}$$

2. Computing the degrees of freedom

a. The degrees of freedom between groups for factor A, df_A, equals $k_A - 1$, where k_A is the number of levels in factor A.

b. The degrees of freedom between groups for factor B, df_B, equals $k_B - 1$, where k_B is the number of levels in factor B.

c. The degrees of freedom between groups for the interaction, $df_{A \times B}$, equals df_A multiplied times df_B.

d. The degrees of freedom within groups equals $N - k_{A \times B}$, where N is the total N of the study and $k_{A \times B}$ is the total number of cells in the study.

3. Computing the mean square

a. *The formula for MS_A is*

$$MS_A = \frac{SS_A}{df_A}$$

b. *The formula for MS_B is*

$$MS_B = \frac{SS_B}{df_B}$$

c. *The formula for $MS_{A \times B}$ is*

$$MS_{A \times B} = \frac{SS_{A \times B}}{df_{A \times B}}$$

d. *The formula for MS_{wn} is*

$$MS_{wn} = \frac{SS_{wn}}{df_{wn}}$$

4. Computing F_{obt}

a. *The formula for F_A is*

$$F_A = \frac{MS_A}{MS_{wn}}$$

b. *The formula for F_B is*

$$F_B = \frac{MS_B}{MS_{wn}}$$

c. *The formula for $F_{A \times B}$ is*

$$F_{A \times B} = \frac{MS_{A \times B}}{MS_{wn}}$$

5. The critical values of F are found in Table 5 of Appendix D.

a. To find F_{crit} to test F_A, use df_A and df_{wn}.

b. To find F_{crit} to test F_B, use df_B and df_{wn}.

c. To find F_{crit} to test $F_{A \times B}$, use $df_{A \times B}$ and df_{wn}.

6. Performing Tukey's *HSD post hoc* comparisons

a. *For each significant main effect, the computational formula for the HSD is*

$$HSD = (q_k)\left(\sqrt{\frac{MS_{\text{wn}}}{n}}\right)$$

where q_k is found in Table 6 for k equal to the number of levels in the factor, MS_{wn} is the denominator of F_{obt}, and n is the number of scores used to compute each mean in the factor. Any means that differ by an amount that is greater than the value of *HSD* are significantly different.

b. *For a significant interaction, the HSD is computed as follows.*

(1) Enter the following table for the design (or number of cells), and obtain the adjusted value of k.

Values of Adjusted k

Design of study	*Number of cell means in study*	*Adjusted value of k*
2×2	4	3
2×3	6	5
2×4	8	6
3×3	9	7
3×4	12	8
4×4	16	10
4×5	20	12

(2) Enter Table 6 for the value of q_k, using the adjusted k and df_{wn} at the appropriate σ.

(3) Compute the value of *HSD* as described in step 6.a, above.

(4) Any unconfounded cell means that differ by an amount that is greater than the value of *HSD* are significantly different.

7. *The computational formula for eta squared is*

$$\eta^2 = \frac{\text{sum of squares between for the factor}}{SS_{\text{tot}}}$$

When η^2 is computed for factor A, factor B, or the A $\times$ B interaction, the sum of squares for the effect is SS_{A}, SS_{B}, or $SS_{\text{A}\times\text{B}}$, respectively.

8. *The computational formula for computing the confidence interval for a single* μ, *using the results of a between-subjects ANOVA, is*

$$\left(\sqrt{\frac{MS_{\text{wn}}}{n}}\right)(-t_{\text{crit}}) + \bar{X} \leq \mu \leq \left(\sqrt{\frac{MS_{\text{wn}}}{n}}\right)(+t_{\text{crit}}) + \bar{X}$$

where t_{crit} is the two-tailed value at the appropriate α with $df = df_{\text{wn}}$, MS_{wn} is from the ANOVA, and $\bar{X}$ and n are from the level or cell being described.

15

Nonparametric Procedures for Frequency Data and Ranked Data

To understand the upcoming chapter:

- From Chapter 2, recall the four types of measurement scales (nominal, ordinal, interval, and ratio).
- From Chapter 12, remember the types of designs that call for either the independent samples t-test or the related samples t-test, and understand the assumptions of each test.
- From Chapter 13 understand why we select either a between-subjects or a within-subjects ANOVA.

Then your goals for this chapter are to learn:

- The type of data that require the use of nonparametric statistics.
- The logic and use of the one-way chi square.
- The logic and use of the two-way chi square.
- The names and interpretation of the nonparametric procedures corresponding to the independent samples and related samples t-test, and those corresponding to the between-subjects and within-subjects ANOVA.

Throughout this book we have performed parametric inferential statistics in which we assumed that our samples represented raw score populations that were normally distributed, that the population variances were homogeneous, and that the dependent variable was measured using an interval or ratio scale. However, we cannot insist on using a parametric procedure when we seriously violate these assumptions. If we do, then the actual probability of a Type I error will be substantially *larger* than our alpha level. Therefore, in such instances we turn to nonparametric procedures.

Nonparametric procedures are never our first choice, because they are less powerful than parametric procedures. But sometimes we must use them. As the name implies, nonparametric procedures do not require that we make assumptions about the parameters of the populations represented by our samples. Therefore, we use nonparametric statistics when we have scores from very skewed or otherwise nonnormal distributions, when the population variance is heterogeneous, or when our dependent scores are from an ordinal (ranked) or nominal (categorical) variable.

Nonparametric procedures are still inferential statistics that we use to decide whether the differences between our samples represent differences we would find in the populations. Therefore, the concepts of H_0 and H_a, sampling distributions, Type I and Type II errors, alpha levels, critical values, and power all apply. We will still ask the same questions that we asked in previous chapters, but here, the characteristics of our data simply dictate that we compute the answers differently.

In this chapter we will first discuss the most common nonparametric procedure, chi square, which is used when we have nominal data. Then we will discuss the nonparametric procedures that are analogous to *t*-tests and ANOVAs, except that they are used for rank-ordered scores or for interval or ratio data that violate the assumptions of *t*-tests or ANOVAs.

CHI SQUARE PROCEDURES

Until now, a score has measured the *amount* of a dependent variable that a subject demonstrates, and we have summarized the scores using the mean. However, sometimes researchers study scores that do not indicate an amount, but rather indicate the *category* that a subject falls in. We summarize these data by indicating the *number* of subjects who give a certain response: how many answer yes, no, or maybe to a question; how many indicate that they are male or female; how many claim to vote Republican, Democratic, or Communist; how many say that they were or were not abused children; and so on. In the above cases, each variable is a nominal, or categorical, variable, and our data consist of the total number, or *frequency,* of subjects falling in each category.

For example, we might find that out of 100 subjects, 40 say yes to a particular question and 60 say no. These numbers indicate how the *frequencies are distributed* across our categories of yes/no. As usual, we want to draw inferences about what we would find in the population: If we were to ask the entire population this question, can we infer that 40% of the population would say yes and 60% would say no? Or would the frequencies be distributed in a different manner? To make inferences about the frequencies that would be found in the population, we perform the procedure called chi square (pronounced "kigh square"). The **chi square procedure** is the nonparametric inferential procedure for testing whether the frequencies in each category in our sample represent certain frequencies in the population.

> ***STAT ALERT*** Whenever you measure the number of subjects that fall in different categories, use the chi square procedure for significance testing.

The symbol for the chi square statistic is χ^2. Theoretically, there is no limit to the number of categories, or levels, we may have in a variable and no limit to the number of variables we may have. We describe a chi square design in the same way we described ANOVAs: when a study has only one variable, we use the one-way chi square; when a study has two variables, we use the two-way chi square; and so on.

ONE-WAY CHI SQUARE: THE GOODNESS OF FIT TEST

We use the one-way chi square when our data consist of the frequencies with which subjects belong to the different categories of *one* variable. As usual, we are examining a relationship, but here the relationship is between the different categories and the frequency with which subjects fall in each category. We are asking, "As we change the categories, do the frequencies with which subjects fall in the categories change in a consistent fashion?"

Here is a study that calls for a one-way chi square. Scientists believe that being right-handed or left-handed is related to brain organization and function. Interestingly, many of history's great geniuses were left-handed. To explore the relationship between the frequencies of left- and right-handedness in geniuses, say that, using an IQ test, we randomly select a sample of 50 geniuses. Then we ask them whether they are left- or right-handed (ambidextrous is not an option). The total numbers of left- and right-handers are the frequencies in the two categories. We can summarize our results as shown in the following diagram.

Handedness

Left-handers	*Right-handers*
$f_o = 10$	$f_o = 40$

$k = 2$
$N = \text{total } f_o = 50$

Each column contains the frequency with which subjects fall in that category. We call this value the *observed frequency,* symbolized by f_o. The sum of the f_o's from all categories must equal N, the total number of subjects in the study. Notice that the symbol k again stands for the number of categories, or levels, in a one-way chi square, and here $k = 2$.

The results of the above study seem pretty straightforward: 10 of our 50 subjects, or 20%, are left-handers, and 40 of them, or 80%, are right-handers. Because this is a random sample, we want to argue that the same distribution of 20% left-handers and 80% right-handers would occur in the population of all geniuses. But, of course, there is the usual problem: sampling error. Maybe, by luck, the subjects in our sample are unrepresentative, so if we could examine the population of all geniuses, we would not find this distribution of right- and left-handers. Maybe our results poorly represent some *other* distribution. As usual, this is our null hypothesis, implying that we are being misled by sampling error.

The question is "What is that 'other distribution' of frequencies that our sample poorly represents?" To answer this question, we create a *model* of the distribution of the frequencies we expect to find in the population when H_0 is true. Recall that H_0 always implies that our study did not work—that it failed to demonstrate the predicted relationship. Therefore, our H_0 model describes the distribution of frequencies we would find in the population if there is not the predicted relationship.

You'll notice that the subtitle of this section is "The Goodness of Fit Test." This is because the one-way χ^2 procedure tests how "good" the "fit" is between our data and the H_0 model. Thus, goodness of fit is merely another way of determining whether our sample data are likely to represent the distribution of frequencies in the population that is described by H_0.

Creating the Statistical Hypotheses for χ^2

Usually, researchers test the H_0 that there is no difference between the frequencies in the categories in the population. This, of course, means that H_0 states that there is no relationship in the population. For our handedness study, say that, for the sake of illustration, we ignore that there are generally more right-handers than left-handers in the world. Therefore, if there is no relationship in the population, then there is no difference between the frequencies of left- and right-handed geniuses. Thus, our H_0 model is that the frequencies of left- and right-handed geniuses in the population are equal. There is no conventional way to write our hypotheses in symbols, so we simply write H_0 as

H_0: all frequencies in the population are equal

H_0 implies that if the observed frequencies (f_o) in our sample are not equal, it is because of sampling error.

Our alternative hypothesis always implies that the study did demonstrate the predicted relationship. Here, our H_a is

H_a: all frequencies in the population are not equal

H_a implies that our observed frequencies represent different frequencies of left- and right-handers in the population of geniuses.

Notice that we can only test whether our sample frequencies are different from those described by H_0, so the one-way χ^2 only tests two-tailed hypotheses.

Computing the Expected Frequencies, f_e

To compute the χ^2 statistic, we must translate our H_0 model into a specific expected frequency for each category. The **expected frequency** is the frequency we would expect in a category if the sample data perfectly represented the distribution of frequencies in the population described by the null hypothesis. The symbol for an expected frequency is f_e.

In our study, H_0 is that the frequencies of left- and right-handedness are equal in the population of geniuses. If our sample perfectly represents this population, then out of the 50 subjects in our study, 25 should be right-handed and 25 should be left-handed. Thus, when H_0 is true, the expected frequency in each category is $f_e = 25$.

For future reference, notice that f_e is actually based on a probability. If the frequencies in the population are equal, then the probability of someone's being left-handed equals the probability of someone's being right-handed. With only two possible categories, the probability of someone's falling in either category is .5. Since probability is the same as relative frequency, we expect .5 of all geniuses to be left-handed and .5 of all geniuses to be right-handed. Therefore, out of the 50 geniuses in our study, we expect to have a frequency of (.5) (50), or 25, in each category. *Thus, the expected frequency in a category is equal to the probability of someone's falling in that category multiplied times the* N *of the study.*

Whenever our H_0 is that the frequencies in the categories are equal, the expected frequency for each category can be computed as the total N in the study divided by the number of categories. Thus,

THE COMPUTATIONAL FORMULA FOR EACH EXPECTED FREQUENCY, WHEN TESTING AN H_0 *OF NO DIFFERENCE, IS*

$$f_e \text{ in each category} = \frac{N}{k}$$

Thus, in our handedness study, with an N of 50 and $k = 2$,

$$f_e \text{ in each category} = \frac{50}{2} = 25$$

Note: Sometimes f_e may contain a decimal. (For example, if we included a third category, ambidextrous, then k would be 3, and for 50 subjects, each f_e would be 16.67.)

As in any statistical test, we must check that our study meets the assumptions of the test.

Assumptions of the One-Way Chi Square

The assumptions of the one-way χ^2 are:

1. We are categorizing subjects along one variable having two or more categories, counting the frequency (the number) of subjects belonging to each category.
2. Each subject is measured only once and can be in one and only one category.

3. Category membership is independent: the fact that a particular subject falls in one category does not influence the probability of any other subject's falling in any category.
4. The computations are based on the responses of all subjects in the study. (That is, we would not count only the number of right-handers. Or, in a different study, if we counted the number of subjects who agreed with some statement, we would also include, as a category, those subjects who disagreed with the statement.)
5. So that our data meet certain theoretical considerations, the f_e in any category should equal at least 5.

Computing χ^2

If the sample perfectly represents the situation where there are no differences in handedness in the population, then we expect to find 25 subjects in each category. In other words, when H_0 is true, our f_o "should" equal our f_e. Any difference between f_o and f_e is chalked up to sampling error. Of course, the greater the difference between the observed frequency and the expected frequency, the less likely it is that the difference is due to sampling error. Therefore, the greater the difference between f_o and f_e, the less likely it is that H_0 is true and that our sample represents an equal distribution of frequencies in the population.

The χ^2 is a way to measure the overall differences between f_o and f_e in the categories in our study. We compute an obtained χ^2, which we'll call χ^2_{obt}.

THE COMPUTATIONAL FORMULA FOR CHI SQUARE, χ^2_{obt}, IS

$$\chi^2_{obt} = \Sigma\left(\frac{(f_o - f_e)^2}{f_e}\right)$$

In English, you find the difference between f_o and f_e in each category, and square that difference. Then divide each squared difference by the f_e for that category. After doing this for all categories, sum the quantities, and the answer is χ^2_{obt}. (Note that because each difference is squared, χ^2 can never be a negative number.)

For our handedness study, we have these frequencies:

Handedness

Left-handers	*Right-handers*
$f_o = 10$ $f_e = 25$	$f_o = 40$ $f_e = 25$

Filling in the formula, we have

$$\chi^2_{obt} = \Sigma\left(\frac{(f_o - f_e)^2}{f_e}\right) = \left(\frac{(10 - 25)^2}{25}\right) + \left(\frac{(40 - 25)^2}{25}\right)$$

After subtracting, we have

$$\chi^2_{obt} = \left(\frac{(-15)^2}{25}\right) + \left(\frac{(15)^2}{25}\right)$$

Squaring then gives

$$\chi^2_{obt} = \left(\frac{225}{25}\right) + \left(\frac{225}{25}\right)$$

After dividing, we have

$$\chi^2_{obt} = 9 + 9 = 18.0$$

so $\chi^2_{obt} = 18.0$.

Interpreting χ^2

As always, to interpret our obtained statistic we must determine its location on the sampling distribution when H_0 is true. The H_0 sampling distribution of χ^2 contains all possible values of χ^2 when H_0 is true (the observed frequencies represent the model described by H_0). Thus, for our handedness study, our χ^2-distribution is the distribution of all possible values of χ^2 when there are two categories and the frequencies in the two categories in the population are equal. We envision the χ^2-distribution as shown in Figure 15.1.

Even though the χ^2-distribution is not at all normal, it is used in the same way as previous sampling distributions. When the data perfectly represent the H_0 model so that each f_o equals the corresponding f_e, χ^2 is zero. The larger the value of χ^2, the larger the differences between the expected and observed frequencies and the less frequently they occur when the observed frequencies do represent our H_0 model. Therefore, the larger the χ^2_{obt}, the less likely it is that our data represent the population frequencies described by H_0. Notice that with chi square we again have two-tailed hypotheses but one region of rejection. If χ^2_{obt} is larger

FIGURE 15.1 Sampling Distribution of χ^2 When H_0 Is True

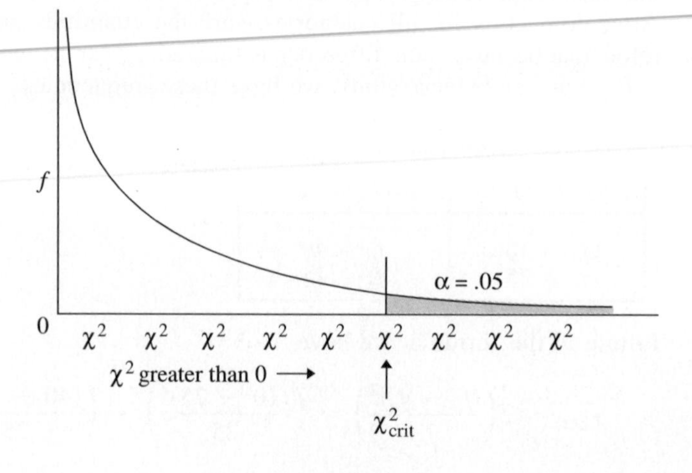

than the critical value, then it is in the region of rejection: it occurs less than 5% of the time when H_0 is true. Then χ^2 is significant, because the observed frequencies are so different from the expected frequencies that the observed frequencies are too unlikely to accept as representing the distribution of frequencies in the population described by H_0.

To determine if our χ^2_{obt} is significant, we compare it to the critical value, symbolized by χ^2_{crit}. As with previous statistics, the χ^2-distribution changes shape as the degrees of freedom change, so to find the appropriate value of χ^2_{crit} for our particular study, we must first have the degrees of freedom.

In a one-way χ^2, the degrees of freedom equals $k - 1$, where k is the number of categories.

To determine the critical value of χ^2, we turn to Table 8 in Appendix D, entitled "Critical Values of Chi Square." Then, for the appropriate degrees of freedom ($k - 1$) and α, we locate the critical value of χ^2. For our handedness study, $k = 2$, so $df = 1$, and with $\alpha = .05$, our $\chi^2_{crit} = 3.84$. Our χ^2_{obt} of 18.0 is larger than the χ^2_{crit} of 3.84, so our results are significant: we reject the H_0 that each f_o represents an equal frequency in the population. We report our results as

$$\chi^2(1) = 18.0, \quad p < .05$$

Notice the df in parentheses.

By rejecting H_0, we accept the H_a that our observed frequencies represent frequencies in the population that are not equal. In fact, as in our samples, we would expect to find about 20% left-handers and 80% right-handers in the population of geniuses. We conclude that we have evidence of a relationship between the categories of handedness and the frequency with which geniuses fall in each.

If our χ^2_{obt} had not been significant, we would have failed to reject H_0. As usual, failing to reject H_0 does not prove that it is true. Therefore, we would *not* be able to say that the distribution of left- and right-handed geniuses in the population was equal. We would simply remain unconvinced that the distribution was unequal.

Other Uses of the Goodness of Fit Test

Instead of testing an H_0 that the frequencies in all categories are distributed equally, we can also test the goodness of fit to other H_0 models, which say that the frequencies are distributed in some other way. For example, we should not have ignored the fact that only about 10% of the general population is left-handed. The better test is to determine whether the distribution of handedness in our sample of geniuses fits this model of the distribution of handedness in the general population.

The null hypothesis is still that our data fit our H_0 model, so we can state our H_0 as

H_0: 10% left-handed, 90% right-handed

For simplicity, we can write our H_a as

H_a: not H_0

H_a implies that our observed frequencies represent a population that does not fit the H_0 model, so the population of geniuses is not 10% left-handed and 90% right-handed.

As usual, we compute our values of f_e based on H_0, but now the new model is that 10% of the population is left-handed and 90% is right-handed. If H_0 is true and our sample is perfectly representative of the general population, then we expect left-handed geniuses to occur 10% of the time. For the 50 geniuses in our study, 10% is 5, so our expected frequency for left-handers is $f_e = 5$. We expect right-handed geniuses to occur 90% of the time, and because 90% of 50 is 45, our expected frequency for right-handers is $f_e = 45$. As usual, according to H_0, any differences between the observed and expected frequencies are due to sampling error in representing this model.

We should *not* perform two χ^2 procedures on the same data, but for the sake of illustration, we will compare our previous handedness frequencies and our new expected frequencies. We have

Handedness

Left-handers	*Right-handers*
$f_o = 10$ $f_e = 5$	$f_o = 40$ $f_e = 45$

$k = 2$
Total $f_o = 50$

We compute χ^2 using the same formula we used in the previous section. Putting the above values in our formula, we have

$$\chi^2_{obt} = \Sigma\left(\frac{(f_o - f_e)^2}{f_e}\right) = \left(\frac{(10 - 5)^2}{5}\right) + \left(\frac{(40 - 45)^2}{45}\right)$$

(Notice that we now have a different value of f_e in each fraction.) Working through the formula, we have

$$\chi^2_{obt} = 5.0 + .56$$

so $\chi^2_{obt} = 5.56$.

With $\alpha = .05$ and $k = 2$, the critical value of χ^2 for $df = 1$ is again 3.84. Because the χ^2_{obt} of 5.56 is larger than the χ^2_{crit} of 3.84, our results fall in the region of rejection. Therefore, we reject H_0 and conclude that the observed frequencies are significantly different from what we would expect if handedness in the population of geniuses was distributed as it is in the general population. Instead, our best guess is that the population of geniuses would be distributed as in our sample data, with 20% left-handers and 80% right-handers.

If our χ^2_{obt} had not been significant, we would have failed to reject H_0 and would simply remain unconvinced that handedness is distributed differently in geniuses than it is in the general population.

Additional Procedures in a One-Way Chi Square

As usual, a graph is a useful way to summarize data, especially if there are more than two categories. We label the Y axis with the frequencies and the X axis with

FIGURE 15.2 Frequencies of Left- and Right-Handed Geniuses

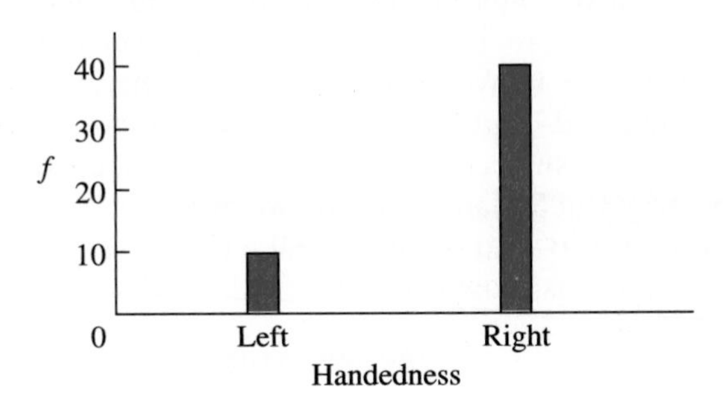

the levels or categories, and then we plot the f_o in each category. Figure 15.2 shows the results of our handedness study. Notice that handedness is a categorical, or nominal, variable, and that when the X variable is a nominal variable, we create a bar graph.

Notably, unlike ANOVA, the one-way chi square usually is not followed by *post hoc* comparisons. A significant χ^2_{obt} indicates that, across all categories of the variable, the frequencies are distributed in a manner that is significantly different from that described by H_0. We then use the observed frequency in each category to estimate the frequencies that would be found in the population. Likewise, we do not compute such measures as eta squared.

THE TWO-WAY CHI SQUARE: THE TEST OF INDEPENDENCE

We use the *two-way* chi square procedure when our data consist of the frequencies with which subjects belong to the categories in each of *two* variables. This is similar to the complete factorial design we saw in the previous chapter. Depending on the number of categories in each variable, the design can be 2×2, 2×3, 4×3, and so on. The procedures for computing χ^2 are the same regardless of the design.

The assumptions of the two-way chi square are the same as for the one-way chi square. (If an f_e is less than 5, we cannot compute χ^2. Instead, we perform Fisher's exact test.[1])

Logic of the Two-Way Chi Square

Here is a study that calls for a two-way chi square. Some psychologists have claimed that they have identified two personality types: Type A and Type B. The Type A personality tends to be a very pressured, hostile individual who never seems to have enough time. The Type B personality tends not to be so time pressured, being more relaxed and mellow. A controversy developed over whether people with Type A

[1]Described in S. Siegel and N. J. Castellan (1988), *Nonparametric Statistics for the Behavioral Sciences,* 2nd ed. (New York: McGraw Hill).

personalities are less healthy, especially when it comes to the big one—having heart attacks. Say that we enter this controversy by randomly selecting a sample of 80 people. Using the appropriate personality test, we determine how many are Type A and how many Type B. We then count the frequency with which Type A and Type B subjects have had heart attacks. We must also count the frequency with which Type A and Type B subjects have *not* had heart attacks (see item 4 in "Assumptions of the One-Way Chi Square"). Therefore, we have two categorical variables: personality type (A or B) and subject's health (heart attack or no heart attack). We can diagram this study as shown below.

		Personality type	
		Type A	*Type B*
Subject's health	*Heart attack*	f_o	f_o
	No heart attack	f_o	f_o

Understand that although this looks like a two-way ANOVA, it is not analyzed like one. Instead of a series of tests for main effects and an interaction, *the two-way χ^2 procedure only tests what is essentially the interaction.* That is, it tests whether the distribution of the frequencies in the categories of one variable depends on which category of the other variable we examine. Because of this, the two-way χ^2 is called a *test of independence:* we determine whether the frequency of subjects falling in a particular category of one variable is independent of the frequency of their falling into a particular category of the other variable.

Thus, in our study, we will test whether the frequencies for having or not having a heart attack are independent of the frequencies for being Type A or Type B. Table 15.1 shows an ideal example of data when the two variables are independent. Here, the frequency of subjects' having or not having a heart attack does not depend on the frequency of subjects' being Type A or Type B. Thus, the two variables are independent.

Another way to view the two-way χ^2 is as a test of whether a correlation exists between the two variables. When the variables are independent, there is no

TABLE 15.1 Observed Frequencies When Personality Type and Heart Attacks Are Independent of Each Other

		Personality type	
		Type A	*Type B*
Subject's health	*Heart attack*	$f_o = 20$	$f_o = 20$
	No heart attack	$f_o = 20$	$f_o = 20$

TABLE 15.2 Observed Frequencies When Personality Type and Heart Attacks Are Dependent on Each Other

		Personality type	
		Type A	*Type B*
Subject's health	*Heart attack*	$f_o = 40$	$f_o = 0$
	No heart attack	$f_o = 0$	$f_o = 40$

correlation. Then, using the categories from one variable does not help us to accurately predict the frequencies for the other variable. In Table 15.1, using Type A or Type B does not help us predict how frequently subjects have or do not have a heart attack (and using the categories of heart attack and no heart attack does not help us predict the frequency of each personality type).

On the other hand, Table 15.2 shows an ideal example of data when the two variables are not independent: they are dependent. Here, the frequency of a heart attack or no heart attack *depends* on personality type. Thus, a correlation exists, because whether subjects are Type A or Type B is a very good predictor of whether they have or have not had a heart attack (and vice versa).

The null hypothesis always says that there is zero correlation in the population, so the null hypothesis in the two-way χ^2 always says that the variables are independent in the population. If, in the sample data, the variables appear to be dependent and correlated, H_0 says that this is due to sampling error. The alternative hypothesis is that the variables are dependent (correlated).

Computing the Expected Frequencies in the Two-Way Chi Square

As usual, our expected frequencies are based on our model of the population described by H_0, so here we compute the f_e in each category based on the idea that the variables are independent. To see how we do this, say that we actually obtained the following data in our heart attack study.

TABLE 15.3 Frequencies as a Function of Personality Type and Subject's Health

		Personality type		
		Type A	*Type B*	
Subject's health	*Heart attack*	$f_o = 25$	$f_o = 10$	row total = 35
	No heart attack	$f_o = 5$	$f_o = 40$	row total = 45
		column total = 30	column total = 50	total = 80 $N = 80$

Notice that, after recording the f_o for each cell, we compute the total of the observed frequencies in each column and in each row. Also, we compute the total of all frequencies, which equals N. (To check your work, note that the sum of the row totals should equal the sum of the column totals, which equals N.)

Now we compute the expected frequency for each cell when the variables are independent. As with the one-way χ^2, the expected frequency is based on the probability of a subject's being located in the cell. First let's compute the probability of a subject's being in the top left-hand cell of heart attack and Type A. Out of a total of 80 subjects in the study, 35 reported having had a heart attack (the row total). Thus, the probability of someone's reporting a heart attack is 35/80, or .438. Similarly, the probability of a subject's being Type A is 30 (the column total) out of 80, or 30/80, which is .375. We want to know the probability of a subject's being Type A *and* reporting a heart attack. Whenever we want to determine the probability of having event A and event B occur simultaneously, we *multiply* the probability of event A times the probability of event B (a discussion of this multiplication rule for independent events is given in Appendix B.) Thus, the probability of a subject's reporting a heart attack *and* being Type A is equal to the probability of reporting a heart attack multiplied times the probability of being Type A, (.438)(.375), which is .164. Thus, the probability of someone's falling in the cell for a heart attack and Type A is .164, if the two variables are independent.

Because probability is the same as relative frequency, we expect .164 of our subjects to fall in this cell if the variables are independent. This means that out of our 80 subjects, we expect .164 times 80, or 13.125 subjects to be in this cell. Therefore, our expected frequency for this cell is $f_e = 13.125$.

Luckily, there is a shortcut formula for calculating each f_e. Above, we multiplied 35/80 times 30/80 and then multiplied the answer times 80. The 35 is the total f_o of the *row* that contains the cell, 30 is the total f_o of the *column* that contains the cell, and 80 is the total N of the study. Using these components, we can construct a formula.

THE COMPUTATIONAL FORMULA FOR COMPUTING THE EXPECTED FREQUENCY IN A CELL OF A TWO-WAY CHI SQUARE IS

$$f_e = \frac{(\text{cell's row total } f_o)\ (\text{cell's column total } f_o)}{N}$$

To find f_e for a cell, we multiply the total observed frequencies for the row containing the cell times the total observed frequencies for the column containing the cell, and then we divide by the N of the study.

Table 15.4 shows the finished diagram for our study, giving the computed f_e for each cell. To check your work, confirm that the sum of the f_e's in each column or row equals the column or row total.

If H_0 is true and the variables are independent, then each observed frequency should equal each corresponding expected frequency. H_0 implies that any difference between f_o and f_e is merely due to sampling error. However, the greater the

TABLE 15.4 Diagram Containing f_o and f_e for Each Cell

Each f_e equals the row total times the column total, divided by N.

		Personality type		
		Type A	*Type B*	
Subject's health	*Heart attack*	$f_o = 25$ $f_e = 13.125$ (35)(30)/80	$f_o = 10$ $f_e = 21.875$ (35)(50)/80	row total = 35
	No heart attack	$f_o = 5$ $f_e = 16.875$ (45)(30)/80	$f_o = 40$ $f_e = 28.125$ (45)(50)/80	row total = 45
		column total = 30	column total = 50	total = 80

value of χ^2_{obt}, the greater the difference between f_o and f_e, so the less likely it is that the data poorly represent variables that are independent.

Computing the Two-Way Chi Square

We compute the χ^2_{obt} for the two-way χ^2 using the same formula we used in the one-way design, which is

$$\chi^2_{obt} = \Sigma\left(\frac{(f_o - f_e)^2}{f_e}\right)$$

Using the data in Table 15.4 from our heart attack study, we have

$$\chi^2_{obt} = \left(\frac{(25 - 13.125)^2}{13.125}\right) + \left(\frac{(10 - 21.875)^2}{21.875}\right) + \left(\frac{(5 - 16.875)^2}{16.875}\right) + \left(\frac{(40 - 28.125)^2}{28.125}\right)$$

As before, in the numerator of each fraction is the observed frequency minus the expected frequency for a cell, and in the denominator is the expected frequency for that cell. Solving each fraction gives

$$\chi^2_{obt} = 10.74 + 6.45 + 8.36 + 5.01$$

so $\chi^2_{obt} = 30.56$.

Although this is a rather large value, such answers are possible. (If you get one, however, it's a good idea to triple-check your computations.)

To evaluate our χ^2_{obt}, we need to find the appropriate χ^2_{crit}, so first we must determine the degrees of freedom.

THE DEGREES OF FREEDOM IN A TWO-WAY CHI SQUARE IS

$$df = (\text{number of rows} - 1)(\text{number of columns} - 1)$$

For our study, df is $(2 - 1)(2 - 1)$, or 1. We again find the critical value of χ^2 in Table 8 in Appendix D. At $\alpha = .05$ and $df = 1$, the χ^2_{crit} is 3.84.

Our χ^2_{obt} of 30.56 is larger than the χ^2_{crit} of 3.84, so our obtained χ^2 is significant. When the two-way χ^2_{obt} is significant, the observed frequencies are too unlikely for us to accept as poorly representing frequencies from variables that are independent. Therefore, we reject our H_0 that the variables are independent and accept the alternative hypothesis: we are confident that the sample represents frequencies from two variables that are dependent in the population. In other words, we conclude that there is a significant correlation such that the frequency of having or not having a heart attack depends on the frequency of being Type A or Type B (and vice versa). We report our results as

$$\chi^2(1) = 30.56, \quad p < .05$$

If our χ^2_{obt} had not been larger than the critical value, we would not have rejected H_0. Therefore, we could not say whether these variables are independent.

STAT ALERT A significant two-way χ^2 indicates that the sample data are likely to represent two variables that are dependent (or correlated) in the population.

Additional Procedures in the Two-Way Chi Square

When we find a significant two-way χ^2_{obt}, there are two procedures we can apply to further describe our data: we graph the data, and we describe the strength of the relationship.

Graphing the two-way chi square We graph the data of a two-way chi square in the same way that we graphed a two-way interaction in the previous chapter, except that here we create a bar graph. Frequency is plotted along the Y axis, and one of the category variables is plotted along the X axis. The other category variable is indicated within the body of the graph. Figure 15.3 shows such a bar graph for our heart attack study. It is interpreted in the same way that we interpreted our table of frequencies: the frequency of subjects having or not having a heart attack depends on whether we are talking about Type A or Type B personalities.

Describing the relationship in a two-way chi square A significant two-way chi square indicates that there is a significant relationship between the two variables. Remember, however, that we don't keep the strength of a significant relationship secret. Instead, as usual, we first think "correlation coefficient."

If we have performed a 2×2 chi square (and it is significant), we describe the strength of the relationship by computing a new correlation coefficient known as the **phi coefficient.** The symbol for the phi coefficient is ϕ, and its value can

FIGURE 15.3 Frequency of Heart Attacks and Personality Type

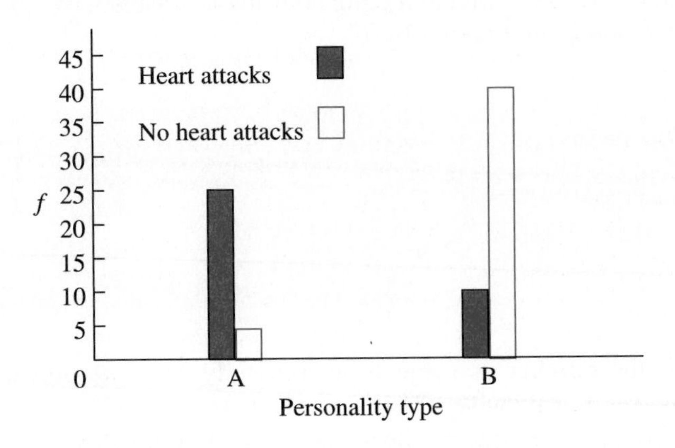

be between 0 and +1.0. You can think of phi as comparing our actual data to the ideal situation when the variables are perfectly dependent. A value of 0 would indicate that our data are not dependent, matching the pattern of independent frequencies shown back in Table 15.1. However, the larger the value of phi, the closer our data come to fitting the ideal pattern of dependent data shown back in Table 15.2.

THE COMPUTATIONAL FORMULA FOR THE PHI COEFFICIENT IS

$$\phi = \sqrt{\frac{\chi^2_{\text{obt}}}{N}}$$

N equals the total number of subjects in the study.

For our heart attack study, χ^2_{obt} was 30.56 and N was 80, so ϕ is

$$\phi = \sqrt{\frac{\chi^2_{\text{obt}}}{N}} = \sqrt{\frac{30.56}{80}} = \sqrt{.382} = .62$$

Thus, on a scale of 0 to +1.0, where +1.0 indicates that the variables are perfectly dependent, we found a correlation of .62 between the frequency of heart attacks and the frequency of personality types.

But remember, the best way to evaluate a relationship is to think "squared correlation coefficient" and compute the proportion of variance accounted for. If we did not take the square root in the above formula, we would have ϕ^2 (phi squared). This is analogous to r^2 or η^2, indicating how much more accurately we can predict scores by using the relationship. In our study above, $\phi^2 = .38$, so we are 38% more accurate in predicting the frequency of heart attacks/no heart attacks when we know personality type (or vice versa).

If we obtain a significant two-way chi square but it is *not* from a 2×2 design, then we do *not* compute the phi coefficient. Instead, we compute the **contingency coefficient,** symbolized by C.

THE COMPUTATIONAL FORMULA FOR THE CONTINGENCY COEFFICIENT, C, IS

$$C = \sqrt{\frac{\chi^2_{\text{obt}}}{N + \chi^2_{\text{obt}}}}$$

N is the number of subjects in the study. We interpret C in the same way we interpret ϕ. Likewise, C^2 is analogous to ϕ^2.

Finally, when one of the variables in a significant two-way χ^2 contains more than two categories, we can perform advanced procedures that are somewhat analogous to *post hoc* comparisons.[2]

NONPARAMETRIC PROCEDURES FOR RANKED DATA

In addition to chi square, there are several other nonparametric procedures you should be aware of. We perform these procedures when we have rank-ordered (ordinal) scores, where subjects' scores are 1st, 2nd, 3rd, and so on. We obtain ranked scores in a study for one of two reasons. First, sometimes we directly measure subjects' scores on the dependent variable using ranked scores. Second, sometimes we initially measure the dependent variable using interval or ratio scores, but the data then violate the assumptions of parametric procedures by not being normally distributed or not having homogeneous variance. Therefore, we transform the scores by assigning them ranks (the highest raw score is ranked 1, the next highest score is ranked 2, and so on). Either way, we then compute one of the following nonparametric inferential statistics to determine whether there are significant differences in the ranked scores for the different conditions of our independent variable.

The Logic of Nonparametric Procedures for Ranked Data

Instead of computing the mean of each condition in the experiment, as we did with t or F, with nonparametric procedures we add the ranked scores in each condition and then examine these sums of ranks. Our symbol for a sum of ranks is ΣR. (So that we will have an accurate ΣR, we handle tied ranks as described in Chapter 7.) Then we compare the observed sum of ranks to an expected sum of ranks. To see the logic of this, say we have the following ranked scores:

[2]Described in S. Siegel and N. J. Castellan (1988), *Nonparametric Statistics for the Behavioral Sciences,* 2nd ed. (New York: McGraw Hill).

Group 1	*Group 2*
1	2
4	3
5	6
8	7
$\Sigma R = 18$	$\Sigma R = 18$

Here there is no difference between the groups, with each group containing both high and low ranks. When the high and low ranks are distributed equally between the two groups, the sums of ranks are equal (here ΣR is 18 in each group). Since there is no difference between these two samples, they may represent the same distribution of ranks in their respective populations. That is, in each population there may be both low ranks and high ranks. Our null hypothesis always states that the populations from our conditions are the same, so here H_0 is that we have the same distribution of ranks in each population. Notice that, for our data to be consistent with H_0, we *expect* the sum of the ranks in each group to equal 18.

But now look at the following two groups of ranked scores:

Group 1	*Group 2*
1	5
2	6
3	7
4	8
$\Sigma R = 10$	$\Sigma R = 26$

Group 1 contains all of the low ranks, and Group 2 contains all of the high ranks. Because these samples are different, they may represent two different populations. Our alternative hypothesis is always that the populations represented by our samples are different, so here H_a says that the distribution of ranks in each population is different, one containing predominantly low ranks and the other containing predominantly high ranks. Notice that when our data is consistent with H_a, the sum of ranks in each sample is different from the expected sum of ranks when H_0 is true. (Here, each ΣR is not equal to 18.)

As usual, the problem is that there is another reason that each observed sum of ranks may not equal the expected sum of ranks. It may be that H_0 is true and the groups represent the same distribution of ranks in the population, but our data reflect sampling error. However, the larger the difference between the expected and observed sum of ranks, the less likely it is that this difference is due to sampling error, and the more likely it is that each sample represents a different distribution of ranks in the population.

Therefore, in each of the following procedures, we compute a statistic that measures the difference between the expected and the observed sum of ranks. If the statistic is a certain size, then we reject H_0 and accept H_a: we are confident that the reason the observed sum of ranks is different from the expected sum of ranks is that the samples represent different distributions of ranks in the population. (If the ranks reflect underlying interval or ratio scores, a significant difference in ranks indicates that the raw score populations are also different.)

Choosing Between the Nonparametric Procedures

Each of the parametric procedures found in previous chapters has a corresponding nonparametric procedure for ranked data. Your first task is to know which nonparametric procedure to choose for the type of research design you are testing. Table 15.5 shows the name of the nonparametric version of each parametric procedure we have previously discussed. The steps in calculating these nonparametric tests are described in the following sections.

Tests for Two Independent Samples: The Mann-Whitney *U* Test and the Rank Sums Test

There are two nonparametric procedures that are analogous to the *t*-test for two independent samples: the Mann-Whitney *U* test and the Rank Sums test. Both are used to test for significant differences between ranked scores measured under two conditions of an independent variable. Which test we use depends on the *n* in each condition.

The Mann-Whitney *U* test for independent samples The Mann-Whitney test is appropriate when the *n* in each condition is equal to or less than 20 and we have two independent samples of ranks. For example, say we measure the reaction times of two groups of subjects to certain symbols. For one group the symbols are printed in black ink, and for the other group the symbols are printed in red ink. We wish to know whether there is a significant difference between reaction times for each colored symbol. However, a raw score population of reaction times tends to be highly positively skewed, so we cannot perform the *t*-test. Therefore, we convert the reaction time scores to ranks. Say that our *n* in each condition is 5 (but we can perform this procedure when the *n*'s are not equal). Table 15.6 gives the reaction times (measured in milliseconds) and their corresponding ranks from our study.

To perform the Mann-Whitney *U* test, do the following.

TABLE 15.5 Parametric Procedures and Their Nonparametric Counterparts

Type of design	*Parametric test*	*Nonparametric test*
Two independent samples	Independent samples *t*-test	Mann-Whitney *U* or Rank Sums test
Two related samples	Related samples *t*-test	Wilcoxon *T* test
Three or more independent samples	Between-subjects ANOVA (*Post hoc* test: protected *t*-test)	Kruskal-Wallis *H* test (*Post hoc* test: Rank Sums test)
Three or more repeated measures samples	Within-Subjects ANOVA (*Post hoc* test: Tukey's *HSD*)	Friedman χ^2 test (*Post hoc* test: Nemenyi's test)

TABLE 15.6 Ranked Data from Two Independent Samples

Red symbols		***Black symbols***	
Reaction time	*Ranked score*	*Reaction time*	*Ranked score*
540	2	760	7
480	1	890	8
600	5	1105	10
590	3	595	4
605	6	940	9
	$\Sigma R = 17$		$\Sigma R = 38$
	$n = 5$		$n = 5$

1. *Assign ranks to all scores in the experiment.* As shown in Table 15.5, assign the rank of 1 to the lowest score in the experiment, regardless of which group it is in. Assign the rank of 2 to the second lowest score in the experiment, and so on.
2. *Compute the sum of the ranks for each group.* Compute ΣR for each group, and note its n, the number of scores in the group.
3. *Compute two versions of the Mann-Whitney U.* First, compute U_1 for Group 1, using the formula

$$U_1 = (n_1)(n_2) + \frac{n_1(n_1 + 1)}{2} - \Sigma R_1$$

where n_1 is the n of Group 1, n_2 is the n of the other group, and ΣR_1 is the sum of ranks from Group 1. We'll call the red symbol group Group 1, so filling in the above formula with the values from Table 15.6, we have

$$U_1 = (5)(5) + \frac{5(5 + 1)}{2} - 17 = 40 - 17 = 23.0$$

Now, compute U_2 for Group 2, using the formula

$$U_2 = (n_1)(n_2) + \frac{n_2(n_2 + 1)}{2} - \Sigma R_2$$

Here, the numerator of the fraction involves n_2 instead of n_1, and we use the sum of ranks from Group 2, ΣR_2. We call the black symbol group Group 2, so filling in the formula we have

$$U_2 = (5)(5) + \frac{5(5 + 1)}{2} - 38 = 40 - 38 = 2.0$$

4. *Determine the Mann-Whitney U_{obt}.* In a two-tailed test, the value of U_{obt} equals the *smaller* of U_1 or U_2. In our example, $U_1 = 23.0$ and $U_2 = 2.0$, so $U_{obt} = 2.0$. In a one-tailed test, we predict that one of the groups has the larger sum of ranks. The corresponding value of U_1 or U_2 from that group becomes our U_{obt}.

5. *Find the critical value of U in Table 9 of Appendix D entitled "Critical Values of Mann-Whitney U."* Choose the appropriate part of the table for either a two-tailed or a one-tailed test. Then, locate U_{crit} using n_1 across the top of the table and n_2 along the left-hand side of the table. For our example, with a two-tailed test and $n_1 = 5$ and $n_2 = 5$, U_{crit} is 2.0.
6. *Compare U_{obt} to U_{crit}.* WATCH OUT! This is a biggie! Unlike any statistic we have yet discussed, the U_{obt} is significant if it is *equal to or less than* U_{crit}. (This is because the *smaller* the U_{obt}, the more likely it is that the group represents a distribution of ranks that is different from the distribution represented by the other group.)

STAT ALERT The Mann-Whitney U_{obt} is significant if it is *less than or equal to* the critical value of *U*.

In our example, $U_{obt} = 2.0$ and $U_{crit} = 2.0$, so our results are significant. Therefore, we conclude that the distribution of ranked scores represented by one sample is significantly different from the distribution represented by the other sample. Because the ranks reflect reaction time scores, we also conclude that the populations of reaction time scores for subjects who see red and black symbols are different. With $\alpha = .05$, the probability that we have made a Type I error is $p < .05$.

7. To describe the effect size, compute eta squared, as shown in the following section.

The rank sums test for independent samples The rank sums test is used to test two independent samples of ranks when the *n* in either condition is *greater* than 20. To illustrate how this statistic is calculated, however, we'll violate this rule and use our previous ranked scores from our reaction time study.

To perform the rank sums test, do the following.

1. *Assign ranks to the scores in the experiment.* As shown in Table 15.5, rank-order all scores in the experiment.
2. *Choose one group and compute the sum of the ranks.* Compute ΣR for one group, and note *n*, the number of scores in the group.
3. *Compute the expected sum of ranks, ΣR_{exp}, for the chosen group.* Use the formula

$$\Sigma R_{exp} = \frac{n(N + 1)}{2}$$

where *n* is the *n* of the chosen group and *N* is the total *N* of the study. We'll compute ΣR_{exp} for the red symbol group, which had $\Sigma R = 17$ and $n = 5$ (*N* is 10). Filling in the formula, we have

$$\Sigma R_{exp} = \frac{n_1(N + 1)}{2} = \frac{5(10 + 1)}{2} = \frac{55}{2} = 27.5$$

Thus, $\Sigma R_{exp} = 27.5$.

4. *Compute the rank sums statistic, symbolized by* z_{obt}. Use the formula

$$z_{obt} = \frac{\Sigma R - \Sigma R_{exp}}{\sqrt{\frac{(n_1)(n_2)(N + 1)}{12}}}$$

where ΣR is the sum of the ranks for the chosen group, ΣR_{exp} is the expected sum of ranks for the chosen group, n_1 and n_2 are the n's of the two groups, and N is the total N of the study.

For our example, we have

$$z_{obt} = \frac{\Sigma R - \Sigma R_{exp}}{\sqrt{\frac{(n_1)(n_2)(N + 1)}{12}}} = \frac{17 - 27.5}{\sqrt{\frac{(5)(5)(10 + 1)}{12}}}$$

$$z_{obt} = \frac{-10.5}{\sqrt{22.92}} = \frac{-10.5}{4.79} = -2.19$$

Thus, $z_{obt} = -2.19$.

5. *Find the critical value of z in the z-tables (Table 1 in Appendix D).* For our alpha level, determine the critical value of z. At $\alpha = .05$, the two-tailed $z_{crit} = \pm 1.96$. (If we had predicted that the sum of ranks of the chosen group would be only greater than or only less than the expected sum of ranks, then we would use the one-tailed value of either $+1.645$ or -1.645.)

6. *Compare* z_{obt} *to* z_{crit}. If the absolute value of z_{obt} is larger than the corresponding z_{crit}, there is a significant difference between the two samples. In our example, $z_{obt} = -2.19$ and $z_{crit} = \pm 1.96$. Therefore, we conclude that the distribution of ranked scores represented by one sample is significantly different from the distribution represented by the other sample. Because the ranks reflect reaction time scores, we also conclude that the populations of reaction time scores for subjects who see red and black symbols are different ($p < .05$).

7. *Describe a significant relationship using eta squared.* Here eta squared is analogous to r^2_{pb} which we discussed in Chapter 12. To compute eta squared, use the formula

$$\eta^2 = \frac{(z_{obt})^2}{N - 1}$$

where z_{obt} is computed in the above rank sums test and N is the total number of subjects.

In our example, z_{obt} is -2.19 and N is 10, so we have $(2.19)^2/9$, or .53. Thus, the color of the symbols accounts for approximately .53 of the variance, or differences, in the ranks. Because the ranks reflect reaction time scores, *approximately* 53% of the differences in reaction time scores are associated with the color of the symbol.

The Wilcoxon T Test for Related Samples

The Wilcoxon test is analogous to the related samples t-test for ranked data. Recall that we have related samples when we match samples or when we have repeated measures. For example, say we perform a study similar to the above reaction time study, but this time we measure the reaction times of the *same* subjects to both the red and black symbols. Table 15.7 gives the data we might obtain.

To determine whether the two samples of scores represent different populations of ranks, we compute the Wilcoxon T_{obt} as follows. (This T_{obt} is not the t_{obt} from Chapters 11 and 12.)

1. *Determine the difference score, D, for each pair of scores.* For each pair of scores, subtract the score in one condition from the score in the other. It makes no difference which score is subtracted from which, but subtract the scores the same way for all pairs. Record the difference scores.
2. *Determine the N of the difference scores, but ignore all difference scores equal to zero.* The N is the total number of nonzero difference scores. In our study, there is one difference of zero (for subject 10), so even though there were ten subjects, $N = 9$.
3. *Assign ranks to the nonzero difference scores.* Ignore the sign (+ or −) of each difference. Assign the rank of 1 to the smallest difference, the rank of 2 to the second smallest difference, and so on. Record ranked scores in a column.
4. *Separate the ranks, using the sign of the difference scores.* Create two columns of ranks, labeled "$R-$" and "$R+$." The $R-$ column contains the ranks you assigned to negative differences in step 3 above. The $R+$ column contains the ranks you assigned to positive differences.
5. *Compute the sums of ranks for the positive and negative difference scores.* Compute ΣR for the column labeled "$R+$." Then compute ΣR for the column labeled "$R-$."

TABLE 15.7 Data for the Wilcoxon Test for Two Related Samples

Subject	*Reaction time to red symbols*	*Reaction time to black symbols*	*Difference, D*	*Ranked scores*	$R-$	$R+$
1	540	760	−220	6	6	
2	580	710	−130	4	4	
3	600	1105	−505	9	9	
4	680	880	−200	5	5	
5	430	500	− 70	3	3	
6	740	990	−250	7	7	
7	600	1050	−450	8	8	
8	690	640	+ 50	2		2
9	605	595	+ 10	1		1
10	520	520	0			
			$N = 9$		$\Sigma R = 42$	$\Sigma R = 3$

6. *Determine the Wilcoxon* T_{obt}. In the two-tailed test, the Wilcoxon T_{obt} is equal to the *smallest* ΣR. In our example, the smallest ΣR equals 3, so $T_{obt} = 3$. In the one-tailed test, we predict whether most differences are positive or negative, depending on our experimental hypotheses. Thus, we predict whether the $R+$ or $R-$ column contains the smaller ΣR. The ΣR that we predict is smallest is our value of T_{obt}. (In our study, say we predicted that red symbols would produce the largest reaction time scores. Given the way we subtracted, we would predict that ΣR for the $R-$ column would be smaller, so our T_{obt} would be 42.)
7. *Find the critical value of T in Table 10 of Appendix D, entitled "Critical Values of the Wilcoxon T."* Find T_{crit} for the appropriate level of alpha and N, the number of nonzero difference scores. In our study, $N = 9$, so for $\alpha = .05$, T_{crit} is 5.0.
8. *Compare* T_{obt} *to* T_{crit}. Again, watch out: the Wilcoxon T is significant if it is *equal to or less than* T_{crit}. The critical value is the largest value that our smallest ΣR can be and still reflect a significant difference.

STAT ALERT The Wilcoxon T_{obt} is significant if it is *less than or equal to* the critical value of T.

In the above example, for our two-tailed test, the T_{obt} of 3.0 is less than the T_{crit} of 6.0, so we have a significant difference. Therefore, we conclude that each sample represents a different distribution of ranks and thus a different population of reaction time scores ($p < .05$).

There is no recognized way to compute η^2 for this procedure.

The Kruskal-Wallis *H* Test for Independent Samples

The Kruskal-Wallis H test is analogous to a between-subjects, one-way ANOVA for ranks. It assumes that the study involves one independent variable and that there are *three* or more independent samples with at least five subjects in each sample. The null hypothesis states that all conditions represent the same distribution of ranks in the population.

As an example, consider a study that explores the relationship between the independent variable of a golfer's height and the dependent variable of the distance he or she hits the ball. We test three groups of novice golfers, classified on the factor of height as either short, medium, or tall. We measure the distance each subject drives the ball in meters. However, say that, based on the F_{max} test discussed in Chapter 13, we cannot assume that the distance scores have homogeneous variance, so we dare not use the parametric ANOVA. Instead, we rank the scores and perform the Kruskal-Wallis H test. Our data are shown in Table 15.8.

To compute the Kruskal-Wallis H, do the following.

1. *Assign ranks, using all scores in the experiment.* Assign the rank of 1 to the lowest score in the experiment, the rank of 2 to the second lowest score in the experiment, and so on.
2. *Compute the sum of the ranks in each condition.* Compute the sum of the ranks, ΣR, in each column. Also note the n in each condition.

TABLE 15.8 Data for the Kruskal-Wallis *H* Test

Height					
Short		*Medium*		*Tall*	
Score	*Rank*	*Score*	*Rank*	*Score*	*Rank*
10	2	24	3	68	14
28	6	27	5	71	15
26	4	35	7	57	10
39	8	44	9	60	12
6	1	58	11	62	13
$\Sigma R_1 = 21$		$\Sigma R_2 = 35$		$\Sigma R_3 = 64$	
$n_1 = 5$		$n_2 = 5$		$n_3 = 5$	

$N = 15$

3. *Compute the sum of squares between groups,* SS_{bn}. Use the formula

$$SS_{bn} = \frac{(\Sigma R_1)^2}{n_1} + \frac{(\Sigma R_2)^2}{n_2} + \cdots + \frac{(\Sigma R_k)^2}{n_k}$$

For each level, square the sum of the ranks and then divide that quantity by the n in the level. (There may be a different n in each level.) After doing this for all k levels in the factor, add the amounts together.

For our example, we have from Table 15.8

$$SS_{bn} = \frac{(21)^2}{5} + \frac{(35)^2}{5} + \frac{(64)^2}{5} = 88.2 + 245 + 819.2$$

so $SS_{bn} = 1152.4$.

4. *Compute the* H_{obt}. Use the formula

$$H_{obt} = \left(\frac{12}{N(N+1)}\right)(SS_{bn}) - 3(N+1)$$

where N is the total N of the study. Divide 12 by $N(N+1)$ and multiply the answer times the SS_{bn}. Then subtract $3(N+1)$.

In our example, we have

$$H_{obt} = \left(\frac{12}{15(15+1)}\right)(1152.4) - 3(15+1) = (.05)(1152.4) - 48$$

$$H_{obt} = 57.62 - 48$$

Thus, the answer is $H_{obt} = 9.62$.

5. *Find the critical value of H in the χ^2 tables (Table 8 in Appendix D).* Values of H have the same sampling distribution as χ^2. The degrees of freedom are

$$df = k - 1$$

where k is the number of levels in the factor.
In our example, k is 3, so $df = 2$. In the χ^2 tables for $\alpha = .05$ and $df = 2$, χ^2_{crit} is 5.99.

6. *Compare the obtained value of H to the critical value of χ^2.* If H_{obt} is *larger* than the critical value found in the χ^2-tables, then H_{obt} is significant. For our study of golfers, the H_{obt} of 9.62 is larger than the χ^2_{crit} of 5.99, so it is significant. This means that at least two of our samples represent different populations of ranks. Because the distance the subjects hit the ball underlies each rank, we conclude that at least two of the populations of distances for short, medium, and tall golfers are not the same ($p < .05$).

7. *Perform* post hoc *comparisons using the rank sums test.* When H_{obt} is significant, we determine which specific conditions differ by performing the rank sums test on every pair of conditions. This is analogous to Fisher's protected t-test, discussed in Chapter 13, and it is used regardless of the n in each group. To perform the procedure, treat each pair of conditions being compared as if they comprised the entire study, then follow the procedure described previously for the rank sums test.

STAT ALERT When performing the rank sums test as a *post hoc* test for H_{obt}, re-rank the scores in the two conditions being compared, using only those scores.

In our example, comparing the ranks of short and medium-height golfers produces a z_{obt} of 1.36, comparing short and tall golfers produces a z_{obt} of 2.62, and comparing medium-height and tall golfers produces a z_{obt} of 2.40. With $\alpha = .05$, from the z-tables we find a z_{crit} of ± 1.96. Therefore, the scores of short and medium-height subjects are not significantly different, but they both differ significantly from the scores of tall subjects. We conclude that our tall golfers represent one population of distances that is different from the population for short and medium-height golfers.

8. *Describe a significant relationship using eta squared.* Use the formula

$$\eta^2 = \frac{H_{obt}}{N - 1}$$

where H_{obt} is the value computed in the Kruskal-Wallis test and N is the total number of subjects. In our study, we have $H_{obt} = 9.62$ and $N = 15$. Substituting into the above formula, we have $\eta^2 = 9.62/14$, or .69. Therefore, the variable of a player's height accounts for approximately .69 of the variance in the distance scores.

The Friedman χ^2 Test for Repeated Measures

The Friedman χ^2 test is analogous to a one-way, repeated measures ANOVA for ranks. It assumes that the study involves one independent variable, or factor, and that the same subjects are repeatedly measured under *three* or more conditions. If there are only three levels of the factor, there must be at least ten subjects in the study. If there are only four levels of the factor, there must be at least five subjects.

As an example, consider a study in which the scores we collect are already ranked. The three levels of our independent variable are the teaching styles of Dr. Highman, Dr. Shyman, and Dr. Whyman. We obtain a random sample of students who have taken courses from all three instructors, and each student rank-orders the three instructors. Table 15.9 gives the data for our study.

To perform the Friedman χ^2 test, follow these steps.

1. *Assign ranks within the scores of each subject.* If the scores are not already ranks, assign the rank of 1 to the lowest score received by Subject 1, assign the rank of 2 to the second lowest score received by Subject 1, and so on. Repeat the process for each subject.
2. *Compute the sum of the ranks, ΣR, in each condition.* Find the sum of the ranks in each column.
3. *Compute the sum of squares between groups, SS_{bn}.* Use the formula

$$SS_{bn} = (\Sigma R_1)^2 + (\Sigma R_2)^2 + \cdots + (\Sigma R_k)^2$$

Square the sum of the ranks in each of the k conditions, and then add the squared sums together. In our example, we have

$$SS_{bn} = (12)^2 + (23)^2 + (25)^2$$

so $SS_{bn} = 1298$.

TABLE 15.9 Data for the Friedman Test

	Rankings for three instructors		
Subject	*Dr. Highman*	*Dr. Shyman*	*Dr. Whyman*
1	1	2	3
2	1	3	2
3	1	2	3
4	1	3	2
5	2	1	3
6	1	3	2
7	1	2	3
8	1	3	2
9	1	3	2
10	2	1	3
$N = 10$	$\Sigma R_1 = 12$	$\Sigma R_2 = 23$	$\Sigma R_3 = 25$

4. *Compute the Friedman χ^2 statistic.* Use the formula

$$\chi^2_{obt} = \left(\frac{12}{(k)(N)(k+1)}\right)(SS_{bn}) - 3(N)(k+1)$$

where N is the number of subjects and k is the number of levels of the factor. First divide 12 by the quantity $(k)(N)(k + 1)$. Then multiply this number times SS_{bn}. Then subtract the quantity $3(N)(k + 1)$.

In our example,

$$\chi^2_{obt} = \left(\frac{12}{(3)(10)(3+1)}\right)(1298) - 3(10)(3+1)$$

$$\chi^2_{obt} = (.10)(1298) - 120 = 129.8 - 120$$

And the survey says: $\chi^2_{obt} = 9.80$.

5. *Find the critical value of χ^2 in the χ^2-tables (Table 8 in Appendix D).* The degrees of freedom is

$$df = k - 1$$

where k is the number of levels in the factor.

For our example, $k = 3$, so for $df = 2$ and $\alpha = .05$, the critical value is 5.99.

6. *Compare χ^2_{obt} to the critical value of χ^2.* If χ^2_{obt} is larger than χ^2_{crit}, the results are significant. For our example, our χ^2_{obt} of 9.80 is larger than the χ^2_{crit} of 5.99, so our results are significant. Thus, we conclude that at least two of the samples represent different populations ($p < .05$).

7. *When the Friedman χ^2 is significant, perform* post hoc *comparisons using Nemenyi's procedure.* This procedure is analogous to Tukey's *HSD* procedure, which we saw in Chapter 13. We compute one value that is the *critical difference.* Any two conditions that differ by more than this critical difference are significantly different. To perform Nemenyi's procedure, follow these steps.

   a. *Compute the critical difference.* Use the formula

$$\text{Critical difference} = \sqrt{\left(\frac{k(k+1)}{6(N)}\right)(\chi^2_{crit})}$$

   where k is the number of levels of the factor, N is the number of subjects, and χ^2_{crit} is the critical value used to test the Friedman χ^2. Multiply k times $k + 1$, and then divide by $6(N)$. Multiply this number times χ^2_{crit}, and then find the square root.

   In our example, $\chi^2_{crit} = 5.99$, $k = 3$, and $N = 10$. We have

$$\text{Critical difference} = \sqrt{\left(\frac{k(k+1)}{6(N)}\right)(\chi^2_{crit})} = \sqrt{\left(\frac{3(3+1)}{6(10)}\right)(5.99)}$$

$$\text{Critical difference} = \sqrt{(.2)(5.99)} = \sqrt{1.198} = 1.09$$

   so our critical difference is ± 1.09.

b. Compute the mean rank for each condition. For each condition, divide the sum of ranks (ΣR) by the number of subjects. In our example, the sums of ranks are 12, 23, and 25 in the three conditions, and N is 10. Therefore, the mean ranks are 1.2, 2.3, and 2.5 for Highman, Shyman, and Whyman, respectively.

c. Compute the differences between all pairs of mean ranks. Subtract each mean rank from the other mean ranks. Any absolute difference between two mean ranks that is greater than the critical difference indicates that the two conditions differ significantly. In our example, the differences between the mean ranks for Dr. Highman and the other two instructors are 1.10 and 1.30, respectively, and the difference between Shyman and Whyman is .20. Our critical difference is 1.09, so only Dr. Highman's ranking is significantly different from those of the other two instructors. Thus, we conclude that if the entire population were to rank the three instructors, Dr. Highman would be ranked superior to the other two instructors!

8. *Describe a significant relationship using eta squared.* Use the formula

$$\eta^2 = \frac{\chi^2_{obt}}{(N)(k) - 1}$$

where χ^2_{obt} is the value computed in the Friedman χ^2 test, N is the number of subjects, and k is the number of levels of the factor. For our example, we have

$$\eta^2 = \frac{\chi^2_{obt}}{(N)(k) - 1} = \frac{9.80}{(10)(3) - 1} = \frac{9.80}{30 - 1} = .34$$

Thus, our instructor variable accounts for .34 of the variability, or differences, in rankings.

FINALLY

Congratulations! If you are reading this, you have read an entire statistics book, and that is an accomplishment! You should be proud of the sophisticated level of your knowledge. You are now familiar with the vast majority of statistical procedures used in psychology and other behavioral sciences. Although you may encounter more complicated research designs, after this experience with introductory statistics you will find that there really is nothing new under the sun.

CHAPTER SUMMARY

1. *Nonparametric procedures* are used when our data do not meet the assumptions of parametric procedures.

2. *Chi square,* χ^2, is used when we have one or more categorical variables and the data are the frequencies with which subjects fall in a category. Chi square assumes that (a) any subject falls in only one category, (b) the probability of a subject's falling in a particular category is independent of any other subject's falling in a particular category, (c) the responses of all subjects are included in the study, and (d) the f_e in any category is at least 5.

3. The *one-way* χ^2 is a goodness of fit test that determines whether the observed frequencies fit the model of the expected frequencies described by H_0. The larger the χ^2_{obt}, the larger the overall differences between the observed and expected frequencies. A significant χ^2_{obt} indicates that the observed frequencies are unlikely to represent the distribution of frequencies in the population described by H_0.

4. In the *two-way* χ^2, H_0 states that the frequencies in the categories of the two variables are independent. The χ^2_{obt} describes the differences between the observed frequencies and the expected frequencies when H_0 is true and the variables are independent. A significant χ^2_{obt} indicates that the observed frequencies are unlikely to represent variables that are independent in the population. Instead, we conclude that the two variables are dependent, or correlated.

5. In a significant 2×2 chi square, the strength of the relationship is described by the *phi correlation coefficient,* ϕ. In a significant two-way chi square that is not 2×2, the strength of the relationship is described by the *contingency coefficient, C.* The larger these coefficients are, the closer the data are to forming a relationship where the variables are perfectly dependent, or correlated. Squaring ϕ or C gives the proportion of variance accounted for, which indicates how much more accurately the frequencies of category membership on one variable can be predicted by knowing subjects' category membership on the other variable.

6. There are two nonparametric versions of the independent samples t-test for ranks. The *Mann-Whitney U test* is performed when the n in each condition is less than 20. The *rank sums test* is performed when the n in either condition is greater than 20.

7. The *Wilcoxon T test* is the nonparametric equivalent of the related samples t-test for ranks.

8. The *Kruskal-Wallis H test* is the nonparametric equivalent of the one-way, between-subjects ANOVA for ranks. The rank sums test is used as the *post hoc* test to identify the specific conditions that differ from each other.

9. The *Friedman* χ^2 *test* is the nonparametric equivalent of the one-way, repeated measures ANOVA for ranks. *Nemenyi's test* is used as the *post hoc* test to identify the specific conditions that differ from each other.

10. *Eta squared* describes the relationship found in experiments involving ranked data.

PRACTICE PROBLEMS

(Answers for odd-numbered problems are provided in Appendix E.)

1. *a.* What do all nonparametric procedures have in common with all parametric procedures?

b. What aspects of the data cause us to use nonparametric inferential procedures?

c. Why, if possible, should a researcher design a study so that the data meet the assumptions of a parametric procedure?

d. Why shouldn't we use parametric procedures for data that clearly violate their assumptions?

2. *a.* When do we use the chi square?

b. When do we use the one-way chi square?

c. When do we use the two-way chi square?

3. *a.* What is the symbol for observed frequency? What does it mean?

b. What is the symbol for expected frequency? What does it mean?

4. *a.* What does a significant one-way chi square indicate?

b. What does a significant two-way chi square indicate?

5. In the general population, the distribution of political party affiliation is 30% Republican, 55% Democratic, and 15% other. We wish to determine whether this distribution is also found among the elderly. In a sample of 100 senior citizens, we find 18 Republicans, 64 Democrats, and 18 other.

a. What are H_0 and H_a?

b. What is f_e for each group?

c. Compute χ^2_{obt}.

d. With $\alpha = .05$, what do you conclude about party affiliation in the population of senior citizens?

6. A survey finds that, given the choice, 34 females prefer males much taller than themselves, and 55 females prefer males only slightly taller than themselves.

a. What are H_0 and H_a for this survey?

b. With $\alpha = .05$, what would you conclude about the preference of females in the population?

c. Describe how you would graph these results.

7. In a study, Foofy counts the students who say they like Professor Demented and those who say they like Professor Randomsampler. She then performs a one-way χ^2 to determine if there is a significant difference between the frequency with which students like each professor.

a. Why is this approach incorrect?

b. How should she analyze the data?

8. What is the phi coefficient, and when is it used?

b. What does the squared phi coefficient indicate?

c. What is the contingency coefficient, and when is it used?

d. What does the squared contingency coefficient indicate?

9. A study similar to the one in problem 5 determines the frequency of the different political party affiliations for male and female senior citizens. The following data are obtained:

		Affiliation		
		Republican	*Democrat*	*Other*
Gender	*Male*	18	43	14
	Female	39	23	18

a. What are H_0 and H_a?

b. What is f_e in each cell?

c. Compute χ^2_{obt}.

d. With $\alpha = .05$, what should we conclude about gender and party affiliation in the population of senior citizens?

e. How consistent is this relationship?

10. The following data reflect the frequency with which subjects voted in the last election and were satisfied with the officials elected:

		Satisfied	
		yes	*no*
Vote	*yes*	48	35
	no	33	52

a. What are H_0 and H_a?

b. What is f_e in each cell?

c. Compute χ^2_{obt}.

d. With $\alpha = .05$, what should we conclude about the correlation coefficient?

e. How consistent is the relationship for these data?

11. What is the nonparametric version of each of the following?

a. A one-way, between-subjects ANOVA

b. An independent samples *t*-test ($n < 20$)

c. A related samples *t*-test

d. An independent samples *t*-test ($n > 20$)

e. A one-way, repeated measures ANOVA

f. Fisher's protected *t*-test

g. Tukey's *HSD* test

12. Select the statistical procedure that should be used to analyze the data from each of the following studies:

a. An investigation of the effects of a new pain reliever on rankings of the emotional content of words describing pain. A randomly selected group of subjects is tested before and after administration of the drug.

b. An investigation of the effects of eight different colors of spaghetti sauce on tastiness scores. A different random sample of subjects tastes each color of sauce, and then the tastiness scores are ranked.

c. An investigation of the effects of increasing amounts of alcohol consumption on reaction-time scores. The scores are ranked, and the same group of subjects is tested after 1, 3, and 5 drinks.

d. An investigation of two levels of the variable of family income. Two random samples of scores are used to rank-order the percentage of subjects' income that they spent on new clothing last year.

13. A study compares the maturity level of a group of students who have completed statistics to a group of students who have not. Maturity scores for college students tend to be skewed. For the following interval scores, answer the questions below.

Nonstatistics	*Statistics*
43	51
52	58
65	72
23	81
31	92
36	64

a. Do the groups differ significantly ($\alpha = .05$)?

b. What do you conclude about maturity scores you expect to find in the population of students who have taken statistics and in the population that hasn't?

14. We wish to compare the attitude scores of subjects when tested in the morning to their attitude scores when tested in the afternoon. From a morning and an afternoon attitude test, we obtain the following interval data. We determine that we have significantly heterogeneous variance. With $\alpha = .05$, determine if there is a significant difference in scores as a function of testing times.

Morning	*Afternoon*
14	36
18	31
20	19
28	48
3	10
34	49
20	20
24	29

15. An investigator evaluated the effectiveness of activity therapy on three types of patients. She collected the following improvement ratings. (In the population, these data form highly skewed distributions.)

Depressed	*Manic*	*Schizophrenic*
16	7	13
11	9	6
12	6	10
20	4	15
21	8	9

a. Which procedure should be used to analyze these data? Why?

b. What should she do first to the data?

c. If the results are significant, what should she do next?

d. Ultimately, what conclusions can be drawn from this study?

16. A therapist evaluates the progress of a sample of clients in a new treatment program after one month, after two months, and again after three months. Such progress data don't have homogeneous variance.

a. What statistical procedure should be used to analyze the data? Why?

b. What is the first thing the therapist must do?

c. If the results are significant, what should the therapist then do?

d. Ultimately, what will the therapist be able to identify?

17. What is the basic logic underlying the testing of H_0 in all nonparametric procedures for ranked data?

SUMMARY OF FORMULAS

A. Summary of chi square formulas

1. *The computational formula for chi square is*

$$\chi^2_{\text{obt}} = \Sigma\left(\frac{(f_o - f_e)^2}{f_e}\right)$$

where f_o is the observed frequency in a cell and f_e is the expected frequency in a cell.

a. Computing expected frequency

(1) *In a one-way chi square,* the expected frequency in a category is equal to the probability of someone's falling in that category

multiplied times the N of the study. *In testing an H_0 of no difference, the computational formula for each expected frequency is*

$$f_e \text{ in each category} = \frac{N}{k}$$

where N is the total N in the study and k is the number of categories.

(2) *In a two-way chi square, the computational formula for finding the expected frequency in each cell is*

$$f_e = \frac{(\text{cell's row total } f_o)(\text{cell's column total } f_o)}{N}$$

b. *Critical values of χ^2 are found in Table 8 of Appendix D.*

(1) In a one-way chi square, the degrees of freedom is

$$df = k - 1$$

where k is the number of categories in the variable.

(2) In a two-way chi square, the degrees of freedom is

$$df = (\text{number of rows} - 1)(\text{number of columns} - 1)$$

2. *The computational formula for the phi coefficient is*

$$\phi = \sqrt{\frac{\chi^2_{obt}}{N}}$$

where N is the total number of subjects in the study.

3. *The computational formula for the contingency coefficient, C, is*

$$C = \sqrt{\frac{\chi^2_{obt}}{N + \chi^2_{obt}}}$$

where N is the total number of subjects in the study.

B. Summary of nonparametric formulas

1. *Formulas for two independent samples*

a. *When N is equal to or less than 20, the computational formula for the Mann-Whitney* U *test for independent samples is*

$$U_1 = (n_1)(n_2) + \frac{n_1(n_1 + 1)}{2} - \Sigma R_1$$

and

$$U_2 = (n_1)(n_2) + \frac{n_2(n_2 + 1)}{2} - \Sigma R_2$$

Here n_1 and n_2 are the n's of the groups. After ranks are assigned based on all scores, ΣR_1 is the sum of ranks in Group 1, and ΣR_2 is the sum of ranks in Group 2.

In a two-tailed test, the value of U_{obt} equals the *smaller* of U_1 or U_2. In a one-tailed test, the value of U_1 or U_2 from the group predicted to have the largest sum of ranks is U_{obt}. Critical values of U are found in Table 9 of Appendix D. (Note that U is significant if it is equal to or less than the critical value.)

b. *When either* n *is greater than 20, the computational formula for the rank sums test for independent samples is*

$$z_{obt} = \frac{\Sigma R - \Sigma R_{exp}}{\sqrt{\frac{(n_1)(n_2)(N + 1)}{12}}}$$

Here n_1 and n_2 are the n's of the two groups, and N is the total N. After ranks are assigned based on all scores, ΣR is the sum of the ranks for the chosen group. ΣR_{exp} is the expected sum of ranks for the chosen group, found using the formula

$$\Sigma R_{exp} = \frac{n(N + 1)}{2}$$

where n is the n of the chosen group. Critical values of z are found in Table 1 of Appendix D.

c. *Eta squared is computed using the formula*

$$\eta^2 = \frac{(z_{obt})^2}{N - 1}$$

2. *For related samples, the computational formula for the Wilcoxon* T *is*

$$T_{obt} = \Sigma R$$

After the difference scores are found and assigned ranks, in the two-tailed test, ΣR is the smaller of the sum of ranks for the positive difference scores or the sum of ranks for the negative difference scores. In the one-tailed test, ΣR is the sum of ranks that is predicted to be the smallest. Critical values of T are found in Table 10 of Appendix D, where N is the number of nonzero difference scores. (Note that T is significant if it is equal to or less than the critical value.)

3. *For three or more independent samples, the computational formula for the Kruskal-Wallis* H *test is*

$$H_{obt} = \left(\frac{12}{N(N + 1)}\right)(SS_{bn}) - 3(N + 1)$$

(continued on next page)

where N is the number of subjects in the study. After ranks are assigned using all scores, SS_{bn} is found using the formula

$$SS_{bn} = \frac{(\Sigma R_1)^2}{n_1} + \frac{(\Sigma R_2)^2}{n_2} + \cdots + \frac{(\Sigma R_k)^2}{n_k}$$

where each n is the number of scores in a level, each ΣR is the sum of ranks for that level, and k is the number of levels of the factor.

Critical values of H are found in Table 8 of Appendix D, for $df = k - 1$, where k is the number of levels in the factor.

a. When H_{obt} is significant, *post hoc* comparisons are performed using the rank sums test, regardless of the size of n.

b. Eta squared is computed using the formula

$$\eta^2 = \frac{H_{obt}}{N - 1}$$

4. *For three or more related samples, the computational formula for the Friedman χ^2 test is*

$$\chi^2_{obt} = \left(\frac{12}{(k)(N)(k + 1)}\right)(SS_{bn}) - 3(N)(k + 1)$$

where N is the number of subjects and k is the number of levels of the factor. After ranks are assigned within the scores of each subject, SS_{bn} is found using the formula

$$SS_{bn} = (\Sigma R_1)^2 + (\Sigma R_2)^2 + \cdots + (\Sigma R_k)^2$$

where each $(\Sigma R)^2$ is the squared sum of ranks for a level. Critical values of χ^2 are found in Table 8 of Appendix D, for $df = k - 1$, where k is the number of levels in the factor.

a. When the Friedman χ^2 is significant, *post hoc* comparisons are performed using Nemenyi's procedure.

(1) Compute the critical difference using the formula

$$\text{Critical difference} = \sqrt{\left(\frac{k(k + 1)}{6(N)}\right)(\chi^2_{crit})}$$

where k is the number of levels of the factor and N is the number of subjects. χ^2_{crit} is the critical value of χ^2 for the appropriate α at $df = k - 1$.

(2) Compute the mean rank in each condition as $\Sigma R/n$.

(3) Any two mean ranks that differ by more than the critical difference are significantly different.

b. Eta squared is found using the formula

$$\eta^2 = \frac{\chi^2_{obt}}{(N)(k) - 1}$$

Interpolation

This appendix presents the procedures for linear interpolation of z-scores, discussed in Chapter 6, and of values of t_{crit}, discussed in Chapter 11.

INTERPOLATING FROM THE z-TABLES

In essence, linear interpolation is a procedure for reading between the lines of a table. We interpolate when we must find an exact proportion that is not shown in the z-table or when we seek a proportion for a z-score that has three decimal places. In any interpolation, carry all computations to four decimal places.

Finding an Unknown z-Score

Say that we seek a target z-score that corresponds to a target proportion of exactly .45 (.4500) of the curve between the mean and z. To interpolate, enter the z-tables and identify the two bracketing proportions—the closest proportions above and below the target proportion. Note their corresponding z-scores. For .4500, the bracketing proportions are .4505 at $z = 1.6500$ and .4495 at $z = 1.6400$. Arrange the values this way:

	Known proportion under curve	*Unknown z-score*
Upper bracket	.4505	1.6500
Target	.4500	?
Lower bracket	.4495	1.6400

Notice the labels. We seek the "unknown" target z-score which corresponds to the "known" target proportion of .4500. The known target proportion is bracketed by .4505 and .4495. The unknown target z-score is bracketed by 1.6500 and 1.6400.

Interpolation is actually quite simple. If you look at the above table, you can see that the target proportion of .4500 is a number that is halfway between .4495 and .4505. That is, the difference between the lower known proportion and our target proportion is one-half of the difference between the two known proportions.

We assume that the z-score corresponding to .4500 is also halfway between the two bracketing z-scores of 1.6400 and 1.6500. The difference between the two bracketing z-scores is .010, and one-half of that is .005. Thus, to go to halfway between 1.6400 and 1.6500, we add .005 to 1.6400. Our answer is that a z-score of 1.6450 corresponds to .4500 of the curve between the mean and z.

The answer will not always be as obvious as in this example, so use the following steps.

Step 1. Determine the difference between the upper known bracket and the lower known bracket. In the above example, .4505 − .4495 = .0010. This is the total distance between the two known proportions.

Step 2. Determine the difference between the known target and the lower known bracket. Above, .4500 − .4495 = .0005.

Step 3. Form a fraction, with the answer from step 2 as the numerator and the answer from step 1 as the denominator. Above, the fraction is .0005/.0010, which equals .5. This tells us that .4500 is one-half of the distance from .4495 to .4505.

Step 4. Find the difference between the two brackets in the unknown column. Above, 1.6500 − 1.6400 = .010. This is the total distance between the two z-scores that bracket the unknown target z-score.

Step 5. Multiply the proportion found in step 3 by the answer found in step 4. Above, (.5)(.010) = .005. This tells us that the unknown target z-score is .005 larger than the lower bracketing z-score.

Step 6. Add the answer in step 5 to the lower bracketing score. Above, .005 + 1.640 = 1.645. This is our unknown target z-score. Thus, .4500 of the normal curve lies between the mean and z = 1.645.

Finding an Unknown Proportion

We can also apply the above steps to find an unknown proportion for a known three-decimal-place z-score. For example, say we have a known target z of 1.382. What is the corresponding proportion between the mean and this z? From the z-table, we see that the upper and lower known brackets around our target z-score are 1.390 and 1.380. We arrange the z-scores and corresponding proportions as shown below.

	Known z-score	***Unknown proportion under curve***
Upper bracket	1.390	.4177
Target	1.382	?
Lower bracket	1.380	.4162

Here the z-scores are in the "known" column and the proportions are in the "unknown" column. To find the target proportion, use the above steps.

Step 1: 1.390 − 1.380 = .010

This is the total difference between the known bracketing z-scores.

Step 2: 1.382 − 1.380 = .002

This is the distance between the lower known bracketing z-score and the target z-score.

Step 3: $\frac{.002}{.010} = .20$

This is the proportion of the distance that the target z-score lies from the lower bracket. A z of 1.382 is .20 of the distance from 1.380 to 1.390.

Step 4: .4177 − .4162 = .0015

This tells us that .0015 is the total distance between the brackets of .4177 and .4162 in the unknown column.

Our known target z-score is .20 of the distance from the lower bracketing z-score to the higher bracketing z-score. Therefore, the proportion we seek is .20 of the distance from the lower bracketing proportion to the upper bracketing proportion.

Step 5: (0.20)(0.0015) = .0003

This tells us that .20 of the distance separating the bracketing proportions in the unknown column is .0003.

Step 6: .4162 + .0003 = .4165

Increasing the lower proportion in the unknown column by .0003 takes us to the point corresponding to .20 of the distance between the bracketing proportions. This point is .4165, which is the proportion that corresponds to the z-score of 1.382.

INTERPOLATING CRITICAL VALUES

In conducting inferential statistical procedures, sometimes we must interpolate between tabled critical values. We apply the same steps described above, except that now we use tabled degrees of freedom and critical values.

For example, say that we are performing a t-test and we seek the critical value corresponding to 35 *df* (with $\alpha = .05$, two-tailed test). The t-tables have values only for 30 *df* and 40 *df*. We obtain the following:

	Known df	*Unknown critical value*
Upper bracket	30	2.042
Target	35	?
Lower bracket	40	2.021

Logically, because 35 *df* is halfway between 30 *df* and 40 *df*, we know that the critical value for 35 *df* must be halfway between 2.042 and 2.021.

To find the unknown target critical value, follow the steps described for z-scores.

Step 1: $40 - 30 = 10$

This is the total distance between the known bracketing *df*'s.

Step 2: $35 - 30 = 5$

Notice a change here: this is the distance between the *upper* bracketing *df* and our target *df*.

Step 3: $\frac{5}{10} = .50$

This is the proportion of the distance that our target *df* lies from the upper known bracket. The *df* of 35 is .50 of the distance from 30 to 40.

Step 4: $2.042 - 2.021 = .021$

This tells us that .021 is the total distance between the bracketing critical values of 2.042 and 2.021 in the unknown column.

Our *df* of 35 is .50 of the distance between the bracketing *df*'s, so the critical value we seek is .50 of the distance between 2.042 and 2.021, or .50 of .021.

Step 5: $(.50)(.021) = .0105$

This tells us that .50 of the distance separating the bracketing critical values is .0105. Because critical values decrease as *df* increases, and we are going from 30 *df* to 35 *df*, we *subtract* .0105 from the larger value, 2.042.

Step 6: $2.042 - .0105 = 2.0315$

Thus, $t = 2.0315$ is the critical value associated with 35 *df* at $\alpha = .05$ for a two-tailed test.

The same logic can be applied to find critical values for any of the other statistical procedures.

PRACTICE PROBLEMS

(Answers for odd-numbered problems are provided in Appendix E.)

1. What is the z-score you must score above to be in the top 25% of scores?
2. Foofy obtains a z-score of 1.909. What proportion of scores are between her score and the mean?
3. For $\alpha = .05$, what is the two-tailed t_{crit} for $df = 50$?
4. For $\alpha = .05$, what is the two-tailed t_{crit} for $df = 55$?

B

Additional Formulas for Computing Probability

This appendix extends the discussion of computing probability in Chapter 9.

THE MULTIPLICATION RULE

When computing the probability of complex events, sometimes we are "satisfied" only if several events occur. We use the multiplication rule when the word "and" links the events that all must occur in order for us to be satisfied. *The following multiplication rule can be used only with independent events.* (For dependent events, a different, more complex rule is needed.)

THE MULTIPLICATION RULE FOR INDEPENDENT RESULTS IS

$$p(\text{A and B}) = p(\text{A}) \times p(\text{B})$$

The multiplication rule states that the probability of several independent events is equal to the probabilities of the individual events *multiplied* together. (When you say "and," think "multiply.") Thus, the probability that we will be satisfied by having both A and B occur is equal to the probability of A multiplied times the probability of B. (If there were three events, then all three probabilities would be multiplied together, and so on.)

We may use the multiplication rule when describing a *series* of independent events. Say that we want to know the probability of obtaining 3 heads on 3 coin tosses. We can restate the problem as the probability of obtaining a head *and* then a head *and* then a head. Thus, by the multiplication rule, the probability of 3 heads is

$$p(\text{head}) \times p(\text{head}) \times p(\text{head}) = .5 \times .5 \times .5 = .125$$

The answer we obtain is the same one we would find by forming a fraction with the number of sequences of heads and tails that would satisfy us divided by the number of possible sequences we could obtain when tossing 3 coins. The multiplication rule simply provides a shorter, less complicated route.

We may also use the multiplication rule if we are satisfied when two or more independent events occur *simultaneously.* For example, the probability of drawing the king of hearts can be restated as the probability of drawing a king *and* a heart simultaneously. Since there are 4 kings, the probability of drawing a king is 4/52, or .0769. Since there are 13 hearts, the probability of drawing a heart is 13/52, or .25. Thus, the probability of drawing a king and a heart is $(.0769 \times .25)$, so $p(\text{king and heart}) = .0192$. (As a check on our answer, there is 1 king of hearts in 52 cards, so the probability of drawing the king of hearts is 1/52, which is again .0192. Amazing!)

THE ADDITION RULE

Sometimes we are satisfied by any *one* of a number of outcomes that may occur in one sample. We use the addition rule when the word "or" links the events that will satisfy us. For example, if we will be satisfied by either A *or* B, then we seek $p(\text{A or B})$. However, there are two versions of the addition rule, depending on whether we are describing mutually exclusive or mutually inclusive events. **Mutually exclusive events** are events that cannot occur together: the occurrence of one event prohibits, or excludes, the occurrence of another. Head and tail, for example, are mutually exclusive on any *one* flip of a coin. Conversely, **mutually inclusive events** are events that can occur together. For example, drawing a king from a deck is mutually inclusive with drawing a heart, because we can draw the king of hearts.

THE ADDITION RULE FOR MUTUALLY EXCLUSIVE EVENTS IS

$$p(\text{A or B}) = p(\text{A}) + p(\text{B})$$

This formula says that the probability of being satisfied by having either A or B occur is equal to the probability of A plus the probability of B. (For mutually exclusive events, when you say "or," think "add.") For example, the probability of randomly drawing a queen or a king is found by adding the probability of a king (which is 4/52, or .0769) and the probability of a queen (which is also .0769). Thus, $p(\text{king or queen}) = .0769 + .0769$, so $p(\text{king or queen}) = .1538$. In essence, we have found that there are a total of 8 cards out of 52 that can satisfy us with a king or a queen, and 8/52 corresponds to a p of .1538.

When events are mutually *inclusive,* we may obtain A, we may obtain B, or we may obtain A and B simultaneously. To see how this can play havoc with our computations, say that we seek the probability of randomly drawing either a king *or* a heart in one draw. We *might* think that since there are 4 kings and there are 13 hearts, there are a total of 17 cards that will satisfy us—right? Wrong! If you count the cards in a deck that will satisfy us, you will find only 16, not

17. The problem is that we counted the king of hearts twice, once as a king and once as a heart. To correct this, we must subtract the "extra" king of hearts.

THE ADDITION RULE FOR MUTUALLY INCLUSIVE EVENTS IS

$$p(\text{A or B}) = p(\text{A}) + p(\text{B}) - [p(\text{A}) \times p(\text{B})]$$

This version of the addition rule states that the probability of obtaining any one of several mutually inclusive events is equal to the sum of the probabilities of the individual events *minus* the probability of those events' occurring simultaneously (minus the probability of A and B). We compute the probability of A and B using the multiplication rule, where $p(\text{A and B}) = p(\text{A}) \times p(\text{B})$. The probability of a king is 4/52, or .0769, and the probability of a heart is 13/52, or .25. Since we have counted the king of hearts both as a king and as a heart, we then subtract the extra king of hearts: the probability of a king and a heart is 1/52, or .0192. Therefore, altogether we have

$$p(\text{king or heart}) = .0769 + .25 - .0192 = .3077$$

This is equivalent to finding the ratio of the number of outcomes that satisfy the event to the total number of possible outcomes. There are 16 out of 52 outcomes that can satisfy king or heart, and 16/52 corresponds to $p = .3077$.

For fun, we can combine the addition and multiplication rules. For example, what is the probability of drawing either the jack of diamonds *or* the king of spades on one draw, *and* then drawing either the 5 *or* the 6 of diamonds on a second draw? In symbols, we have

$$p[(\text{A or B}) \text{ and } (\text{C or D})]$$

Because these events are all mutually exclusive, we have

$$p[(\text{A or B}) \text{ and } (\text{C or D})] = [p(\text{A}) + p(\text{B})] \times [p(\text{C}) + p(\text{D})]$$

If we are sampling with replacement, the answer is .00148. If we are sampling without replacement, the answer is .00151. (Come on, try it.)

THE BINOMIAL EXPANSION

In Chapter 9 we computed the probability of obtaining 7 heads in 7 coin tosses. In doing so, we listed all of the possible sequences of heads and tails that might occur, and then we counted the number of outcomes that would satisfy us. We could use the same technique to determine the probability of obtaining 1 head in 3 coin tosses. If we list all of the possible sequences of head and tails we might obtain with 3 tosses, we have

head	head	head	tail	tail	tail
head	head	tail	tail	tail	head
head	tail	head	head	tail	tail
tail	head	head	tail	head	tail

There are a total of 8 possible combinations we might obtain, 3 of which contain 1 head. Thus, the probability of obtaining 1 head in 3 tosses is 3/8, or .375.

Instead of listing all of the possible outcomes and then counting those that satisfy us, we can use a mathematical formula, called the *binomial expansion,* for computing such probabilities. A "binomial" situation exists when one of only two possible outcomes occurs on each occasion and the two outcomes are mutually exclusive. Then the binomial expansion can be used to compute the probability of obtaining a certain number of one of the outcomes in some total number of tries. In statistical terms, we find the probability of a certain *combination* of N events taken r at a time. Because either a head or a tail occurs on each toss of the coin, the binomial expansion can be used to determine the probability of obtaining some number of heads in a certain number of tosses. We will use the symbol p_C to stand for the probability of the particular combination that satisfies us.

THE FORMULA FOR THE BINOMIAL EXPANSION IS

$$p_C = \left(\frac{N!}{r!(N - r)!}\right)(p^r)(q^{N-r})$$

N stands for the total number of tries or occasions, and r stands for the number of events that satisfy us. Thus, to find the probability of obtaining 1 head in 3 tries, we use $N = 3$ and $r = 1$. The symbol p stands for the probability of the desired event, and we raise it to the r power (multiply it times itself r times). Here heads is the desired event, so $p = .5$. Since $r = 1$, we have $.5^1$. The symbol q stands for the probability of the event that is not desired, and it is raised to the $N - r$ power. Tails is the undesirable event, so $q = .5$. Since $N - r$ equals 2, we have $.5^2$. Thus, filling in the binomial expansion, we have

$$p_C = \left(\frac{3!}{1!(3 - 1)!}\right)(.5^1)(.5^2)$$

The exclamation point (!) is the symbol for *factorial,* meaning that you multiply the number times all whole numbers less than it down to 1. Thus, 3! equals 3 times 2 times 1, for an answer of 6. The quantity 1! is (1)(1), or 1, and the quantity (3 − 1)! is (2)!, which is (2)(1), or 2. Now the formula becomes

$$p_C = \left(\frac{6}{1(2)}\right)(.5^1)(.5^2)$$

Any number raised to the first power is that number, so $.5^1$ equals .5. (In a different problem, if you had to raise p or q to the zero power, the answer would be 1.) Because $.5^2$ is .25, we have

$$p_C = \left(\frac{6}{1(2)}\right)(.5)(.25)$$

Multiplying 1 times 2 yields 2, divided into 6 is 3. Multiplying .5 times .25 gives .125, so we have

$$p_C = 3(.125)$$

and $p_C = .375$. Thus, the probability of obtaining 1 head in 3 coin tosses is, as we found initially, equal to .375. Notice that this is not the probability of obtaining at least 1 head. Rather, it is the probability of obtaining precisely 1 head (and 2 tails) in 3 coin tosses.

We can also use the binomial expansion when we can classify various events as either "yes" or "no." For example, say we are playing with dice, and we want to determine the probability of showing a two on 4 out of 6 rolls of one die. The desired event of a two is the "yes." We want it to happen 4 times, so $r = 4$. Its probability on any single roll is p, which equals 1/6, or .167. Showing any other number on the die is a "no," and the probability of any other number on 1 roll of the die is q, which equals 5/6, or .83. Thus, filling in the binomial expansion, we have

$$p_C = \left(\frac{N!}{r!(N-r)!}\right)(p^r)(q^{N-r}) = \frac{6!}{4!(6-4)!}(.167^4)(.83^2)$$

which becomes

$$p_C = \left(\frac{720}{24(2)}\right)(.00078)(.689)$$

which equals

$$p_C = \left(\frac{720}{48}\right)(.00054)$$

Thus,

$$p_C = 15(.00054) = .0081$$

so the probability of rolling a die 6 times and showing a two on 4 of the rolls equals .0081.

PRACTICE PROBLEMS

(Answers for odd-numbered problems are provided in Appendix E.)

1. *a.* When we state a question in terms of the probability of this "and" that, what mathematical procedure do we employ?

b. When we state the question in terms of the probability of this "or" that, what do we do?

c. When we phrase a question using "or," what characteristics of the events must we consider?

2. Which of the following events are mutually inclusive and which are mutually exclusive?

   a. Being male or female

   b. Being sunny and raining

   c. Being tall and weighing a lot

   d. Being age 17 and being a registered voter

3. Researchers have found that for every 100 people, 34 have an IQ above 116 and the rest are below 116, and 40 are introverted, 35 are extroverted, and 25 are in-between.

   a. What is the probability of randomly selecting two people with an IQ above 116?

   b. What is the probability of selecting either an introverted or an extroverted subject?

   c. What is the probability of selecting someone with an IQ above 116 who is introverted?

   d. What is the probability of selecting either someone with an IQ above 116 or someone who is introverted?

   e. What is the probability of selecting someone in-between introverted and extroverted, and then selecting either someone who has an IQ above 116 or who is introverted?

4. You want to determine the probability of obtaining three heads in a row in a coin toss.

   a. Determine the probability by making a fraction of the total number of ways three heads can occur and the total number of possible combinations of heads and tails.

   b. Determine the probability using the multiplication rule.

5. *a.* When do we use the binomial expansion?

   b. Out of 5 coin tosses, what is the probability of obtaining 4 heads?

   c. What is the probability of obtaining 1 head in 5 coin tosses?

   d. Why is the answer in part *c* the same as in part *b*?

6. When rolling dice, what is the probability of each of the following?

   a. Getting a 4 or a 5 rolling one die

   b. Getting a 4 twice in a row rolling one die

   c. Getting a 5 on only one die when rolling two dice at once

   d. Getting three 1's in 5 rolls of one die

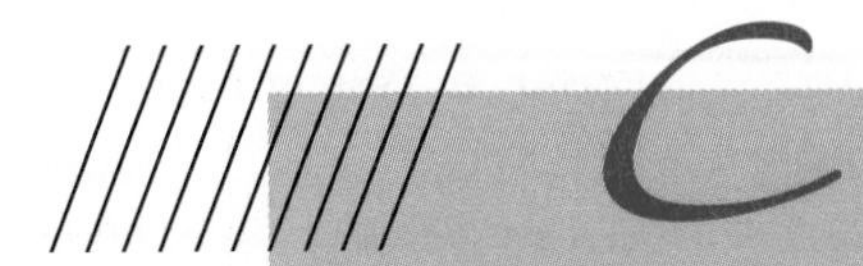

One-Way Within-Subjects Analysis of Variance

This appendix contains the formulas for performing a one-way within-subjects ANOVA, discussed in Chapter 13. Most often this procedure is applied in an experiment involving a repeated measures design, so we will focus on this use of the procedure. A one-way ANOVA for repeated measures is similar to the two-way ANOVA, discussed in Chapter 14, so read that chapter first.

ASSUMPTIONS OF THE REPEATED MEASURES ANOVA

In a repeated measures ANOVA, the same subjects are measured repeatedly under all levels of one factor, and there must be an equal number of scores in each condition. The other assumptions of the repeated measures ANOVA are the same as those for the between-subjects ANOVA: (1) subjects are randomly selected and the dependent variable is a ratio or interval variable, (2) the populations of scores represented by the conditions are normally distributed and the mean is the appropriate measure of central tendency, and (3) the population variances are equal or homogeneous.

LOGIC OF THE ONE-WAY REPEATED MEASURES *F*-RATIO

Let's look at an example of a repeated measures study. Say we are interested in whether subjects' form of dress influences how comfortable they feel in a social setting. On three consecutive days we ask each subject to act as a "greeter" for other subjects participating in a different experiment. On the first day our subjects dress casually, on the second day they dress semiformally, and on the third day they dress formally. We test the very unpowerful N of five subjects. Labeling our independent variable of type of dress as factor A, we can diagram the design as follows.

		Factor A: Type of Dress		
		Level A_1 *Casual*	*Level* A_2 *Semiformal*	*Level* A_3 *Formal*
Subjects Factor	1	X	X	X
	2	X	X	X
	3	X	X	X
	4	X	X	X
	5	X	X	X
		$\bar{X}_{A_1}$	$\bar{X}_{A_2}$	$\bar{X}_{A_3}$

At the end of each day, our subjects answer several questions that describe how comfortable they felt greeting people. Therefore, first we will test the five subjects under level A_1 (casual) and obtain their comfort scores, then we will test the same five subjects under level A_2 (semiformal) and obtain their scores, and then we will test them again under level A_3 (formal). We then find the mean of each column of scores under factor A. As usual, we are testing whether the means from the different levels of the factor represent different population μ's. Therefore, the hypotheses are the same as in a between-subjects design.

H_0: $\mu_1 = \mu_2 = \mu_3$

H_a: not all μ's are equal

Notice that although this is a one-way ANOVA, we can view it as a two-way ANOVA, with factor A as one factor and the different subjects as a second factor, here with five levels. Then factor A is the type of dress, factor B is the subjects factor, and the interaction is between subjects and type of dress.

ELEMENTS OF THE REPEATED MEASURES ANOVA

Previously when computing F_{obt} we needed a mean square within groups (MS_{wn}). This was our estimate of the error variance (σ^2_{error}), the inherent variability of any of the populations being represented. We computed MS_{wn} using the differences between the scores in each cell and the mean of the cell. However, as you can see from the above diagram, each cell contains only one score, because there is only one subject. Therefore, the mean of each cell *is* the score in the cell, and the differences within a cell are always zero. Obviously, we cannot compute MS_{wn} in the usual way.

However, in a repeated measures ANOVA, the mean square for the interaction between factor A and the subjects (abbreviated $MS_{A \times subs}$) does reflect the inherent variability of scores. Recall that an interaction indicates that the effect of one factor changes across the levels of the other factor. It is because of the inherent variability, or differences, between subjects that the effect of type of dress will change across the different subjects. Therefore, $MS_{A \times subs}$ is our estimate of error variance (σ^2_{error}), and it is used as the denominator of the repeated measures F-ratio.

As usual, MS_A describes the difference between the means in factor A, and it is our estimate of variability due to error plus variability due to treatment. Thus, the F-ratio for a repeated measures factor is

	Sample	***Estimates***	***Population***
$F_{obt} =$	$\dfrac{MS_A}{MS_{A \times subs}}$	$\rightarrow$ $\rightarrow$	$\dfrac{\sigma^2_{error} + \sigma^2_{treat}}{\sigma^2_{error}}$

As usual, if the data represent the situation where H_0 is true and all μ's are equal, then both the numerator and the denominator will contain only σ^2_{error}, so F_{obt} will equal 1.0. However, the larger the F_{obt}, the less likely it is that the differences between the means merely reflect sampling error in representing one population μ. If F_{obt} is significant, then at least two of the means from factor A represent different values of μ.

COMPUTING THE ONE-WAY REPEATED MEASURES ANOVA

Say that we obtained these data:

		Factor A: Type of Dress			
		Level A_1 casual	*Level A_2 semiformal*	*Level A_3 formal*	
	1	4	9	1	$\Sigma X_{sub} = 14$
	2	6	12	3	$\Sigma X_{sub} = 21$
Subjects	*3*	8	4	4	$\Sigma X_{sub} = 16$
	4	2	8	5	$\Sigma X_{sub} = 15$
	5	10	7	2	$\Sigma X_{sub} = 19$
		$\Sigma X = 30$ $\Sigma X^2 = 220$ $n_1 = 5$ $\bar{X}_1 = 6$	$\Sigma X = 40$ $\Sigma X^2 = 354$ $n_2 = 5$ $\bar{X}_2 = 8$	$\Sigma X = 15$ $\Sigma X^2 = 55$ $n_3 = 5$ $\bar{X}_3 = 3$	Total: $\Sigma X_{tot} = 30 + 40 + 15 = 85$ $\Sigma X^2_{tot} = 220 + 354 + 55 = 629$ $N = 15$ $k = 3$

The first step is to compute the ΣX, the $\bar{X}$, and the ΣX^2 for each level of factor A (each column). Then compute ΣX_{tot} and ΣX^2_{tot}. Also compute ΣX_{sub}, which is the ΣX for each subject's scores (each horizontal row). Notice that the n's and N are based on the number of *scores,* not the number of subjects.

Then follow these steps.

Step 1 is to compute the total sum of squares (SS_{tot}).

THE COMPUTATIONAL FORMULA FOR SS_{tot} IS

$$SS_{tot} = \Sigma X^2_{tot} - \left(\frac{(\Sigma X_{tot})^2}{N}\right)$$

Filling in this formula from our example data, we have

$$SS_{tot} = 629 - \left(\frac{85^2}{15}\right)$$

$$SS_{tot} = 629 - 481.67$$

so

$$SS_{tot} = 147.33$$

Note that we call the quantity $(\Sigma X_{tot})^2/N$ the *correction* in the following computations. (Here the correction is 481.67.)

Step 2 is to compute the sum of squares for factor A, SS_A.

THE COMPUTATIONAL FORMULA FOR THE SUM OF SQUARES BETWEEN GROUPS FOR FACTOR A IS

$$SS_A = \Sigma\left(\frac{(\text{sum of scores in the column})^2}{n \text{ of scores in the column}}\right) - \left(\frac{(\Sigma X_{tot})^2}{N}\right)$$

This formula says to take the sum of X in each level (column) of factor A, square the sum, and divide by the n of the level. After doing this for all levels, add the results together and subtract the correction.

In our example, we have

$$SS_A = \left(\frac{(30)^2}{5} + \frac{(40)^2}{5} + \frac{(15)^2}{5}\right) - 481.67$$

so

$$SS_A = 545 - 481.67$$

Thus,

$$SS_A = 63.33$$

Step 3 is to find the sum of squares for subjects, SS_{subs}.

THE COMPUTATIONAL FORMULA FOR THE SUM OF SQUARES FOR SUBJECTS, SS_{subs}, IS

$$SS_{subs} = \frac{(\Sigma X_{sub1})^2 + (\Sigma X_{sub2})^2 + \cdots + (\Sigma X_n)^2}{k} - \frac{(\Sigma X_{tot})^2}{N}$$

This says to take ΣX_{sub}, the sum for each subject, and square it. Then add the squared sums together. Next divide by k, where k is the number of levels of factor A. Finally, subtract the correction.

In our example, we have

$$SS_{subs} = \frac{(14)^2 + (21)^2 + (16)^2 + (15)^2 + (19)^2}{3} - 481.67$$

so

$$SS_{subs} = 493 - 481.67$$

Thus,

$$SS_{subs} = 11.33$$

Step 4 is to find the sum of squares for the interaction, $SS_{A \times subs}$. To do this, subtract the sums of squares for the other factors from the total.

THE COMPUTATIONAL FORMULA FOR THE INTERACTION OF FACTOR A BY SUBJECTS, $SS_{A \times subs}$, IS

$$SS_{A \times subs} = SS_{tot} - SS_A - SS_{subs}$$

In our example,

$$SS_{A \times subs} = 147.33 - 63.33 - 11.33$$

so

$$SS_{A \times subs} = 72.67$$

Step 5 is to determine the degrees of freedom.

THE DEGREES OF FREEDOM BETWEEN GROUPS FOR FACTOR A IS

$$df_A = k_A - 1$$

k_A is the number of levels of factor A. (In the example, there are three levels of type of dress, so df_A is 2.)

THE DEGREES OF FREEDOM FOR THE INTERACTION IS

$$df_{A\times subs} = (k_A - 1)(k_{subs} - 1)$$

k_A is the number of levels of factor A, and k_{subs} is the number of subjects. In the example, there are three levels of factor A and 5 subjects, so $df_{A\times subs} = (2)(4) = 8$.

Compute df_{subs} and df_{tot} to check the above *df*. The $df_{subs} = k_{subs} - 1$, where k_{subs} is the number of subjects. The $df_{tot} = N - 1$, where N is the total number of *scores* in the experiment. The df_{tot} is equal to the sum of all the other *df*'s.

Step 6 is to place the sum of squares and the *df*'s in the summary table. For our example, we have the following table.

Summary Table of One-Way Repeated Measures ANOVA

Source	*Sum of squares*	*df*	*Mean square*	*F*
Subjects	11.33	4		
Factor A (dress)	63.33	2	MS_A	F_A
Interaction (A × subjects)	72.67	8	$MS_{A\times subs}$	
Total	147.33	14		

Because we have only one factor of interest here (type of dress), we find only the F_{obt} for factor A.

Step 7 is to find the mean squares for factor A and the interaction.

THE MEAN SQUARE FOR FACTOR A, MS_A, EQUALS

$$MS_A = \frac{SS_A}{df_A}$$

In our example,

$$MS_A = \frac{SS_A}{df_A} = \frac{63.33}{2} = 31.67$$

THE MEAN SQUARE FOR THE INTERACTION BETWEEN FACTOR A AND SUBJECTS, $MS_{A\times subs}$, IS

$$MS_{A\times subs} = \frac{SS_{A\times subs}}{df_{A\times subs}}$$

In our example,

$$MS_{A\times subs} = \frac{SS_{A\times subs}}{df_{A\times subs}} = \frac{72.67}{8} = 9.08$$

Step 8 is to find F_{obt}.

THE COMPUTATIONAL FORMULA FOR THE REPEATED MEASURES F-RATIO IS

$$F_{obt} = \frac{MS_A}{MS_{A\times subs}}$$

In our example,

$$F_{obt} = \frac{MS_A}{MS_{A\times subs}} = \frac{31.67}{9.08} = 3.49$$

Thus, our finished summary table is as follows:

Source	*Sum of Squares*	*df*	*Mean Square*	*F*
Subjects	11.33	4		
Factor A (Dress)	63.33	2	31.67	3.49
Interaction (A × Subjects)	72.67	8	9.08	
Total	147.33	14		

Step 9 is to find the critical value of *F* in Table 5 of Appendix D. Use df_A as the degrees of freedom between groups and $df_{A\times subs}$ as the degrees of freedom within groups. In the example, for $\alpha = .05$, $df_A = 2$, and $df_{A\times subs} = 8$, the F_{crit} is 4.46.

INTERPRETING THE REPEATED MEASURES *F*

At this point, we interpret the above F_{obt} in exactly the same way we would a between-subjects F_{obt}. Since F_{obt} in the above example is *not* larger than F_{crit}, it is not significant. Thus, we do not have evidence that the means from at least two of our levels of type of dress represent different populations of comfort scores. Had F_{obt} been significant, it would indicate that at least two of the level means differ significantly. Then, for *post hoc* comparisons, graphing, eta squared, and confidence intervals, we would follow the procedures discussed in Chapter 13. However, in any of those formulas, in place of the term MS_{wn} we would use the above $MS_{A\times subs}$.

Note that in discussing the related samples *t*-test in Chapter 12, we said that it is more powerful than the independent samples *t*-test, because the variability in the scores is less. For the same reason, the repeated measures ANOVA is more powerful than a between-subjects ANOVA for the same data. In the repeated measures ANOVA, we delete some of the variability in the raw scores by separating (and then ignoring) the sum of squares for the subjects factor. By removing the differences due to subjects from our calculations, we obtain an $MS_{A \times subs}$ that is smaller than the MS_{wn} we would compute in a between-subjects ANOVA. Therefore, we obtain a larger F_{obt}, which is more likely to be significant, so we have greater power. The results in the above example were not significant because of the relatively small *N*, given the amount of variability in the scores. However, this design was still more powerful—more likely to produce significant results—than a comparable between-subjects design.

PRACTICE PROBLEMS

(Answers for odd-numbered problems are provided in Appendix E.)

1. You read in a research report that the repeated measures factor of a subject's weight gain led to a decrease in his or her mood.
   a. What does this tell you about the design?
   b. What does it tell you about the results?

2. Which of these relationships suggest using a repeated measures design?
   a. Examining the improvement in language ability as children grow older
   b. Measuring subjects' reaction when the experimenter surprises them by unexpectedly shouting, under three levels of volume of shouting
   c. Comparing the dating strategies of males and females
   d. Comparing memory ability under the conditions of subjects' consuming different amounts of alcoholic beverages

3. In a study on the influence of practice on performing a task requiring eye-hand coordination, we test subjects when they have had no practice, after 1 hour of practice, and again after 2 hours of practice. We obtain the following data, with higher scores indicating better performance. (See next page.)
   a. What are H_0 and H_A?
   b. Complete the ANOVA summary table.
   c. With $\alpha = .05$, what do you conclude about F_{obt}?
   d. Perform the appropriate *post hoc* comparisons.
   e. What should you conclude about this relationship?
   f. What is the effect size in this study?

Amount of Practice

Subjects	*Zero*	*One hour*	*Two hours*
S1	4	3	6
S2	3	5	5
S3	1	4	3
S4	3	4	6
S5	1	5	6
S6	2	6	7
S7	2	4	5
S8	1	3	8

4. You conduct a study in which you measure 21 students' degree of positive attitude toward statistics at 4 equally spaced intervals during the semester. The mean score for each level is as follows: time 1, 62.50; time 2, 64.68; time 3, 69.32; time 4, 72.00. You obtain the following sums of squares:

Source	***Sum of Squares***	***df***	***Mean Square***	***F***
Subjects	402.79			
Factor A	189.30			
A × Subjects	688.32			
Total	1280.41			

a. What are H_0 and H_A?

b. Complete the ANOVA summary table.

c. With $\alpha = .05$, what do you conclude about F_{obt}?

d. Perform the appropriate *post hoc* comparisons.

e. What is the effect size in this study?

f. What do you conclude about this relationship?

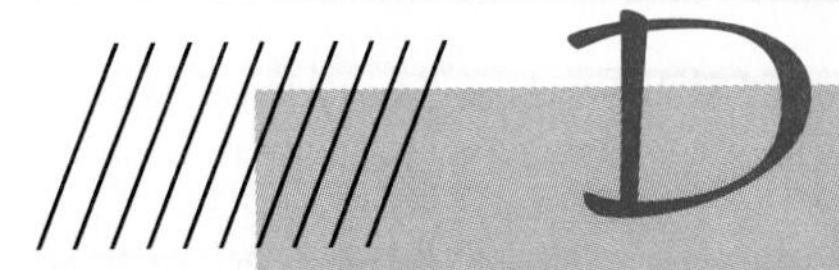

Statistical Tables

Table 1 Proportions of Area Under the Standard Normal Curve: The *z*-Tables

Column (A) lists *z*-score values. Column (B) lists the proportion of the area between the mean and the *z*-score value. Column (C) lists the proportion of the area beyond the *z*-score. *Note:* Because the normal distribution is symmetrical, areas for negative *z*-scores are the same as those for positive *z*-scores.

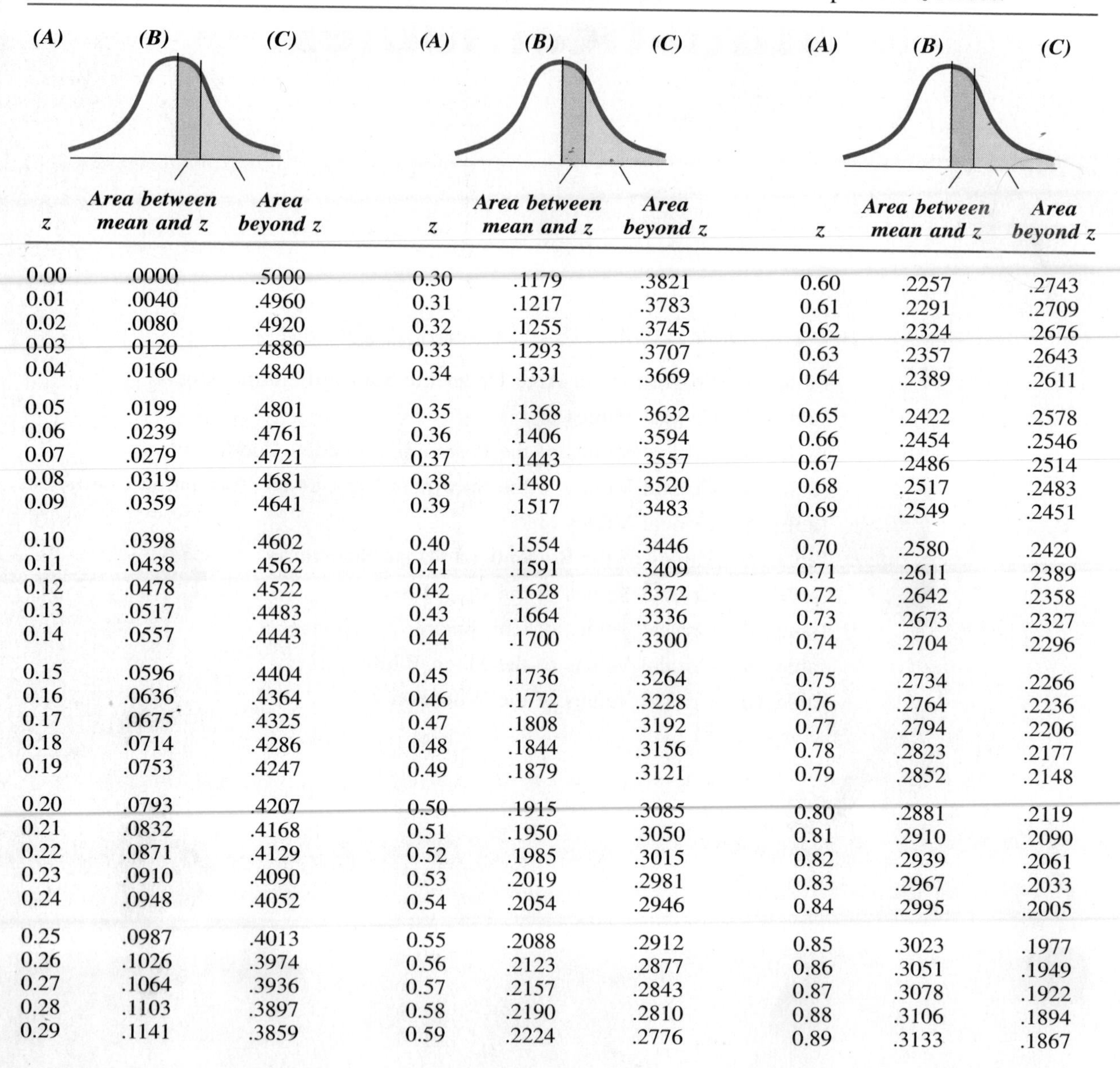

(A) *z*	(B) Area between mean and *z*	(C) Area beyond *z*	(A) *z*	(B) Area between mean and *z*	(C) Area beyond *z*	(A) *z*	(B) Area between mean and *z*	(C) Area beyond *z*
0.00	.0000	.5000	0.30	.1179	.3821	0.60	.2257	.2743
0.01	.0040	.4960	0.31	.1217	.3783	0.61	.2291	.2709
0.02	.0080	.4920	0.32	.1255	.3745	0.62	.2324	.2676
0.03	.0120	.4880	0.33	.1293	.3707	0.63	.2357	.2643
0.04	.0160	.4840	0.34	.1331	.3669	0.64	.2389	.2611
0.05	.0199	.4801	0.35	.1368	.3632	0.65	.2422	.2578
0.06	.0239	.4761	0.36	.1406	.3594	0.66	.2454	.2546
0.07	.0279	.4721	0.37	.1443	.3557	0.67	.2486	.2514
0.08	.0319	.4681	0.38	.1480	.3520	0.68	.2517	.2483
0.09	.0359	.4641	0.39	.1517	.3483	0.69	.2549	.2451
0.10	.0398	.4602	0.40	.1554	.3446	0.70	.2580	.2420
0.11	.0438	.4562	0.41	.1591	.3409	0.71	.2611	.2389
0.12	.0478	.4522	0.42	.1628	.3372	0.72	.2642	.2358
0.13	.0517	.4483	0.43	.1664	.3336	0.73	.2673	.2327
0.14	.0557	.4443	0.44	.1700	.3300	0.74	.2704	.2296
0.15	.0596	.4404	0.45	.1736	.3264	0.75	.2734	.2266
0.16	.0636	.4364	0.46	.1772	.3228	0.76	.2764	.2236
0.17	.0675	.4325	0.47	.1808	.3192	0.77	.2794	.2206
0.18	.0714	.4286	0.48	.1844	.3156	0.78	.2823	.2177
0.19	.0753	.4247	0.49	.1879	.3121	0.79	.2852	.2148
0.20	.0793	.4207	0.50	.1915	.3085	0.80	.2881	.2119
0.21	.0832	.4168	0.51	.1950	.3050	0.81	.2910	.2090
0.22	.0871	.4129	0.52	.1985	.3015	0.82	.2939	.2061
0.23	.0910	.4090	0.53	.2019	.2981	0.83	.2967	.2033
0.24	.0948	.4052	0.54	.2054	.2946	0.84	.2995	.2005
0.25	.0987	.4013	0.55	.2088	.2912	0.85	.3023	.1977
0.26	.1026	.3974	0.56	.2123	.2877	0.86	.3051	.1949
0.27	.1064	.3936	0.57	.2157	.2843	0.87	.3078	.1922
0.28	.1103	.3897	0.58	.2190	.2810	0.88	.3106	.1894
0.29	.1141	.3859	0.59	.2224	.2776	0.89	.3133	.1867

Table 1 (cont.) Proportions of Area Under the Standard Normal Curve: The z-Tables

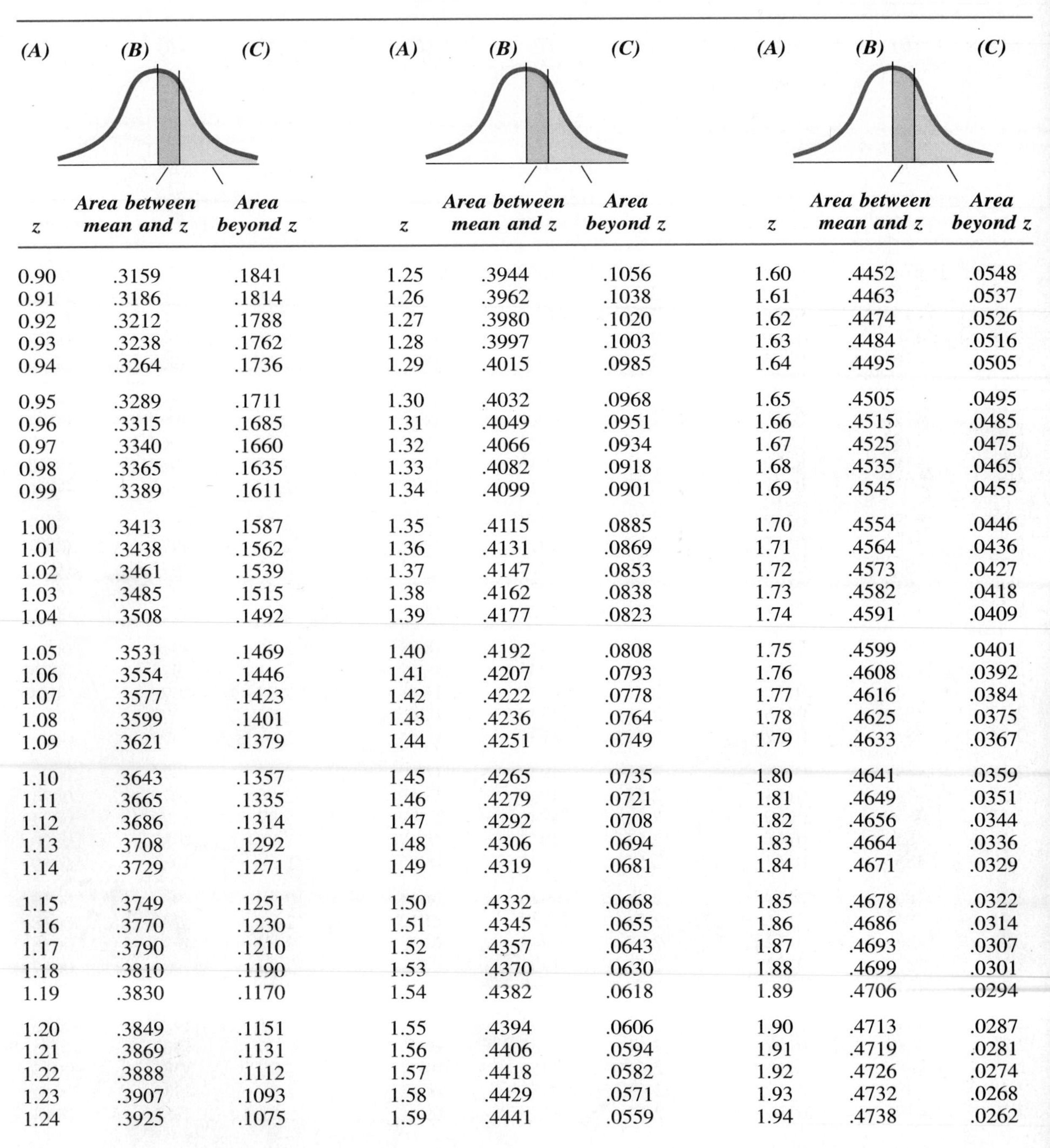

(A) z	(B) Area between mean and z	(C) Area beyond z	(A) z	(B) Area between mean and z	(C) Area beyond z	(A) z	(B) Area between mean and z	(C) Area beyond z
0.90	.3159	.1841	1.25	.3944	.1056	1.60	.4452	.0548
0.91	.3186	.1814	1.26	.3962	.1038	1.61	.4463	.0537
0.92	.3212	.1788	1.27	.3980	.1020	1.62	.4474	.0526
0.93	.3238	.1762	1.28	.3997	.1003	1.63	.4484	.0516
0.94	.3264	.1736	1.29	.4015	.0985	1.64	.4495	.0505
0.95	.3289	.1711	1.30	.4032	.0968	1.65	.4505	.0495
0.96	.3315	.1685	1.31	.4049	.0951	1.66	.4515	.0485
0.97	.3340	.1660	1.32	.4066	.0934	1.67	.4525	.0475
0.98	.3365	.1635	1.33	.4082	.0918	1.68	.4535	.0465
0.99	.3389	.1611	1.34	.4099	.0901	1.69	.4545	.0455
1.00	.3413	.1587	1.35	.4115	.0885	1.70	.4554	.0446
1.01	.3438	.1562	1.36	.4131	.0869	1.71	.4564	.0436
1.02	.3461	.1539	1.37	.4147	.0853	1.72	.4573	.0427
1.03	.3485	.1515	1.38	.4162	.0838	1.73	.4582	.0418
1.04	.3508	.1492	1.39	.4177	.0823	1.74	.4591	.0409
1.05	.3531	.1469	1.40	.4192	.0808	1.75	.4599	.0401
1.06	.3554	.1446	1.41	.4207	.0793	1.76	.4608	.0392
1.07	.3577	.1423	1.42	.4222	.0778	1.77	.4616	.0384
1.08	.3599	.1401	1.43	.4236	.0764	1.78	.4625	.0375
1.09	.3621	.1379	1.44	.4251	.0749	1.79	.4633	.0367
1.10	.3643	.1357	1.45	.4265	.0735	1.80	.4641	.0359
1.11	.3665	.1335	1.46	.4279	.0721	1.81	.4649	.0351
1.12	.3686	.1314	1.47	.4292	.0708	1.82	.4656	.0344
1.13	.3708	.1292	1.48	.4306	.0694	1.83	.4664	.0336
1.14	.3729	.1271	1.49	.4319	.0681	1.84	.4671	.0329
1.15	.3749	.1251	1.50	.4332	.0668	1.85	.4678	.0322
1.16	.3770	.1230	1.51	.4345	.0655	1.86	.4686	.0314
1.17	.3790	.1210	1.52	.4357	.0643	1.87	.4693	.0307
1.18	.3810	.1190	1.53	.4370	.0630	1.88	.4699	.0301
1.19	.3830	.1170	1.54	.4382	.0618	1.89	.4706	.0294
1.20	.3849	.1151	1.55	.4394	.0606	1.90	.4713	.0287
1.21	.3869	.1131	1.56	.4406	.0594	1.91	.4719	.0281
1.22	.3888	.1112	1.57	.4418	.0582	1.92	.4726	.0274
1.23	.3907	.1093	1.58	.4429	.0571	1.93	.4732	.0268
1.24	.3925	.1075	1.59	.4441	.0559	1.94	.4738	.0262

Table 1 (cont.) Proportions of Area Under the Standard Normal Curve: The z-Tables

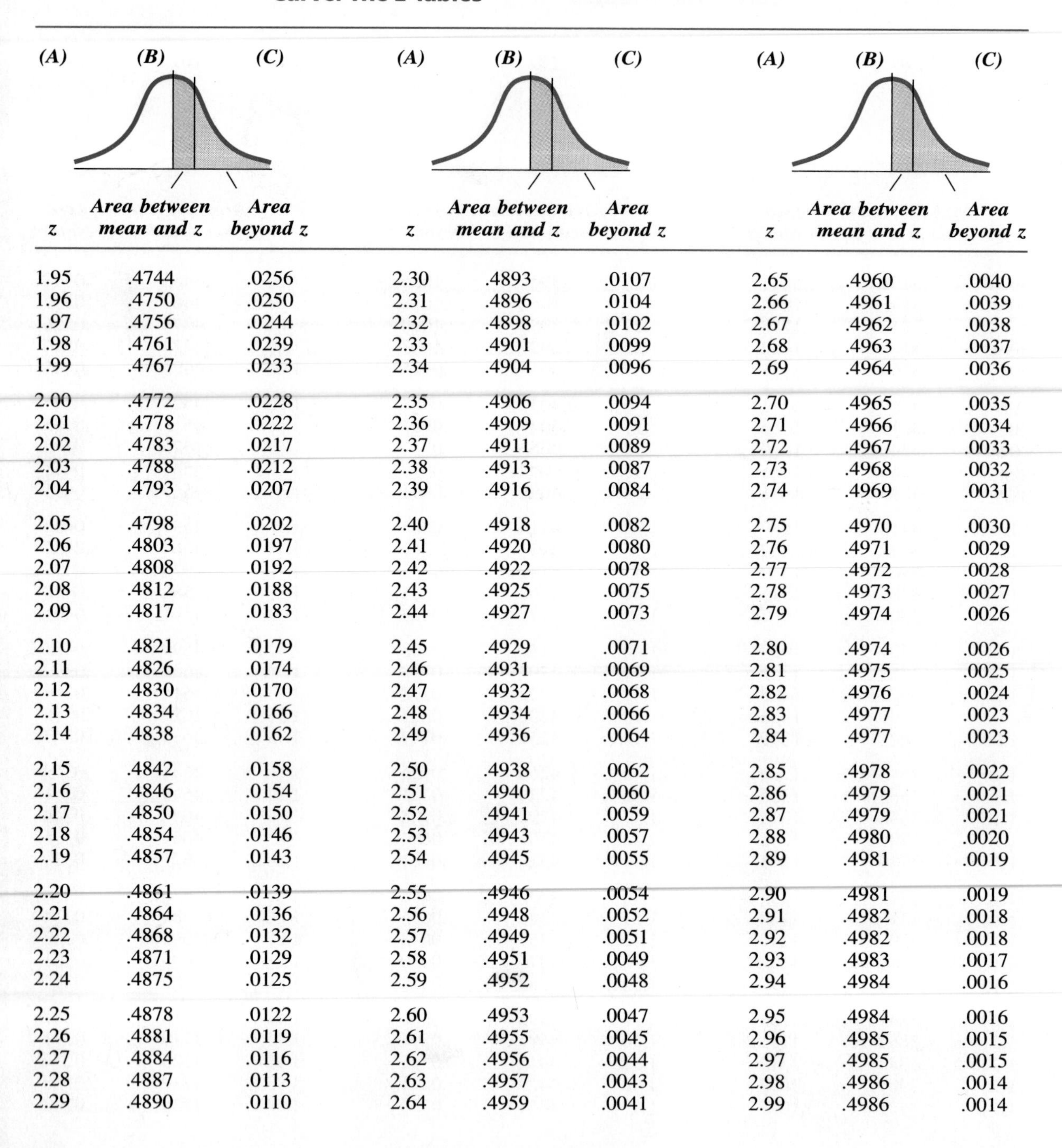

(A) z	(B) Area between mean and z	(C) Area beyond z	(A) z	(B) Area between mean and z	(C) Area beyond z	(A) z	(B) Area between mean and z	(C) Area beyond z
1.95	.4744	.0256	2.30	.4893	.0107	2.65	.4960	.0040
1.96	.4750	.0250	2.31	.4896	.0104	2.66	.4961	.0039
1.97	.4756	.0244	2.32	.4898	.0102	2.67	.4962	.0038
1.98	.4761	.0239	2.33	.4901	.0099	2.68	.4963	.0037
1.99	.4767	.0233	2.34	.4904	.0096	2.69	.4964	.0036
2.00	.4772	.0228	2.35	.4906	.0094	2.70	.4965	.0035
2.01	.4778	.0222	2.36	.4909	.0091	2.71	.4966	.0034
2.02	.4783	.0217	2.37	.4911	.0089	2.72	.4967	.0033
2.03	.4788	.0212	2.38	.4913	.0087	2.73	.4968	.0032
2.04	.4793	.0207	2.39	.4916	.0084	2.74	.4969	.0031
2.05	.4798	.0202	2.40	.4918	.0082	2.75	.4970	.0030
2.06	.4803	.0197	2.41	.4920	.0080	2.76	.4971	.0029
2.07	.4808	.0192	2.42	.4922	.0078	2.77	.4972	.0028
2.08	.4812	.0188	2.43	.4925	.0075	2.78	.4973	.0027
2.09	.4817	.0183	2.44	.4927	.0073	2.79	.4974	.0026
2.10	.4821	.0179	2.45	.4929	.0071	2.80	.4974	.0026
2.11	.4826	.0174	2.46	.4931	.0069	2.81	.4975	.0025
2.12	.4830	.0170	2.47	.4932	.0068	2.82	.4976	.0024
2.13	.4834	.0166	2.48	.4934	.0066	2.83	.4977	.0023
2.14	.4838	.0162	2.49	.4936	.0064	2.84	.4977	.0023
2.15	.4842	.0158	2.50	.4938	.0062	2.85	.4978	.0022
2.16	.4846	.0154	2.51	.4940	.0060	2.86	.4979	.0021
2.17	.4850	.0150	2.52	.4941	.0059	2.87	.4979	.0021
2.18	.4854	.0146	2.53	.4943	.0057	2.88	.4980	.0020
2.19	.4857	.0143	2.54	.4945	.0055	2.89	.4981	.0019
2.20	.4861	.0139	2.55	.4946	.0054	2.90	.4981	.0019
2.21	.4864	.0136	2.56	.4948	.0052	2.91	.4982	.0018
2.22	.4868	.0132	2.57	.4949	.0051	2.92	.4982	.0018
2.23	.4871	.0129	2.58	.4951	.0049	2.93	.4983	.0017
2.24	.4875	.0125	2.59	.4952	.0048	2.94	.4984	.0016
2.25	.4878	.0122	2.60	.4953	.0047	2.95	.4984	.0016
2.26	.4881	.0119	2.61	.4955	.0045	2.96	.4985	.0015
2.27	.4884	.0116	2.62	.4956	.0044	2.97	.4985	.0015
2.28	.4887	.0113	2.63	.4957	.0043	2.98	.4986	.0014
2.29	.4890	.0110	2.64	.4959	.0041	2.99	.4986	.0014

Table 1 (cont.) Proportions of Area Under the Standard Normal Curve: The z-Tables

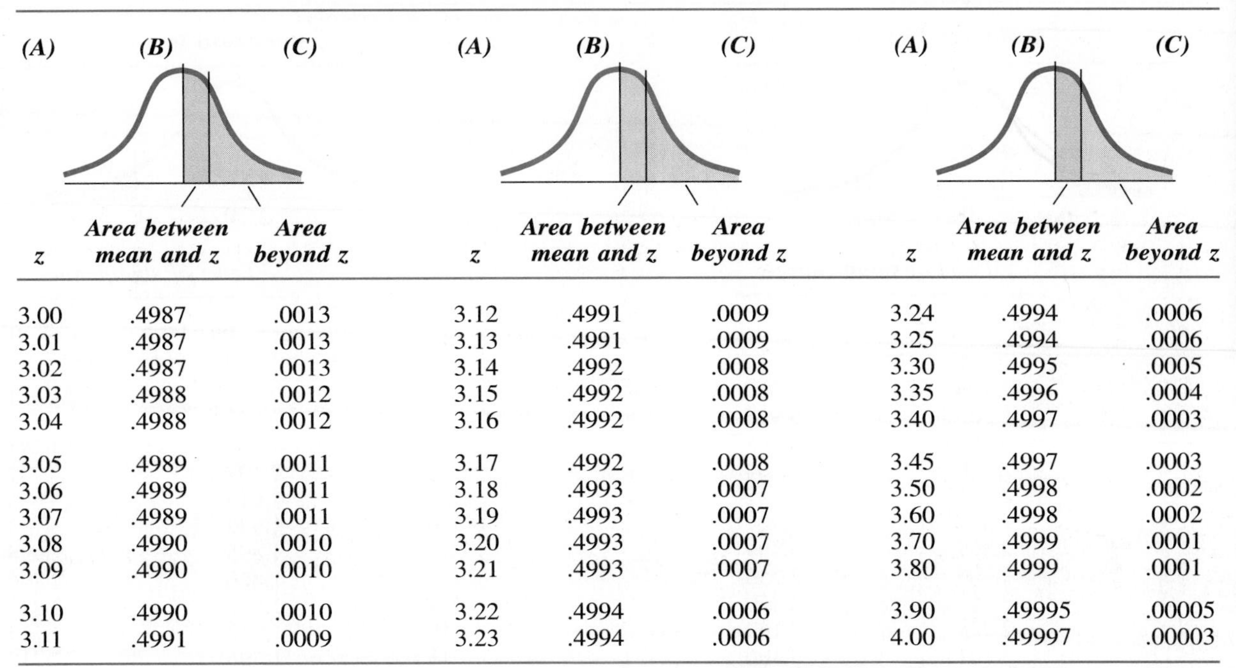

(A)	*(B)*	*(C)*	*(A)*	*(B)*	*(C)*	*(A)*	*(B)*	*(C)*
z	***Area between mean and z***	***Area beyond z***	*z*	***Area between mean and z***	***Area beyond z***	*z*	***Area between mean and z***	***Area beyond z***
3.00	.4987	.0013	3.12	.4991	.0009	3.24	.4994	.0006
3.01	.4987	.0013	3.13	.4991	.0009	3.25	.4994	.0006
3.02	.4987	.0013	3.14	.4992	.0008	3.30	.4995	.0005
3.03	.4988	.0012	3.15	.4992	.0008	3.35	.4996	.0004
3.04	.4988	.0012	3.16	.4992	.0008	3.40	.4997	.0003
3.05	.4989	.0011	3.17	.4992	.0008	3.45	.4997	.0003
3.06	.4989	.0011	3.18	.4993	.0007	3.50	.4998	.0002
3.07	.4989	.0011	3.19	.4993	.0007	3.60	.4998	.0002
3.08	.4990	.0010	3.20	.4993	.0007	3.70	.4999	.0001
3.09	.4990	.0010	3.21	.4993	.0007	3.80	.4999	.0001
3.10	.4990	.0010	3.22	.4994	.0006	3.90	.49995	.00005
3.11	.4991	.0009	3.23	.4994	.0006	4.00	.49997	.00003

Table 2 Critical Values of *t*: The *t*-Tables

Note: Values of $-t_{crit}$ = values of $+t_{crit}$.

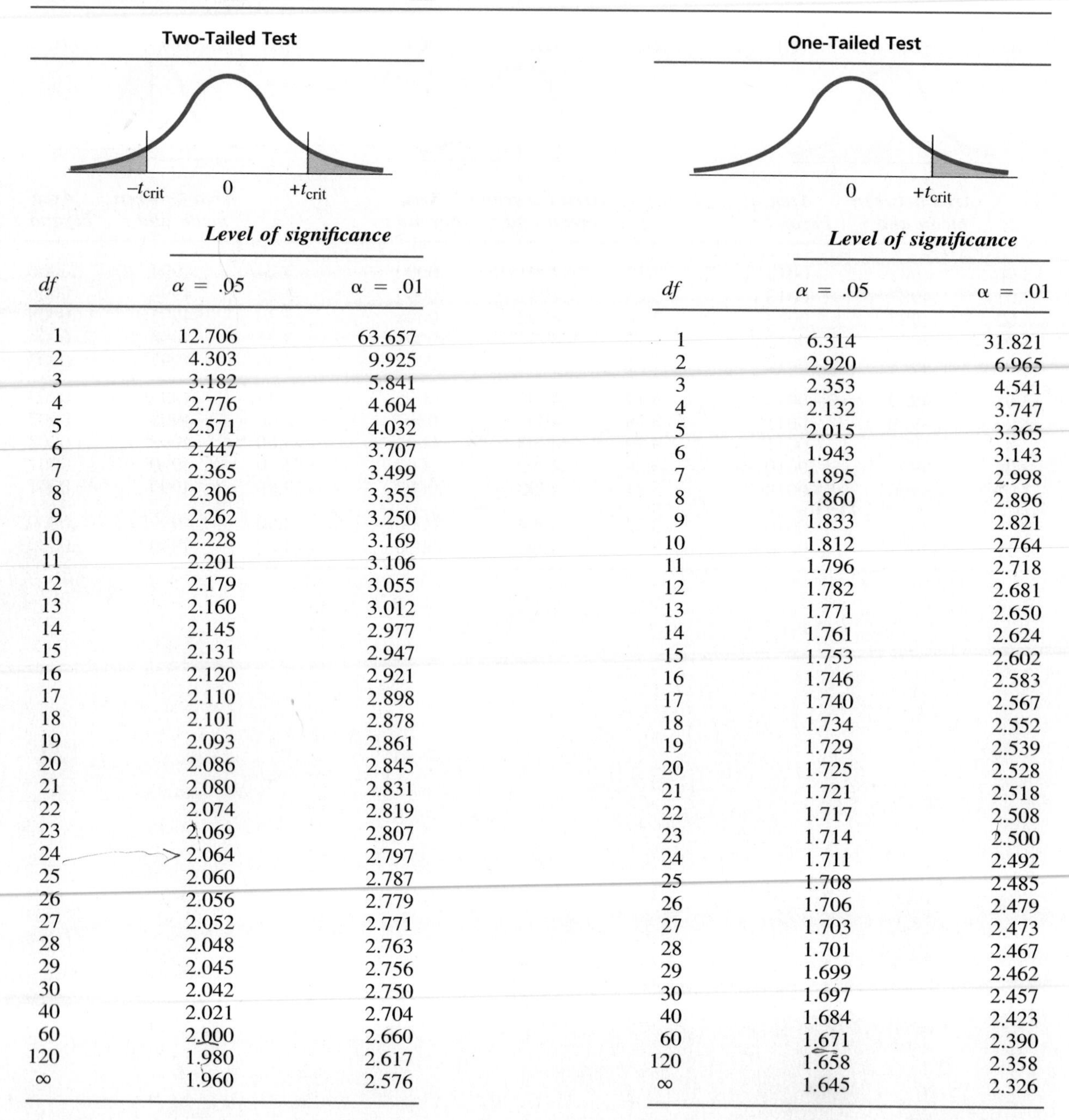

	Two-Tailed Test		One-Tailed Test	
	Level of significance		*Level of significance*	
df	α = .05	α = .01	α = .05	α = .01
1	12.706	63.657	6.314	31.821
2	4.303	9.925	2.920	6.965
3	3.182	5.841	2.353	4.541
4	2.776	4.604	2.132	3.747
5	2.571	4.032	2.015	3.365
6	2.447	3.707	1.943	3.143
7	2.365	3.499	1.895	2.998
8	2.306	3.355	1.860	2.896
9	2.262	3.250	1.833	2.821
10	2.228	3.169	1.812	2.764
11	2.201	3.106	1.796	2.718
12	2.179	3.055	1.782	2.681
13	2.160	3.012	1.771	2.650
14	2.145	2.977	1.761	2.624
15	2.131	2.947	1.753	2.602
16	2.120	2.921	1.746	2.583
17	2.110	2.898	1.740	2.567
18	2.101	2.878	1.734	2.552
19	2.093	2.861	1.729	2.539
20	2.086	2.845	1.725	2.528
21	2.080	2.831	1.721	2.518
22	2.074	2.819	1.717	2.508
23	2.069	2.807	1.714	2.500
24	2.064	2.797	1.711	2.492
25	2.060	2.787	1.708	2.485
26	2.056	2.779	1.706	2.479
27	2.052	2.771	1.703	2.473
28	2.048	2.763	1.701	2.467
29	2.045	2.756	1.699	2.462
30	2.042	2.750	1.697	2.457
40	2.021	2.704	1.684	2.423
60	2.000	2.660	1.671	2.390
120	1.980	2.617	1.658	2.358
∞	1.960	2.576	1.645	2.326

Table 3 Critical Values of the Pearson Correlation Coefficient: The *r*-Tables

Two-Tailed Test

$-r_{crit}$ 0 $+r_{crit}$

df (*no. of pairs* − 2)	*Level of significance* α = .05	α = .01
1	.997	.9999
2	.950	.990
3	.878	.959
4	.811	.917
5	.754	.874
6	.707	.834
7	.666	.798
8	.632	.765
9	.602	.735
10	.576	.708
11	.553	.684
12	.532	.661
13	.514	.641
14	.497	.623
15	.482	.606
16	.468	.590
17	.456	.575
18	.444	.561
19	.433	.549
20	.423	.537
21	.413	.526
22	.404	.515
23	.396	.505
24	.388	.496
25	.381	.487
26	.374	.479
27	.367	.471
28	.361	.463
29	.355	.456
30	.349	.449
35	.325	.418
40	.304	.393
45	.288	.372
50	.273	.354
60	.250	.325
70	.232	.302
80	.217	.283
90	.205	.267
100	.195	.254

One-Tailed Test

0 $+r_{crit}$

df (*no. of pairs* − 2)	*Level of significance* α = .05	α = .01
1	.988	.9995
2	.900	.980
3	.805	.934
4	.729	.882
5	.669	.833
6	.622	.789
7	.582	.750
8	.549	.716
9	.521	.685
10	.497	.658
11	.476	.634
12	.458	.612
13	.441	.592
14	.426	.574
15	.412	.558
16	.400	.542
17	.389	.528
18	.378	.516
19	.369	.503
20	.360	.492
21	.352	.482
22	.344	.472
23	.337	.462
24	.330	.453
25	.323	.445
26	.317	.437
27	.311	.430
28	.306	.423
29	.301	.416
30	.296	.409
35	.275	.381
40	.257	.358
45	.243	.338
50	.231	.322
60	.211	.295
70	.195	.274
80	.183	.256
90	.173	.242
100	.164	.230

From Table VI of R. A. Fisher and F. Yates, *Statistical Tables for Biological, Agricultural and Medical Research,* 6th ed. London: Longman Group Ltd., 1974 (previously published by Oliver and Boyd Ltd., Edinburgh).

Table 4 Critical Values of the Spearman Rank-Order Correlation Coefficient: The r_s-Tables

Note: To interpolate the critical value for an *N* not given, find the critical values for the *N* above and below your *N*, add them together, and then divide the sum by 2.

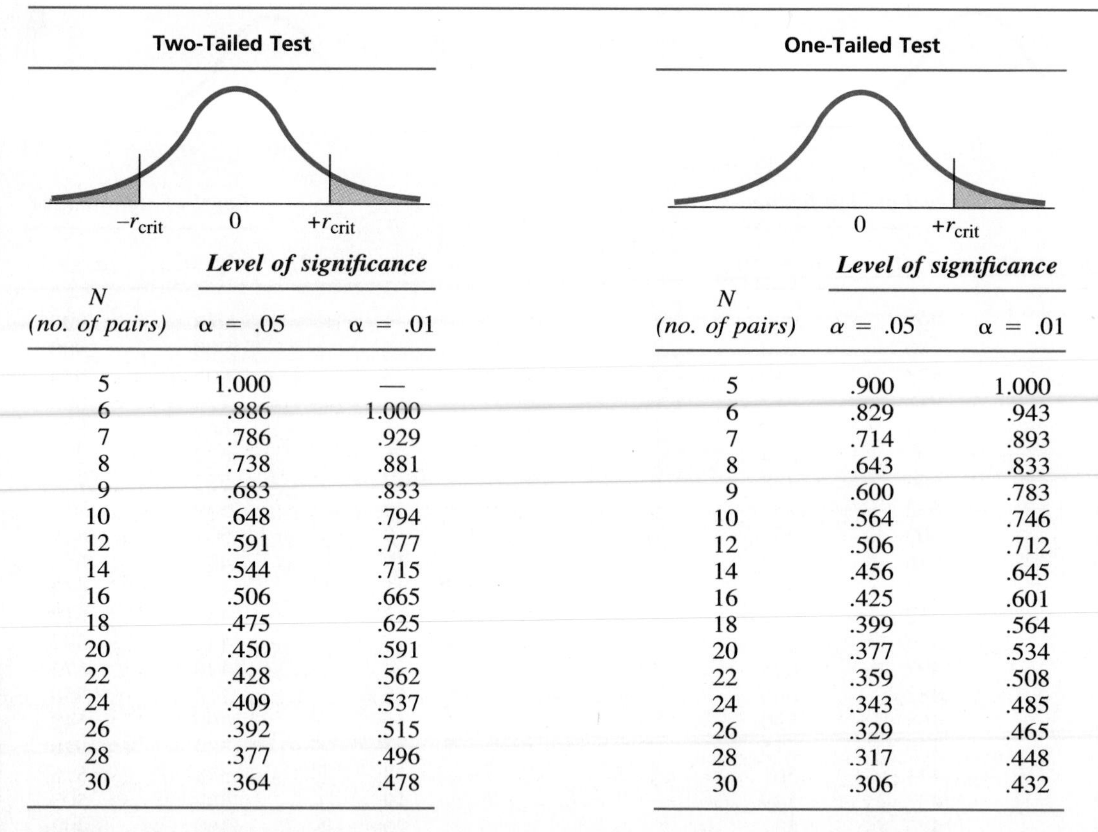

Two-Tailed Test

N (*no. of pairs*)	*Level of significance* α = .05	α = .01
5	1.000	—
6	.886	1.000
7	.786	.929
8	.738	.881
9	.683	.833
10	.648	.794
12	.591	.777
14	.544	.715
16	.506	.665
18	.475	.625
20	.450	.591
22	.428	.562
24	.409	.537
26	.392	.515
28	.377	.496
30	.364	.478

One-Tailed Test

N (*no. of pairs*)	*Level of significance* α = .05	α = .01
5	.900	1.000
6	.829	.943
7	.714	.893
8	.643	.833
9	.600	.783
10	.564	.746
12	.506	.712
14	.456	.645
16	.425	.601
18	.399	.564
20	.377	.534
22	.359	.508
24	.343	.485
26	.329	.465
28	.317	.448
30	.306	.432

From E. G. Olds (1949), The 5 Percent Significance Levels of Sums of Squares of Rank Differences and a Correction, *Ann. Math. Statist.*, **20,** 117–118, and E. G. Olds (1938), Distribution of Sums of Squares of Rank Differences for Small Numbers of Individuals, *Ann. Math. Statist.*, **9,** 133–148. Reprinted with permission of the Institute of Mathematical Statistics.

Table 5 Critical Values of *F*: The *F*-Tables

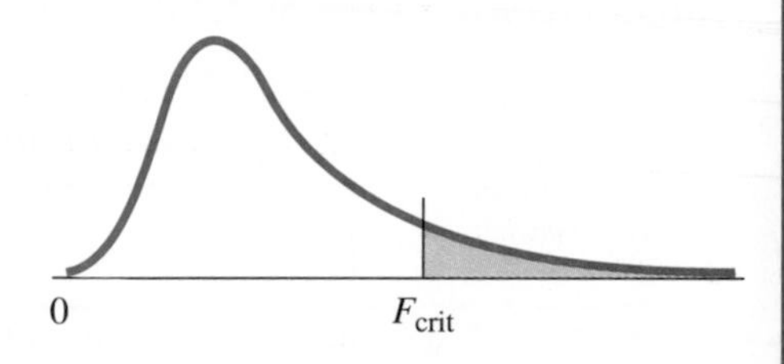

Critical values for α = .05 are in **dark numbers.**
Critical values for α = .01 are in light numbers.

Degrees of freedom within groups (degrees of freedom in denominator of F ratio)	*α*	*Degrees of freedom between groups (degrees of freedom in numerator of F ratio)* 1	2	3	4	5	6	7	8	9	10	11	12	14	16	20
1	**.05**	**161**	**200**	**216**	**225**	**230**	**234**	**237**	**239**	**241**	**242**	**243**	**244**	**245**	**246**	**248**
	.01	4,052	4,999	5,403	5,625	5,764	5,859	5,928	5,981	6,022	6,056	6,082	6,106	6,142	6,169	6,208
2	**.05**	**18.51**	**19.00**	**19.16**	**19.25**	**19.30**	**19.33**	**19.36**	**19.37**	**19.38**	**19.39**	**19.40**	**19.41**	**19.42**	**19.43**	**19.44**
	.01	98.49	99.00	99.17	99.25	99.30	99.33	99.34	99.36	99.38	99.40	99.41	99.42	99.43	99.44	99.45
3	**.05**	**10.13**	**9.55**	**9.28**	**9.12**	**9.01**	**8.94**	**8.88**	**8.84**	**8.81**	**8.78**	**8.76**	**8.74**	**8.71**	**8.69**	**8.66**
	.01	34.12	30.82	29.46	28.71	28.24	27.91	27.67	27.49	27.34	27.23	27.13	27.05	26.92	26.83	26.69
4	**.05**	**7.71**	**6.94**	**6.59**	**6.39**	**6.26**	**6.16**	**6.09**	**6.04**	**6.00**	**5.96**	**5.93**	**5.91**	**5.87**	**5.84**	**5.80**
	.01	21.20	18.00	16.69	15.98	15.52	15.21	14.98	14.80	14.66	14.54	14.45	14.37	14.24	14.15	14.02
5	**.05**	**6.61**	**5.79**	**5.41**	**5.19**	**5.05**	**4.95**	**4.88**	**4.82**	**4.78**	**4.74**	**4.70**	**4.68**	**4.64**	**4.60**	**4.56**
	.01	16.26	13.27	12.06	11.39	10.97	10.67	10.45	10.27	10.15	10.05	9.96	9.89	9.77	9.68	9.55
6	**.05**	**5.99**	**5.14**	**4.76**	**4.53**	**4.39**	**4.28**	**4.21**	**4.15**	**4.10**	**4.06**	**4.03**	**4.00**	**3.96**	**3.92**	**3.87**
	.01	13.74	10.92	9.78	9.15	8.75	8.47	8.26	8.10	7.98	7.87	7.79	7.72	7.60	7.52	7.39
7	**.05**	**5.59**	**4.47**	**4.35**	**4.12**	**3.97**	**3.87**	**3.79**	**3.73**	**3.68**	**3.63**	**3.60**	**3.57**	**3.52**	**3.49**	**3.44**
	.01	12.25	9.55	8.45	7.85	7.46	7.19	7.00	6.84	6.71	6.62	6.54	6.47	6.35	6.27	6.15
8	**.05**	**5.32**	**4.46**	**4.07**	**3.84**	**3.69**	**3.58**	**3.50**	**3.44**	**3.39**	**3.34**	**3.31**	**3.28**	**3.23**	**3.20**	**3.15**
	.01	11.26	8.65	7.59	7.01	6.63	6.37	6.19	6.03	5.91	5.82	5.74	5.67	5.56	5.48	5.36
9	**.05**	**5.12**	**4.26**	**3.86**	**3.63**	**3.48**	**3.37**	**3.29**	**3.23**	**3.18**	**3.13**	**3.10**	**3.07**	**3.02**	**2.98**	**2.93**
	.01	10.56	8.02	6.99	6.42	6.06	5.80	5.62	5.47	5.35	5.26	5.18	5.11	5.00	4.92	4.80
10	**.05**	**4.96**	**4.10**	**3.71**	**3.48**	**3.33**	**3.22**	**3.14**	**3.07**	**3.02**	**2.97**	**2.94**	**2.91**	**2.86**	**2.82**	**2.77**
	.01	10.04	7.56	6.55	5.99	5.64	5.39	5.21	5.06	4.95	4.85	4.78	4.71	4.60	4.52	4.41
11	**.05**	**4.84**	**3.98**	**3.59**	**3.36**	**3.20**	**3.09**	**3.01**	**2.95**	**2.90**	**2.86**	**2.82**	**2.79**	**2.74**	**2.70**	**2.65**
	.01	9.65	7.20	6.22	5.67	5.32	5.07	4.88	4.74	4.63	4.54	4.46	4.40	4.29	4.21	4.10
12	**.05**	**4.75**	**3.88**	**3.49**	**3.26**	**3.11**	**3.00**	**2.92**	**2.85**	**2.80**	**2.76**	**2.72**	**2.69**	**2.64**	**2.60**	**2.54**
	.01	9.33	6.93	5.95	5.41	5.06	4.82	4.65	4.50	4.39	4.30	4.22	4.16	4.05	3.98	3.86
13	**.05**	**4.67**	**3.80**	**3.41**	**3.18**	**3.02**	**2.92**	**2.84**	**2.77**	**2.72**	**2.67**	**2.63**	**2.60**	**2.55**	**2.51**	**2.46**
	.01	9.07	6.70	5.74	5.20	4.86	4.62	4.44	4.30	4.19	4.10	4.02	3.96	3.85	3.78	3.67
14	**.05**	**4.60**	**3.74**	**3.34**	**3.11**	**2.96**	**2.85**	**2.77**	**2.70**	**2.65**	**2.60**	**2.56**	**2.53**	**2.48**	**2.44**	**2.39**
	.01	8.86	6.51	5.56	5.03	4.69	4.46	4.28	4.14	4.03	3.94	3.86	3.80	3.70	3.62	3.51
15	**.05**	**4.54**	**3.68**	**3.29**	**3.06**	**2.90**	**2.79**	**2.70**	**2.64**	**2.59**	**2.55**	**2.51**	**2.48**	**2.43**	**2.39**	**2.33**
	.01	8.68	6.36	5.42	4.89	4.56	4.32	4.14	4.00	3.89	3.80	3.73	3.67	3.56	3.48	3.36
16	**.05**	**4.49**	**3.63**	**3.24**	**3.01**	**2.85**	**2.74**	**2.66**	**2.59**	**2.54**	**2.49**	**2.45**	**2.42**	**2.37**	**2.33**	**2.28**
	.01	8.53	6.23	5.29	4.77	4.44	4.20	4.03	3.89	3.78	3.69	3.61	3.55	3.45	3.37	3.25

Table 5 (cont.) Critical Values of F: The F-Tables

Degrees of freedom within groups (degrees of freedom in denominator of F ratio)		*Degrees of freedom between groups (degrees of freedom in numerator of F ratio)*														
	α	1	2	3	4	5	6	7	8	9	10	11	12	14	16	20
17	.05	4.45	3.59	3.20	2.96	2.81	2.70	2.62	2.55	2.50	2.45	2.41	2.38	2.33	2.29	2.23
	.01	8.40	6.11	5.18	4.67	4.34	4.10	3.93	3.79	3.68	3.59	3.52	3.45	3.35	3.27	3.16
18	.05	4.41	3.55	3.16	2.93	2.77	2.66	2.58	2.51	2.46	2.41	2.37	2.34	2.29	2.25	2.19
	.01	8.28	6.01	5.09	4.58	4.25	4.01	3.85	3.71	3.60	3.51	3.44	3.37	3.27	3.19	3.07
19	.05	4.38	3.52	3.13	2.90	2.74	2.63	2.55	2.48	2.43	2.38	2.34	2.31	2.26	2.21	2.15
	.01	8.18	5.93	5.01	4.50	4.17	3.94	3.77	3.63	3.52	3.43	3.36	3.30	3.19	3.12	3.00
20	.05	4.35	3.49	3.10	2.87	2.71	2.60	2.52	2.45	2.40	2.35	2.31	2.28	2.23	2.18	2.12
	.01	8.10	5.85	4.94	4.43	4.10	3.87	3.71	3.56	3.45	3.37	3.30	3.23	3.13	3.05	2.94
21	.05	4.32	3.47	3.07	2.84	2.68	2.57	2.49	2.42	2.37	2.32	2.28	2.25	2.20	2.15	2.09
	.01	8.02	5.78	4.87	4.37	4.04	3.81	3.65	3.51	3.40	3.31	3.24	3.17	3.07	2.99	2.88
22	.05	4.30	3.44	3.05	2.82	2.66	2.55	2.47	2.40	2.35	2.30	2.26	2.23	2.18	2.13	2.07
	.01	7.94	5.72	4.82	4.31	3.99	3.76	3.59	3.45	3.35	3.26	3.18	3.12	3.02	2.94	2.83
23	.05	4.28	3.42	3.03	2.80	2.64	2.53	2.45	2.38	2.32	2.28	2.24	2.20	2.14	2.10	2.04
	.01	7.88	5.66	4.76	4.26	3.94	3.71	3.54	3.41	3.30	3.21	3.14	3.07	2.97	2.89	2.78
24	.05	4.26	3.40	3.01	2.78	2.62	2.51	2.43	2.36	2.30	2.26	2.22	2.18	2.13	2.09	2.02
	.01	7.82	5.61	4.72	4.22	3.90	3.67	3.50	3.36	3.25	3.17	3.09	3.03	2.93	2.85	2.74
25	.05	4.24	3.38	2.99	2.76	2.60	2.49	2.41	2.34	2.28	2.24	2.20	2.16	2.11	2.06	2.00
	.01	7.77	5.57	4.68	4.18	3.86	3.63	3.46	3.32	3.21	3.13	3.05	2.99	2.89	2.81	2.70
26	.05	4.22	3.37	2.98	2.74	2.59	2.47	2.39	2.32	2.27	2.22	2.18	2.15	2.10	2.05	1.99
	.01	7.72	5.53	4.64	4.14	3.82	3.59	3.42	3.29	3.17	3.09	3.02	2.96	2.86	2.77	2.66
27	.05	4.21	3.35	2.96	2.73	2.57	2.46	2.37	2.30	2.25	2.20	2.16	2.13	2.08	2.03	1.97
	.01	7.68	5.49	4.60	4.11	3.79	3.56	3.39	3.26	3.14	3.06	2.98	2.93	2.83	2.74	2.63
28	.05	4.20	3.34	2.95	2.71	2.56	2.44	2.36	2.29	2.24	2.19	2.15	2.12	2.06	2.02	1.96
	.01	7.64	5.45	4.57	4.07	3.76	3.53	3.36	3.23	3.11	3.03	2.95	2.90	2.80	2.71	2.60
29	.05	4.18	3.33	2.93	2.70	2.54	2.43	2.35	2.28	2.22	2.18	2.14	2.10	2.05	2.00	1.94
	.01	7.60	5.42	4.54	4.04	3.73	3.50	3.33	3.20	3.08	3.00	2.92	2.87	2.77	2.68	2.57
30	.05	4.17	3.32	2.92	2.69	2.53	2.42	2.34	2.27	2.21	2.16	2.12	2.09	2.04	1.99	1.93
	.01	7.56	5.39	4.51	4.02	3.70	3.47	3.30	3.17	3.06	2.98	2.90	2.84	2.74	2.66	2.55
32	.05	4.15	3.30	2.90	2.67	2.51	2.40	2.32	2.25	2.19	2.14	2.10	2.07	2.02	1.97	1.91
	.01	7.50	5.34	4.46	3.97	3.66	3.42	3.25	3.12	3.01	2.94	2.86	2.80	2.70	2.62	2.51
34	.05	4.13	3.28	2.88	2.65	2.49	2.38	2.30	2.23	2.17	2.12	2.08	2.05	2.00	1.95	1.89
	.01	7.44	5.29	4.42	3.93	3.61	3.38	3.21	3.08	2.97	2.89	2.82	2.76	2.66	2.58	2.47
36	.05	4.11	3.26	2.86	2.63	2.48	2.36	2.28	2.21	2.15	2.10	2.06	2.03	1.98	1.93	1.87
	.01	7.39	5.25	4.38	3.89	3.58	3.35	3.18	3.04	2.94	2.86	2.78	2.72	2.62	2.54	2.43
38	.05	4.10	3.25	2.85	2.62	2.46	2.35	2.26	2.19	2.14	2.09	2.05	2.02	1.96	1.92	1.85
	.01	7.35	5.21	4.34	3.86	3.54	3.32	3.15	3.02	2.91	2.82	2.75	2.69	2.59	2.51	2.40
40	.05	4.08	3.23	2.84	2.61	2.45	2.34	2.25	2.18	2.12	2.07	2.04	2.00	1.95	1.90	1.84
	.01	7.31	5.18	4.31	3.83	3.51	3.29	3.12	2.99	2.88	2.80	2.73	2.66	2.56	2.49	2.37
42	.05	4.07	3.22	2.83	2.59	2.44	2.32	2.24	2.17	2.11	2.06	2.02	1.99	1.94	1.89	1.82
	.01	7.27	5.15	4.29	3.80	3.49	3.26	3.10	2.96	2.86	2.77	2.70	2.64	2.54	2.46	2.35

Table 5 (cont.) Critical Values of *F*: The *F*-Tables

Degrees of freedom within groups (degrees of freedom in denominator of F ratio)	α	*Degrees of freedom between groups (degrees of freedom in numerator of F ratio)* 1	2	3	4	5	6	7	8	9	10	11	12	14	16	20
44	**.05**	**4.06**	**3.21**	**2.82**	**2.58**	**2.43**	**2.31**	**2.23**	**2.16**	**2.10**	**2.05**	**2.01**	**1.98**	**1.92**	**1.88**	**1.81**
	.01	7.24	5.12	4.26	3.78	3.46	3.24	3.07	2.94	2.84	2.75	2.68	2.62	2.52	2.44	2.32
46	**.05**	**4.05**	**3.20**	**2.81**	**2.57**	**2.42**	**2.30**	**2.22**	**2.14**	**2.09**	**2.04**	**2.00**	**1.97**	**1.91**	**1.87**	**1.80**
	.01	7.21	5.10	4.24	3.76	3.44	3.22	3.05	2.92	2.82	2.73	2.66	2.60	2.50	2.42	2.30
48	**.05**	**4.04**	**3.19**	**2.80**	**2.56**	**2.41**	**2.30**	**2.21**	**2.14**	**2.08**	**2.03**	**1.99**	**1.96**	**1.90**	**1.86**	**1.79**
	.01	7.19	5.08	4.22	3.74	3.42	3.20	3.04	2.90	2.80	2.71	2.64	2.58	2.48	2.40	2.28
50	**.05**	**4.03**	**3.18**	**2.79**	**2.56**	**2.40**	**2.29**	**2.20**	**2.13**	**2.07**	**2.02**	**1.98**	**1.95**	**1.90**	**1.85**	**1.78**
	.01	7.17	5.06	4.20	3.72	3.41	3.18	3.02	2.88	2.78	2.70	2.62	2.56	2.46	2.39	2.26
55	**.05**	**4.02**	**3.17**	**2.78**	**2.54**	**2.38**	**2.27**	**2.18**	**2.11**	**2.05**	**2.00**	**1.97**	**1.93**	**1.88**	**1.83**	**1.76**
	.01	7.12	5.01	4.16	3.68	3.37	3.15	2.98	2.85	2.75	2.66	2.59	2.53	2.43	2.35	2.23
60	**.05**	**4.00**	**3.15**	**2.76**	**2.52**	**2.37**	**2.25**	**2.17**	**2.10**	**2.04**	**1.99**	**1.95**	**1.92**	**1.86**	**1.81**	**1.75**
	.01	7.08	4.98	4.13	3.65	3.34	3.12	2.95	2.82	2.72	2.63	2.56	2.50	2.40	2.32	2.20
65	**.05**	**3.99**	**3.14**	**2.75**	**2.51**	**2.36**	**2.24**	**2.15**	**2.08**	**2.02**	**1.98**	**1.94**	**1.90**	**1.85**	**1.80**	**1.73**
	.01	7.04	4.95	4.10	3.62	3.31	3.09	2.93	2.79	2.70	2.61	2.54	2.47	2.37	2.30	2.18
70	**.05**	**3.98**	**3.13**	**2.74**	**2.50**	**2.35**	**2.23**	**2.14**	**2.07**	**2.01**	**1.97**	**1.93**	**1.89**	**1.84**	**1.79**	**1.72**
	.01	7.01	4.92	4.08	3.60	3.29	3.07	2.91	2.77	2.67	2.59	2.51	2.45	2.35	2.28	2.15
80	**.05**	**3.96**	**3.11**	**2.72**	**2.48**	**2.33**	**2.21**	**2.12**	**2.05**	**1.99**	**1.95**	**1.91**	**1.88**	**1.82**	**1.77**	**1.70**
	.01	6.96	4.88	4.04	3.56	3.25	3.04	2.87	2.74	2.64	2.55	2.48	2.41	2.32	2.24	2.11
100	**.05**	**3.94**	**3.09**	**2.70**	**2.46**	**2.30**	**2.19**	**2.10**	**2.03**	**1.97**	**1.92**	**1.88**	**1.85**	**1.79**	**1.75**	**1.68**
	.01	6.90	4.82	3.98	3.51	3.20	2.99	2.82	2.69	2.59	2.51	2.43	2.36	2.26	2.19	2.06
125	**.05**	**3.92**	**3.07**	**2.68**	**2.44**	**2.29**	**2.17**	**2.08**	**2.01**	**1.95**	**1.90**	**1.86**	**1.83**	**1.77**	**1.72**	**1.65**
	.01	6.84	4.78	3.94	3.47	3.17	2.95	2.79	2.65	2.56	2.47	2.40	2.33	2.23	2.15	2.03
150	**.05**	**3.91**	**3.06**	**2.67**	**2.43**	**2.27**	**2.16**	**2.07**	**2.00**	**1.94**	**1.89**	**1.85**	**1.82**	**1.76**	**1.71**	**1.64**
	.01	6.81	4.75	3.91	3.44	3.14	2.92	2.76	2.62	2.53	2.44	2.37	2.30	2.20	2.12	2.00
200	**.05**	**3.89**	**3.04**	**2.65**	**2.41**	**2.26**	**2.14**	**2.05**	**1.98**	**1.92**	**1.87**	**1.83**	**1.80**	**1.74**	**1.69**	**1.62**
	.01	6.76	4.71	3.88	3.41	3.11	2.90	2.73	2.60	2.50	2.41	2.34	2.28	2.17	2.09	1.97
400	**.05**	**3.86**	**3.02**	**2.62**	**2.39**	**2.23**	**2.12**	**2.03**	**1.96**	**1.90**	**1.85**	**1.81**	**1.78**	**1.72**	**1.67**	**1.60**
	.01	6.70	4.66	3.83	3.36	3.06	2.85	2.69	2.55	2.46	2.37	2.29	2.23	2.12	2.04	1.92
1000	**.05**	**3.85**	**3.00**	**2.61**	**2.38**	**2.22**	**2.10**	**2.02**	**1.95**	**1.89**	**1.84**	**1.80**	**1.76**	**1.70**	**1.65**	**1.58**
	.01	6.66	4.62	3.80	3.34	3.04	2.82	2.66	2.53	2.43	2.34	2.26	2.20	2.09	2.01	1.89
∞	**.05**	**3.84**	**2.99**	**2.60**	**2.37**	**2.21**	**2.09**	**2.01**	**1.94**	**1.88**	**1.83**	**1.79**	**1.75**	**1.69**	**1.64**	**1.57**
	.01	6.64	4.60	3.78	3.32	3.02	2.80	2.64	2.51	2.41	2.32	2.24	2.18	2.07	1.99	1.87

Table 6 Values of the Studentized Range Statistic, q_k

For a one-way ANOVA, or a comparison of the means from a main effect, the value of k is the number of means in the factor.

To compare the means from an interaction, find the appropriate design (or number of cell means) in the table below and obtain the adjusted value of k. Then use adjusted k as k to find the value of q_k.

Values of Adjusted k

Design of study	*Number of cell means in study*	*Adjusted value of k*
2 × 2	4	3
2 × 3	6	5
2 × 4	8	6
3 × 3	9	7
3 × 4	12	8
4 × 4	16	10
4 × 5	20	12

Values of q_k for α = .05 are **dark numbers** and for α = .01 are light numbers.

Degrees of freedom within groups (degrees of freedom in denominator of F ratio)		*k = number of means being compared*										
	α	*2*	*3*	*4*	*5*	*6*	*7*	*8*	*9*	*10*	*11*	*12*
1	**.05**	**18.00**	**27.00**	**32.80**	**37.10**	**40.40**	**43.10**	**45.40**	**47.40**	**49.10**	**50.60**	**52.00**
	.01	90.00	135.00	164.00	186.00	202.00	216.00	227.00	237.00	246.00	253.00	260.00
2	**.05**	**6.09**	**8.30**	**9.80**	**10.90**	**11.70**	**12.40**	**13.00**	**13.50**	**14.00**	**14.40**	**14.70**
	.01	14.00	19.00	22.30	24.70	26.60	28.20	29.50	30.70	31.70	32.60	33.40
3	**.05**	**4.50**	**5.91**	**6.82**	**7.50**	**8.04**	**8.48**	**8.85**	**9.18**	**9.46**	**9.72**	**9.95**
	.01	8.26	10.60	12.20	13.30	14.20	15.00	15.60	16.20	16.70	17.10	17.50
4	**.05**	**3.93**	**5.04**	**5.76**	**6.29**	**6.71**	**7.05**	**7.35**	**7.60**	**7.83**	**8.03**	**8.21**
	.01	6.51	8.12	9.17	9.96	10.60	11.10	11.50	11.90	12.30	12.60	12.80
5	**.05**	**3.64**	**4.60**	**5.22**	**5.67**	**6.03**	**6.33**	**6.58**	**6.80**	**6.99**	**7.17**	**7.32**
	.01	5.70	6.97	7.80	8.42	8.91	9.32	9.67	9.97	10.20	10.50	10.70
6	**.05**	**3.46**	**4.34**	**4.90**	**5.31**	**5.63**	**5.89**	**6.12**	**6.32**	**6.49**	**6.65**	**6.79**
	.01	5.24	6.33	7.03	7.56	7.97	8.32	8.61	8.87	9.10	9.30	9.49
7	**.05**	**3.34**	**4.16**	**4.69**	**5.06**	**5.36**	**5.61**	**5.82**	**6.00**	**6.16**	**6.30**	**6.43**
	.01	4.95	5.92	6.54	7.01	7.37	7.68	7.94	8.17	8.37	8.55	8.71
8	**.05**	**3.26**	**4.04**	**4.53**	**4.89**	**5.17**	**5.40**	**5.60**	**5.77**	**5.92**	**6.05**	**6.18**
	.01	4.74	5.63	6.20	6.63	6.96	7.24	7.47	7.68	7.87	8.03	8.18
9	**.05**	**3.20**	**3.95**	**4.42**	**4.76**	**5.02**	**5.24**	**5.43**	**5.60**	**5.74**	**5.87**	**5.98**
	.01	4.60	5.43	5.96	6.35	6.66	6.91	7.13	7.32	7.49	7.65	7.78

Table 6 (cont.) Values of the Studentized Range Statistic, q_k

Degrees of freedom within groups (degrees of freedom in denominator of F ratio)	α	k = number of means being compared										
		2	3	4	5	6	7	8	9	10	11	12
10	**.05**	**3.15**	**3.88**	**4.33**	**4.65**	**4.91**	**5.12**	**5.30**	**5.46**	**5.60**	**5.72**	**5.83**
	.01	4.48	5.27	5.77	6.14	6.43	6.67	6.87	7.05	7.21	7.36	7.48
11	**.05**	**3.11**	**3.82**	**4.26**	**4.57**	**4.82**	**5.03**	**5.20**	**5.35**	**5.49**	**5.61**	**5.71**
	.01	4.39	5.14	5.62	5.97	6.25	6.48	6.67	6.84	6.99	7.13	7.26
12	**.05**	**3.08**	**3.77**	**4.20**	**4.51**	**4.75**	**4.95**	**5.12**	**5.27**	**5.40**	**5.51**	**5.62**
	.01	4.32	5.04	5.50	5.84	6.10	6.32	6.51	6.67	6.81	6.94	7.06
13	**.05**	**3.06**	**3.73**	**4.15**	**4.45**	**4.69**	**4.88**	**5.05**	**5.19**	**5.32**	**5.43**	**5.53**
	.01	4.26	4.96	5.40	5.73	5.98	6.19	6.37	6.53	6.67	6.79	6.90
14	**.05**	**3.03**	**3.70**	**4.11**	**4.41**	**4.64**	**4.83**	**4.99**	**5.13**	**5.25**	**5.36**	**5.46**
	.01	4.21	4.89	5.32	5.63	5.88	6.08	6.26	6.41	6.54	6.66	6.77
16	**.05**	**3.00**	**3.65**	**4.05**	**4.33**	**4.56**	**4.74**	**4.90**	**5.03**	**5.15**	**5.26**	**5.35**
	.01	4.13	4.78	5.19	5.49	5.72	5.92	6.08	6.22	6.35	6.46	6.56
18	**.05**	**2.97**	**3.61**	**4.00**	**4.28**	**4.49**	**4.67**	**4.82**	**4.96**	**5.07**	**5.17**	**5.27**
	.01	4.07	4.70	5.09	5.38	5.60	5.79	5.94	6.08	6.20	6.31	6.41
20	**.05**	**2.95**	**3.58**	**3.96**	**4.23**	**4.45**	**4.62**	**4.77**	**4.90**	**5.01**	**5.11**	**5.20**
	.01	4.02	4.64	5.02	5.29	5.51	5.69	5.84	5.97	6.09	6.19	6.29
24	**.05**	**2.92**	**3.53**	**3.90**	**4.17**	**4.37**	**4.54**	**4.68**	**4.81**	**4.92**	**5.01**	**5.10**
	.01	3.96	4.54	4.91	5.17	5.37	5.54	5.69	5.81	5.92	6.02	6.11
30	**.05**	**2.89**	**3.49**	**3.84**	**4.10**	**4.30**	**4.46**	**4.60**	**4.72**	**4.83**	**4.92**	**5.00**
	.01	3.89	4.45	4.80	5.05	5.24	5.40	5.54	5.56	5.76	5.85	5.93
40	**.05**	**2.86**	**3.44**	**3.79**	**4.04**	**4.23**	**4.39**	**4.52**	**4.63**	**4.74**	**4.82**	**4.91**
	.01	3.82	4.37	4.70	4.93	5.11	5.27	5.39	5.50	5.60	5.69	5.77
60	**.05**	**2.83**	**3.40**	**3.74**	**3.98**	**4.16**	**4.31**	**4.44**	**4.55**	**4.65**	**4.73**	**4.81**
	.01	3.76	4.28	4.60	4.82	4.99	5.13	5.25	5.36	5.45	5.53	5.60
120	**.05**	**2.80**	**3.36**	**3.69**	**3.92**	**4.10**	**4.24**	**4.36**	**4.48**	**4.56**	**4.64**	**4.72**
	.01	3.70	4.20	4.50	4.71	4.87	5.01	5.12	5.21	5.30	5.38	5.44
∞	**.05**	**2.77**	**3.31**	**3.63**	**3.86**	**4.03**	**4.17**	**4.29**	**4.39**	**4.47**	**4.55**	**4.62**
	.01	3.64	4.12	4.40	4.60	4.76	4.88	4.99	5.08	5.16	5.23	5.29

From B. J. Winer, *Statistical Principles in Experimental Design,* McGraw-Hill, 1962; abridged from H. L. Harter, D. S. Clemm, and E. H. Guthrie, The probability integrals of the range and of the studentized range, WADC Tech. Rep. 58–484, Vol. 2, 1959, Wright Air Development Center, Table II.2, pp. 243–281. Reproduced with permission of McGraw-Hill, Inc.

Table 7 Critical Values of the F_{max} Test

Critical values for α = .05 are **dark numbers** and for α = .01 are light numbers.

Note: n = number of scores in each condition or cell.

		k = number of samples in the study										
n − 1	*α*	*2*	*3*	*4*	*5*	*6*	*7*	*8*	*9*	*10*	*11*	*12*
4	**.05**	**9.60**	**15.50**	**20.60**	**25.20**	**29.50**	**33.60**	**37.50**	**41.40**	**44.60**	**48.00**	**51.40**
	.01	23.20	37.00	49.00	59.00	69.00	79.00	89.00	97.00	106.00	113.00	120.00
5	**.05**	**7.15**	**10.80**	**13.70**	**16.30**	**18.70**	**20.80**	**22.90**	**24.70**	**26.50**	**28.20**	**29.90**
	.01	14.90	22.00	28.00	33.00	38.00	42.00	46.00	50.00	54.00	57.00	60.00
6	**.05**	**5.82**	**8.38**	**10.40**	**12.10**	**13.70**	**15.00**	**16.30**	**17.50**	**18.60**	**19.70**	**20.70**
	.01	11.10	15.50	19.10	22.00	25.00	27.00	30.00	32.00	34.00	36.00	37.00
7	**.05**	**4.99**	**6.94**	**8.44**	**9.70**	**10.80**	**11.80**	**12.70**	**13.50**	**14.30**	**15.10**	**15.80**
	.01	8.89	12.10	14.50	16.50	18.40	20.00	22.00	23.00	24.00	26.00	27.00
8	**.05**	**4.43**	**6.00**	**7.18**	**8.12**	**9.03**	**9.78**	**10.50**	**11.10**	**11.70**	**12.20**	**12.70**
	.01	7.50	9.90	11.70	13.20	14.50	15.80	16.90	17.90	18.90	19.80	21.00
9	**.05**	**4.03**	**5.34**	**6.31**	**7.11**	**7.80**	**8.41**	**8.95**	**9.45**	**9.91**	**10.30**	**10.70**
	.01	6.54	8.50	9.90	11.10	12.10	13.10	13.90	14.70	15.30	16.00	16.60
10	**.05**	**3.72**	**4.85**	**5.67**	**6.34**	**6.92**	**7.42**	**7.87**	**8.28**	**8.66**	**9.01**	**9.34**
	.01	5.85	7.40	8.60	9.60	10.40	11.10	11.80	12.40	12.90	13.40	13.90
12	**.05**	**3.28**	**4.16**	**4.79**	**5.30**	**5.72**	**6.09**	**6.42**	**6.72**	**7.00**	**7.25**	**7.48**
	.01	4.91	6.10	6.90	7.60	8.20	8.70	9.10	9.50	9.90	10.20	10.60
15	**.05**	**2.86**	**3.54**	**4.01**	**4.37**	**4.68**	**4.95**	**5.19**	**5.40**	**5.59**	**5.77**	**5.93**
	.01	4.07	4.90	5.50	6.00	6.40	6.70	7.10	7.30	7.50	7.80	8.00
20	**.05**	**2.46**	**2.95**	**3.29**	**3.54**	**3.76**	**3.94**	**4.10**	**4.24**	**4.37**	**4.49**	**4.59**
	.01	3.32	3.80	4.30	4.60	4.90	5.10	5.30	5.50	5.60	5.80	5.90
30	**.05**	**2.07**	**2.40**	**2.61**	**2.78**	**2.91**	**3.02**	**3.12**	**3.21**	**3.29**	**3.36**	**3.39**
	.01	2.63	3.00	3.30	3.40	3.60	3.70	3.80	3.90	4.00	4.10	4.20
60	**.05**	**1.67**	**1.85**	**1.96**	**2.04**	**2.11**	**2.17**	**2.22**	**2.26**	**2.30**	**2.33**	**2.36**
	.01	1.96	2.20	2.30	2.40	2.40	2.50	2.50	2.60	2.60	2.70	2.70

From Table 31 of E. Pearson and H. Hartley, *Biometrika Tables for Statisticians,* Vol 1, 3rd ed. Cambridge: Cambridge University Press, 1966. Reprinted with permission of the Biometrika trustees.

Table 8 Critical Values of Chi Square: The χ^2-Tables

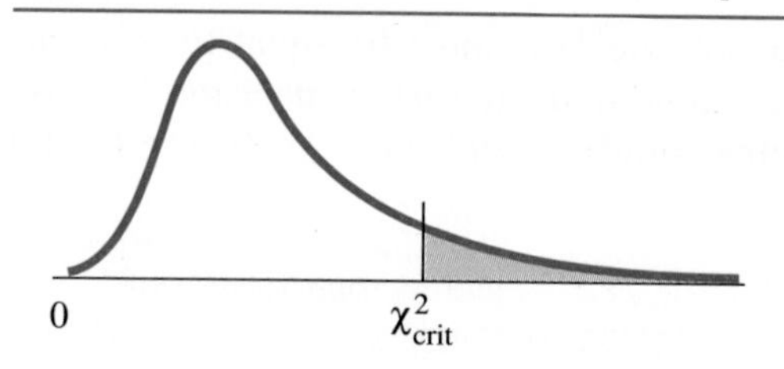

	Level of significance	
df	α = .05	α = .01
1	3.84	6.64
2	5.99	9.21
3	7.81	11.34
4	9.49	13.28
5	11.07	15.09
6	12.59	16.81
7	14.07	18.48
8	15.51	20.09
9	16.92	21.67
10	18.31	23.21
11	19.68	24.72
12	21.03	26.22
13	22.36	27.69
14	23.68	29.14
15	25.00	30.58
16	26.30	32.00
17	27.59	33.41
18	28.87	34.80
19	30.14	36.19
20	31.41	37.57
21	32.67	38.93
22	33.92	40.29
23	35.17	41.64
24	36.42	42.98
25	37.65	44.31
26	38.88	45.64
27	40.11	46.96
28	41.34	48.28
29	42.56	49.59
30	43.77	50.89
40	55.76	63.69
50	67.50	76.15
60	79.08	88.38
70	90.53	100.42

From Table IV of R. A. Fisher and F. Yates, *Statistical Tables for Biological, Agricultural and Medical Research,* 6th ed. London: Longman Group Ltd., 1974 (previously published by Oliver and Boyd Ltd., Edinburgh).

Table 9 Critical Values of the Mann-Whitney *U*

To be significant, the U_{obt} must be equal to or be *less than* the critical value. (Dashes in the table indicate that no decision is possible.) Critical values for $\alpha = .05$ are **dark numbers** and for $\alpha = .01$ are light numbers.

Two-Tailed Test

n_2 (no. of scores in Group 2)	α	n_1 (no. of scores in Group 1) 1	2	3	4	5	6	7	8	9
1	**.05**	—	—	—	—	—	—	—	—	—
	.01	—	—	—	—	—	—	—	—	—
2	**.05**	—	—	—	—	—	—	—	**0**	**0**
	.01	—	—	—	—	—	—	—	—	—
3	**.05**	—	—	—	—	**0**	**1**	**1**	**2**	**2**
	.01	—	—	—	—	—	—	—	—	0
4	**.05**	—	—	—	**0**	**1**	**2**	**3**	**4**	**4**
	.01	—	—	—	—	—	0	0	1	1
5	**.05**	—	—	**0**	**1**	**2**	**3**	**5**	**6**	**7**
	.01	—	—	—	—	0	1	1	2	3
6	**.05**	—	—	**1**	**2**	**3**	**5**	**6**	**8**	**10**
	.01	—	—	—	0	1	2	3	4	5
7	**.05**	—	—	**1**	**3**	**5**	**6**	**8**	**10**	**12**
	.01	—	—	—	0	1	3	4	6	7
8	**.05**	—	**0**	**2**	**4**	**6**	**8**	**10**	**13**	**15**
	.01	—	—	—	1	2	4	6	7	9
9	**.05**	—	**0**	**2**	**4**	**7**	**10**	**12**	**15**	**17**
	.01	—	—	0	1	3	5	7	9	11
10	**.05**	—	**0**	**3**	**5**	**8**	**11**	**14**	**17**	**20**
	.01	—	—	0	2	4	6	9	11	13
11	**.05**	—	**0**	**3**	**6**	**9**	**13**	**16**	**19**	**23**
	.01	—	—	0	2	5	7	10	13	16
12	**.05**	—	**1**	**4**	**7**	**11**	**14**	**18**	**22**	**26**
	.01	—	—	1	3	6	9	12	15	18
13	**.05**	—	**1**	**4**	**8**	**12**	**16**	**20**	**24**	**28**
	.01	—	—	1	3	7	10	13	17	20
14	**.05**	—	**1**	**5**	**9**	**13**	**17**	**22**	**26**	**31**
	.01	—	—	1	4	7	11	15	18	22
15	**.05**	—	**1**	**5**	**10**	**14**	**19**	**24**	**29**	**34**
	.01	—	—	2	5	8	12	16	20	24
16	**.05**	—	**1**	**6**	**11**	**15**	**21**	**26**	**31**	**37**
	.01	—	—	2	5	9	13	18	22	27
17	**.05**	—	**2**	**6**	**11**	**17**	**22**	**28**	**34**	**39**
	.01	—	—	2	6	10	15	19	24	29
18	**.05**	—	**2**	**7**	**12**	**18**	**24**	**30**	**36**	**42**
	.01	—	—	2	6	11	16	21	26	31
19	**.05**	—	**2**	**7**	**13**	**19**	**25**	**32**	**38**	**45**
	.01	—	0	3	7	12	17	22	28	33
20	**.05**	—	**2**	**8**	**13**	**20**	**27**	**34**	**41**	**48**
	.01	—	0	3	8	13	18	24	30	36

*n*₁ (no. of scores in Group 1)										
10	*11*	*12*	*13*	*14*	*15*	*16*	*17*	*18*	*19*	*20*
—	**—**	**—**	**—**	**—**	**—**	**—**	**—**	**—**	**—**	**—**
—	—	—	—	—	—	—	—	—	—	—
0	**0**	**1**	**1**	**1**	**1**	**1**	**2**	**2**	**2**	**2**
—	—	—	—	—	—	—	—	—	0	0
3	**3**	**4**	**4**	**5**	**5**	**6**	**6**	**7**	**7**	**8**
0	0	1	1	1	2	2	2	2	3	3
5	**6**	**7**	**8**	**9**	**10**	**11**	**11**	**12**	**13**	**13**
2	2	3	3	4	5	5	6	6	7	8
8	**9**	**11**	**12**	**13**	**14**	**15**	**17**	**18**	**19**	**20**
4	5	6	7	7	8	9	10	11	12	13
11	**13**	**14**	**16**	**17**	**19**	**21**	**22**	**24**	**25**	**27**
6	7	9	10	11	12	13	15	16	17	18
14	**16**	**18**	**20**	**22**	**24**	**26**	**28**	**30**	**32**	**34**
9	10	12	13	15	16	18	19	21	22	24
17	**19**	**22**	**24**	**26**	**29**	**31**	**34**	**36**	**38**	**41**
11	13	15	17	18	20	22	24	26	28	30
20	**23**	**26**	**28**	**31**	**34**	**37**	**39**	**42**	**45**	**48**
13	16	18	20	22	24	27	29	31	33	36
23	**26**	**29**	**33**	**36**	**39**	**42**	**45**	**48**	**52**	**55**
16	18	21	24	26	29	31	34	37	39	42
26	**30**	**33**	**37**	**40**	**44**	**47**	**51**	**55**	**58**	**62**
18	21	24	27	30	33	36	39	42	45	48
29	**33**	**37**	**41**	**45**	**49**	**53**	**57**	**61**	**65**	**69**
21	24	27	31	34	37	41	44	47	51	54
33	**37**	**41**	**45**	**50**	**54**	**59**	**63**	**67**	**72**	**76**
24	27	31	34	38	42	45	49	53	56	60
36	**40**	**45**	**50**	**55**	**59**	**64**	**67**	**74**	**78**	**83**
26	30	34	38	42	46	50	54	58	63	67
39	**44**	**49**	**54**	**59**	**64**	**70**	**75**	**80**	**85**	**90**
29	33	37	42	46	51	55	60	64	69	73
42	**47**	**53**	**59**	**64**	**70**	**75**	**81**	**86**	**92**	**98**
31	36	41	45	50	55	60	65	70	74	79
45	**51**	**57**	**63**	**67**	**75**	**81**	**87**	**93**	**99**	**105**
34	39	44	49	54	60	65	70	75	81	86
48	**55**	**61**	**67**	**74**	**80**	**86**	**93**	**99**	**106**	**112**
37	42	47	53	58	64	70	75	81	87	92
52	**58**	**65**	**72**	**78**	**85**	**92**	**99**	**106**	**113**	**119**
39	45	51	56	63	69	74	81	87	93	99
55	**62**	**69**	**76**	**83**	**90**	**98**	**105**	**112**	**119**	**127**
42	48	54	60	67	73	79	86	92	99	105

Table 9 (cont.) Critical Values of the Mann-Whitney U

One-Tailed Test

n_2 (no. of scores in Group 2)	α	n_1 (no. of scores in Group 1) 1	2	3	4	5	6	7	8	9
1	**.05**	—	—	—	—	—	—	—	—	—
	.01	—	—	—	—	—	—	—	—	—
2	**.05**	—	—	—	—	**0**	**0**	**0**	**1**	**1**
	.01	—	—	—	—	—	—	—	—	—
3	**.05**	—	—	**0**	**0**	**1**	**2**	**2**	**3**	**3**
	.01	—	—	—	—	—	—	0	0	1
4	**.05**	—	—	**0**	**1**	**2**	**3**	**4**	**5**	**6**
	.01	—	—	—	—	0	1	1	2	3
5	**.05**	—	**0**	**1**	**2**	**4**	**5**	**6**	**8**	**9**
	.01	—	—	—	0	1	2	3	4	5
6	**.05**	—	**0**	**2**	**3**	**5**	**7**	**8**	**10**	**12**
	.01	—	—	—	1	2	3	4	6	7
7	**.05**	—	**0**	**2**	**4**	**6**	**8**	**11**	**13**	**15**
	.01	—	—	0	1	3	4	6	7	9
8	**.05**	—	**1**	**3**	**5**	**8**	**10**	**13**	**15**	**18**
	.01	—	—	0	2	4	6	7	9	11
9	**.05**	—	**1**	**3**	**6**	**9**	**12**	**15**	**18**	**21**
	.01	—	—	1	3	5	7	9	11	14
10	**.05**	—	**1**	**4**	**7**	**11**	**14**	**17**	**20**	**24**
	.01	—	—	1	3	6	8	11	13	16
11	**.05**	—	**1**	**5**	**8**	**12**	**16**	**19**	**23**	**27**
	.01	—	—	1	4	7	9	12	15	18
12	**.05**	—	**2**	**5**	**9**	**13**	**17**	**21**	**26**	**30**
	.01	—	—	2	5	8	11	14	17	21
13	**.05**	—	**2**	**6**	**10**	**15**	**19**	**24**	**28**	**33**
	.01	—	0	2	5	9	12	16	20	23
14	**.05**	—	**2**	**7**	**11**	**16**	**21**	**26**	**31**	**36**
	.01	—	0	2	6	10	13	17	22	26
15	**.05**	—	**3**	**7**	**12**	**18**	**23**	**28**	**33**	**39**
	.01	—	0	3	7	11	15	19	24	28
16	**.05**	—	**3**	**8**	**14**	**19**	**25**	**30**	**36**	**42**
	.01	—	0	3	7	12	16	21	26	31
17	**.05**	—	**3**	**9**	**15**	**20**	**26**	**33**	**39**	**45**
	.01	—	0	4	8	13	18	23	28	33
18	**.05**	—	**4**	**9**	**16**	**22**	**28**	**35**	**41**	**48**
	.01	—	0	4	9	14	19	24	30	36
19	**.05**	**0**	**4**	**10**	**17**	**23**	**30**	**37**	**44**	**51**
	.01	—	1	4	9	15	20	26	32	38
20	**.05**	**0**	**4**	**11**	**18**	**25**	**32**	**39**	**47**	**54**
	.01	—	1	5	10	16	22	28	34	40

n_1 *(no. of scores in Group 1)*										
10	*11*	*12*	*13*	*14*	*15*	*16*	*17*	*18*	*19*	*20*
—	**—**	**—**	**—**	**—**	**—**	**—**	**—**	**—**	**0**	**0**
—	—	—	—	—	—	—	—	—	—	—
1	**1**	**2**	**2**	**2**	**3**	**3**	**3**	**4**	**4**	**4**
—	—	—	0	0	0	0	0	0	1	1
4	**5**	**5**	**6**	**7**	**7**	**8**	**9**	**9**	**10**	**11**
1	1	2	2	2	3	3	4	4	4	5
7	**8**	**9**	**10**	**11**	**12**	**14**	**15**	**16**	**17**	**18**
3	4	5	5	6	7	7	8	9	9	10
11	**12**	**13**	**15**	**16**	**18**	**19**	**20**	**22**	**23**	**25**
6	7	8	9	10	11	12	13	14	15	16
14	**16**	**17**	**19**	**21**	**23**	**25**	**26**	**28**	**30**	**32**
8	9	11	12	13	15	16	18	19	20	22
17	**19**	**21**	**24**	**26**	**28**	**30**	**33**	**35**	**37**	**39**
11	12	14	16	17	19	21	23	24	26	28
20	**23**	**26**	**28**	**31**	**33**	**36**	**39**	**41**	**44**	**47**
13	15	17	20	22	24	26	28	30	32	34
24	**27**	**30**	**33**	**36**	**39**	**42**	**45**	**48**	**51**	**54**
16	18	21	23	26	28	31	33	36	38	40
27	**31**	**34**	**37**	**41**	**44**	**48**	**51**	**55**	**58**	**62**
19	22	24	27	30	33	36	38	41	44	47
31	**34**	**38**	**42**	**46**	**50**	**54**	**57**	**61**	**65**	**69**
22	25	28	31	34	37	41	44	47	50	53
34	**38**	**42**	**47**	**51**	**55**	**60**	**64**	**68**	**72**	**77**
24	28	31	35	38	42	46	49	53	56	60
37	**42**	**47**	**51**	**56**	**61**	**65**	**70**	**75**	**80**	**84**
27	31	35	39	43	47	51	55	59	63	67
41	**46**	**51**	**56**	**61**	**66**	**71**	**77**	**82**	**87**	**92**
30	34	38	43	47	51	56	60	65	69	73
44	**50**	**55**	**61**	**66**	**72**	**77**	**83**	**88**	**94**	**100**
33	37	42	47	51	56	61	66	70	75	80
48	**54**	**60**	**65**	**71**	**77**	**83**	**89**	**95**	**101**	**107**
36	41	46	51	56	61	66	71	76	82	87
51	**57**	**64**	**70**	**77**	**83**	**89**	**96**	**102**	**109**	**115**
38	44	49	55	60	66	71	77	82	88	93
55	**61**	**68**	**75**	**82**	**88**	**95**	**102**	**109**	**116**	**123**
41	47	53	59	65	70	76	82	88	94	100
58	**65**	**72**	**80**	**87**	**94**	**101**	**109**	**116**	**123**	**130**
44	50	56	63	69	75	82	88	94	101	107
62	**69**	**77**	**84**	**92**	**100**	**107**	**115**	**123**	**130**	**138**
47	53	60	67	73	80	87	93	100	107	114

From the *Bulletin of the Institute of Educational Research,* 1, No. 2, Indiana University, with permission of the publishers.

Table 10 Critical Values of the Wilcoxon T

To be significant, the T_{obt} must be equal to or *less than* the critical value. (Dashes in the table indicate that no decision can be made.) In the table, N is the number of nonzero differences that occurred when T_{obt} was calculated.

Two-Tailed Test

N	$\alpha = .05$	$\alpha = .01$	N	$\alpha = .05$	$\alpha = .01$
5	—	—	28	116	91
6	0	—	29	126	100
7	2	—	30	137	109
8	3	0	31	147	118
9	5	1	32	159	128
10	8	3	33	170	138
11	10	5	34	182	148
12	13	7	35	195	159
13	17	9	36	208	171
14	21	12	37	221	182
15	25	15	38	235	194
16	29	19	39	249	207
17	34	23	40	264	220
18	40	27	41	279	233
19	46	32	42	294	247
20	52	37	43	310	261
21	58	42	44	327	276
22	65	48	45	343	291
23	73	54	46	361	307
24	81	61	47	378	322
25	89	68	48	396	339
26	98	75	49	415	355
27	107	83	50	434	373

Table 10 (cont.) Critical Values of the Wilcoxon T

One-Tailed Test

N	$\alpha = .05$	$\alpha = .01$	N	$\alpha = .05$	$\alpha = .01$
5	0	—	28	130	101
6	2	—	29	140	110
7	3	0	30	151	120
8	5	1	31	163	130
9	8	3	32	175	140
10	10	5	33	187	151
11	13	7	34	200	162
12	17	9	35	213	173
13	21	12	36	227	185
14	25	15	37	241	198
15	30	19	38	256	211
16	35	23	39	271	224
17	41	27	40	286	238
18	47	32	41	302	252
19	53	37	42	319	266
20	60	43	43	336	281
21	67	49	44	353	296
22	75	55	45	371	312
23	83	62	46	389	328
24	91	69	47	407	345
25	100	76	48	426	362
26	110	84	49	446	379
27	119	92	50	466	397

From F. Wilcoxon and R. A. Wilcox, *Some Rapid Approximate Statistical Procedures.* New York: Lederle Laboratories, 1964. Reproduced with the permission of the American Cyanamid Company.

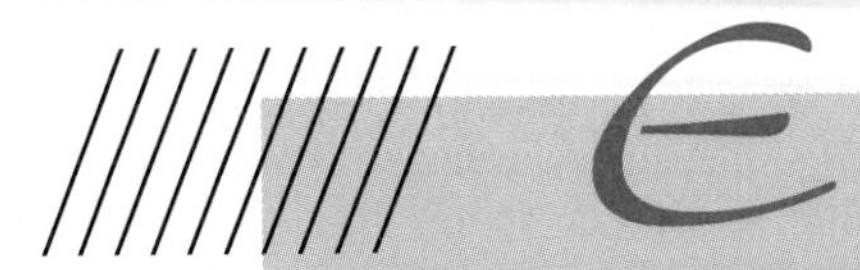

Answers to Odd-Numbered Practice Problems

Chapter 1

1. Researchers need to learn statistics to conduct their own research and to understand the research of others.
3. *Statistical notation* refers to the symbols we use for the mathematical operations we perform, for the order in which we perform operations, and for the answers we obtain.
5. *a.* To two more decimal places than are found in the original scores.
   b. If the number in the third decimal place is 5 or greater than 5, round up the number in the second decimal place. If the number in the third decimal place is less than 5, round down by not changing the number in the second decimal place.
7. Perform squaring and taking a square root first, then multiplication and division, and then addition and subtraction.
9. A proportion is a decimal number that indicates a fraction of the total. To transform a number to a proportion, divide the number by the total.
11. *a.* 5/15 = .33 *b.* 10/50 = .20
    c. 1/1000 = .001
13. To transform a percentage to a proportion, divide the percentage by 100.
15. *a.* 13.75 *b.* 10.04 *c.* 10.05 *d.* .08 *e.* 1.00
17. $Q = (8 + -2)(64+4) = (6)(68) = 408$
19. $D = (-3.25)(3) = -9.75$
21. *a.* (.60)40 = 24; (.60)35 = 21; (.60)60 = 36
    b. (.60)135 = 81
    c. 115/135 = .85, multiplied by 100 is 85%
23. A data point is the "dot" placed on a graph when plotting a pair of *X* and *Y* scores.

Chapter 2

1. A relationship exists between two variables when certain scores on one variable are associated with certain scores on the other variable, and as the scores on one variable change, the scores on the other variable tend to change in a consistent fashion.
3. Because of individual differences, all subjects who obtain the same *X* score will not obtain the same *Y* score, so there will be inconsistency in the association of *X* and *Y* scores.
5. The independent variable is the overall causal variable the researcher is interested in; the conditions are the specific amounts or categories of the independent variable under which subjects are tested.
7. Random sampling is a method of selecting a sample in which every score in the population has an equal chance of being selected for the sample, and every sample has an equal chance of being selected.
9. Descriptive statistics are used to organize, summarize, and describe the characteristics of a sample of scores.
11. *a.* A statistic is a number that describes a characteristic of a sample of scores. A parameter is a number that describes a characteristic of a population of scores.
    b. The symbols for statistics are letters from the English alphabet. The symbols for population parameters are letters from the Greek alphabet.
13. The problem is that a statistical analysis cannot prove anything.
15. Even though she has a random sample, it may not be representative of all college students. Perhaps,

by chance, she only selected those few students who prefer sauerkraut juice.

17. The design of the study and the scale of measurement used to measure the variables.

19. Samples A and D.

21. A relationship is indicated in Study A and Study C. In each, as the scores on one variable change, the scores on the other variable change in a consistent fashion. This does not occur in Study B, so no relationship is indicated.

23. Because each relationship suggests there is something about the way nature operates so that as the amount of the *X* variable changes, the amount of the *Y* variable that occurs also changes in a consistent fashion.

25.

Variable	*Qualitative or Quantitative*	*Continuous, Discrete or Dichotomous*	*Type of Measurement Scale*
gender	qualitative	dichotomous	nominal
academic major	qualitative	discrete	nominal
time	quantitative	continuous	interval
restaurant ratings	quantitative	discrete	ordinal
speed	quantitative	continuous	ratio
money	quantitative	discrete	ratio
position in line	quantitative	discrete	ordinal
change in weight	quantitative	continuous	interval

Chapter 3

1. *a.* *N* is the number of scores in a sample.
b. *f* is frequency, the number of times a score occurs.
c. *rel. f* is relative frequency, the proportion of time certain scores occur.
d. *cf* is cumulative frequency, the number of times scores at or below a certain score occur.

3. Previously the graph showed the relationship in which as subjects' scores on the *X* variable change their scores on the *Y* variable also change. A frequency distribution shows the relationship in which as subjects' *X* scores change, the frequency of each score (shown on *Y*) also changes.

5. It means that the score is either a high or low extreme score relative to the middle scores in the distribution, and that it occurs with a relatively low frequency.

7. *a.* The middle IQ score has the highest frequency in a symmetrical distribution; the higher and lower scores have lower frequencies, and the highest and lowest scores have a relatively very low frequency.
b. The agility scores form a symmetrical distribution containing two distinct "humps" where there are two scores that occur more frequently than the surrounding scores.
c. The memory scores form an asymmetrical distribution in which there are some very infrequent, extremely low scores, but there are not correspondingly infrequent high scores.

9. *a.* A percentile is the percent of all scores at or below a particular score.
b. A percentile is the percent of all scores below a particular score.

11. *a.* Simple frequency is the number of times a score occurs in a sample; relative frequency is the proportion of time the score occurs.
b. Cumulative frequency is the number of times that scores at or below the score occur; percentile is the percent of the time that scores at or below the score occur.

13. *a.* 70, 72, 60, 85, 45.
b. Because .20 of the scores are below 60, it is at the 20th percentile.
c. Because .50 of the curve is to the left of 70, and .20 is to the left of 60, then .50 − .20, or .30, is between 60 and 70; so .30 of the students scored between 60 and 70.
d. With .25 of the scores between 80 and 70, and with .50 of the scores below 70, in total (.50 + .25) .75 of the scores are below 80; so it is at the 75th percentile.

15.

Score	*f*	*rel. f*	*cf*
53	1	.05	18
52	3	.17	17
51	2	.11	14
50	5	.28	12
49	4	.22	7
48	0	.00	3
47	3	.17	3

17. The score at the 50th percentile is:

$$\text{Score} = 49.5 + \left(\frac{9-7}{5}\right)(1) = 49.90$$

Chapter 4

1. It indicates where on a variable most of the scores in a distribution tend to be located.

3. The mode is the most frequently occurring score. It is used with scores from a nominal variable.

5. The mean is the average of the scores—the mathematical center of a distribution. It is used with symmetrical, unimodal distributions of interval or ratio scores.

7. Because, as the mathematical center of the distribution, the mean is not located near most of the scores.

9. *a.* $\Sigma_X = 638$, $N = 11$, $\overline{X} = 58$
b. The mode is 58.

11. $\Sigma X = 460$, $N = 20$, $\overline{X} = 23.00$

13. *a.* mean
b. median (these ratio scores are skewed)
c. mode (this is a nominal variable)
d. median (this is an ordinal variable)

15. Deviations convey (1) whether a score is above or below the mean and (2) how far the score is from the mean.

17. *a.* The subject with -5; it is farthest below the mean.
b. Again, the subject with -5; it is farthest into the tail where the lowest-frequency scores occur.
c. The subject with 0; this score equals the mean score, which, in a normal distribution, is the highest-frequency score.
d. The subject with $+3$; it is farthest above the mean score.

19. She is correct *unless* the variable is something on which it is undesirable to have a high score. In that case, being below the mean and having a negative deviation is best.

21. Mean errors do not change until there has been 5 hours of sleep deprivation. Mean errors then increase as a function of increasing sleep deprivation.

23. *a.* line graph; income on *Y* axis, age on *X* axis; find median income per age group (income is skewed).
b. bar graph; positive votes on *Y* axis, presence or absence of a wildlife refuge on *X* axis; find mean number of votes.
c. line graph; running speed on *Y* axis, amount of carbohydrates consumed on *X* axis; find mean running speed if normally distributed.
d. bar graph; alcohol abuse on *Y* axis, ethnic group on *X* axis; find mean rate of alcohol abuse per group.

Chapter 5

1. The distribution's shape, its central tendency, and its variability.

3. *a.* The range is the difference or distance between the highest and lowest scores in the distribution.
b. Because it includes only the most extreme scores (which are often the least frequent scores). Thus it does not summarize most of the differences in a distribution.

5. *a.* The variance is the average of the squared deviations around the mean.
b. The variance equals the squared standard deviation, and the standard deviation equals the square root of the variance.

7. Because a sample value too often tends to be smaller than the population value. Thus, in calculating the unbiased estimate of the population, our final division involves the quantity $N-1$, a slightly smaller quantity that results in a slightly larger, generally more accurate estimate.

9. *a.* Range $= 9 - 0 = 9$, so the error scores spanned 9 different scores.
b. Because $\Sigma_X = 100$, $\Sigma_X{}^2 = 668$, and $N = 20$, $S_X^2 = (668 - 500)/20 = 8.40$. The average squared deviation of error scores from the mean error score is 8.40.
c. Because $S_X = \sqrt{8.4} = 2.90$, the average amount individual error scores differed from the mean error score is 2.90.

11. About 160 people. The score of 2.90 is one standard deviation below the mean of 5 $(5 - 2.90 = 2.10)$. Since a total of 50% of all scores are below the mean and about 34% of the scores are between 2.10 and the mean, 50% − 34%, or 16%, of the scores are below the score of 2.10. And 16% of 1000 is $(.16)(1000) = 160$.

13. *a.* The sample tends to be normally distributed, so we would expect the population to be normal.
b. Because $\overline{X} = 1297/17 = 76.29$, we would estimate the typical score, μ, to be 76.29.
c. We would estimate the population variance to be $(99{,}223 - 98{,}953.47)/16 = 16.85$.
d. We would estimate the standard deviation to be $\sqrt{16.85} = 4.10$.
e. Between 72.19 $(76.29 - 4.10)$ and 80.39 $(76.29 + 4.10)$.

15. *a.* Pluto. Because his standard deviation is smaller, his scores are generally closer to the mean and closer to each other, so he tends to be a more consistent student.

b. Also Pluto, because his scores are closer to the mean of 60, so it more accurately describes all of his scores.

17. You should predict the mean score of 65 for each student. We measure "average error" using variance, so your error will be $S_X^2 = 6^2 = 36$.

19. She is computing the proportion of variance in exam scores accounted for by the relationship with study time.

better offer, because her income will be closer to the average cost of living in that city.

21. Convert $\bar{X}$ to a z-score. First, $\sigma_{\bar{X}}$ equals $6/\sqrt{50}$, or .849. Then $z = (18 - 19.4)/.849 = -1.65$. From the z-tables, .0495 of the curve is below this score. Out of 1,000 samples, you would expect (.0495)(1000) = 49.5 sample means to be below 18.

Chapter 6

1. A z-score indicates the distance, measured in standard deviation units, that a score is above or below the mean.

3. A z-distribution is the distribution that results when we transform a distribution of raw scores into z-scores.

5. Because z-scores standardize or equate different distributions so that they can be compared and graphed on the same set of axes.

7. *a.* He should consider the size of each class's standard deviation.
b. Small. A small S_X, will give him a large positive z-score, placing him at the top of his class.
c. Large. With a large S_X, he will have a small negative z and still be close to the mean.

9. *a.* It is our model of the perfect normal z-distribution.
b. It is used as a model of any normal distribution of raw scores after they have been transformed to z-scores.
c. The raw scores should be at least approximately normally distributed, the scores should be from a continuous interval or ratio variable, and the sample should be large.

11. $\Sigma X = 103$, $\Sigma X^2 = 931$, and $N = 12$, so $S_X = 1.98$ and $\bar{X} = 8.58$.
a. For $X = 10$, $z = (10 - 8.58)/1.98 = +.72$.
b. For $X = 6$, $z = (6 - 8.58)/1.98 = -1.30$.

13. *a.* $z = +1.0$ *b.* $z = -2.8$
c. $z = -.70$ *d.* $z = -2.0$

15. *a.* .4706 *b.* .0107 *c.* .3944 + .4970 = .8914 *d.* .0250 + .0250 = .05

17. From the z-table, the 25th percentile is at approximately $z = -.67$. The cutoff score is then $X = (-.67)(10) + 75 = 68.3$.

19. For City A, her salary has a z of (27,000 − 50,000)/15,000 = −1.53. For City B, her salary has a z of (12,000 − 14,000)/1000 = −2.0. City A is the

Chapter 7

1. *a.* In experiments, the researcher manipulates one variable and measures subjects' responses on another variable; in correlational studies, the researcher merely measures subjects' responses on two variables.
b. In experiments, the researcher computes the mean of the dependent scores (Y scores) for each condition of the independent variable (each X score); in correlational studies the researcher examines the entire relationship over all X-Y pairs by computing a correlation coefficient.

3. In correlational research, we don't necessarily know which variable occurred first, nor have we controlled or eliminated any other variables that might potentially cause scores to change.

5. *a.* A scatterplot is a graph of the individual data points formed from a set of X-Y pairs.
b. A regression line is the summary straight line that best fits through the scatterplot.

7. *a.* As the X scores increase, the Y scores tend to increase.
b. As the X scores increase, the Y scores tend to decrease.
c. As the X scores increase, the Y scores do not tend to only increase or only decrease.

9. *a.* The scatterplot has a circular or horizontal elliptical shape.
b. The variability in Y at each X is equal to the overall variability across all Y scores in the data.
c. The Y scores are not at all relatively close to the regression line.
d. Knowing X does not improve accuracy in predicting Y.

11. He is drawing the causal inference that more people cause fewer bears. It may not be the number of people that affects the bear population so much as the number of hunters, or the amount of pesticides used, or the noise level associated with more people.

13. *a.* With $r = -.73$, the scatterplot is skinnier.

b. With $r = -.73$, there is less variability in Y at each X.

c. With r = $-.73$, the Y scores hug the regression line more closely.

d. No. He thought a positive r was better than a negative r. Consider the absolute value.

15. *a.* When a small range of X and/or Y scores is measured, the coefficient underestimates the actual strength of the relationship that would be found with unrestricted data.

b. By selecting subjects who all have very similar scores on a particular X or Y variable.

c. By selecting a sample that is likely to produce a wide range of X and Y scores.

17. *a.* ρ stands for the Pearson correlation coefficient in the population.

b. ρ is estimated from an r calculated on a random and representative sample.

c. ρ indicates the strength and type of linear relationship found between all pairs of X and Y scores in the population.

19. Because a correlation coefficient equal to ± 1.0 indicates a perfect linear relationship, so a value beyond ± 1.0 would indicate better than perfect, which is not possible.

21. First compute r_{pb}. For those with degrees, $\overline{Y}_2 = 8.6$; for those without degrees, $\overline{Y}_1 = 5.2$; $S_Y = 3.208$, $p = .50$, and $q = .50$. $r_{pb} = (1.06)(.50) = .53$. Looking at those without college degrees and then at those with degrees, this is a positive linear relationship with an intermediate degree of association.

23. To answer this question, compute r. $\Sigma X = 38$, $\Sigma X^2 = 212$, $(\Sigma X)^2 = 1444$, $\Sigma Y = 68$, $\Sigma Y^2 = 552$, $(\Sigma Y)^2 = 4624$, $\Sigma XY = 317$, and $N = 9$. $r = (2853 - 2584)/\sqrt{(464)(344)} = +.67$. This is a positive linear relationship of intermediate strength, so a nurse's "burnout" score will allow reasonably accurate prediction of her absenteeism.

25. Compute r_s: $\Sigma D^2 = 312$; $r_s = 1 - (1872/990) = -.89$. There is a strong negative relationship between these variables, so that the most dominant tend to weigh the most, and less dominant weigh less.

Chapter 8

1. The linear regression line is the line that summarizes a scatterplot by, on average, passing through the center of the Y scores at each X.

3. Y' is the symbol for the predicted Y score for a given X, computed from the linear regression equation.

5. *a.* The Y-intercept indicates the value of Y where the regression line crosses the Y axis.

b. The slope indicates the direction of the regression line and the degree to which it is slanted.

7. *a.* It is called the standard error of the estimate.

b. It is a standard deviation, indicating the "average" amount that the Y scores at each X deviate from their corresponding values of Y'.

c. It indicates the "average" amount that the actual scores differ from the predicted Y' scores, so it is the "average" error.

9. *a.* $S_{Y'}$ is inversely related to the absolute value of r.

b. $S_{Y'}$ is at its maximum (equal to S_Y) when $r = 0$, because when there is no relationship the amount that the Y scores deviate from the Y' scores equals the overall spread in the data.

c. $S_{Y'}$ is at its minimum (equal to 0) when $r = \pm 1.0$, because when there is a perfect linear relationship there are no differences between Y scores and the corresponding Y' scores.

11. *a.* r^2 is called the coefficient of determination, or the proportion of variance in Y that is accounted for by the relationship with X.

b. r^2 can be interpreted as the proportional improvement in accuracy when we use the relationship with X to predict Y scores, compared to using the overall mean of Y to predict Y scores.

13. *a.* Foofy. The positive r indicates that the higher the statistics grade, the higher the test score.

b. The relationship does not account for 83% of the variance (and $S_{Y'}$ is rather large), so predictions based on this relationship will not be very accurate, as with the case of Bubbles and Foofy.

15. *a.* He should use multiple correlation and multiple regression procedures, simultaneously considering both a subject's concentration and visualization abilities when predicting memory ability.

b. With a multiple r of $+.67$, r^2 is .45: he will be 45% more accurate in predicting memory ability by considering both concentration and visualization abilities than if these predictors were not considered.

17. *a.* Compute r: $\Sigma X = 45$, $\Sigma X^2 = 259$, $(\Sigma X)^2 = 2025$, $\Sigma Y = 89$, $\Sigma Y^2 = 887$, $(\Sigma Y)^2 = 7921$, $\Sigma XY = 460$, and $N = 10$, so $r = (4600 - 4005)/\sqrt{(565)(949)} = +.81$.

b. $b = (4600 - 4005)/565 = +1.05$ and $a = 8.9 - (1.05)(4.5) = 4.18$, so $Y' = (+1.05)X + 4.18$.

c. Using the completed regression equation, for subjects with an attraction score of 9, the predicted anxiety score is $Y' = (+1.05)9 + 4.18 = 13.63$.

d. Compute $S_{Y'}$; $S_Y = 3.081$, so, $S_{Y'} = 3.081\sqrt{1 - .81^2} = 1.81$. Our "average error" is 1.81 when we use Y' to predict each anxiety score.

19. First, square each coefficient: $.20^2 = .04$. Then, as if you had performed regression, knowing the relationship and a student's class ranking allows you to be 4% more accurate in predicting his or her studiousness scores. Likewise, using the relationship with a subject's gender also allows you to account for 4% of the variance in studiousness scores.

Chapter 9

1. *a.* It is the expected relative frequency of the event.
b. It is based on the relative frequency of the event in the population.

3. *a.* $p = 1/6 = .167$.
b. $p = 13/52 = .25$.
c. $p = 1/4 = .25$.
d. $p = 0$: after selecting the ace the first time, it would no longer be in the deck to select again.

5. No. The sex of a child is an independent event, so the sex of previous children does not influence the probability that a child will be a boy or a girl.

7. The p of a hurricane here is $160/200 = .80$. The uncle is looking at an unrepresentative sample, consisting of the past 13 years. Poindexter uses the gambler's fallacy, failing to realize that this p is based on the long run, and in the next few years there may not be a hurricane.

9. *a.* Dependent: You are less likely to golf in rain, snow, hurricanes, and so on.
b. The answer depends on the amount of money or credit you have. If you are rich, the events are independent; if you are poor, they are probably dependent.
c. Dependent: Weight loss depends on calories consumed.
d. Independent: Your chances of winning are the same, whether you use the same or different numbers.

11. *a.* $z = (27 - 43)/8 = -2.0$; $p = .0228$.
b. $z = (51 - 43)/8 = +1.0$; $p = .1587$.
c. $z = (42 - 43)/8 = -.13$; $z = (44 - 43)/8 = +.13$; $p = .0517 + .0517 = .1034$.
d. $z = (33 - 43)/8 = -1.25$; $z = (49 - 43)/8 = +.75$; $p = .1056 + .2266 = .3322$.

13. Transform 24 to z: $\sigma_{\overline{X}} = 12/\sqrt{30} = 2.19$; $z = (24 - 18)/2.19 = +2.74$; $p = .0031$.

15. Either the sample is somewhat unrepresentative of the population because of the luck of who was selected for the sample, or the sample represents some other population.

17. No. With a $z = +2.74$, this mean falls beyond the critical value of $+1.96$. It is too unlikely for it to be accepted as being representative of this population.

19. *a.* The $\overline{X} = 321/9 = 35.67$; $\sigma_{\overline{X}} = 5/\sqrt{9} = 1.67$. Then $z = (35.67 - 30)/1.67 = +3.40$. With a critical value of ± 1.96, conclude that the football players do not represent this population.
b. Football players, as represented by your sample, form a population different from non–football players, having a μ of about 35.67.

21. *a.* For Fred's sample we'd estimate $\mu = 26$, and for Ethel's, $\mu = 18$.
b. The population with $\mu = 26$ is most likely to produce a sample with $\overline{X} = 26$, and the population with $\mu = 18$ is most likely to produce a sample with $\overline{X} = 18$.

Chapter 10

1. *a.* Sampling error is the difference between a sample statistic and the population parameter it represents; it occurs by luck in drawing that sample.
b. We never know whether a sample (1) poorly represents one population parameter because of sampling error, or (2) represents some other population parameter.

3. *a.* In a real relationship, there is something in nature that ties different X scores to different Y scores so that a relationship is produced. Sampling error can also produce scores that by chance pair up to produce a relationship.
b. Not necessarily. We mean only that there is a relationship involving some variables reflected by our scores, even if we have not correctly identified the variables.

5. *a.* They are more powerful than nonparametric procedures.
b. Parametric procedures are robust, which means that violating their assumptions somewhat does

not result in a large error in accurately determining the probability of a Type I error.

7. Experimental hypotheses describe the predicted relationship that we may or may not demonstrate in our experiment.

9. We use a two-tailed test when we predict a relationship but do not predict the direction in which scores will change. We use a one-tailed test when we do predict the direction the scores will change.

11. *a.* The experiment will demonstrate that changing the independent variable from a week other than finals week to finals week increases the dependent variable of amount of pizza consumed; the experiment will not demonstrate an increase.
b. The experiment will demonstrate that changing the independent variable from not performing breathing exercises to performing them changes the dependent variable of blood pressure; the experiment will not demonstrate a change.
c. The experiment will demonstrate that changing the independent variable by increasing hormone levels changes the dependent variable of pain sensitivity; the experiment will not demonstrate a change.
d. The experiment will demonstrate that changing the independent variable by increasing amount of light will decrease the dependent variable of frequency of dreams; the experiment will not demonstrate a decrease.

13. *a.* α stands for our criterion probability; it determines the size of the region of rejection and it is the theoretical probability of a Type I error.
b. α defines our criterion for deciding whether a result is significant; the smaller the α, the larger the absolute value of z_{crit}, and the larger z_{obt} must be to be significant.

15. *a.* The term *beneficial* implies only higher scores, so he should use a one-tailed test.
b. H_0: $\mu \leq 50$, H_a: $\mu > 50$
c. $\sigma_{\overline{X}} = 12/\sqrt{49} = 1.71$; $z_{obt} = (54.63 - 50)/1.71 = +2.71$
d. $z_{crit} = +1.645$
e. Because z_{obt} is beyond z_{crit}, his results are significant: he has evidence of a relationship in the population, so changing from the condition of no music to the condition of music results in test scores' changing from a μ of 50 to a μ of around 54.63.

17. *a.* The probability of a Type I error is $p < .05$. The error would be concluding that music influences scores, when really it does not.
b. By rejecting H_0, we have no chance of making a Type II error. It would be concluding that music does not influence scores, when really it does.

19. *a.* Power is the probability of rejecting a false H_0 (the probability of not making a Type II error).
b. So they can detect relationships when they exist and thus learn something about nature.
c. In a one-tailed test the critical value is smaller than in a two-tailed test; so our obtained value is more likely to be larger than the critical value, and thus is more likely to be significant.

21. Although she is correct about it being easier to reject H_0, she is incorrect about Type I errors. For a given α, the total size of the region of rejection is the same regardless of whether a one- or a two-tailed test is used. Since α is the probability of making a Type I error over the long run, such errors are equally likely using either type of test.

23. *a.* She is correct; that is what $p < .0001$ indicates.
b. She is incorrect. In both studies, the researchers decided the results were too unlikely to reflect sampling error from the H_0 population; they merely defined too unlikely differently.
c. The probability of a Type I error is less in Study B.

Chapter 11

1. *a.* The *t*-test and the *z*-test.
b. Compute z if the true standard deviation of the raw score population (σ_X) is known; compute t if σ_X must be estimated by s_X.

3. *a.* $s_{\overline{X}}$ is the estimated standard error of the mean; $\sigma_{\overline{X}}$ is the true standard error of the mean.
b. Both are used as a standard deviation to locate a sample mean on the sampling distribution of means.

5. The *df* of 80 are .33 of the distance between the *df* at 60 and 120, so the target t_{crit} is .33 of the distance from 2.000 to 1.980: $2.000 - 1.980 = .020$, so $(.020)(.33) = .0066$, and thus $2.000 - .0066$ equals the target t_{crit} of 1.993.

7. *a.* Power is the probability of rejecting H_0 when it really is false (the probability of not making a Type II error).
b. Because we don't know if we were likely to reject H_0, even if it really was false.
c. They initially design the study to maximize power.

9. *a.* H_0: $\mu = 68.5$; H_a: $\mu \neq 68.5$

b. $s_X^2 = 130.5$; $s_{\bar{X}} = \sqrt{130.5/10} = 3.61$; $t_{obt} = (78.5 - 68.5)/3.61 = +2.77$

c. With $df = 9$, $t_{crit} = \pm 2.262$.

d. Using this book rather than other books produces a significant improvement in exam scores: $t_{obt}(9) = 2.77$, $p < .05$.

e. $(3.61)(-2.262) + 78.5 \leq \mu \leq (3.61)(+2.262) + 78.5 = 70.33 \leq \mu \leq 86.67$

11. *a.* H_0: $\mu = 50$; H_a: $\mu \neq 50$.

b. $t_{obt} = (53.25 - 50)/8.44 = +.39$.

c. For $df = 7$, $t_{crit} = \pm 2.365$.

d. $t(7) = +.39$, $p > .05$.

e. The results are not significant, so do not compute the confidence interval.

f. She has no evidence that strong arguments change people's attitudes toward this issue.

13. Disagree. Everything Poindexter said was meaningless, because he failed to first perform significance testing to eliminate the possibility that his correlation was merely a fluke resulting from sampling error.

15. *a.* H_0: $\rho = 0$; H_a: $\rho \neq 0$.

b. With $df = 70$, $r_{crit} = \pm .232$.

c. $r(70) = +.38$, $p < .05$.

d. The correlation is significant, so he should conclude that the relationship exists in the population, and he should estimate that ρ is approximately $+.38$.

e. The regression equation and r^2 should be computed.

17. *a.* r_{pb}

b. H_0: $\rho_{pb} = 0$; H_a: $\rho_{pb} \neq 0$.

c. For $df = 40$, $r_{crit} = \pm .304$.

d. The r_{pb} is significant, so she expects the ρ_{pb} to be approximately .33. She should expect it to be a positive relationship only if left-handers are assigned a lower score than right-handers on the variable of handedness.

e. $(r_{pb})^2 = .11$. The relationship accounts for only 11% of the variance in personality scores, so the results are not very useful.

Chapter 12

1. *a.* The independent samples *t*-test and the related samples *t*-test.

b. Whether the scientist created independent samples or related samples.

3. Create a related samples design if possible, because the related samples *t*-test is more powerful.

5. *a.* $s_{\bar{X}_1 - \bar{X}_2}$ is the standard error of the difference—the standard deviation of the sampling distribution of differences between means for independent samples.

b. $s_{\bar{D}}$ is the standard error of the mean difference, the standard deviation of the sampling distribution of $\bar{D}$ for related samples.

7. She should graph the results, compute the appropriate confidence interval, and compute the effect size.

9. It indicates a range of values of μ_D, one of which our $\bar{D}$ is likely to represent.

11. *a.* H_0: $\mu_1 - \mu_2 = 0$; H_a: $\mu_1 - \mu_2 \neq 0$.

b. $s^2_{pool} = 23.695$; $s_{\bar{X}_1 - \bar{X}_2} = 1.78$; $t_{obt} = (43 - 39)/1.78 = +2.25$.

c. With $df = (15 - 1) + (15 - 1) = 28$, $t_{crit} = \pm 2.048$.

d. The results are significant: in the population, hot baths (with μ about 43) produce different relaxation scores than cold baths (with μ about 39).

e. $(1.78)(-2.048) + 4 \leq \mu_1 - \mu_2 \leq (1.78)(+2.048) + 4 = .35 \leq \mu_1 - \mu_2 \leq 7.65$

f. $r^2_{pb} = (2.25)^2/[(2.25)^2 + 28] = .15$, so bath temperatures do not have a very large effect.

g. Label the *X* axis bath temperature; label the *Y* axis mean relaxation score; plot the data point for cold baths at a *Y* of 39 and for hot baths at a *Y* of 43; connect the data points with a straight line.

13. *a.* She should retain H_0, because in her one-tailed test the signs of t_{obt} and t_{crit} are different.

b. She probably did not subtract her sample means in the same way that she subtracted the μ's in her hypotheses.

15. *a.* H_0: $\mu_D \leq 0$; H_a: $\mu_D > 0$.

b. $t_{obt} = (2.63 - 0)/.75 = +3.51$.

c. With $df = 7$, $t_{crit} = +1.895$, so $t(7) = +3.51$, $p < .05$.

d. $.86 \leq \mu_D \leq 4.40$.

e. The $\bar{X}$ of 15.5, the $\bar{X}$ of 18.13.

f. $r^2_{pb} = (3.51)^2/[(3.51)^2 + 7] = .64$; they are on average about 64% more accurate.

g. Subjects exposed to high amounts of sunshine exhibit a significantly higher well-being score than do subjects exposed to low amounts, with the μ of the difference scores between .86 and 4.40.

17. *a.* H_0: $\mu_D \leq 0$; H_a: $\mu_D > 0$.

b. $\bar{D} = 1.2$, $s^2_D = 1.289$, $s_{\bar{D}} = .359$; $t_{obt} = (1.2 - 0)/.359 = +3.34$

c. With $df = 9$, $t_{crit} = +1.833$.

d. The results are significant. In the population,

children exhibit more aggressive acts after watching the show (with μ about 3.9), than they do before the show (with μ about 2.7).

e. $(.359)(-2.262) + 1.2 \leq \mu_D \leq (.359)(+2.262) + 1.2 = .39 \leq \mu_D \leq 2.01$

f. $r^2_{pb} = (3.34)^2/[(3.34)^2 + 9] = .55$, so violent television shows are an important variable to consider here.

19. a. Select subjects with a severe phobia and give them substantial therapy, so that you are likely to see a large reduction in anxiety scores.

b. Give all subjects the identical therapy, from the same therapist with the same attitudes, and so on.

Chapter 13

1. a. Analysis of variance.

b. A study that contains one independent variable.

c. An independent variable.

d. A condition of the independent variable.

e. All samples are independent.

f. All samples are related, because we employed either a repeated measures or a related samples design.

3. a. Error variance is the inherent variability, or differences, between scores in each population. Treatment variance is the variability between scores in the different populations, created by the different levels of the factor.

b. The mean square within groups estimates the error variance; the mean square between groups estimates the error variance plus the treatment variance.

5. a. H_0: $\mu_1 = \mu_2 = \mu_3 = \mu_4$

b. H_a: not all μ's are equal.

c. H_0 maintains that all μ's represented by the levels are the same; H_a maintains that not all μ's represented by the levels are the same.

7. a. The MS_{bn} is less than the MS_{wn}; either term is a poor estimate of σ^2_{error} and H_0 is assumed to be true.

b. He made a computational error because F_{obt} cannot be a negative number.

9. a. It is necessary when F_{obt} is significant and k is greater than 2. The F_{obt} indicates only that two or more sample means differ significantly; therefore, *post hoc* tests are used to determine which specific levels produced significant differences.

b. It is not necessary when F_{obt} is not significant, because we are not convinced there are any differences to be found. It's also not necessary when $k = 2$, because there is only one possible difference between means in the study.

11. a. Both describe the effect size, or the proportion of variance in dependent scores accounted for by changing the levels of the independent variable.

b. η^2 is a descriptive statistic for describing effect size in the sample data; ω^2 is used to estimate the effect size in the population.

13. a.

Source	***Sum of squares***	***df***	***Mean square***	***F***
Between	134.80	3	44.93	17.08
Within	42.00	16	2.63	
Total	176.80	19		

b. With $df = 3$ and 16, $F_{crit} = 3.24$, so F_{obt} is significant, $p < .05$.

c. For $k = 4$ and $df_{wn} = 16$, $q_k = 4.05$, so $HSD = (4.05)(\sqrt{2.63/5}) = 2.94$: $\bar{X}_4 = 4.4$, $\bar{X}_6 = 10.8$, $\bar{X}_8 = 9.40$, $\bar{X}_{10} = 5.8$.

d. This is an inverted U-shaped function, in which only ages 4 and 10 and ages 6 and 8 do not differ significantly.

e. Because $\eta^2 = 134.8/176.8 = .76$; this relationship accounts for 76% of the variance, so it is a very important relationship.

f. Label the X axis with the independent variable of age and the Y axis with the mean creativity score. Plot the mean score for each condition, and connect adjacent data points with straight lines.

15. a. F_{max} is used for determining whether we can assume homogeneity of variance when performing the independent samples t-test or the between-subjects ANOVA.

b. $F_{max} = 43.68/9.50 = 4.598$. With $k = 3$ and $n - 1 = 15$, the critical value is 3.54. These variances differ significantly, so do not assume homogeneity.

c. Technically, we should not perform ANOVA. We should perform a nonparametric procedure.

Chapter 14

1. a. She can use either a t-test or a one-way ANOVA.

b. She can use either a two-way between-subjects

ANOVA, a two-way within-subjects ANOVA; or a two-way mixed design ANOVA.

c. The procedure to use depends on, respectively, whether both factors are tested using independent samples, both factors are tested using related samples (usually with repeated measures), or one factor involves independent samples and one involves related samples.

3. a. It is the overall effect on dependent scores of changing the levels of one factor, ignoring the other factor.

b. It is the combined effect of factor A and factor B, in which the effect of changing the levels of one factor is not consistent for each level of the other factor; the effect of one factor depends on the level of the other factor being examined.

c. Because by saying that the influence of factor A depends on each level of factor B and vice versa, we contradict any conclusions about an overall consistent influence of A or B.

5. a. H_0: $\mu_{A_1} = \mu_{A_2}$, H_A: Not all μ_A's are equal.

b. H_0: $\mu_{B_1} = \mu_{B_2}$, H_a: Not all μ_B's are equal.

c. H_0: The cell means do not represent μ's such that an interaction exists in the population; H_a: The cell means do represent μ's such that an interaction exists in the population.

7. Study 1: For A, means are 7 and 9; for B, means are 3 and 13. Apparently there are effects for A and B but not for A $\times$ B.

Study 2: For A, means are 7.5 and 7.5; for B, means are 7.5 and 7.5. There is no effect for either A or B, but there is an effect for A $\times$ B.

Study 3: For A, means are 8 and 8; for B, means are 11 and 5. There is no effect for A, but there are effects for B and A $\times$ B.

9. Perform Tukey's *post hoc* comparisons on each main effect and the interaction, graph each main effect and interaction and compute its η^2; where appropriate, compute confidence intervals for the μ represented by a cell or level mean.

11. a. We can maximize the differences between the levels of each main effect or the cells of the interaction, minimize the variability of scores in each cell, and maximize n.

b. The power of the *post hoc* comparisons.

13. The complexity of the interaction increases dramatically with three or more factors, becoming virtually uninterpretable. Yet, if it is significant, it will contradict the conclusions drawn from any main effects and so we must focus our interpretation of the study on this confusing interaction.

15. a.

Source	*Sum of squares*	*df*	*Mean square*	*F*
Between groups				
Factor A	7.20	1	7.20	1.19
Factor B	115.20	1	115.20	19.04
Interaction	105.80	1	105.80	17.49
Within groups	96.80	16	6.05	
Total	325.00			

For each factor, df = 1 and 16, so $F_{crit} = 4.49$: factor B and the interaction are significant, $p < .05$.

b. For factor A, $\bar{X}_1 = 8.9$, $\bar{X}_2 = 10.1$; for factor B, $\bar{X}_1 = 11.9$, $\bar{X}_2 = 7.1$; for the interaction, $\bar{X}_{A_1B_1} = 9.0$, $\bar{X}_{A_1B_2} = 8.8$, $\bar{X}_{A_2B_1} = 14.8$, $\bar{X}_{A_2B_2} = 5.4$.

c. Because factor A is not significant and factor B contains only two levels, such tests are unnecessary for the main effects. For A $\times$ B, adjusted $k = 3$, so $q_k = 3.65$, $HSD = (3.65)(\sqrt{6.05/5}) = 4.02$: the only significant differences are between males and females tested early, and between females tested early and females tested late.

d. We can conclude only that a relationship exists between gender and test scores when testing is done early in the day, and that both early and late testing produce a relationship with test scores for females, $p < .05$.

e. For B, $\eta^2 = 115.2/325 = .35$; for A $\times$ B, $\eta^2 = 105.8/325 = .33$.

Chapter 15

1. a. Both types of procedures test whether, due to sampling error, the data poorly represent the absence of the predicted relationship in the population.

b. We use nonparametric procedures with ordinal or nominal data or with interval or ratio data that are either very skewed or do not have homogeneous population variances.

c. Parametric procedures are more powerful (less likely to produce a Type II error).

d. The probability of a Type I error will then be greater than the alpha level we have selected.

3. a. Observed frequency, f_o, refers to the number of subjects who fall in a category or cell.

b. Expected frequency, f_e, refers to the expected

number of subjects who fall in a category or cell if the data perfectly represent the distribution described by H_0.

5. *a.* H_0: the elderly population is 30% Republican, 55% Democrat, and 15% other; H_a: affiliations in the elderly population are not distributed this way.
   b. For Republicans, $f_e = (.30)(100) = 30$; for Democrats, $f_e = (.55)(100) = 55$; and for others, $f_e = (.15)(100) = 15$.
   c. $\chi^2_{obt} = 4.80 + 1.47 + .60 = 6.87$
   d. For $df = 2$, $\chi^2_{crit} = 5.99$, so the results are significant: party membership in the population of senior citizens is different from party membership in the general population, and it is distributed as in our samples, $p < .05$.
7. *a.* The frequency with which students dislike each professor also must be included.
   b. She can perform a separate one-way χ^2 on the data for each professor to test for a difference between the frequency for "like" and "dislike," or she can perform a two-way χ^2 to determine if whether students like or dislike one professor is correlated with whether they like or dislike the other professor.
9. *a.* H_0: gender and political party affiliation are independent in the population; H_a: gender and political party affiliation are dependent in the population.
   b. For males, Republican $f_e = (75)(57)/155 = 27.58$, Democrat $f_e = (75)(66)/155 = 31.94$, and other $f_e = (75)(32)/155 = 15.48$. For females, Republican $f_e = (80)(57)/155 = 29.42$, Democrat $f_e = (80)(66)/155 = 34.06$, and other $f_e = (80)(32)/155 = 16.52$.
   c. $\chi^2_{obt} = 3.33 + 3.83 + .14 + 3.12 + 3.59 + .133 = 14.14$
   d. With $df = 2$, $\chi^2_{crit} = 5.99$, so the results are significant: in the population, frequency of political party affiliation depends on gender, $p < .05$.
   e. $C = \sqrt{14.14/(155 + 14.14)} = .29$, indicating a somewhat consistent relationship.
11. *a.* The Kruskal-Wallis H-test
   b. The Mann-Whitney U test
   c. The Wilcoxon T test
   d. The rank sums test
   e. The Friedman χ^2 test
   f. The rank sums test
   g. Nemenyi's procedure
13. *a.* Yes. Because these are independent groups, perform the Mann-Whitney test. $U_1 = 32$ and $U_2 = 4$; therefore, $U_{obt} = 4$. $U_{crit} = 5$, so the two groups of ranks differ significantly, as do the groups of underlying maturity scores, $p < .05$.
   b. Return to the raw scores: for students who have not taken statistics, $\overline{X} = 41.67$, so you would expect μ to be around 41.67. For statistics students, $\overline{X} = 69.67$, so you would expect their μ to be around 69.67.
15. *a.* She should use the Kruskal-Wallis H-test, because this is a nonparametric one-way, between-subjects design.
   b. She should assign ranks to the 20 scores, assigning a 1 to the lowest score in the study, a 2 to the next lowest, and so on.
   c. She should perform the *post hoc* comparisons: for each possible pair of conditions, she should rerank the scores and perform the rank sums test.
   d. She will determine which types of patients have significantly different improvement ratings.
17. If the sample is perfectly representative of the distribution of ranks described by H_0, then the sum of the ranks in a group should equal the expected sum of ranks. The greater the difference between the observed and the expected sums of ranks, the less likely that H_0 is true. When the statistic is significant, the observed ranks are too unlikely for us to accept them as representing the distribution of ranks described by H_0.

Appendix A

1. The target z-score is between $z = .670$ at .2514 of the curve and $z = .680$ at .2483. With .2500 at .0014/.0031 of the distance between .2514 and .2483, the corresponding z-score is .00452 above .67, at .67452.
3. The df of 50 is bracketed by $df = 40$ with $t_{crit} = 2.021$, and $df = 60$ with $t_{crit} = 2.000$. Because 50 is at .5 of the distance between 40 and 60, the target t_{crit} is .5 of the .021 between the brackets, which is 2.0105.

Appendix B

1. *a.* With "and" we multiply the individual probabilities times each other.
   b. With "or" we add the individual probabilities together.
   c. We must consider whether the events are mutually inclusive or mutually exclusive.
3. *a.* The probability is the same as that of first selecting one such person and then selecting another. The

p(above 116) = .34, so p(two people above 116) = (.34)(.34) = .1156.

b. These are mutually exclusive events, so with p(introverted) = .40 and p(extroverted) = .35, p(introverted or extroverted) = .40 + .35 = .75.

c. With p(above 116) = .34 and p(introverted) = .40, p(above 116 and introverted) = (.34)(.40) = .136.

d. These are mutually inclusive events, so with p(above 116) = .34, p(introverted) = .40, and p(above 116 and introverted) = .136, p(above 116 or introverted) = (.34 + .40) − .136 = .604.

e. The p(in-between) = .25, and from part *d*, p(above 116 or introverted) = .604. Therefore, p(in-between and then above 116 or introverted) = (.25)(.604) = .151.

5. a. We use the binomial expansion when we want to find the probability of obtaining a sequence of events in which each event involves one of two mutually exclusive possibilities.

b. p(4 heads) = $(5!/4!(1))\ (.5^4)(.5^1)$ = (120/24)(.03125) = .15625.

c. p(1 head) = $(5!/(1!(4!))\ (.5^1)(.5^4)$ = (120/24)(.03125) = .15625.

d. Obtaining 1 head out of 5 tosses is equivalent to obtaining 4 tails; since p(4 tails) = p(4 heads), the answer in part *c* is the same as in part *b*.

Appendix C

1. a. It tells you that the researcher tracked subjects' weight gain by repeatedly weighing the same group of people.

b. It tells you that, on some occasions when subjects were weighed, if their weight had increased, their mood significantly decreased.

3. a. H_0: $\mu_1 = \mu_2 = \mu_3$; H_a: not all μ's are equal.

b. $SS_{tot} = 477 - 392.04$; $SS_A = 445.125 - 392.04$; and $SS_{subs} = 1205/3 - 392.04$.

Source	*Sum of Squares*	*df*	*Mean Square*	*F*
Subjects	9.63	7		
Factor A	53.08	2	26.54	16.69
A × Subjects	22.25	14	1.59	
Total	84.96	23		

c. With $df_A = 2$ and $df_{A \times subs} = 14$, the F_{crit} is 3.74. The F_{obt} is significant.

d. The $q_k = 3.70$ and $HSD = 1.65$. The means for zero, one, and two hours are 2.13, 4.25, and 5.75, respectively. Significant differences occurred between zero and one hour and between zero and two hours, but not between one and two hours.

e. Eta squared (η^2) = 53.08/84.96 = .62.

f. The variable of amount of practice is important in determining performance scores, but although 1 or 2 hours of practice significantly improved performance compared to no practice, 2 hours was not significantly better than 1 hour.

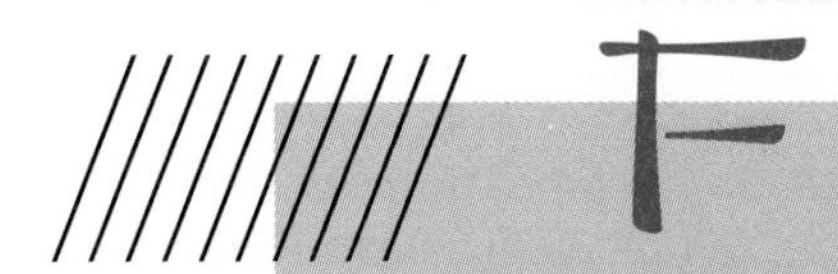

Introduction to Statistics Software Packages

This appendix introduces three computerized data analysis packages: HMSTAT, MYSTAT, and SPSS/pc+ Studentware Plus for MS-DOS. Each package already "knows" how to perform statistics. Essentially, all you must do is simply tell the computer the specific procedure you wish to perform and enter the data. The computer then spits out the statistical answer.

The packages differ in cost, difficulty, hardware requirements, and statistical procedures performed. HMSTAT is the simplest, and it performs all of the procedures described in this textbook. MYSTAT and Studentware are more complicated. They accept larger sets of data and a greater number of different variables at once, and they perform procedures beyond those discussed in this textbook.

A NOTE TO USERS OF ALL PROGRAMS

The following instructions should be read while you are sitting at your computer with the statistical package in hand. For the package you select, work through these instructions from the beginning, starting with the simpler routines. (Do not jump in at a later, complex procedure.)

The instructions for HMSTAT completely describe the program. MYSTAT comes with a 153-page instruction book, and SPSS Studentware comes with a 604-page instruction book. Refer to those instructions for detailed coverage and tutorials.

Throughout the following, commands you must type into the computer are displayed in **bold** print. After you've given a command, the computer usually won't proceed until you press the Enter key.

When performing an inferential procedure, the computer often provides a value for *p,* which is the probability of making a Type I error if you call the result significant. A result is significant if your α is *larger than* the value the computer prints. For example, the computer might print $p = .03$. If you have set α at .05, then this result is significant, because you will accept a probability as large as .05 of making a Type I error. Thus .03 is smaller than your accepted level. However, if the computer prints $p = .051$, you would not consider this result significant, as this represents a larger probability than your α of .050.

Remember: (1) Always make sure what you have typed is exactly correct before you press the Enter key (and remember to press Enter). (2) Always check your data. Inputting an incorrect score is the only thing that will cause the computer to compute an incorrect answer.

HMSTAT

Your instructor may have selected the version of this textbook that includes a computer disk containing HMSTAT.

HMSTAT is a menu-driven program that can accept and store data and perform the statistical procedures discussed in this textbook. It can be used by students who have no experience with computers. It can be run on any IBM or IBM-compatible personal computer with at least 256K of memory, MS-DOS 2.0 or higher, and a monitor (color or monochrome). A printer is optional.

Be sure you have read "A Note to Users of All Programs" found at the beginning of this appendix.

Getting Started

To run HMSTAT:

1. Boot your computer.

   a. If you are using a floppy disk system, turn on the computer with your MS-DOS disk in drive A.

   b. If you are using a machine with a hard disk, MS-DOS is probably already installed on drive C. Just turn on the computer.

2. Back up your disk. Make a copy of the program disk. Refer to your MS-DOS manual for instructions on copying disks. Use the copy as your working disk.

3. Start the program.

   a. If your machine does not have MS-DOS installed on a hard drive, the MS-DOS disk must remain in drive A while you use the program.

      1) Insert your HMSTAT working disk into drive B.

      2) Type **B:** and press Enter.

      3) Type **RUN** and press Enter.

      4) You will see the HMSTAT copyright screen.

   b. If you are using a hard disk system, you can run HMSTAT from the program disk.

      1) Insert your working disk into drive A.

      2) Type **A:** and press Enter.

      3) Type **RUN** and press Enter.

      4) You will see the HMSTAT copyright screen.

c. If you are using a hard disk system, you may install the program in a subdirectory called HMSTAT on your hard drive.

1) Insert the program disk in drive A.

2) Type **A:** and press Enter.

3) Type **HDCOPY** and press Enter.

4) To start HMSTAT from the hard drive:

a) at the C: prompt, type **CD HMSTAT** and press Enter.

b) Type **RUN** and press Enter.

5) You will see the HMSTAT title page.

4. When at the HMSTAT title page, press any key to get to the Main Menu screen.

5. Perform the desired statistical procedure.

6. Exit the program.

a. When you have completed the desired procedures select number **14** (and Enter) to exit the program.

b. To exit the program at any point during a procedure, simultaneously press the keys labeled "Ctrl" and "Break."

The Main Menu

As shown in Screen 1, the main menu provides a numbered list of the major procedures that the program performs. To select a procedure, at the bottom of the Main Menu screen, type the number of the procedure you want (and Enter). (Selecting number **1,** entitled "Instructions," provides a brief review of saving and printing your data and results.) To perform the procedure, simply answer the questions on the screen. After completing a procedure, the program will return to the main menu. Then you can select another procedure or exit the program.

SCREEN 1 HMSTAT Main Menu Screen

```
==================================== HM STAT ====================================
 1. Instructions                         |  8. Two-Way Analysis of Variance
 2. Frequency Tables and Percentile      |  9. Chi Square
 3. One Group Descriptive Statistics     | 10. Mann-Whitney U & Rank Sums Test
 4. Correlation and Regression           | 11. Wilcoxon T Test
 5. Single Sample t Test and z Test      | 12. Kruskal-Wallis H Test
 6. Independent & Related Samples t      | 13. Friedman Chi Square Test
 7. One-Way Analysis of Variance         | 14. Exit Program and Return to DOS
---------------------------------------------------------------------------------
  © Copyright 1992 David W. Abbott, Ph.D.                       (407) 823-2547
=================================================================================

TYPE THE NUMBER OF YOUR CHOICE AND PRESS <Enter> KEY -->
```

Common Routines

All procedures in HMSTAT provide data correction and file saving prior to performing the calculations, give printed copies of your results, and (for inferential procedures) perform significance testing.

Data Input and Correction

Before you type in any data, identify each score by assigning the corresponding subject a number. For example, the first score can be called the score for subject 1, the second score can be called the score for subject 2, and so on.

Among the initial questions you are asked is whether you are inputting data from the keyboard or from a saved file. To input data that has not been previously saved, type **K** (and hit Enter). After you have typed in your last score, you are asked if you want to check the scores that were entered. To do so, type **Y.** You are given the option of displaying the data on the computer screen (type **S**) or having the data printed (type **P**). You are then asked if you want to change the data. If you do, type **Y** and identify the scores you wish to change using the corresponding subject numbers. Follow the instructions on the screen to complete the data correction procedure.

File Saving and Retrieval

After entering the data, you have the option (**Y** or **N**) of saving the data. (You also have the option of seeing a directory of existing files.) The name you give to the data file being saved can contain a maximum of eleven characters, and it must be different from that of any other file. Saved files can then be retrieved anytime you enter the procedure: when the computer asks whether you want to input data using the keyboard or data from a saved file, type **F.** You then can check and print the file and compute your statistics with the data it contains.

Printing Your Results

After completing a procedure, you have the option of printing the results. If you answer yes (**Y**), you can provide a title for the printout that may contain up to eighty characters.

Significance Testing

With inferential procedures, the program either provides the two-tailed critical value for your *df* or it indicates the probability of making a Type I error. All values of p the computer gives are for a two-tailed test. If you are performing a one-tailed test, the actual p for your results will be one-half the value the computer prints. Therefore, if your one-tailed α is .05, your result is significant if the computer prints a p that is no larger than .10 (and your obtained value has the predicted positive or negative sign).

The following sections describe how to perform the specific procedures provided by HMSTAT.

Frequency Tables and Percentiles

As we did in Chapter 3, this procedure creates an ungrouped frequency table showing simple, relative, and cumulative frequencies and percentiles for a set of data. The procedure can be performed on a maximum of twenty scores, and it will not list any score having zero frequency (as we did in Chapter 3). As an example, say that you collected the following age scores: 25, 21, 18, 17, 17, 17, 24, 19, 19, 32, 24, 23, 41, 16, 20, 19.

At the Main Menu screen, type **2.** The screen will then show

HOW MANY SCORES IN THE GROUP? (min = 2, max = 20)

Our sample has sixteen scores, so type **16** (Enter). Then, inputting the data from the keyboard (**K**) produces

FOR SUBJECT 1: X =

From the above scores, type **25** (Enter). The computer then asks for the score for subject 2, and so on until you've entered the score for all sixteen subjects.

Once you've done the data correction and file-saving routines, you'll see Screen 2, which shows each score, its frequency (labeled "f"), its relative frequency (labeled "REL f"), its cumulative frequency (labeled "CUM f") and its percentile (labeled "%").

SCREEN 2 HMSTAT Output for Frequency Table

SCORE	f	REL f	CUM f	%
41	1	.06	16	100
32	1	.06	15	94
25	1	.06	14	88
24	2	.13	13	81
23	1	.06	11	69
21	1	.06	10	63
20	1	.06	9	56
19	3	.19	8	50
18	1	.06	5	31
17	3	.19	4	25
16	1	.06	1	6

DO YOU WANT A PRINTED COPY OF THE RESULTS? (Y/N)

One-Group Descriptive Statistics

As we did in Chapters 4 and 5, this procedure computes the mean, median, and mode (if the data are unimodal) for a set of data, as well as the range, the sample variance and standard deviation, and the estimated population variance and standard deviation. (You must provide the appropriate symbols for these statistics, because the program cannot print them.) The sample can have a maximum *N* of 50.

As an example, say you want to compute these statistics for these age scores: 25, 21, 18, 19, 17, 17, 24, 19, 19, 32, 24, 23, 41, 16, 20, 19. At the Main Menu screen, type **3.** Then enter the data for each subject, as described above. Once you've done the data correction and file-saving routines, you'll see Screen 3. You are given the measures of central tendency; the estimated population standard deviation and the estimated population variance; the sample standard deviation and the sample variance; and lastly, the lowest score, the highest score, and the range.

SCREEN 3 HMSTAT Output for One-Group Descriptive Statistics

```
                         ONE GROUP DESCRIPTIVE STATISTICS

THE MEAN = 22.125 ;   MEDIAN = 19.5 ;   MODE = 19

THE ESTIMATED POPULATION STANDARD DEVIATION = 6.438168
                         ESTIMATED VARIANCE = 41.45

THE SAMPLE STANDARD DEVIATION = 6.233729
              SAMPLE VARIANCE = 38.85938

LOWEST SCORE = 16   HIGHEST = 41   RANGE = 25


DO YOU WANT A PRINTED COPY OF THE RESULTS?  (Y/N)
```

Correlation and Regression

As we did in Chapters 7 and 8, this procedure computes the Pearson correlation coefficient (r), the linear regression equation, the Spearman correlation coefficient (r_s), and the Point-Biserial correlation coefficient (r_{pb}). It also performs related procedures, including significance testing of the coefficients (as in Chapter 11). For each procedure you must have between a minimum of five and a maximum of fifty pairs of scores.

To get into this procedure, at the Main Menu screen type **4.** The screen will show

HOW MANY SCORES IN THE GROUP? (min = 5, max = 50)

Say that you have a set of data containing 10 *X*-*Y* pairs: type **10.** Then select the correlation coefficient you wish to compute.

The Pearson Correlation Coefficient To compute the Pearson correlation coefficient, type **P** (Enter). As an example, say you are examining the relationship between annual income and the number of years of schooling, based on the data in Table F.1. Inputting the data from the keyboard (**K**) produces:

SUBJECT 1 X =

TABLE F.1 Data Set for the Pearson Correlation Coefficient Procedure

Subject	X Annual income (in thousands)	Y Number of years of schooling
1	38	18
2	10	15
3	6	10
4	19	9
5	3.5	14
6	0	13
7	12.3	11
8	13.5	13
9	65	21
10	22	17

SCREEN 4 HMSTAT Data Input Screen for Pearson Correlation

```
SUBJECT 1     X = 38          Y = 18
SUBJECT 2     X = 10          Y = 15
SUBJECT 3     X = 6           Y = 10
SUBJECT 4     X = 19          Y = 9
SUBJECT 5     X = 3.5         Y = 14
SUBJECT 6     X = 0           Y = 13
SUBJECT 7     X = 12.3        Y = 11
SUBJECT 8     X = 13.5        Y = 13
SUBJECT 9     X = 65          Y = 21
SUBJECT 10    X = 22          Y = 17
```

Income is your X variable, so for subject 1 enter the income score (**38**). After you hit Enter the screen will show

SUBJECT 1   X = 38   Y =

Now enter the first subject's Y score (**18**). When you're finished with all the subjects, your screen should look like Screen 4. Once you've done the data correction and file-saving routines, you'll see Screen 5, showing the Pearson r, the df in parentheses (here it's 8), the obtained value (here it's +.752), and the value of p for significance testing. You'll also see a scatterplot of the data, plus the mean and sample standard deviation for the X and Y scores, the standard error of the estimate ("Std Err Estimate"), and r^2 ("r sqr"). At the bottom of the screen is the computed linear regression equation. After indicating whether you want a printed copy of the results, you are given the option of using the regression equation to compute a predicted value of Y' for any value of X. To do so, type **Y** and then provide the known value of X you want to use to predict Y.

SCREEN 5 HMSTAT Output for Pearson Correlation and Regression

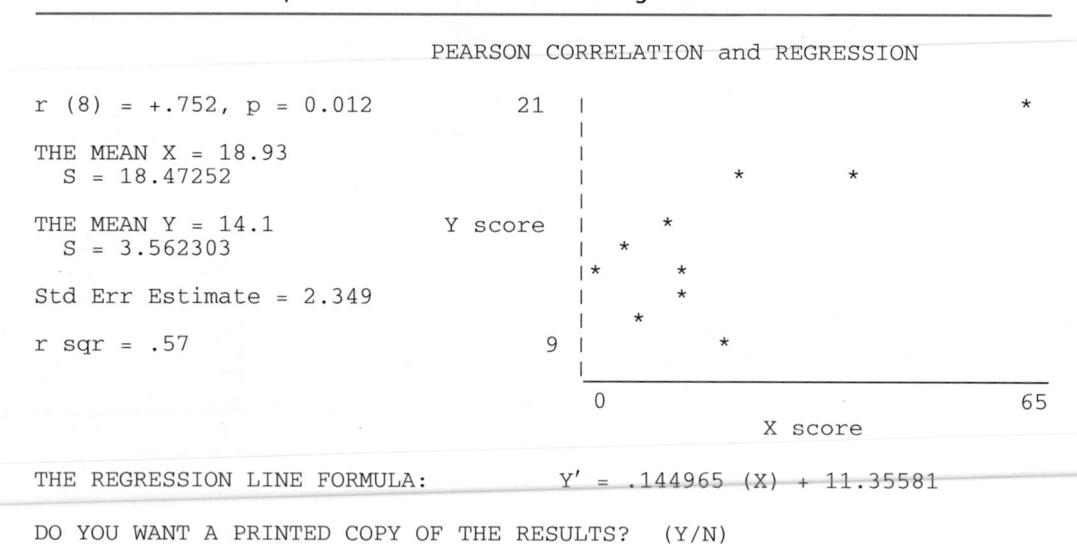

```
                                PEARSON CORRELATION and REGRESSION

r (8)  = +.752, p = 0.012          21  I                                 *
                                       I
THE MEAN X = 18.93                     I
  S = 18.47252                         I            *         *
                                       I
THE MEAN Y = 14.1              Y score I      *
  S = 3.562303                         I   *
                                       I*     *
Std Err Estimate = 2.349               I      *
                                       I   *
r sqr = .57                         9  I          *
                                       I_____________________________
                                       0                            65
                                                  X score

THE REGRESSION LINE FORMULA:          Y' = .144965 (X) + 11.35581

DO YOU WANT A PRINTED COPY OF THE RESULTS?  (Y/N)
```

The Spearman Rank-Order Correlation Coefficient After indicating the number of subjects you have, type **S** to compute the Spearman correlation coefficient (r_S). Then follow the same procedures as for the Pearson r above. As an example, say that you have ranked ten students separately on their class participation and on how much they watch the clock, obtaining the data in Table F.2. The computer automatically resolves any tied ranks (as we did in Chapter 7). Enter each student's class-participation rank at "X =" and his or her clock-watching rank at "Y =." After the data correction and file-saving prompts, you'll see Screen 6, which contains a scatterplot, the obtained Spearman r, the two-tailed critical value (here it's .648), and the mean and sample standard deviation for each variable.

TABLE F.2 Data Set for the Spearman Rank-Order Correlation Coefficient Procedure

Student	X *Participation*	Y *Clock-watching*
1	2	8
2	7	6
3	4	5
4	3	10
5	1	9
6	5	4
7	10	2
8	9	3
9	8	1
10	6	7

SCREEN 6 HMSTAT Output for Spearman Rank-Order Correlation

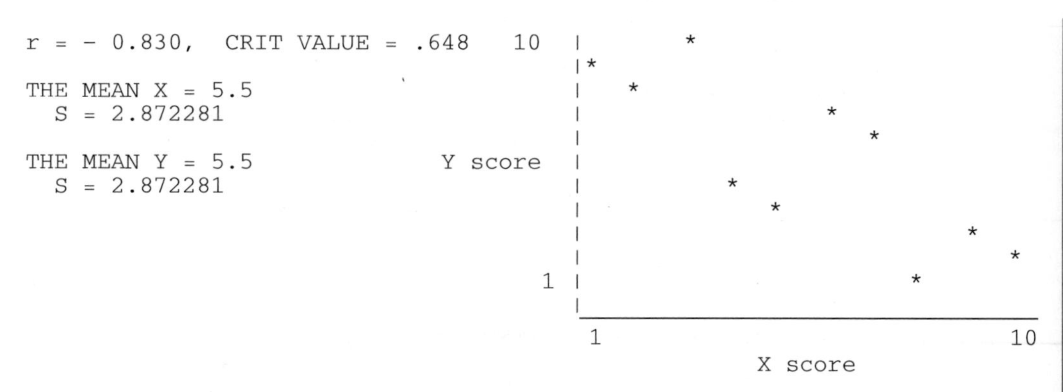

DO YOU WANT A PRINTED COPY OF THE RESULTS? (Y/N)

The Point-Biserial Coefficient After indicating the number of subjects you have, type **PB** to compute the point-biserial correlation coefficient (r_{pb}). For example, say you are examining the relationship between gender and body weight. You use the number 1 to indicate male and the number 2 to indicate female. You obtain the data in Table F.3. For this procedure, the dichotomous variable must be the X variable.

TABLE F.3 Data Set for the Point-Biserial Correlation Coefficient Procedure

Subject	X *Gender*	Y *Body weight*
1	1	200
2	2	120
3	2	135
4	1	180
5	2	125
6	1	160
7	1	165
8	2	140
9	1	150
10	2	130

Input the data as in the above correlations. You'll see Screen 7 (as shown at the top of the next page). Notice that here, instead of the critical value, you are given the p level for significance testing.

SCREEN 7 HMSTAT Output for Point-Biserial Correlation

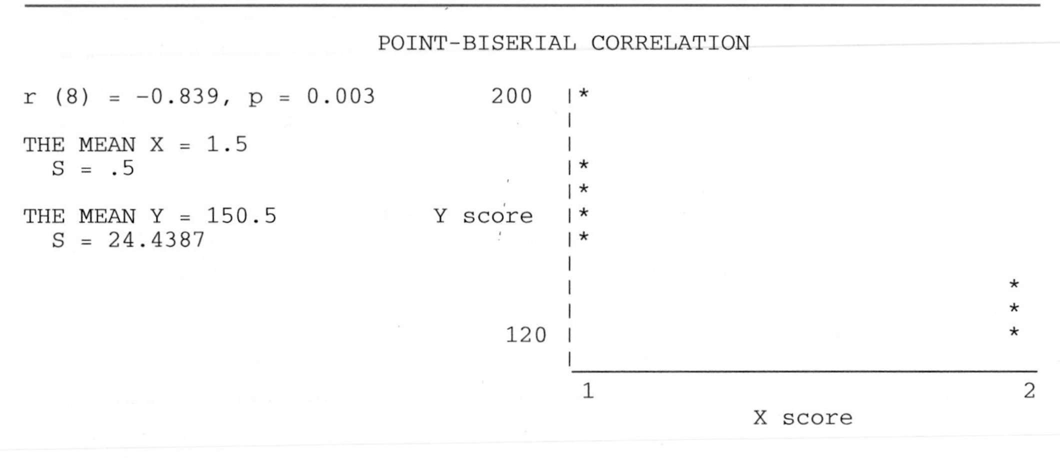

DO YOU WANT A PRINTED COPY OF THE RESULTS? (Y/N)

Single-Sample *t*-Test and *z*-Test

This procedure performs a *z*-test (as we did in Chapter 10) or a single-sample *t*-test (as we did in Chapter 11). As an example, say that you teach fifteen people how to speed-read using a new technique and then measure the number of words they read in a one-minute period. You obtain the following data: 53, 143, 78, 75, 85, 72, 69, 75, 82, 90, 108, 69, 41, 98, 80. From previous research, you know that in the population of people without speed-reading training, $\mu = 75$ words per minute.

At the Main Menu screen, type **5.** Then enter the data as described in the procedure for frequency tables and percentiles (p. 547). Once you've done the data correction and file-saving routines, you'll see the top four lines of Screen 8,

SCREEN 8 HMSTAT Output for Single Sample *t*-Test and *z*-Test

```
                    SINGLE SAMPLE t TEST AND z TEST

THE MEAN = 81.2

ESTIMATED POP STANDARD DEV = 23.5954

ESTIMATED POP VARIANCE = 556.7429

DO YOU WANT A PRINTED COPY OF THE RESULTS?  (Y/N) n

DO YOU WANT A t TEST FOR THE POPULATION MEAN?  (Y/N) y

ENTER THE VALUE OF THE HYPOTHESIZED POPULATION MEAN:  75

t (14)  =  1.018, p = 0.327

DO YOU WANT A CONFIDENCE INTERVAL FOR THE POPULATION MEAN?  (Y/N) y

THE 95% CONFIDENCE INTERVAL:   68.16246  TO  94.23753

DO YOU WANT A z TEST OR CONFIDENCE INTERVAL?  (Y/N)
```

which show the sample mean and the estimated population standard deviation and variance. After indicating whether you want a printed copy of the results, you can perform the one-sample *t*-test. If you do not know the true population standard deviation, type **Y** and then indicate the population mean that your null hypothesis states your sample represents (here, type **75**). The complete statistical result appears, and you have the option of computing a confidence interval for the population mean.

Say that you know the population μ is 75 and the population standard deviation, σ_X, is 7.5. You can then perform the *z*-test. Input these values when prompted, and you are given the results of the *z*-test and the 95% confidence interval.

Independent and Related Samples *t*-Tests

As we did in Chapter 12, this procedure performs the independent and the related samples *t*-tests. Type **6** at the Main Menu screen to select this procedure.

Independent Samples *t*-test Type **I** to perform the independent samples *t*-test. As an example, say that you are testing subjects' ability to remember a list of words under the conditions of either consuming alcohol or not consuming alcohol. Table F.4 shows the data.

TABLE F.4 Data Set for the Independent Samples *t*-Test Procedure
The scores represent the number of words recalled.

Sample 1: Alcohol	*Sample 2: No alcohol*
9	20
15	15
12	25
17	28
14	12
22	16
11	24
7	19
10	18
13	29
17	14
8	11
15	19
11	16
13	21

First indicate the number of scores in group 1 (type **15**). Then indicate how many scores are in group 2. This may be the same or a different *n* as in group 1 (here it is the same, 15). Then enter the data as described in the procedure for frequency tables and percentiles, above, except enter all scores from group 1 first and then the scores from group 2. After the data correction and file-saving routines, you'll see Screen 9 (on the next page), which shows (1) the mean and standard deviation for each group, with a bar graph of the means; (2) the results of the

SCREEN 9 HMSTAT Output for Independent Samples *t*-Test

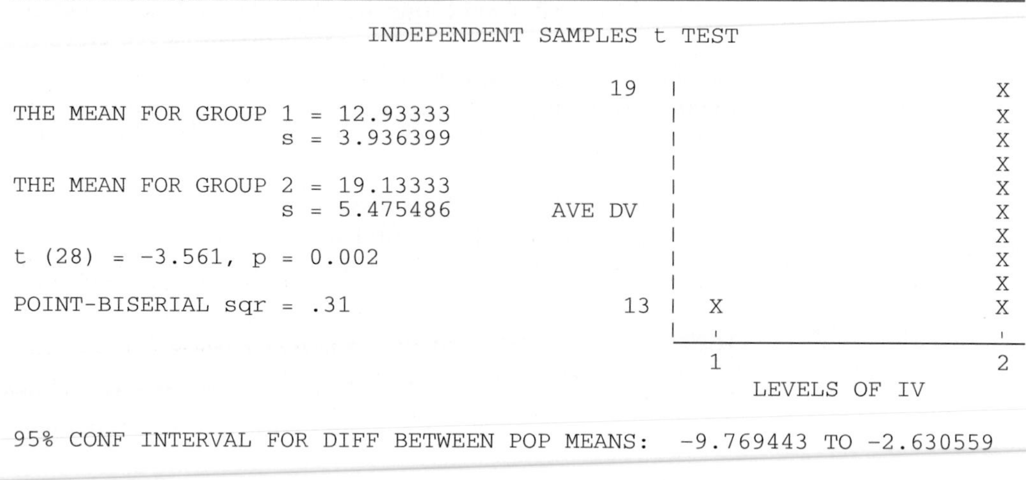

```
                    INDEPENDENT SAMPLES t TEST

                                      19  I                    X
THE MEAN FOR GROUP 1 = 12.93333           I                    X
                   s = 3.936399           I                    X
                                          I                    X
THE MEAN FOR GROUP 2 = 19.13333           I                    X
                   s = 5.475486   AVE DV  I                    X
                                          I                    X
t (28) = -3.561, p = 0.002                I                    X
                                          I                    X
POINT-BISERIAL sqr = .31              13  I  X                 X
                                          I  ____________________
                                             1                 2
                                              LEVELS OF IV

95% CONF INTERVAL FOR DIFF BETWEEN POP MEANS:  -9.769443 TO -2.630559

DO YOU WANT A PRINTED COPY OF THE RESULTS?  (Y/N)
```

two-tailed *t*-test; (3) the proportion of variance accounted for [as calculated by the squared point-biserial correlation coefficient]; and (4) the 95% confidence interval for the difference between the population means.

Related Samples *t*-test Type **R** to perform the related samples *t*-test. As an example, suppose that you again are studying the influence of consuming alcohol on memory, but this time you use a repeated measures design and measure each subject's recall under both conditions. The results are shown in Table F.5.

TABLE F.5 Data Set for the Related Samples *t*-Test Procedure
The scores represent the number of words recalled.

Subject	*Alcohol*	*No alcohol*
1	19	20
2	15	15
3	12	25
4	17	28
5	17	12
6	22	16
7	24	11
8	7	19
9	10	18
10	13	22
11	17	14
12	11	8
13	15	19
14	11	16
15	20	21

First indicate the number of pairs of scores you have (here, type **15**). Then enter the data as described for the independent samples *t*-test. *Note:* The computer pairs each subject's scores for the two conditions, so be sure that the first score in group 2 is for the same subject as the first score in group 1, and so on. After inputting the data, you are given the same information as in the independent samples *t*-test although the specific values will be different (see Screen 9).

One-Way Analysis of Variance

This procedure computes a one-way, random-groups ANOVA; descriptive statistics; and Tukey's *HSD* (as we did in Chapter 13). You must have equal *n*'s for all levels of your factor. As an example, suppose that you create a condition of either high, medium, or low anxiety in your subjects and then measure their performance on an exam, obtaining the results in Table F.6.

TABLE F.6 Data Set for the One-way Analysis of Variance Procedure
The scores represent the number of questions correctly answered.

High anxiety	*Medium anxiety*	*Low anxiety*
9	20	13
15	15	21
12	25	11
17	28	16
14	12	15
22	16	19
11	24	11
7	19	8
10	18	17
13	29	14

At the Main Menu screen, type **7.** Then indicate the number of groups you have (**3**) and the number of scores per group (**10**). Then enter the data in for the independent samples *t*-test, entering the scores for each level. After the data collection and file saving routines, you will see Screen 10 (on the next page), which shows (1) the mean and sample standard deviation for each of the three groups (levels), with a bar graph of the means; (2) the SS, MS, and *df* for the between-group variance and the within-group variance (use these values when creating your ANOVA summary table); (3) the description for your obtained *F* and the value of eta squared; and (4) if the *F* is significant, the value of Tukey's *HSD* (any difference between two means that is larger than the *HSD* value is a significant difference). To compute the confidence interval for the mean of a level, enter the data from that level into the single-sample *t*-test procedure described above.

SCREEN 10 HMSTAT Output for One-Way Analysis for Variance

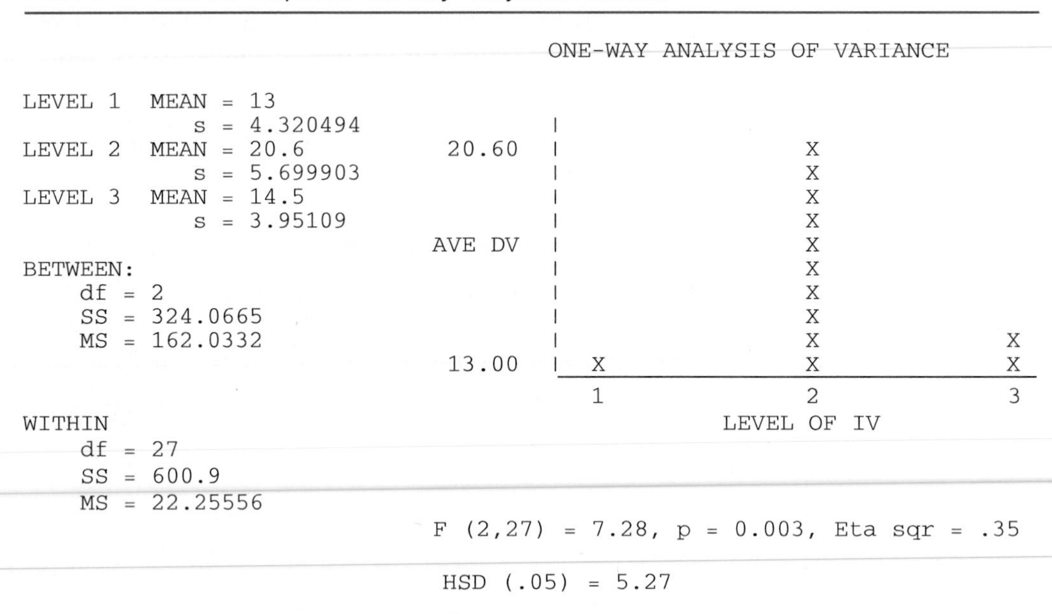

```
                                            ONE-WAY ANALYSIS OF VARIANCE

LEVEL 1  MEAN = 13
            s = 4.320494              I
LEVEL 2  MEAN = 20.6          20.60   I               X
            s = 5.699903              I               X
LEVEL 3  MEAN = 14.5                  I               X
            s = 3.95109               I               X
                             AVE DV   I               X
BETWEEN:                              I               X
    df = 2                            I               X
    SS = 324.0665                     I               X
    MS = 162.0332                     I               X              X
                              13.00   I   X           X              X
                                      ------------------------------------
                                          1           2              3
WITHIN                                            LEVEL OF IV
    df = 27
    SS = 600.9
    MS = 22.25556
                             F (2,27) = 7.28, p = 0.003, Eta sqr = .35

                             HSD (.05) = 5.27

DO YOU WANT A PRINTED COPY OF THE RESULTS?  (Y/N)
```

Two-Way Analysis of Variance

This procedure computes a two-way, between-subjects ANOVA as we did in Chapter 14. Note that you can have no more than nine cells, and you must have equal *n*'s for all cells. The program computes F_{obt}, the value of Tukey's *HSD,* eta squared, and descriptive statistics for each main effect and for the interaction. As an example, say that factor A is amount of sleep subjects are given (with three levels—0 hours, 4 hours, and 8 hours) and factor B is amount of caffeine subjects consume (with two levels—0 mg and 100 mg). The dependent variable is the subjects' exam performance. The results are given in Table F.7.

At the Main Menu, type **8**. You must then indicate the number of levels of factor A you have: type **3** (max = 4). Then indicate that there are **2** levels of factor B (max = 4) and **10** scores in each cell (max = 50).

As you enter the data, the computer identifies the scores in each cell by using the level numbers for both factor A and factor B that you used to create the cell.

TABLE F.7 Data Set for the Two-way Analysis of Variance Procedure
The scores represent the number of questions correctly answered.

		Factor A: Sleep		
		A_1: *0 hours*	A_2: *4 hours*	A_3: *8 hours*
Factor B: Caffeine consumption	B_1: *0 mg*	9	20	13
		15	15	21
		12	25	11
		17	28	16
		14	12	15
		22	16	19
		11	24	11
		7	19	8
		10	18	17
		13	29	14
	B_2: *100 mg*	11	17	11
		17	16	19
		14	11	9
		19	20	14
		16	15	13
		24	17	17
		13	9	9
		9	25	6
		12	22	15
		15	18	12

Thus, the screen will initially show

ENTER SCORE NUMBER 1 FOR CELL A1 B1:

From the way Table F.7 is labeled, you can see that cell A_1B_1 is the cell for 0 hours sleep (A_1) and 0 mg caffeine (B_1), so first type in the scores **9, 15, 12, . . . 13**. Then enter the scores for cell A_1B_2 (0 hours sleep and 100 mg caffeine), cell A_2B_1 (4 hours sleep and 0 mg caffeine), and so on.

After entering and saving the data, you will see Screen 11 (see top of next page), which shows the *df,* SS, MS, and the complete description of the results for each factor and the interaction. When the F_{obt} is significant, eta squared ("eta sqr") is also provided. After indicating whether you want a printed copy, you can see the descriptive statistics and Tukey's *HSD* for the main effect of A (type **A**), the main effect of B (type **B**), and the interaction (type **AB**). In each, you are given the level means or cell means, respectively, a graph of the means, and the value of Tukey's *HSD.* Follow the prompts on the bottom of the screen to cycle through and review the ANOVA data.

SCREEN 11 HMSTAT Output for Two-Way Analysis for Variance

```
FOR FACTOR A:                              df = 2
                                           SS = 342.5333
                                           MS = 171.2666
                                    F (2,54) = 8.31, p = .001, Eta sqr = .22

FOR FACTOR B:                              df = 1
                                           SS = 21.59998
                                           MS = 21.59998
                                    F (1,54) = 1.05, p = .311

FOR THE AB INTERACTION:                    df = 2
                                           SS = 83.19998
                                           MS = 41.59999
                                    F (2,54) = 2.02, p = .141

FOR THE WITHIN GROUPS:                     df = 54
                                           SS = 1113.4
                                           MS = 20.61852

DO YOU WANT A PRINTED COPY OF THE RESULTS? (Y/N)
```

Chi Square

This procedure computes the one-way and two-way chi-square and related descriptive statistics as we did in Chapter 15. To select this procedure, type **9** at the Main Menu screen.

One-Way Chi Square Type **O** to perform the one-way chi square procedure. As an example, say you are counting the frequency of graffiti in restrooms in different types of university buildings. You find 352 pieces of graffiti in the academic buildings, 935 pieces in the dorms, and 145 pieces in the dining halls.

First indicate the number of groups you have (max = 12): here, type **3.** Then indicate whether the expected frequencies are equal: if your null hypothesis is that all frequencies should be equal, type **Y.** (If your null hypothesis is that there is not an equal split between the groups, type **N.** You'll type in the expected frequencies later.) Then input the observed frequency for each group. When all observed frequencies are entered, you'll either be given, or must type in, the expected frequencies. You can then perform data correction, but you cannot save the data. You'll see Screen 12, which shows the results of the chi square test and a graph of the observed and expected frequencies.

SCREEN 12 HMSTAT Output for One-Way Chi Square

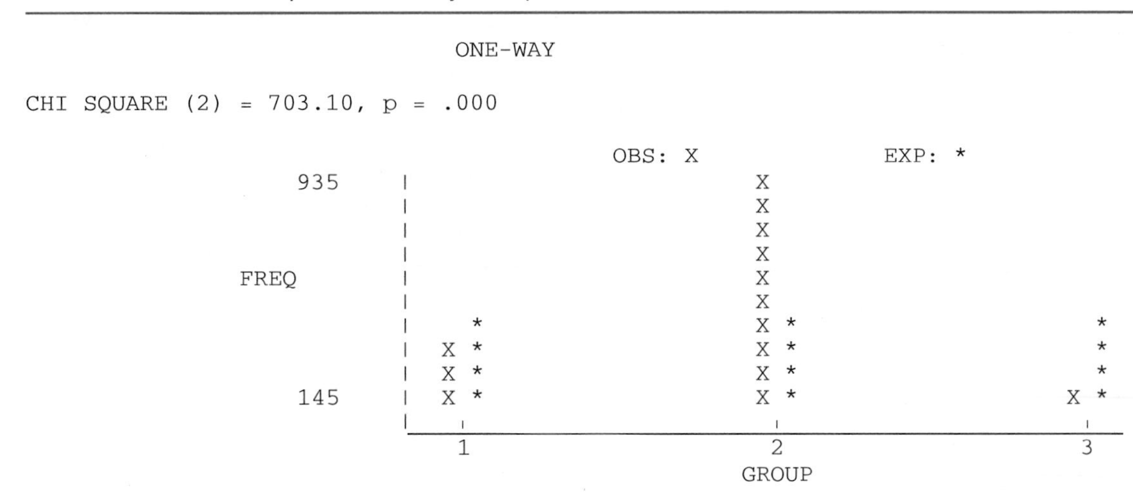

DO YOU WANT A PRINTED COPY OF THE RESULTS? (Y/N)

Two-Way Chi Square Type **T** to compute a two-way chi square. As an example, suppose you are testing whether the frequency of graffiti in restrooms differs among the different buildings described above and whether there are any gender differences. The number of graffiti in the restrooms at each location is shown in Table F.8.

TABLE F.8 Data Set for Two-way Chi Square Procedure
The scores represent the number of graffiti.

		Location		
		Academic buildings	*Dormitories*	*Dining halls*
Restroom type	*Male*	255	98	42
	Female	190	200	50

First indicate how many rows are in your frequency table (max = 3): here, type **2.** Then indicate how many columns are in your frequency table (max = 6): here, type **3.** Then input the observed frequency for, first, the cell formed by "COL 1" and "ROW 1," then the cell formed by "COL 2" and "ROW 1," and so on. You can then correct the data, but you cannot save it. You'll see a screen similar to Screen 12. The only difference is that with a significant 2 × 2 chi square you are given the value of the phi correlation coefficient, and with a significant chi square that is larger than a 2 × 2 you are given the value of the C correlation coefficient.

Mann-Whitney *U* and Rank Sums Test

This procedure computes the Mann-Whitney U and the rank sums tests for independent samples as we did in Chapter 15. The procedure automatically determines which test to perform, based on the number of scores in each condition. Here we will illustrate a Mann-Whitney U test. Assume that we have two groups, with ranked scores as follows.

Group 1: 9, 4, 7, 3, 6. Group 2: 1, 8, 2, 5, 10.

Type **10** at the Main Menu screen. Then enter the data as described previously for the independent samples t-test. You will see Screen 13, which shows the n's and the U values for each group. For a two-tailed test, the smaller U value is U_{obt} (here $U_{obt} = 11$). U_{obt} is significant if it is equal to or less than U_{crit}. The group identified as containing the more frequent high scores is used for identifying the U_{obt} in a one-tailed test, but then you must obtain the critical value from Table 9 in Appendix D.

SCREEN 13 HMSTAT Output for Mann-Whitney *U* Test

```
                          Mann-Whitney U TEST

FOR GROUP 1: N = 5 , U = 11

FOR GROUP 2: N = 5 , U = 14

GROUP 1 HAS MORE FREQUENT HIGH SCORES

CRITICAL VALUE AT .05 LEVEL = 2


DO YOU WANT A PRINTED COPY OF THE RESULTS?  (Y/N)
```

Wilcoxon *T* Test

This procedure computes the Wilcoxon T test for related samples as we did in Chapter 15. As an example, suppose you measured the speed with which people react to a red traffic light and to a green traffic light, collecting the data in Table F.9.

TABLE F.9 Data Set for the Wilcoxon *T* Test Procedure

Subject	*Reaction time to red traffic light (in seconds)*	*Reaction time to green traffic light (in seconds)*
1	1.9	2.0
2	2.7	1.5
3	2.4	2.6
4	1.7	2.8
5	1.3	1.2
6	2.2	1.6
7	1.1	0.9
8	0.7	1.75
9	1.0	2.9
10	1.4	2.25

Type **11** at the Main Menu screen. Then indicate how many scores you have, and enter the data as described for the related samples *t*-test. You'll see Screen 14. The T_{obt} is significant if it is equal to or less than the critical value. (In the example, $T_{obt} = 19$, but the critical value is only 8, so the results are not significant.)

SCREEN 14 HMSTAT Output for Wilcoxon *T* Test

```
              Wilcoxon TEST FOR RELATED SAMPLES

NOTE: DIFFERENCE SCORES COMPUTED AS D = X1 - X2

THE SUM OF THE POSITIVE RANKS = 19

THE SUM OF THE NEGATIVE RANKS = 36

N = 10 ,  T = 19

CRITICAL VALUE AT .05 LEVEL = 8


DO YOU WANT A PRINTED COPY OF THE RESULTS?  (Y/N)
```

Kruskal-Wallis *H* Test

This procedure computes the Kruskal-Wallis *H* Test for independent samples as we did in Chapter 15. As an example, say that you are examining three groups of people who are classified as either underweight, of average weight, or overweight. For each group, you measure their annual income (in thousands of dollars), as shown in Table F.10.

TABLE F.10 Data Set for the Kruskal-Wallis *H* Test Procedure
The scores represent annual income (in thousands of dollars).

Underweight	*Average weight*	*Overweight*
9	20	13
15	15	21
12	25	11
17	28	16
14	12	15
22	16	19
11	24	11
7	19	8
10	18	17
13	29	14

Type **12** at the Main Menu screen. Indicate how many subjects are in each group (for this example, type **10**) and levels you have (here, type **3**), and enter the data as described previously for the one-way analysis of variance. Then you'll see Screen 15 (at top of next page), which shows (1) the sum of ranks for each group (for computing the means of ranks), (2) the results of the statistical test, and (3) eta squared. To perform *post hoc* comparisons, re-enter the data from each pair of conditions into the Mann-Whitney U and rank sums test (procedure 10).

SCREEN 15 HMSTAT Output for Kruskal-Wallis *H* Test

```
                    Kruskal-Wallis H TEST

SUM OF THE RANKS:

GROUP 1 = 106

GROUP 2 = 222.5

GROUP 3 = 136.5

H (2) = 9.42, p = .010, Eta sqr = .32


DO YOU WANT A PRINTED COPY OF THE RESULTS?  (Y/N)
```

Friedman Chi Square Test

This procedure computes the Friedman chi square test for repeated measures as we did in Chapter 15. You must have between three and six levels of your independent variable. If you have three levels, you must have a minimum of ten subjects; if you have four levels, you need only five subjects (max = 50). As an example, say you have ten subjects taste each of three types of colas and rank-order their taste preferences for them, producing the data in Table F.11.

TABLE F.11 Data set for the Friedman Chi Square procedure
The scores represent preference rankings.

Subject #	*Cola 1*	*Cola 2*	*Cola 3*
1	1	2	3
2	2	1	3
3	1	3	2
4	1	2	3
5	2	3	1
6	1	2	3
7	2	1	3
8	1	2	3
9	3	1	2
10	1	2	3

Type **13** at the Main Menu screen. Then indicate how many subjects and levels you have. Enter the data by providing the ranks for one subject for all levels of the independent variable. For example, typing in the data for our first subject would produce this screen:

ENTER THE SCORE FOR SUBJECT 1 AT LEVEL 1: **1**
ENTER THE SCORE FOR SUBJECT 1 AT LEVEL 2: **2**
ENTER THE SCORE FOR SUBJECT 1 AT LEVEL 3: **3**

After entering and correcting the data, you'll see Screen 16, which shows (1) the sum of the ranks for each group (for computing the mean ranks), (2) the results of the statistical test, and (3) eta squared.

SCREEN 16 HMSTAT Output for Friedman Chi Square Test

```
                    Friedman CHI SQUARE TEST

SUM OF RANKS FOR LEVEL 1 = 15

SUM OF RANKS FOR LEVEL 2 = 19

SUM OF RANKS FOR LEVEL 3 = 26

CHI SQUARE (2) = 6.20, p = .044, Eta sqr = .21


DO YOU WANT A PRINTED COPY OF THE RESULTS?  (Y/N)
```

MYSTAT 2.1 FOR MS-DOS

MYSTAT accepts up to fifty variables and thirty-two thousand scores, and it performs all the major analyses discussed in this textbook. It also performs more advanced procedures, such as three-way ANOVAs. MYSTAT can run on any IBM-compatible personal computer with at least 256K RAM, 320K of either hard- or floppy-disk storage space, MS-DOS 2.0 or higher, and a monitor. A printer is optional. You will also need one or two formatted, double-sided, high-density floppy disks for saving data.

Be sure you have read the Note to Users of All Programs found at the beginning of this appendix. The following provides a basic introduction to MYSTAT.

Installing MYSTAT

1. Boot your computer as described under Getting Started in the discussion of HMSTAT (p. 544).

2. Format your floppy disk(s). Use high-density disks.

   a. If you are using a dual disk-drive system, you will need to format two floppy disks.

   b. If you are using a hard disk system, format one floppy disk.

3. Install MYSTAT.

   a. If you are using a dual disk-drive system, at the MS-DOS prompt, remove the MS-DOS disk and put in the original MYSTAT disk. Type **A:** and press Enter. Type **TWODRIVE**, press Enter, and follow the on-screen instructions. Use one of the blank formatted disks to create a MYSTAT program disk and the other to create a MYSTAT data disk.

b. If you are using a hard disk system, insert the original MYSTAT disk in drive A. Type **A:** and press Enter. Type **HARDDISK C:\MYSTAT**, press Enter, and follow the instructions. Use a blank formatted disk as a MYSTAT data disk.

4. Start the program.

a. If you are using a dual disk-drive system, at the MS-DOS prompt, remove the MS-DOS disk from drive A and insert the MYSTAT program disk. Put the MYSTAT data disk in drive B. Type **MYSTAT** and press Enter. You will see the MYSTAT copyright screen.

b. If you are using a hard disk system, at the MS-DOS prompt, put the MYSTAT data disk in drive A, type **CD\MYSTAT**, and press Enter. Then type **MYSTAT** (and Enter). You will see the MYSTAT copyright screen.

5. At the MYSTAT copyright screen, press Enter, and you will see the main command screen (see Screen 17). Perform the desired statistical procedures (selected topics are discussed below).

6. Exit the program. After completing your analysis, type **QUIT** at the main command screen prompt and press Enter.

Getting Started

There are several general aspects to MYSTAT.

The Main Command Screen As shown in Screen 17, the main command screen provides a list of the thirty-six major commands that MYSTAT uses. Below the list of commands is the MYSTAT prompt (>). This is called the command line. All commands are input at the command line. The first time you start MYSTAT, type **DEMO** at the command line to get a brief introduction to MYSTAT and its capabilities.

SCREEN 17 MYSTAT Main Command Screen

```
MYSTAT --- An Instructional Version of SYSTAT

| DEMO   | EDIT    | MENU    | PLOT      | STATS    | MODEL     |
| HELP   |         | NAMES   | BOX       | TABULATE | CATEGORY  |
| SYSTAT | USE     | LIST    | HISTOGRAM | TTEST    | ANOVA     |
|        | SAVE    | FORMAT  | STEM      | PEARSON  | COVARIATE |
|        | PUT     | NOTE    | TPLOT     |          | ESTIMATE  |
|        | SUBMIT  |         |           |          |           |
|        |         |         |           |          |           |
| QUIT   | OUTPUT  | SORT    | CHARSET   | SIGN     |           |
|        |         | RANK    |           | WILCOXON |           |
|        |         | WEIGHT  |           | FRIEDMAN |           |

>
If you are a new user, type DEMO and then press the [Enter] key.
```

On-line Help To obtain information about commands, type **HELP** at the command line, followed by the name of the specific command you are inquiring about. For example, type **HELP PEARSON** (Enter) to obtain information about the procedure for computing a Pearson *r.* Enter then returns you to the main command screen.

Printing You may obtain a printed copy of a data file or analysis by pressing the Prt Sc (Print Screen) key when the information is displayed on your screen.

The MYSTAT Editor

At the main command screen, type **EDIT** on the command line (and Enter) to enter the editor. The editor is used to enter, edit, and save data. The top, boxed-in part of the screen is the Edit window, where data are displayed. The lower portion of the screen is the command area, where you enter commands.

Pressing the Esc key allows you to move between the Edit window and the command area. Typing **QUIT** in the command area returns you to the main command screen.

Entering Your Data

To enter data, follow these steps.

1. At the main command screen, type **EDIT** in the command line (and Enter) to enter the editor.
2. The first cell in the empty Edit window will be highlighted. (If the Edit window is not empty, press the Esc key and you'll move to the command area. Type **NEW** (and Enter), and the Edit window should now be empty. Press Esc and you will be in the first cell of the empty Edit window.) The editor regards each row as a subject, or "case," and each column as a variable.
3. Name each variable using the first row in the Edit window.
   a. Each variable name must be surrounded by single quotation marks.
   b. Each variable name must start with a letter and contain a maximum of eight characters.
   c. If you are using letters to categorize subjects on a variable, the variable name must end with a dollar sign ($).
4. Enter the data.
   a. If the score is a number, type the number.
   b. If a score is actually a letter or word (up to twelve characters), it must be surrounded by single quotation marks.
   c. Enter a decimal point (.) in place of any missing scores.
5. Check your data using the arrow keys to highlight any score you want to change, and then type in the corrected value (and Enter).
6. Save your entered data by moving to the command area (press the Esc key). Then type the word **SAVE** followed by the appropriate disk drive letter and

the name of your file (for example, **SAVE b:Test**). File names must begin with a letter, be a maximum of eight characters long, and be different from other file names.

7. Type **QUIT** to return to the main command screen so you can perform the desired statistical analyses.

To familiarize yourself with the details of entering data—and to create some files for use in later procedures—complete the following three examples.

Data File Example 1 Say you have identified the age and gender of sixteen students as shown in Table F.12.

To name the variables, in the first highlighted cell of the Edit window, type **'AGE'** (Enter). The cursor highlights the cell one column to the right. Type **'GENDER$'** (Enter). After naming the variables, press the Home key (or use the arrow keys) to highlight the first row ("case") under the AGE variable. Now enter the data. Type **25,** the first subject's age (and Enter). The screen will show 25.000, and the case under GENDER$ will be highlighted. Subject 1 is male, so type **'M'** (Enter). The cursor will highlight case 2 under AGE, so type **21** (and Enter) and then **'F'**, and so on. Screen 18 shows the first eight cases entered into the editor screen. When you get to case 16, the screen will scroll upward. This is OK—the data you entered is still there.

Check your data using the arrow keys or the Page Up and Page Down keys to move through the data. Highlight an incorrect score and type in the corrected value (and Enter). Then save the data on your data disk. We'll assume the disk is in drive B. First move the cursor to the command area by pressing the Esc key. Say we wish to name the file AGEGEN1. Type **SAVE B:AGEGEN1**.

To create another data file, be sure you're in the command area of the editor, and then type **NEW**. The Edit window should be empty. Press Esc and the cursor will position itself for you to enter the first variable name. Type **QUIT** in the

TABLE F.12 Age and Gender Distribution for a Hypothetical Class

Age	*Gender* (M = male, F = female)
25	M
21	F
18	F
17	F
17	F
17	M
24	M
19	F
19	F
32	M
24	F
23	F
41	F
16	F
20	M
19	M

SCREEN 18 MYSTAT Editor: First Eight Cases in the AGEGEN1 Data File

```
MYSTAT Editor
  Case          AGE      GENDER$
    1         25.000            m
    2         21.000            f
    3         18.000            f
    4         17.000            f
    5         17.000            f
    6         17.000            m
    7         24.000            m
    8         19.000            f
    9
   10
   11
   12
   13
   14
   15
```

command area to return to the main command screen so you can perform the desired statistical analyses.

Data File Example 2 Say that we want to input the data shown in Table F.1 (page 549). To label the variable of annual income, type **'ANNINC'** in the first highlighted cell of the Edit window. To label the variable of years of schooling, type **'YEARSSCH'**. Press the Home key (or use the arrow keys) to move to the first case under ANNINC. Type **38**. Then type **18** under YEARSSCH. The cursor moves down to case 2, so type **10** (Enter), **15** (Enter), and so on. After entering all the data from Table F.1, the screen should look like Screen 19. If not, correct your data as described in Data File Example 1. Save your data by pressing Esc and typing **SAVE B:CORR**.

SCREEN 19 MYSTAT Editor: Ten Cases Showing ANNINC and YEARSSCH data

```
MYSTAT Editor
  Case      ANNINC      YEARSSCH
    1         38.000       18.000
    2         10.000       15.000
    3          6.000       10.000
    4         19.000        9.000
    5          3.500       14.000
    6           .000       13.000
    7         12.300       11.000
    8         13.500       13.000
    9         65.000       21.000
   10         22.000       17.000
   11
   12
   13
   14
   15
```

Data File Example 3 Say that you have nominal or categorical data—for example, counting the type of bicycle (mountain, racing, or touring) you see people riding in the park and the frequency of each gender (male or female) you see riding. You can code the type of bike using this scheme: mountain bike is 1, racing bike is 2, and touring bike is 3. For the variable of gender, female is 1 and male is 2. You obtain the data shown in Table F.13. To enter the data, label the variable of bike type by typing **'BIKETYPE'** in the first highlighted cell of the Edit window. Label the variable of gender as **'GENDER'**. Press the Home key (or use the arrow keys) to move to the first case under BIKETYPE, and type **1**. Under GENDER, also type **1**. The cursor moves down to case 2, so type **1** and **1**, and so on. Complete the process so that the screen looks like Table F.13. Check and correct your data as described in Example 1, and then save the data by pressing Esc and typing **SAVE B:BIKE**.

TABLE F.13 Categorical Data Set
The categorical independent variables are bicycle type (mountain bike = 1, racing bike = 2, touring bike = 3) and gender (female = 1, male = 2).

BIKE TYPE	GENDER
1	1
1	1
2	2
2	1
1	2
1	2
1	1
3	1
1	2
2	1
1	2
2	2
1	1
1	1
1	1
3	2
1	1
2	2
1	1
2	2
1	2
1	1
2	2
3	1
1	1
1	1
2	2
1	1
1	2
2	2

Performing a Procedure

All statistical analyses are performed from the main command screen. To perform an analysis, you must first tell MYSTAT which saved data file you want. This is done with the USE command. For example, the first step in performing any analysis on the file created in Example 1 (AGEGEN1) is for you to type **USE B:AGEGEN1** (Enter). The variable names used in the file will appear. Press Enter again to return to the main command screen.

Then type in the appropriate commands for the selected procedure, as described below. Be sure to press the Enter key after each command.

Frequency Tables, Percentiles, and Histograms

This procedure creates an ungrouped frequency table showing simple, relative, and cumulative frequencies and percentiles. It will not list any score having zero frequency, as we did in Chapter 3. We'll use the data from Example 1 above, called AGEGEN1.

At the command line in the main command screen, type

USE B:AGEGEN1
TABULATE AGE /LIST

The screen will look like Screen 20. The age scores are shown in the far right-hand column, arranged in ascending order. Also shown are each score's (1) frequency, under COUNT; (2) cumulative frequency, under CUM COUNT; (3) percent of *N*, under PCT (which multiplied by 100 gives its relative frequency), and (4) percentile, under CUM PCT. To get a histogram of the data, you type **HISTOGRAM** on the command line. (See the MYSTAT instructions for additional types of graphs you can obtain.)

SCREEN 20 MYSTAT Output for Frequency Table

COUNT	CUM COUNT	PCT	CUM PCT	AGE
1	1	6.3	6.3	16.000
3	4	18.8	25.0	17.000
1	5	6.3	31.3	18.000
3	8	18.8	50.0	19.000
1	9	6.3	56.3	20.000
1	10	6.3	62.5	21.000
1	11	6.3	68.8	23.000
2	13	12.5	81.3	24.000
1	14	6.3	87.5	25.000
1	15	6.3	93.8	32.000
1	16	6.3	100.0	41.000

One-Group Descriptive Statistics This procedure computes measure of central tendency and measures of variability for any numeric variable in your data file (as we did in Chapters 4 and 5). We will again use the AGEGEN1 data file.

At the command line in the main command screen, type

USE B:AGEGEN1
STATS AGE /MEAN,MIN,MAX,RANGE,VARIANCE

You are given the *N*, the minimum and maximum scores, the range, the mean, and the estimated population variance.

Correlation and Regression

MYSTAT computes the Pearson *r*, the linear regression equation, and the Spearman r_s (as we did in Chapters 7 and 8). As an example, we'll use the data file from Example 2 above, called CORR.

The Pearson Correlation Coefficient At the command line of the main command screen, type

USE B:CORR
PEARSON

Your output will look like Screen 21. This is a "correlation matrix," in which the computer has correlated every variable with every other variable. The Pearson correlation coefficient between income and years of school is under ANNINC in the row labeled YEARSSCH (.754.) (If we had included a third variable in the CORR data file—say, subjects' ages—then we'd have a 3 × 3 matrix, also obtaining the *r*'s between age and income and between age and years in school.) To see a scatterplot for your data, type **PLOT ANNINC*YEARSSCH**. (The first variable you name becomes the *Y* variable.)

SCREEN 21 MYSTAT Output for the Pearson Correlation Coefficient

```
PEARSON CORRELATION MATRIX

                      ANNINC        YEARSSCH

        ANNINC         1.000
      YEARSSCH         0.754             1.000

NUMBER OF OBSERVATIONS:     10
```

The test of significance is performed using the regression procedure.

Linear Regression MYSTAT performs linear regressions for two variables, as well as multiple regression. As an example, we'll again use our income and years-in-school data.

At the command line of the main command screen, type

USE B:CORR
MODEL YEARSSCH = CONSTANT + ANNINC
ESTIMATE

SCREEN 22 MYSTAT Output for Linear Regression

```
DEP VAR:YEARSSCH       N:     10     MULTIPLE R:   .754     SQUARED MULTIPLE R:   .568
ADJUSTED SQUARED MULTIPLE R:    .514       STANDARD ERROR OF ESTIMATE:           2.618

 VARIABLE     COEFFICIENT        STD ERROR       STD COEF TOLERANCE       T     P(2 TAIL)

CONSTANT             11.343          1.187           0.000  .             9.558      0.000

 ANNINC               0.145          0.045           0.754  .100E+01      3.242      0.012

                              ANALYSIS OF VARIANCE

  SOURCE   SUM-OF-SQUARES        DF  MEAN-SQUARE        F-RATIO        P

REGRESSION          72.061        1       72.061        10.512        0.012
  RESIDUAL          54.839        8        6.855
```

You'll see Screen 22. Although much of this information is beyond the introductory level, you can find the components of the regression equation as follows. In the row labeled CONSTANT, under the word COEFFICIENT is the value of *a,* the *Y* intercept of the regression line (here $a = 11.343$). Below that, in the row labeled ANNINC, is the value of *b,* the slope of the regression line (here $b = 0.145$). Use these to form the regression equation for predicting a subject's income based on his or her number of years in school.

The ANOVA summary table at the bottom of the screen essentially tests the significance of the Pearson *r* calculated from these data. The far-right column, labeled P gives the two-tailed *p* level at which the *r* is significant.

The Spearman Correlation Coefficient For illustration, we'll convert the above scores into rank-ordered variables and compute a Spearman r_S.

At the command line in the main command screen, type

USE B:CORR
SAVE B:CORRRANK
RANK
USE B:CORRRANK
PEARSON

(The RANK command separately rank-orders the income and schooling data, and the ranked data is then saved as CORRRANK. If we had already ranked the subjects, we could create a data file by entering their ranked scores into CORRANK directly.)

You will again see a correlation matrix, as in Screen 21. Although you entered PEARSON, the value in the matrix (here it's .511) is actually the Spearman correlation between the ranks of subjects' annual income and years in school.

Independent and Related Samples *t*-Tests

MYSTAT computes the independent samples and the related samples *t*-test (as we did in Chapter 12), as well as relevant descriptive statistics.

Independent Samples *t*-Test First create the data file. As an example, let's use the data in Table F.4. As described on page 553, the independent variable is alcohol consumption, and it has two levels (alcohol consumed and no alcohol consumed). The dependent variable measures word recognition.

You must identify for the computer the condition under which a score belongs. You could, for example, use the following coding scheme: alcohol group = 1; no alcohol group = 2. Label which alcohol condition a subject belongs under by naming the first variable **'ALCCOND'**. Then name the second variable **'WORDREC'** for the subjects' scores. The first subject in Table F.4 is in the alcohol group, so at case 1, type **1** under ALCOND. This subject scored 9, so type **9** under WORDREC. For case 2, type **1** under ALCOND and **15** under WORDREC. Continue in this way for the first 15 cases (the first column of Table F.4). For case 16, enter the first score from the "no alcohol" conditon—so type **2** under ALCOND and **20** under WORDREC. Likewise, type **2** and the subject's score for all subjects in this conditon. Once all data are entered, type **SAVE B:TWOSAMP**.

At the command line in the main command screen, type

USE B:TWOSAMP
TTEST WORDREC*ALCCOND

You will see Screen 23. First you are given the mean and the estimated population standard deviation for each group. The POOLED VARIANCES row provides the t_{obt} (labeled T), the *df*, and the value of *p* for significance testing.

SCREEN 23 MYSTAT Output for the Independent Samples *t*-Test

```
INDEPENDENT SAMPLES T-TEST ON  WORDREC          GROUPED BY   ALCCOND

     GROUP           N       MEAN              SD
       1.000        15         12.933           3.936
       2.000        15         19.133           5.475

SEPARATE VARIANCES T =     3.561 DF =     25.4 PROB = .002
  POOLED VARIANCES T =     3.561 DF =       28 PROB = .001
```

Related Samples *t*-Test The data file for the related samples test is created differently than the data file for the independent samples test. As an example, we'll use the data in Table F.5. As described on page 554, in this study you tested the same subjects under both an alcohol and a no alcohol condition. Here we will identify each condition as a variable in the data file, so once you are in the editor, name the first column **'ALCGRP'** (for the alcohol condition) and the second column **'NOALCGRP'** (for the no alcohol condition). Then enter each subject's scores in the appropriate column (for the first subject, type **19** in the first column and **20** in the second). When you have entered all data, type **SAVE B:TWOSAMP**.

At the command line of the main command screen, type

USE B:TWOSAMP
TTEST ALCGRP,NOALCGRP

You will see Screen 24, which shows (1) the mean difference and the standard deviation of the difference scores and (2) the t_{obt}, *df*, and exact value of *p*.

SCREEN 24 MYSTAT Output for the Related Samples *t*-Test

```
PAIRED SAMPLES T-TEST ON    ALCGRP     VS NOALCGRP      WITH     15 CASES

MEAN DIFFERENCE  =     -2.267
SD DIFFERENCE  =        7.526
T =       1.166  DF  =    14  PROB  =  .263
```

Analysis of Variance

MYSTAT performs a one-way and two-way random-groups ANOVA (as we did in Chapters 13 and 14). It will also perform ANOVAs involving more than two factors. See the MYSTAT Instruction book.

One-Way Analysis of Variance First create the data file. Let's use the data in Table F.6 (page 555). The independent variable of anxiety has three levels—high, medium, and low—and the dependent variable is subjects' exam scores. As in the independent samples *t*-test performed above, we use one variable column in the data file to identify the level of the factor in which a subject was tested. Say that you use this coding scheme: High = 3, Medium = 2, and Low = 1. When in the editor, name the first variable **'ANXLEVEL'** and the second variable **'EXAMSCOR'**. The first subject in Table F.6 is in the high-anxiety group, so for case 1, type **3** under ANXLEVEL and **9** under EXAMSCOR. For case 2, type **3** under ANXLEVEL and **15** under EXAMSCOR. Continue in the same way for the first ten cases (the first column in Table F.6). For case 11, enter the first score from the medium-anxiety group—type **2** under ANXLEVEL and **20** under EXAMSCOR. Likewise, type **2** and the subject's exam score for all of the mild-anxiety group. (The first fifteen cases are shown in Screen 25.) For case 21

SCREEN 25 MYSTAT Editor: First Fifteen Cases of the ONEWAY Data File

```
MYSTAT Editor
   Case    ANXLEVEL          EXAMSCOR
     1            3.000             9.000
     2            3.000            15.000
     3            3.000            12.000
     4            3.000            17.000
     5            3.000            14.000
     6            3.000            22.000
     7            3.000            11.000
     8            3.000             7.000
     9            3.000            10.000
    10            3.000            13.000
    11            2.000            20.000
    12            2.000            15.000
    13            2.000            25.000
    14            2.000            28.000
    15            2.000            12.000
```

through 30, type **1** under ANXLEVEL to indicate the low-anxiety group and enter the appropriate score under EXAMSCOR. Save the data by typing

SAVE B:ONEWAY

At the command line in the main command screen, type

USE B:ONEWAY
CATEGORY ANXLEVEL = 3
ANOVA EXAMSCOR
ESTIMATE

(The CATEGORY command tells the computer our factor is ANXLEVEL, having three levels. The ANOVA command tells it to perform the ANOVA using EXAMSCOR as the dependent variable.)

You will see the ANOVA summary table shown in Screen 26. Note that in the SOURCE column, instead of "between" and "within" as the names of the components you have ANXLEVEL and ERROR. The value of p for significance testing is also included.

SCREEN 26 MYSTAT Output for the One-Way ANOVA

```
DEP VAR:EXAMSCOR        N:   30    MULTIPLE R:  .592     SQUARED MULTIPLE R:   .350

                          ANALYSIS OF VARIANCE

   SOURCE    SUM-OF-SQUARES    DF   MEAN-SQUARE      F-RATIO        P

  ANXLEVEL          324.067     2       162.033        7.281      0.003

     ERROR          600.900    27        22.256
```

Two-Way Analysis of Variance First create the data file. As an example, we'll use the data from Table F.7 (page 557). As described on pages 555 and 556, in this study you have two independent variables: amount of sleep (with three levels) and caffeine consumption (with two levels). The dependent variable is exam scores.

As in the one-way ANOVA performed above, you must identify the level of the factor under which a subject was tested. You must label two variables in the data file for each factor, for use in identifying the level (cell) a subject was tested under. In the editor, name the first variable **'SLEEP'** and identify the three levels of sleep using this scheme: 0 hours = 1, 4 hours = 2, and 8 hours = 3. Name the second variable **'CAFFEINE'**, and use this scheme: 0 mg = 1 and 100 mg = 2. The third column contains the dependent scores; it is named **'EXAMSCOR'**. The first subject was tested under 0 hours sleep and 0 mg caffeine (see Table F.7), so for case 1 type **1** under SLEEP and **1** under CAFFEINE; then type **9**, the subject's score, under EXAMSCOR. Repeat this procedure for all ten scores in the A_1B_1 cell. For case 11, move to cell A_1B_2. The scores here form the 0 hours sleep, 100 mg caffeine cell so for each subject, enter **1** under SLEEP, **2** under CAFFEINE, and the appropriate score under EXAMSCOR. The first fifteen cases are shown in Screen 27.

SCREEN 27 MYSTAT Editor: First Fifteen Cases of the TWOWAY Data File

```
MYSTAT Editor
  Case      SLEEP       CAFFEINE      EXAMSCOR
     1          1.000        1.000         9.000
     2          1.000        1.000        15.000
     3          1.000        1.000        12.000
     4          1.000        1.000        17.000
     5          1.000        1.000        14.000
     6          1.000        1.000        22.000
     7          1.000        1.000        11.000
     8          1.000        1.000         7.000
     9          1.000        1.000        10.000
    10          1.000        1.000        13.000
    11          1.000        2.000        11.000
    12          1.000        2.000        17.000
    13          1.000        2.000        14.000
    14          1.000        2.000        19.000
    15          1.000        2.000        16.000
```

At case 21, begin entering the scores from cell A_2B_1. These scores form the 4 hours sleep, 0 mg caffeine cell so for each subject, type **2** under SLEEP, **1** under ANXLEVEL, and the appropriate score under EXAMSCOR. Continue this procedure for the A_2B_2, A_3B_1, and A_3B_2 cells, respectively, until all data are entered and you have completed sixty cases. Type

SAVE B:TWOWAY

At the command line in the main command screen, type

USE B:TWOWAY
CATEGORY SLEEP = 3, CAFFEINE = 2
ANOVA EXAMSCOR
ESTIMATE

(The CATEGORY command tells the computer your two factors and the number of levels in each, and the ANOVA command identifies the dependent variable.)

You will now see the ANOVA summary table in Screen 28 (on the next page). The first two rows of the table describe the main effects, the third row describes the interaction, and the fourth row describes the within-groups, or error, component.

SCREEN 28 MYSTAT Output for the Two-Way ANOVA

```
DEP VAR:EXAMSCOR       N:    60    MULTIPLE R:   .535     SQUARED MULTIPLE R:   .287

                              ANALYSIS OF VARIANCE

   SOURCE     SUM-OF-SQUARES  DF   MEAN-SQUARE      F-RATIO       P

    SLEEP           342.533   2        171.267        8.306      0.001
  CAFFEINE           21.600   1         21.600        1.048      0.311
   SLEEP*
  CAFFEINE           83.200   2         41.600        2.018      0.143

    ERROR          1113.400  54         20.619
```

Chi Square

MYSTAT computes the one-way and two-way chi square (as we did in Chapter 15).

One-Way Chi Square As an example, we'll use the bicycle data we entered in Data File Example 3, above, which we saved as BIKE. At the command line in the main command screen, type

USE B:BIKE
TABULATE BIKETYPE

The TABULATE command counts the frequency of the different bike type scores, so you can use it to test whether the frequency of the scores 1, 2, and 3 differ significantly. You'll see Screen 29, which shows a frequency table at the top and, in the row labeled CHI SQUARE, the equivalent of the obtained one-way chi square value, the *df,* and the value of *p* for significance testing.

SCREEN 29 MYSTAT Output for the One-Way Chi Square

```
TABLE OF VALUES FOR BIKETYPE
FREQUENCIES
            1.000       2.000        3.000      TOTAL

               18           9            3         30

TEST STATISTIC                              VALUE          DF          PROB
  LIKELIHOOD RATIO CHI-SQUARE              12.040           2          .002
```

Two-Way Chi Square As an example, we'll perform a two-way chi square using Data File Example 3, above. At the command line in the main command screen, type

USE B:BIKE
TABULATE BIKETYPE*GENDER

You'll see Screen 30, which shows (1) a two-way frequency table (at the top of the screen) and (2) the obtained value of the two-way chi square (with its *df* and *p*). Included in the list of coefficients are the phi and contingency coefficients: select the appropriate one for the type of two-way design you are analyzing.

SCREEN 30 MYSTAT Output for the Two-Way Chi Square

```
TABLE OF BIKETYPE          (ROWS)  BY    GENDER         (COLUMNS)
FREQUENCIES
               1.000         2.000        TOTAL

    1.000   |    12            6 |          18
    2.000   |     2            7 |           9
    3.000   |     2            1 |           3

TOTAL            16           14            30

WARNING: MORE THAN ONE-FIFTH OF FITTED CELLS ARE SPARSE (FREQUENCY < 5)
SIGNIFICANCE TESTS ARE SUSPECT

TEST STATISTIC                              VALUE        DF          PROB
  LIKELIHOOD RATIO CHI-SQUARE               5.187         2          .075

COEFFICIENT                                 VALUE     ASYMPTOTIC STD ERROR
  PHI                                       .4082
  CRAMER V                                  .4082
  CONTINGENCY                               .3780
  GOODMAN-KRUSKAL GAMMA                     .4412          .27860
  KENDALL TAU-B                             .2572          .17758
  STUART TAU-C                              .2667          .18152
  SPEARMAN RHO                              .2662          .18065
  SOMERS D      (COLUMN DEPENDENT)          .2469          .17423
  LAMBDA        (COLUMN DEPENDENT)          .3571          .17181
  UNCERTAINTY   (COLUMN DEPENDENT)          .1251          .10377
```

Wilcoxon *T* Test

MYSTAT computes the Wilcoxon *T* test for related samples (as we did in Chapter 15). As an example, we'll convert the file created in the independent samples *t*-test performed above (saved as TWOSAMPLE) into rank-ordered variables, and perform a Wilcoxon *T* test.

At the command line in the main command screen, type

USE B:TWOSAMPLE
SAVE B:WILKT
RANK
USE B:WILKT
WILCOXON

Screen 31 appears (on next page). The obtained *T* is in the matrix below the label Z = (SUM OF . . . It is the one nonzero value under the ALCGRP in the row labeled NOALCGRP (here, .114). The two-tailed value of p is located in the bottom matrix, labeled TWO-SIDED PROBABILITIES. . . under ALCGRP in the row labeled NOALCGRP (here, .909).

SCREEN 31 MYSTAT Output for the Wilcoxon *T* Test

```
WILCOXON SIGNED RANKS TEST RESULTS

 COUNTS OF DIFFERENCES (ROW VARIABLE GREATER THAN COLUMN)

                  ALCGRP      NOALCGRP

  ALCGRP                  0             8
 NOALCGRP                 7             0

 Z = (SUM OF SIGNED RANKS)/SQUARE ROOT(SUM OF SQUARED RANKS)

                  ALCGRP      NOALCGRP

  ALCGRP               .000
 NOALCGRP              .114          .000

 TWO-SIDED PROBABILITIES USING NORMAL APPROXIMATION

                  ALCGRP      NOALCGRP

  ALCGRP              1.000
 NOALCGRP              .909         1.000
```

SPSS/PC+ STUDENTWARE PLUS 1.0 FOR MS-DOS

SPSS/PC+ Studentware Plus 1.0 for MS-DOS accepts up to fifty variables and an unlimited number of subjects ("cases"). It can perform the major analyses presented in this textbook, as well as more advanced procedures. It runs on an IBM-compatible personal computer with at least 512K RAM (640K recommended), a 10 MB (minimum) hard drive, MS-DOS 3.0 or higher, and a monitor. A printer is optional.

Be sure you have read the Note to Users of All Programs found at the beginning of this appendix. The following provides a basic introduction to Studentware.

Installing Studentware

1. Boot your computer as described under Getting Started in the discussion of HMSTAT.

2. Install Studentware: At the MS-DOS prompt, insert the Studentware disk labeled "Inst/SW1" in the A drive and type **A:INSTALL** (Enter). Follow the instructions. When given the option, install the Social Sciences Data Files.

3. Start the program. If you selected the default installation option, at the MS-DOS prompt type **CD\SPSS\DATA** and press Enter. Then type **SPSSSW.** You will see the Studentware windows screen.

4. Perform the desired statistical procedures. Selected coverage is described below.

5. Exit the program. When you are finished, type **FINISH.** (Yes, you really must type the period!) In the Studentware scratch pad, press F10 and then hit Enter.

Getting Started

Studentware is extremely sensitive to the format of the commands it uses. Some commands require the use of punctuation such as periods, slashes, parentheses, quotation marks, or apostrophes. If these punctuation marks are not entered, the program will not work.

The Windows Screen The windows screen is divided into two large windows with a message line on the bottom. The top window, called the "listing window," displays the results of your analyses. The lower window, called the "scratch pad," is where you type in data and your commands. You move between the listing window and the scratch pad by pressing F2 (Enter).

Messages The message line at the bottom of the screen displays diverse information and asks questions. Simply follow the directions. You'll also see a screen containing an error message if you present the program with an impossible task. Pressing Enter should return you to the scratch pad. Usually the problem is that you have not entered a command precisely. Use the arrow keys to move to any incorrect commands and correct them.

Help Functions On-screen help is available by pressing the F1 key. Among other things, you are provided a glossary of statistical terms, a menu of Studentware commands and subcommands, and explanations and examples of their use.

Executing Commands You will type in data file and program commands in columns, and the commands must be executed in order. Therefore, after you have entered your information, always move the cursor back to the first line of your commands (using the arrow keys or Page Up key). The final step is to execute, always by pressing the F10 key and Enter.

"More" Whenever you see the word MORE in the upper right-hand corner of the screen, this means that more information is available. Press the spacebar or Enter to advance to the next screen.

SCREEN 32 Studentware Output for Frequency Table

AGE age of the subject in years MORE

Value Label	Value	Frequency	Percent	Valid Percent	Cum Percent
	16.00	1	6.3	6.3	6.3
	17.00	3	18.8	18.8	25.0
	18.00	1	6.3	6.3	31.3
	19.00	3	18.8	18.8	50.0
	20.00	1	6.3	6.3	56.3
	21.00	1	6.3	6.3	62.5
	23.00	1	6.3	6.3	68.8
	24.00	2	12.5	12.5	81.3
	25.00	1	6.3	6.3	87.5
	32.00	1	6.3	6.3	93.8
	41.00	1	6.3	6.3	100.0
	Total	16	100.0	100.0	

Press the spacebar again and the frequency histogram shown in Screen 33 will appear. Unlike in Chapter 3, here the scores are presented along the left-hand *Y*-axis, and the length of each horizontal bar along the *X*-axis indicates each score's frequency. Press the spacebar twice and you will be in the scratch pad area of the Studentware Windows screen. There is now information in the listing window: press F2 (Enter) and use the arrow keys to browse through the contents of the listing window. Press F2 (Enter) to return to the scratch pad.

SCREEN 33 Studentware Output for the BARCHART Subcommand

Correlation Coefficients

Studentware computes the Pearson and the Spearman correlation coefficients (as we did in Chapter 7). For this example, let's create a file using the data back in Table F.1 (p. 547). Type

data list free
/anninc yearssch.
begin data.

On the next line, begin entering the data in Table F.1, starting with

38 18

and so on. When all data are entered, type

end data.
variable labels anninc "annual income in thousands of dollars"
/yearssch "years of schooling".
save outfile 'corr.sys'.

Move the cursor back to the first line and press F10. This saves the data using the file name CORR.SYS.

The Pearson Correlation Coefficient To compute the Pearson correlation coefficient on the above data, type

get file 'corr.sys'.
correlation anninc with yearssch.
plot plot anninc with yearssch.

These commands identify the variables in the data file to be correlated and plotted. Move the cursor back to anywhere on the get file line and press F10. Page through the screens until, as in Screen 34, you obtain the Pearson correlation coefficient between the variables ANNINC and YEARSSCH (here $r = .7536$). The * after the obtained r indicates that it is statistically significant at the one-tailed, .01 alpha level. Page through the screens further to obtain a scatterplot of the data. Press the space bar again until you return to the windows screen.

SCREEN 34 Studentware Output for the Pearson Correlation Coefficient

```
Correlations:    YEARSSCH

  ANNINC         .7536*

N of cases:      10          1-tailed Signif:  * - .01   ** - .001

" . " is printed if a coefficient cannot be computed
```

The Spearman Rank-Order Correlation Studentware will transform data to ranked scores and compute the Spearman correlation coefficient. As an example, we'll use the CORR.SYS data file created above.

At the windows screen, type

get file 'corr.sys'.
rank anninc yearssch
/rank into ranninc ryrssch.
list.
correlation ranninc ryrssch.
save outfile 'scorr.sys'.

(The **rank** commands cause the scores in ANNINC and YEARSSCH to be ranked and then labeled as RANNINC and RYRSSCH. The **correlation** command then produces the correlation between these ranked variables.) Move the cursor back to the first line and press F10. Paging through the screens, you'll see the transformed variables and scores and then the correlation matrix between RANNINC and RYRSSCH, as in Screen 35. (Here the Spearman r_s is .5106, and it is not significant in a one-tailed test at the .01 or .001 level. Check Table 4 in Appendix D for the .05 and/or two-tailed critical values.)

SCREEN 35 Studentware Output for the Spearman Correlation Coefficient

```
Correlations:  RANNINC     RYRSSCH

  RANNINC      1.0000       .5106
  RYRSSCH       .5106      1.0000

N of cases:     10          1-tailed Signif:   * - .01    ** - .001

" . " is printed if a coefficient cannot be computed
```

Independent and Related Samples *t*-tests

Studentware computes the independent and related samples *t*-tests (as we did in Chapter 12) and provides descriptive statistics and significance testing for each.

Independent Samples *t*-test For this example, we'll use the data in Table F.4. As described on page 553, the independent variable is alcohol consumption, and it has two levels (alcohol consumed and no alcohol consumed). The dependent variable is word recall. You must identify for the computer the condition in which a score belongs, so use the following coding scheme: alcohol group = 1; no alcohol group = 2. We will identify which alcohol condition a subject is in using the first variable, named ALCCOND, and enter dependent scores in the second variable named WORDREC. To create this file, type

data list free
/alccond wordrec.
begin data.

Then type in the data. The first two subjects in Table F.4 are in the alcohol group and scored 9 and 15, respectively, so enter

1 9
1 15

and so on, for the first fifteen rows of data you enter (from the first column in Table F.4). On the sixteenth row, begin entering the scores from the no alcohol condition (column 2 in Table F.4). Type

2 20
2 15

and so on. Then type

end data.
variable labels alccond "alcohol consumption: 1 = alcohol, 2 = no alcohol"
/wordrec "number of words recalled".
save outfile 'ittest.sys'.

Move the cursor back to the first line and press F10 (Enter) to save this file under ITTEST.SYS. Once back at the windows screen, type into the scratch pad:

get file 'ittest.sys'.
ttest groups alccond (1,2)
/variables wordrec.

(These commands tell the computer to perform the *t*-test using our system of 1's and 2's in ALCCOND to form the groups and the scores in WORDREC as the dependent variable.) Move the cursor to the get file line and press F10. Page through the screens until you see Screen 36, which shows the *N* (Number of Cases), the mean, the estimated population standard deviation (SD), and the standard error of the mean (SE of Mean) in each condition. The bottom of the screen, in the row labeled "Equal" shows the *t*-test results: here the obtained value is −3.56, with *df* = 28, and the results are significant in a two-tailed test at an α of .001. The standard error of the difference between the means is given,

SCREEN 36 Studentware Output for the Independent Samples *t*-Test

```
                                                                    MORE
t-tests for independent samples of ALCCOND  alcohol consumption 1: = alcohol,

                             Number
         Variable          of Cases    Mean          SD          SE of Mean
       ---------------------------------------------------------------------
         WORDREC number of words recalled

        ALCCOND 1.00             15     12.9333      3.936          1.016
        ALCCOND 2.00             15     19.1333      5.475          1.414
       ---------------------------------------------------------------------

        Mean Difference = -6.2000
        Levene's Test for Equality of Variances: F = 1.552 P = .223

     t-test for Equality of Means                                  95%
  Variances  t-value   df     2-Tail Sig     SE of Diff        CI for Diff
  -------------------------------------------------------------------------
  Equal      -3.56    28          .001          1.741        (-9.768, -2.632)
  Unequal    -3.56    25.42       .001          1.741        (-9.787, -2.613)
```

and in the far right-hand column is the confidence interval for the difference between the two means.

Related Samples *t*-Test For this example, we'll create a file using the data in Table F.5. As described on page 554, in this study you tested the same subjects under both an alcohol and a no alcohol condition. Here you enter the scores from each condition as a separate variable, so name them ALC (for the alcohol condition) and NOALC (for the no alcohol condition). At the windows screen, type

data list free
/alc noalc.
begin data.

Then enter each subject's pair of scores, first for the alcohol condition and then for the no alcohol condition. For the first two subjects in Table F.5, type

19 20
15 15

and so on. Do likewise for all subjects' scores. Then type

end data.
variable labels alc "subjects who consumed alcohol"
/noalc "subjects who did not consume alcohol".
save outfile 'rttest.sys'.

Move the cursor back to the first line and press F10 (Enter) to save this file under RTTEST.SYS. Once back at the windows screen, type

get file 'rttest.sys'.
ttest pairs alc noalc.

(The **ttest pairs** command tells the computer to perform a related samples *t*-test, comparing the scores in ALC to those in NOALC. Move the cursor back to the get file line, and then press F10 (Enter). Page through the screens until you see Screen 37, which shows *N*, information on the correlation between the scores in

SCREEN 37 Studentware Output for the Related Samples *t*-Test

```
                    - - - t-tests for paired samples - - -

                    Number of          2-tail
Variable              pairs     Corr     Sig         Mean        SD      SE of Mean

ALC   subjects who consumed alcohol                 15.3333     4.746       1.225
                      15     -.116   .680
NOALC  subject who did not consume alc              17.6000     5.316       1.373

        Paired Differences
Mean          SD         SE of Mean          t-value            df    2-tail Sig

-2.2667     7.526       1.943               -1.17              14    .263
95% CI (-6.435, 1.902)
```

the two conditions, and, for each condition, the mean and estimated population standard deviation and the standard error of the mean. At the bottom of the screen is the mean difference, the estimated population standard deviation of the difference scores, the standard error of the mean difference, and the obtained t, df, and exact value of p.

Analysis of Variance

Studentware performs the one-way or two-way between-subjects ANOVA (as we did in Chapters 13 and 14). It will also perform ANOVAs involving more than two factors. See the Studentware Instruction book. Within-subjects (repeated measures) ANOVAs cannot be performed.

One-Way Analysis of Variance First create the data file, using the data in Table F.6. As described on page 555, the independent variable of anxiety has three levels (high, medium, and low), and the dependent variable is subjects' exam scores. As in the independent samples t-test above, use one variable in the data file to identify the level of the factor under which a subject was tested. Say you use this coding scheme: High = 3, Medium = 2, and Low = 1. To create this file, type

data list free
/anxlevel examscor.
begin data.

Then type in the data. The first two subjects in Table F.6 are in the high-anxiety group and scored 9 and 15, respectively, so enter

3 9
3 15

Continue in this way for the first ten rows of data you enter (the first column in Table F.6). On the eleventh row, begin entering the scores from the medium-anxiety condition in Table F.6. Type

2 20
2 15

and so on. Follow the same scheme for the low-anxiety group, identifying each with a **1** in the first column. Then type

end data.
variable labels anxlevel "anxiety level: 3=high, 2=medium, 1=low"
/examscor "number of correct answers".
save outfile 'oneway.sys'.

Move the cursor back to the first line and press F10 (Enter) to save this file under ONEWAY.SYS. Once back at the windows screen, type into the scratch pad:

get file 'oneway.sys'.
means examscor by anxlevel.
oneway examscor by anxlevel (1,3).

Move the cursor back to the get file line and press F10 (Enter). The **means** command uses the scores in EXAMSCOR to produce a screen containing the mean, estimated population standard deviation, and *N* for all data in the study (labeled "For Entire Population") and for each level of ANXLEVEL.

The **oneway** command performs a one-way ANOVA on the dependent scores in the variable named EXAMSCOR for the independent variable identified as ANXLEVEL. The **(1,3)** indicates that you have levels 1 through 3. Page through the screens until you see Screen 38, which shows the ANOVA summary table (including *p*).

SCREEN 38 Studentware Output for the One-Way ANOVA

```
- - - - - - - - - - O N E W A Y - - - - - - - - - -

Variable     EXAMSCOR    number of correct answers

By Variable  ANXLEVEL    anxiety level: 3 = high, 2 = medium, 1 = low

                         Analysis of Variance

                                Sum of         Mean            F        F
Source              D.F.        Squares        Squares         Ratio    Prob.

Between Groups        2         324.0667       162.0333        7.2806   .0030

Within Groups        27         600.9000        22.2556

Total                29         924.9667
```

Two-Way Analysis of Variance As an example, we'll create a file from the data shown in Table F.7 (page 557). As described on pages 556 and 557, in this study you have two independent variables: amount of Sleep (with three levels) and caffeine consumption (with two levels). The dependent variable is exam scores.

As in the one-way ANOVA performed above, you must identify the level of the factor under which a subject was tested. With two factors, create two variables for identifying the cell a subject was tested under. Here, name the first variable SLEEP; in the first column you'll code the three levels of sleep using this scheme: zero hours = 1, four hours = 2, and eight hours = 3. Name the second variable CAFFEINE (the second column) and code it using this scheme: 0 mg = 1 and 100 mg = 2. The third variable (column) contains the dependent scores; it is named EXAMSCOR. To create the file, type

data list free
/sleep caffeine examscor.
begin data.

In Table F.7, the first subject (scoring 9) was tested under 0 hours sleep (coded as level 1) and 0 mg caffeine (coded as level 1), so in the first row type

1 1 9

In subsequent rows, repeat this procedure for all ten scores in the A_1B_1 cell of Table F.7. At row 11, move to cell A_1B_2, where the scores are for 0 hours sleep (coded as 1) and 100 mg caffeine (coded as 2). For subject 1 here, enter

1 2 11

Continue in this way for all the scores in this cell. At row 21, begin entering the scores from cell A_2B_1. These scores are from the 4 hours sleep, 0 mg caffeine cell. For subject 1, type

2 1 20

And so on. Continue this procedure for the A_2B_2, A_3B_1, and A_3B_2 cells, respectively, until all sixty scores are entered. Then type

end data.
variable labels sleep "# of hours of sleep:1=0hrs, 2=4hrs,3=8hrs"
/caffeine "caffeine consumption: 1=0mg, 2=100mg"
/examscor "# of questions correctly answered".
save outfile 'twoway.sys'.

Move the cursor back to the first line and press F10 (Enter) to save this file under TWOWAY.SYS. Once you are at the Windows screen, type

get file 'twoway.sys'.
means examscor by sleep by caffeine.
anova examscor by sleep (1,3) caffeine (1,2).

Move the cursor back to the get file line and press F10 (Enter). The **means** command produces the descriptive statistics on the scores in the variable named EXAMSCOR for the variables SLEEP and CAFFEINE. Paging through the screens will bring you to Screen 39. The statistics describing the main effect for

SCREEN 39 Studentware Output for the MEANS Command

```
Summaries of EXAMSCOR  # of questions correctly answered
By levels of SLEEP     # of hours of sleep: 1 = 0hrs, 2 = 4hrs,
             CAFFEINE  caffeine consumption: 1 = 0mg, 2 = 100mg

Variable         Value  Label                    Mean        Std Dev       Cases

For Entire Population                          15.4333        5.1433          60

SLEEP             1.00                         14.0000        4.3286          20
  CAFFEINE        1.00                         13.0000        4.3205          10
  CAFFEINE        2.00                         15.0000        4.3205          10

SLEEP             2.00                         18.8000        5.4348          20
  CAFFEINE        1.00                         20.6000        5.6999          10
  CAFFEINE        2.00                         17.0000        4.7610          10

SLEEP             3.00                         13.5000        3.9802          20
  CAFFEINE        1.00                         14.5000        3.9511          10
  CAFFEINE        2.00                         12.5000        3.9511          10

  Total Cases =        60
```

each level of sleep are in the rows labeled SLEEP. For example, the main effect mean for sleep level 1 (0 hours) is 14.00, and the estimated population standard deviation is 4.3286. Under each level of sleep are summaries of the cells for the levels of caffeine under that sleep level. For example, the mean in the cell combining level 1 of sleep (0 hours) and level 1 of caffeine (0 mg) is 13.0, and the estimated population standard deviation for the cell is 4.3205. You are not given a summary of the main effect for the second variable (caffeine).

The ANOVA command performs the ANOVA using the dependent variable of EXAMSCOR and the independent variables of SLEEP and CAFFEINE. As in Screen 40, the output is an ANOVA summary table showing the results for first the main effects and then the interaction (including p).

SCREEN 40 Studentware Output for the Two-Way ANOVA

```
           * * * A N A L Y S I S   O F   V A R I A N C E * * *

            EXAMSCOR # of questions correctly answered
       BY   SLEEP    # of hours of sleep: 1 = 0hrs, 2 = 4hrs,
            CAFFEINE caffeine consumption: 1 = 0mg, 2 = 100mg
```

Source of Variation	Sum of Squares	DF	Mean Square	F	Signif of F
Main Effects	364.133	3	121.378	5.887	.001
SLEEP	342.533	2	171.267	8.306	.001
CAFFEINE	21.600	1	21.600	1.048	.311
2-way Interactions	83.200	2	41.600	2.018	.143
SLEEP CAFFEINE	83.200	2	41.600	2.018	.143
Explained	447.333	5	89.467	4.339	.002
Residual	1113.400	54	20.619		
Total	1560.733	59	26.453		

Chi Square

Studentware performs one-way and two-way chi squares (as we did in Chapter 15). As an example, we'll use the categorical data from Table F.13. As described on page 568, the data reflect the type of bicycle each subject is riding and the subject's gender. To create the data file, type

data list free
/biketype gender.
begin data.

Then type in the scores as in Table F.13. For example, for the first three subjects, type

1 1
1 1
2 2

After entering all of the data, type

end data.
variable labels biketype "bicycle type: 1=mount, 2=race, 3=tour"
/gender "gender of rider: 1=female, 2=male".
save outfile 'chi.sys'.

Move the cursor back to the data list free line and press F10 (Enter) to save this file under CHI.SYS.

One-Way Chi Square After creating the data file, to perform the one-way chi-square for the variable of type of bicycle, go to the windows screen and type

get file 'chi.sys'.
npar tests chisquare=biketype.

Move the cursor back to the get file line and press F10 (Enter). The **npar** command tells Studentware to do a nonparametric test, and **chisquare=biketype** performs a chi square on the variable of bike type. Paging through the screens will bring you to Screen 41. For each of the three categories of bike type, you are given the observed frequency, the expected frequency (the program assumes your null hypothesis is an equal distribution of the frequencies in the population), and the residual or difference between the observed and expected frequencies. Then you are given the obtained value of chi square (here 11.4), the *df,* and, under significance, the value of *p*.

SCREEN 41 Studentware Output for the One-Way Chi Square

```
- - - - - Chi-square Test

    BIKETYPE  type of bicycle: 1 = mount, 2 = race, 3 = tour

                   Cases
     Category  Observed  Expected  Residual

        1.00         18     10.00      8.00
        2.00          9     10.00     -1.00
        3.00          3     10.00     -7.00
                     --
       Total         30

       Chi-Square             D.F.           Significance
          11.400                2                  .003
```

Two-Way Chi Square After creating the data file, to perform the two-way chi-square, go to the windows screen and type

get file 'chi.sys'.
crosstabs biketype by gender
/statistics chisq.

Move the cursor back to the get file line and press F10 (Enter). The **crosstabs** command creates a gender-by-bike type (2 × 3) frequency table. The **statistics chisq** command produces Screen 42 (on next page). In the row labeled "Pearson" is the obtained value of the two-way chi square (here 5.00), the *df,* and the value of *p*.

SCREEN 42 Studentware Output for the Two-Way Chi Square

```
                                                                  MORE
      Chi-Square                    Value              DF      Significance
--------------------          -----------            ----      ------------

Pearson                          5.00000               2           .08208
Likelihood Ratio                 5.18709               2           .07475
Mantel-Haenszel test for         1.15079               1           .28338
      linear association

Minimum Expected Frequency -    1.400
Cells with Expected Frequency < 5 -       4 OF      6 ( 66.7%)


Number of Missing Observations: 0
```

Mann-Whitney *U* Test

Studentware computes the Mann-Whitney *U* test and the rank sums test for independent samples (as we did in Chapter 15). As an example, say we have two groups of five subjects each and have ranked the relative performance of all subjects as follows:

Group 1: 9, 4, 7, 3, 6
Group 2: 1, 8, 2, 5, 10

As in the independent samples *t*-test performed above, use one variable in the data file to identify the condition under which a subject was tested. Thus, to create the data file, type

data list free
/grp rank.
begin data.

Then enter the data by first identifying a subject's condition and then entering the score. For example, for the first two subjects, type

1 9
1 4

and so on. Then type

end data.
variable labels grp "group"
/rank "ranked score".
save outfile 'mann.sys'.

Move the cursor back to the data list free line and press F10 (plus Enter) to save this file under MANN.SYS. Once back at the Windows screen, type

get file 'mann.sys'.
npar tests m-w=rank by grp(1,2).

Move the cursor back to the get file line, and then press F10 (Enter). The **npar** command with **m-w** selects the Mann-Whitney *U* test, using the variable you labeled "rank" for dependent scores and the variable you labeled "group" to identify the conditions. The output is in Screen 43, which shows (1) the mean rank and *n* (number of cases) in each condition, (2) the obtained Mann-Whitney *U* (here 11.0) and the two-tailed *p* (here .8413), and (3) the obtained *z* and *p* level for the rank sums test.

SCREEN 43 Studentware Output for the Mann-Whitney *U* and Rank Sums Test

```
- - - - - Mann-Whitney U - Wilcoxon Rank Sum W Test

     RANK         ranked score
  by GRP          group

     Mean Rank      Cases

          5.80          5 GRP = 1.00
          5.20          5 GRP = 2.00
                       --
                       10 Total

                                 EXACT                Corrected for Ties
          U             W       2-tailed P               Z        2-tailed P
        11.0          29.0        .8413              -.3133          .7540
```

Wilcoxon *T* Test

Studentware computes the Wilcoxon *T* test for related samples using ranked data (as we did in Chapter 15). Create the data file in the same way as for the related samples *t*-test performed above, entering the scores from each condition as a separate variable (column). For example, to create a file using the data in Table F.5 (p. 554), which compares subjects tested under both an alcohol and a no alcohol condition, go to the windows screen and type

data list free
/alc noalc.
begin data.

Now enter the data as for the related samples *t*-test. Then type

end data.
variable labels alc "group who consumed alcohol"
/noalc "group who did not consume alcohol".
save outfile 'wil.sys'.

Move the cursor back to the first line and press F10 (plus Enter) to save this file under WIL.SYS. Once back at the Windows screen, type

get file 'wil.sys'.
rank alc noalc
/rank into ralc rnoalc
npar tests wilcoxon=ralc with rnoalc.

Move the cursor back to the get file line, and then press F10 (Enter). The **rank** commands cause the scores in ALC and NOALC to be ranked, then labeled as RALC and RNOALC. The **npar** command with **wilcoxon** selects the Wilcoxon T test, comparing the conditions "ralc" and "rnoalc." The output is in Screen 44, which shows (1) the mean rank and n in each condition and (2) the value of z, which is the obtained T (here $-.1136$), and the two-tailed p (here .9096).

SCREEN 44 Studentware Output for Wilcoxon *T* Test

```
- - - - - Wilcoxon Matched-pairs Signed-ranks Test

     RALC       group who consumed alcohol
with RNOALC     group who did not consume alcohol

     Mean Rank     Cases

          7.25          8   - Ranks  (RNOALC Lt RALC)
          8.86          7   + Ranks  (RNOALC Gt RALC)
                        0     Ties   (RNOALC Eq RALC)
                       --
                       15     Total

        Z =      -.1136            2-tailed P =   .9096
```

Kruskal-Wallis *H* Test

Studentware computes the Kruskal-Wallis H test—the one-way, between-subjects ANOVA for ranks (as we did in Chapter 15). Create the data file in the same way as described for the one-way ANOVA performed above, using one variable in the data file to identify the level of the factor under which a subject was tested. As an example, create the data file from the data in Table F.10 (page 561). As described, you have the annual income scores from subjects in one of three conditions: underweight (coded 1), average weight (coded 2), or overweight (coded 3). Create the file by typing

data list free
/grp anninc.
begin data.

Then enter the data for the first subject by typing

1 9

Continue, ending with the final subject:

3 14

Now type

end data.
variable labels grp "weight group: 1=under, 2=average, 3=over"
/anninc "annual income in thousands of dollars".
save outfile 'kruskal.sys'.

Move the cursor back to the first line and press F10 (Enter) to save this file under KRUSKAL.SYS. Once back at the windows screen, type

get file 'kruskal.sys'.
npar tests k-w=anninc by grp(1,3).

Move the cursor back to the get file line and press F10 (Enter). The **npar** command with **k-w** selects the Kruskal-Wallis *H* test, using the variable you labeled "anninc" for dependent scores and the variable you labeled "grp" to identify the conditions. The output is in Screen 45, which shows (1) the mean rank and *n* for each condition and (2) the obtained chi square value (here 9.4187) and the two-tailed *p* (here .0090).

SCREEN 45 Studentware Output for Kruskal-Wallace *H* Test

```
- - - - - Kruskal-Wallis 1-way ANOVA

   ANNINC      annual income in thousands of dollars
by GRP        weight group: 1=under, 2=average, 3=over

   Mean Rank     Cases

      10.60        10     GRP =     1
      22.25        10     GRP =     2
      13.65        10     GRP =     3
                   --
                   30     Total

                                                 Corrected for Ties
      CASES     Chi-Square   Significance   Chi-Square  Significance
         30         9.4187          .0090       9.4481         .0089
```

Friedman Chi Square Test

Studentware performs the Friedman chi square test—the one-way, within-subjects ANOVA for ranks (as we did in Chapter 15). Create the data file in the same way as described for the related samples *t*-test performed above, using one variable in the data file for each level of your factor. As an example, create the data file from the data in Table F.11 (page 562). As described, here you have subjects' ranked preferences for three colas. Create the file by typing

data list free
/cola1 cola2 cola3.
begin data.

Then enter the data, using one row per subject. Thus, for subject one, type

1 2 3

and so on. Then type

end data.
save outfile 'fri.sys'.

Move the cursor back to the first line and press F10 (Enter) to save this file under FRI.SYS. Once back at the windows screen, type:

get file 'fri.sys'.
npar tests friedman = cola1 cola2 cola3.

Move the cursor back to the get file line and press F10 (Enter). The **npar** command with **friedman** selects the Friedman chi square test, comparing the ranks in the variables (conditions) labeled "cola1," "cola2," and "cola3." The output is in Screen 46, which shows (1) the mean rank and n in each condition and (2) the obtained chi square value (here 6.2000), the *df,* and the two-tailed p.

SCREEN 46 Studentware Output for Friedman Chi Square Test

```
- - - - - Friedman Two-way ANOVA

     Mean Rank   Variable

          1.50   COLA1
          1.90   COLA2
          2.60   COLA3

          Cases          Chi-Square            D.F.        Significance
             10              6.2000               2               .0450
```

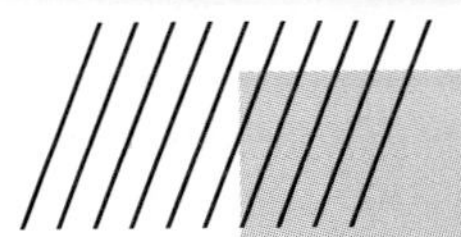

Glossary

Alpha The Greek letter α, which symbolizes the criterion, the size of the region of rejection of a sampling distribution, and the theoretical probability of making a Type I error

Alternative hypothesis The statistical hypothesis describing the population parameters that the sample data represent if the predicted relationship does exist; symbolized by H_a

Analysis of variance The parametric procedure for determining whether significant differences exist in an experiment containing two or more sample means; abbreviated ANOVA

ANOVA Abbreviation of analysis of variance

Bar graph A graph in which a free-standing vertical bar is centered over each score on the X axis; used with nominal or ordinal scores

Beta The Greek letter β, which symbolizes the theoretical probability of making a Type II error

Between-subjects ANOVA The type of ANOVA that is performed when a study involves between-subjects factors

Between-subjects factor The type of factor created when an independent variable is studied using independent samples in all conditions

Bimodal distribution A symmetrical frequency polygon with two distinct humps where there are relatively high frequency scores and with center scores that technically have the same frequency

Cell In a two-way ANOVA, the combination of one level of one factor with one level of the other factor

Central limit theorem A statistical principle that defines the mean, standard deviation, and shape of a theoretical sampling distribution

χ^2-distribution The sampling distribution of all possible values of χ^2 that occur when the samples represent the distribution of frequencies described by the null hypothesis

Chi square procedure The nonparametric inferential procedure for testing whether the frequencies of category membership in the sample represent the predicted frequencies in the population

Coefficient of alienation The proportion of variance not accounted for by a relationship; computed by subtracting the squared correlation coefficient from 1

Coefficient of determination The proportion of variance accounted for by a relationship; computed by squaring the correlation coefficient

Collapsing In a two-way ANOVA, averaging together all scores from all levels of one factor in order to calculate the main effect means for the other factor

Complete factorial design A two-way ANOVA design in which all levels of one factor are combined with all levels of the other factor

Condition An amount or category of the independent variable that creates the specific situation under which subjects' scores on the dependent variable are measured

Confidence interval for a single population μ A statistically defined range of values of μ, any one of which is likely to be represented by the sample mean

Confidence interval for the difference between two μ's A statistically defined range of differences between two population μ's, any one of which is likely to be represented by the difference between the two sample means

Confounded comparison In a two-way ANOVA, a comparison of two cells that differ along more than one factor

Contingency coefficient The statistic that describes the strength of the relationship in a two-way chi square when there are more than two categories for either variable; symbolized by C

Continuous scale A measurement scale that allows for fractional amounts of the variable being measured

Correlation coefficient A number, computed from the pairs of X scores and Y scores in a set of data, that summarizes and describes the type of relationship present and the strength of that relationship

Correlational study A procedure in which subjects' scores on two variables are simply measured, without manipulation of any variables, to determine whether there is a relationship

Criterion The probability that provides the basis for deciding whether a sample is too unlikely to have occurred by chance and thus is unrepresentative of a particular population

Criterion variable The variable in a relationship whose unknown scores are predicted through use of the known scores on the predictor variable

Critical value The value of the sample statistic that marks the edge of the region of rejection in a sampling distribution; values that fall beyond it fall in the region of rejection

Cumulative frequency The frequency of those scores at or below a particular score; symbolized by cf

Cumulative frequency distribution A distribution of scores organized to show the frequency of the scores at or below each score

Curvilinear relationship See *Nonlinear relationship*

Data point A dot plotted on a graph to represent a pair of X and Y scores

Degree of association See *Strength of a relationship*

Degrees of freedom The number of scores in a sample that are free to vary, and thus the number that is used to calculate an estimate of the population variability; symbolized by df

Dependent events Events for which the probability of one is influenced by the occurrence of the other

Dependent samples See *Related samples*

Dependent variable In an experiment, the variable that is measured under each condition of the independent variable

Descriptive statistics Procedures for organizing and summarizing data so that the important characteristics can be described and communicated

Deviation The distance that separates a score from the mean and thus indicates how much the score differs from the mean

Dichotomous variable A discrete variable that has only two possible amounts or categories

Discrete scale A measurement scale that allows for measurement only in whole-number amounts

Distribution An organized set of data

Effect size The proportion of variance accounted for in an experiment, which indicates how consistently differences in the dependent scores are "caused" by changes in the independent variable

Empirical knowledge Knowledge obtained through observation of events

Empirical probability distribution A probability distribution based on observations of the relative frequency of events

Error variance The inherent variability within a population, estimated in ANOVA by the mean square within groups

Estimated standard error of the mean An estimate of the standard deviation of the sampling distribution of means, used in calculating the single-sample *t*-test; symbolized by $s_{\bar{X}}$

Eta The correlation coefficient used to describe a linear or nonlinear relationship containing two or more levels of a factor; symbolized by η

Eta squared The proportion of variance in the dependent variable that is accounted for by changing the levels of a factor, and thus the measurement of effect size, in the sample; symbolized by η^2

Expected frequency In chi square, the frequency expected in a category if the sample data perfectly represent the distribution of frequencies in the population described by the null hypothesis; symbolized by f_e

Experiment A research procedure in which one variable is actively changed or manipulated, the scores on another variable are measured, and all other variables are kept constant, to determine whether there is a relationship

Experimental hypotheses Two statements made before a study is begun, describing the predicted relationship that may or may not be demonstrated by the study

Experiment-wise error rate The probability of making a Type I error when comparing all means in an experiment

Extreme scores The scores that are relatively far above and below the middle score of any distribution

Factor In ANOVA, an independent variable

***F*-distribution** The sampling distribution of all possible values of *F* that occur when the null hypothesis is true and all conditions represent one population μ

Fisher's protected *t*-test The *post hoc* procedure performed with ANOVA to compare means from a factor in which all levels do not have equal *n*

***F*-ratio** In ANOVA, the ratio of the mean square between groups to the mean square within groups

Frequency The number of times each score occurs within a set of data; also called simple frequency; symbolized by *f*

Frequency polygon A graph that shows interval or ratio scores (*X* axis) and their frequencies (*Y* axis), using data points connected by straight lines

Friedman χ^2-test The nonparametric version of the one-way repeated measures ANOVA for ranked scores

***F* statistic** In ANOVA, the statistic used to compare all sample means for a factor to determine whether two or more sample means represent different population means; equal to the *F*-ratio

Grouped distribution A distribution formed by combining different scores to make small groups whose total frequencies, relative frequencies, or cumulative frequencies can then be manageably reported

Heterogeneity of variance A characteristic of data describing populations represented by samples in a study that do not have the same variance

Heteroscedasticity An unequal spread of *Y* scores around the regression line (that is, around the values of *Y*′)

Histogram A graph similar to a bar graph but with adjacent bars touching, used to plot the frequency distribution of a small range of interval or ratio scores

Homogeneity of variance A characteristic of data describing populations represented by samples in a study that have the same variance

Homogeneity of variance test A test performed before a t-test or ANOVA is conducted, to determine whether the estimated variances for two samples are significantly different from each other; also called the F_{max} test

Homoscedasticity An equal spread of Y scores around the regression line (that is, around the values of Y')

Incomplete factorial design A two-way ANOVA design in which not all levels of the two factors are combined

Independent events Events for which the probability of one is not influenced by the occurrence of the other

Independent samples Samples created by selecting each subject for one sample, without regard to the subjects selected for any other sample

Independent variable In an experiment, a variable that is changed or manipulated by the experimenter; a variable hypothesized to cause a change in the dependent variable

Individual differences Variations in individuals' traits, backgrounds, genetic make-up, etc., that influence their behavior in a given situation and thus the strength of a relationship

Inferential statistics Procedures for determining whether sample data represent a particular relationship in the population

Interaction effect The effect produced by the concurrent manipulation of two independent variables such that the influence of changing the levels of one factor depends on which level of the other factor is present

Interval scale A measurement scale in which each score indicates an actual amount and there is an equal unit of measurement between consecutive scores, but in which zero is simply another point on the scale (not zero amount)

Kruskal-Wallis *H*-test The nonparametric version of the one-way between-subjects ANOVA for ranked scores

Level In ANOVA, each condition of the factor (independent variable); also called treatment

Linear regression The procedure for describing the best-fitting straight line that summarizes a linear relationship

Linear regression equation The equation that defines the straight line summarizing a linear relationship by describing the value of Y' at each X

Linear regression line The straight line that summarizes the scatterplot of a linear relationship by, on average, passing through the center of all Y scores

Linear relationship A correlation between the X scores and Y scores in a set of data in which the Y scores tend to change in only one direction as the X scores increase, forming a slanted straight regression line on a scatterplot

Line graph A graph in which X scores from an interval or ratio variable are plotted by connecting the data points with straight lines

Main effect In a two-way ANOVA, the effect on the dependent scores of changing the levels of one factor while collapsing over other factors in the study

Mann-Whitney *U* test The nonparametric version of the independent samples t-test for ranked scores when n is less than or equal to 20

Mean The score located at the mathematical center of a distribution

Mean square In ANOVA, an estimated population variance, symbolized by MS

Mean square between groups In ANOVA, the variability in scores that occurs between the levels in a factor or the cells in an interaction

Mean square within groups In ANOVA, the variability in scores that occurs in the conditions, or cells; also known as the error term

Measure of central tendency A score that summarizes the location of a distribution on a variable by indicating where the center of the distribution tends to be located

Measures of variability Measures that summarize and describe the extent to which scores in a distribution differ from one another

Median The score located at the 50th percentile; symbolized by Mdn; also called the median score

Modal score See *Mode*

Mode The most frequently occurring score in a sample; also called the modal score

Negative linear relationship A linear relationship in which the Y scores tend to decrease as the X scores increase

Negatively skewed distribution A frequency polygon with low frequency, extreme low scores but without corresponding low frequency, extreme high ones, so that its only pronounced tail is in the direction of the lower scores

Nemenyi's procedure The *post hoc* procedure performed with the Friedman χ^2-test

Nominal scale A measurement scale in which each score is used simply for identification and does not indicate an amount

Nonlinear relationship A relationship in which the Y scores change their direction of change as the X scores change, forming a curved regression line; also called a curvilinear relationship

Nonparametric statistics Inferential procedures that do not require stringent assumptions about the parameters of the raw score population represented by the sample data; usually used with scores most appropriately described by the median or the mode

Nonsignificant Describes results that are considered likely to result from chance sampling error when the predicted relationship does not exist; it indicates failure to reject the null hypothesis

Normal curve model The most common model of how nature operates; it is based on the normal curve and describes a normal distribution of a population of scores

Null hypothesis The statistical hypothesis describing the population parameters that the sample data represent if the predicted relationship does not exist; symbolized by H_0

Observed frequency In chi square, the frequency with which subjects fall in a category of a variable; symbolized by f_o

Omega squared In ANOVA, an estimate of the proportion of variance in the population that would be accounted for by a relationship; symbolized by ω^2

One-tailed test The test used to evaluate a statistical hypothesis that predicts that scores will only increase or only decrease (that is, the region of rejection falls in only one tail of the sampling distribution)

One-way ANOVA The analysis of variance performed when an experiment has only one independent variable

One-way chi square The chi square procedure performed in testing whether the sample frequencies of category membership along one variable represent the predicted distribution of frequencies in the population

Ordinal scale A measurement scale in which scores indicate only rank order or a relative amount

Parameter See *Population parameter*

Parametric statistics Inferential procedures that require certain assumptions about the parameters of the raw score population represented by the sample data; usually used with scores most appropriately described by the mean

Pearson correlation coefficient The correlation coefficient that describes the linear relationship between two interval or ratio variables; symbolized by r

Percent A proportion multiplied times 100

Percentile A cumulative percentage; the percentage of all scores in the sample that are at or below a particular score

Phi coefficient The statistic that describes the strength of the relationship in a two-way chi square when there are only two categories for each variable; symbolized by ϕ

Point-biserial correlation coefficient The correlation coefficient that describes the linear relationship between scores from one continuous interval or ratio variable and one dichotomous variable; symbolized by r_{pb}

Pooled variance The weighted average of the sample variances in a two-sample experiment; symbolized by s^2_{pool}

Population The infinitely large group of all possible scores that would be obtained if the behavior of every individual of interest in a particular situation could be measured

Population parameter A number that describes a characteristic of a population of scores, symbolized by a letter from the Greek alphabet; also called a parameter

Positive linear relationship A linear relationship in which the Y scores tend to increase as the X scores increase

Positively skewed distribution A frequency polygon with low frequency, extreme high scores but without corresponding low frequency, extreme low ones, so that its only pronounced tail is in the direction of the higher scores

***Post hoc* comparisons** In ANOVA, statistical procedures used to compare all possible pairs of sample means in a significant effect, to determine which means differ significantly from each other

Power The probability that a statistical test will detect a true relationship and allow the rejection of a false null hypothesis

Predicted *Y* score In linear regression, the best description and prediction of the Y scores at a particular X, based on the linear relationship summarized by the regression line; symbolized by Y'

Predictor variable The variable for which known scores in a relationship are used to predict unknown scores on another variable

Probability A mathematical statement indicating the likelihood that an event will occur when a particular population is randomly sampled; symbolized by p

Probability distribution The probability of every possible event in a population, derived from the relative frequency of every possible event in that population

Proportion A decimal number between 0 and 1 that indicates a fraction of a total

Proportion of the area under the curve The proportion of the total area beneath the normal curve at certain scores, which represents the relative frequency of those scores

Proportion of variance accounted for The proportion of the error in predicting scores that is eliminated when, instead of using the mean of Y, we use the relationship with the X variable to predict Y scores; the proportional improvement in predicting Y scores thus achieved

Qualitative variable A variable that reflects a quality or category

Quantitative variable A variable that reflects a quantity or amount

Random sampling A method of selecting samples whereby all members of the population have the same chance of being selected for a sample and all samples have the same chance of being selected

Range The distance between the highest and lowest scores in a set of data

Rank sums test The nonparametric version of the independent samples t-test for ranked scores when n is greater than 20; also the *post hoc* procedure performed with the Kruskal-Wallis H-test

Ratio scale A measurement scale in which each score indicates an actual amount, there is an equal unit of measurement, and there is a true zero (zero amount)

Rectangular distribution A symmetrical frequency polygon shaped like a rectangle; it has no discernible tails because its extreme scores do not have relatively low frequencies

Region of rejection That portion of a sampling distribution containing values considered too unlikely to occur by chance, found in the tail or tails of the distribution

Regression line The line drawn through the long dimension of a scatterplot that best fits the center of the scatterplot, thereby visually summarizing the scatterplot and indicating the type of relationship that is present

Related samples Samples created by matching each subject in one sample with a subject in the other sample or by repeatedly measuring the same subject under all conditions; also called dependent samples

Relationship A correlation between two variables whereby a change in one variable is accompanied by a consistent change in the other

Relative frequency The proportion of time a score occurs in a distribution, which is equal to the proportion of the total number of scores that the score's simple frequency represents; symbolized by *rel. f*

Relative frequency distribution A distribution of scores, organized to show the proportion of time each score occurs in a set of data

Relative standing A description of a particular score derived from a systematic evaluation of the score using the characteristics of the sample or population in which it occurs

Repeated measures design A related samples research design in which the same subjects are measured repeatedly under all conditions of an independent variable

Representative sample A sample whose characteristics accurately reflect those of the population

Research design The way in which a study is laid out so as to demonstrate a relationship

Restriction of range In correlation, improper limitation of the range of scores obtained on one or both variables, leading to an underestimate of the strength of the relationship between the two variables

Robust procedure A procedure that results in only a negligible amount of error in estimating the probability of a Type I error, even if the assumptions of the procedure are not perfectly met; describes parametric procedures

Sample A relatively small subset of a population, intended to represent the population; a subset of the complete group of scores found in any particular situation

Sample statistic A number that describes a characteristic of a sample of scores, symbolized by a letter from the English alphabet; also called a statistic

Sampling distribution of a correlation coefficient A frequency distribution showing all possible values of the coefficient that occur when samples of a particular size are drawn from a population whose correlation coefficient is zero

Sampling distribution of differences between the means A frequency distribution showing all possible differences between two means that occur when two independent samples of a particular size are drawn from the population of scores described by the null hypothesis

Sampling distribution of mean differences A frequency distribution showing all possible mean differences that occur when the difference scores from two related samples of a particular size are drawn from the population of difference scores described by the null hypothesis

Sampling distribution of means A frequency distribution showing all possible sample means that occur when samples of a particular size are drawn from the raw score population described by the null hypothesis

Sampling error The variation, due to random chance, between a sample statistic and the population parameter it represents

Sampling with replacement A sampling procedure in which a previously selected sample is returned to the population before any additional samples are selected

Sampling without replacement A sampling procedure in which previously selected samples are not returned to the population before additional samples are selected

Scatterplot A graph of the individual data points from a set of *X*-*Y* pairs

Semi-interquartile range The average distance between the median and the scores at the 25th and 75th percentiles (the quartiles), used to describe highly skewed distributions

Significant Describes results that are too unlikely to accept as resulting from chance sampling error when the predicted relationship does not exist; it indicates rejection of the null hypothesis

Simple frequency distribution A distribution of scores, organized to show the number of times each score occurs in a set of data

Single-sample *t*-test The parametric procedure used to test the null hypothesis for a single-sample experiment when the standard deviation of the raw score population must be estimated

Skewed distribution A frequency polygon similar in shape to a normal distribution except that it is not symmetrical and it has only one pronounced tail

Slope A number that indicates how much a linear regression line slants and in which direction it slants; used in computing predicted *Y* scores; symbolized by b

Spearman rank-order correlation coefficient The correlation coefficient that describes the linear relationship between pairs of ranked scores; symbolized by r_s

Squared sum of *X* A result calculated by adding all scores and then squaring their sum; symbolized by $(\Sigma X)^2$

Standard deviation The statistic that communicates the average of the deviations of the scores from the mean in a set of data, computed by obtaining the square root of the variance

Standard error of the difference The estimated standard deviation of the sampling distribution of differences between the means of independent samples in a two-sample experiment; symbolized by $s_{\bar{X}_1-\bar{X}_2}$

Standard error of the estimate A standard deviation indicating the amount that the actual *Y* scores in a sample differ from, or are spread out around, their corresponding *Y′* scores; symbolized by $S_{Y'}$

Standard error of the mean The standard deviation of the sampling distribution of means; used in the *z*-test (symbolized by $\sigma_{\bar{X}}$) and in the single-sample *t*-test (symbolized by $s_{\bar{X}}$)

Standard error of the mean difference The standard deviation of the sampling distribution of mean differences between related samples in a two-sample experiment; symbolized by $s_{\bar{D}}$

Standard normal curve A theoretical perfect normal curve, which serves as a model of the perfect normal *z*-distribution

Standard scores See *z-score*

Statistic See *Sample statistic*

Statistical hypotheses Two statements (H_0 and H_a) that describe the population parameters the sample statistics will represent if the predicted relationship exists or does not exist

Strength of a relationship The extent to which one value of *Y* within a relationship is consistently associated with one and only one value of *X*; also called the degree of association

Subjects The individuals who are measured in a sample

Sum of squares The sum of the squared deviations of a set of scores around a statistic

Sum of the deviations around the mean The sum of all differences between the scores and the mean; symbolized as $\sigma(X - \bar{X})$

Sum of the squared *X*'s A result calculated by squaring each score in a sample and adding the squared scores; symbolized by ΣX^2

Sum of *X* The sum of the scores in a sample; symbolized by ΣX

Tail (of a distribution) The far-left or far-right portion of a frequency polygon, containing relatively low frequency, extreme scores

***t*-distribution** The sampling distribution of all possible values of t that occur when samples of a particular size represent the raw score population(s) described by the null hypothesis

Theoretical probability distribution A probability distribution based on a theoretical model of the relative frequencies of events in a population

Tied rank The situation that occurs when two subjects in a sample receive the same rank-order score on a variable

Total area under the curve The area beneath the normal curve, which represents the total frequency of all scores

Transformation A systematic mathematical procedure for converting a set of scores into a different but equivalent set of scores

Treatments The conditions of the independent variable; also called levels

Treatment variance In ANOVA, the variability between scores from different populations that would be created by the different levels of a factor

***t*-test for independent samples** The parametric procedure used for significance testing of sample means from two independent samples

***t*-test for related samples** The parametric procedure used for significance testing of sample means from two related or dependent samples

Tukey's HSD test The *post hoc* procedure performed with ANOVA to compare means from a factor in which all levels have equal n

Two-tailed test The test used to evaluate a statistical hypothesis that predicts a relationship but not whether scores will increase or decrease (that is, the region of rejection falls in both tails of the sampling distribution)

Two-way chi square The chi square procedure performed in testing whether, in the population, frequency of category membership on one variable is independent of frequency of category membership on the other variable

Two-way interaction effect In a two-way ANOVA, the effect of the combination of the levels of one factor with the levels of the other factor

Type I error A statistical decision-making error in which a large amount of sampling error causes rejection of the null hypothesis when the null hypothesis is true (that is, when the predicted relationship does not exist)

Type II error A statistical decision-making error in which the closeness of the sample statistic to the population parameter described by the null hypothesis causes the null hypothesis to be retained when it is false (that is, when the predicted relationship does exist)

Type of relationship The form of the correlation between the X scores and the Y scores in a set of data, determined by the overall direction in which the Y scores change as the X scores change

Unconfounded comparisons In a two-way ANOVA, comparisons between cells that differ along only one factor

Unimodal distribution A distribution whose frequency polygon has only one hump and thus has only one score qualifying as the mode

Variable Anything that, when measured, can produce two or more different scores

Variance A measure of the variability of the scores in a set of data, computed as the average of the squared deviations of the scores around the mean

Wilcoxon *T* test The nonparametric version of the related samples t-test for ranked scores

Within-subjects ANOVA The type of ANOVA performed when a study involves within-subjects factors

Within-subjects factor The type of factor created when an independent variable is studied using related samples in all conditions, either because subjects are matched or repeatedly measured

Y-intercept The value of Y at the point where the linear regression line intercepts the Y axis; used in computing predicted Y scores; symbolized by a

z-distribution The distribution of z-scores produced by transforming all raw scores in a distribution into z-scores

z-score The statistic that describes the location of a raw score in terms of its distance from the mean when measured in standard deviation units; symbolized by z; also known as a standard score because it allows comparison of scores on different kinds of variables by equating, or standardizing, the distributions

z-table The table that gives the proportion of the total area under the standard normal curve for any two-decimal z-score

z-test The parametric procedure used to test the null hypothesis for a single-sample experiment when the true standard deviation of the raw score population is known

Index

List of Symbols

Chapter 3

N	number of scores in the data
f	frequency
cf	cumulative frequency
rel. f	*relative frequency*

Chapter 4

X	scores
Y	scores
k	constant
ΣX	sum of X
Mdn	median
$\bar{X}$	sample mean of Xs
$X - \bar{X}$	deviation
μ	mu; population mean

Chapter 5

ΣX^2	sum of squared Xs
$(\Sigma X)^2$	squared sum of Xs
S_X	sample standard deviation
S_X^2	sample variance
σ_X	population standard deviation
σ_X^2	population variance
s_X	estimated population standard deviation
s_X^2	estimated population variance
df	degrees of freedom

Chapter 6

$\pm$	plus or minus
z	z-score
$\sigma_{\bar{X}}$	standard error of the mean

Chapter 7

$\bar{Y}$	sample mean of Ys
ΣY	sum of Ys
ΣY^2	sum of squared Ys
$(\Sigma Y)^2$	squared sum of Ys
ΣXY	sum of cross products of X and Y
D	difference score
r	Pearson correlation coefficient
r_s	Spearman correlation coefficient
r_{pb}	point-biserial correlation coefficient
ρ	rho; population correlation coefficient

Chapter 8

Y'	Y prime; predicted value of Y
$S_{Y'}^2$	variance of the Y scores around Y'
$S_{Y'}$	standard error of the estimate
b	slope of the regression line
a	Y-intercept of the regression line
r^2	coefficient of determination
$1 - r^2$	coefficient of alienation

Chapter 9

p	probability
$p(A)$	probability of event A

Chapter 10

$>$	greater than
$<$	less than
$\geq$	greater than or equal to
$\leq$	less than or equal to
$\neq$	not equal to
H_a	alternative hypothesis
H_0	null hypothesis
z_{obt}	obtained value of z-test
z_{crit}	critical value of z-test
α	alpha; theoretical probability of a Type I error
β	beta; theoretical probability of a Type II error
$1 - \beta$	power